Power Electronics and Power Systems

The Power Electronics and Power Systems book series encompasses power electronics, electric power restructuring, and holistic coverage of power systems. The series comprises advanced textbooks, state-of-the-art titles, research monographs, professional books, and reference works related to the areas of electric power transmission and distribution, energy markets and regulation, electronic devices, electric machines and drives, computational techniques, and power converters and inverters. The series features leading international scholars and researchers within authored books and edited compilations. All titles are peer reviewed prior to publication to ensure the highest quality content.

To inquire about contributing to the series, please contact:

Dr. Joe Chow

Administrative Dean of the College of Engineering and Professor of Electrical, Computer and Systems Engineering

Rensselaer Polytechnic Institute

Jonsson Engineering Center, Office 7012

110 8th Street

Troy, NY USA

Tel: 518-276-6374

chowj@rpi.edu

Hanoch Lev-Ari • Aleksandar M. Stanković

Dynamic Phasors in Energy Processing Systems

Springer

Hanoch Lev-Ari
Electrical and Computer Engineering
Northeastern University
Boston, MA, USA

Aleksandar M. Stanković
Applied Energy Division
SLAC National Accelerator Laboratory
Menlo Park, CA, USA

ISSN 2196-3185 ISSN 2196-3193 (electronic)
Power Electronics and Power Systems
ISBN 978-3-031-50019-0 ISBN 978-3-031-50020-6 (eBook)
https://doi.org/10.1007/978-3-031-50020-6

This Springer imprint is published by the registered company Springer Nature Switzerland AG
The registered company address is: Gewerbestrasse 11, 6330 Cham, Switzerland

To my wife Aviva and our son Edan
– Hanoch Lev-Ari

To my wife Tina and our sons Michael, Gabriel, and Raphael
– Aleksandar M. Stanković

Preface

The book aims to explore representations of rich and diverse signals in electric energy systems (EES) that are simple and useful for tasks ranging from protection and monitoring to estimation and control. EES play a focal role in engineered energy conversion at levels ranging from personal devices, vehicles such as cars, airplanes, and ships, to regions and even whole continents. EES offer efficient, large, and precisely controlled energy flows, together with the ability to be customized for new sources and loads. This book provides a unifying modeling framework allowing consistent multi-scale analysis, design, and integration with physical and cyber environments.

The scale and rate of change in EES are momentous, and old problems are recast while new ones are spawned. The book aims to bring about a dialogue among practitioners and researchers in the energy industry while serving as a reference for educational ventures, including courses at various levels. To that order, the book introduces a solid analytical platform with adjustable levels of mathematical rigor and numerous references connecting with research in other fields.

We present a set of tools that enable frequency-selective modeling, simulation, and control. The switching mode of operation present in modern EES gives rise to multiple frequency components in signals under study. The modeling concepts described here provide mechanisms to quantify the dynamics of harmonics in power networks by developing a unified analysis framework for power electronic converters, electric machines and drives, and other power system components. A precise characterization of EES transients is of primary engineering interest, particularly for emerging control and protection strategies that utilize signal processing and microcontrollers. Aspiring to be reasonably self-contained, the book includes background material, technical details, and data for test scenarios in the Appendices.

Boston, MA, USA — Hanoch Lev-Ari
Menlo Park, CA, USA — Aleksandar M. Stanković

Contents

Prologue

Signals observed in electric energy systems are rich and kaleidoscopic. The advent of new technologies related to renewable sources and electronic loads has increased the number and variability of transients. This book aims to explore representations of such signals that are simple and useful for tasks ranging from protection and monitoring to estimation and control.

Simplicity has advantages that transcend aesthetics, and has gained a good reputation in the theoretical realm and in industrial applications. It is often linked to sparsity or parsimony of the objects used in engineering studies. Perhaps the first example that comes to mind is the phasor representation of a sinusoid with known frequency. It represents a continuous-time signal by two numbers (magnitude and phase) while achieving perfect signal reconstruction when necessary. This general approach has blossomed in the mathematically beautiful and practically exciting area of compressed sensing. It establishes that a compressive signal can be reconstructed from a few measured samples, much more limited in number than required by general sampling theorems.

We rely on decades of observations and physics-based modeling of power systems and power electronics to derive the basis vectors used in our representations. One could in principle determine these vectors by learning directly from data, as in some forms of machine learning (ML). While we do not pursue such paths in this book, we nevertheless feel that our results may prove useful for methods coming from data science and ML. We envision signal-informed architectures in ML that will harmonize strengths of physics- and data-driven approaches.

Electric energy systems (EES) play a focal role in engineered energy conversion at levels ranging from personal devices, vehicles such as cars, airplanes, and ships, to regions and even whole continents. EES offer efficient, large and precisely controlled energy flows, together with the ability to be customized for new sources and loads. EES technology is a confluence of many developments, ranging from materials to controls to communications and computation. Models are the glue that

H. Lev-Ari, A. M. Stanković, *Dynamic Phasors in Energy Processing Systems*,
Power Electronics and Power Systems, https://doi.org/10.1007/978-3-031-50020-6_1

holds the EES workflow together, and they differ in temporal and spatial scales, in their intended uses, and in their ownership.

In the multifaceted enterprise of EES operation and control, heterogeneous models are used to match different goals and uses. In a large system with many owners and operators, models are a key means of exchanging information among the various parties. They represent evolving summaries of accumulated knowledge about the system and starting points for future explorations [1]. However, present EES lack a unifying modeling framework that would allow consistent multi-scale analysis, design, and integration with physical and cyber environments. This book aims to make a contribution along this direction.

In EES, the dielectric and thermal properties of materials used in components limit the magnitudes and allowable variations of voltages and currents (and of their products, such as power). Furthermore, Maxwell's equations, which describe and couple voltages and currents, involve time derivatives. Since the low-frequency approximation of Maxwell's equations is most relevant for engineered energy conversion, periodic and even sinusoidal shapes are natural candidates. Such waveform families are closed under time differentiation (with reasonable technical requirements such as smoothness on a coarse scale). Moreover, control interventions needed to match the variable operational conditions (such as load changes) are evident for sinusoidal signals, as they often depend on a single parameter (such as the phase angle). In addition, system components – such as sources, transmission networks, and loads – place constraints on voltages, currents, and power they can handle. The need to provide backward compatibility with legacy resources and the necessity to coordinate bandwidth allocation with other electrical systems, such as communications, result in fairly tightly regulated waveforms. Consequently, typical waveforms are repetitive or close-to-periodic, with the basic (average) frequency selected to satisfy the constraints imposed by the materials and to exercise control actions at appropriate time scales.

The historical dichotomy between DC and AC systems is less relevant today, when switching is the key means of actuation in energy conversion. Power electronics, the key enabling technology for energy efficiency, relies on switching actions. The resulting waveforms introduce multiple time scales into energy conversion. Power converters are hybrid systems whose uninterrupted operation is made possible by feedback control, leading to cyber connections in networked systems. In many applications, AC waveforms are advantageous because of the ease of voltage/current level changes using transformers and the superior size/weight metrics of electric machines that require temporal variation of the magnetic flux (even DC machines have mechanical contacts to disrupt the rotor current).

The system-theoretic concept of a *convergent system* nicely encapsulates a large class of EES of interest [2], such as components of power systems – transformers, transmission lines, inverters, electric machines, loads – and their interconnections in standard operation. Informally, a system is called convergent if it has a unique, globally asymptotically stable solution, called the steady-state solution. While each of these attributes can be suitably modified, collectively they capture the behavior commonly observed in EES. Convergent systems naturally extend the class of asymptotically stable linear time-invariant systems into the realm of engineered nonlinear systems. In particular, they admit analysis in the frequency domain [3] using nonlinear frequency response functions and Bode plots. For example, T-periodic inputs to convergent systems produce T-periodic steady-state solutions. Similarly, the property of being a convergent system is preserved under series, parallel and feedback interconnections (with some reasonable technical assumptions [2]).

Modeling Electric Energy Systems

Many elements in energy processing systems – such as power electronic converters, electric drives, and power transmission and distribution systems – have undergone profound changes in recent years. This process is primarily driven by a desire to reduce the environmental impact and to increase efficiency, with a concomitant reduction in energy costs, power losses (and cooling requirements), and component size. A primary means for efficiency improvement has been an increased reliance on a switching mode of operation, made possible by advances in power semiconductor technology.

Energy processing systems have a large number of components connected in a hierarchical, multi-layered structure. These systems are built with efficiency as the primary concern, especially in portable and stand-alone applications. The components exhibit various types of nonlinearities due to properties of materials, geometries of associated electromagnetic fields, and switching modes of operation. The continuous operation of these systems is made possible by feedback control. The control structure becomes critical for overall resilience and reliability when sudden and potentially detrimental transients are triggered by abrupt changes in system components or the environment.

The introduction of switching subsystems in energy networks has profound repercussions for system dynamics: (1) constitutive relationships for components are altered in terms of time constants and complexity (e.g., with both continuous and discrete variables); (2) the emergence of new dynamical couplings between components that are spatially quite distant adds a new dimension to the problems of analysis and control. Analytical complexities associated with hybrid dynamics are thus compounded with issues related to system size and the need to precisely quantify system performance during transients. The changing nature of sources and loads and the emergence of new types of subsystems – such as all-electronic microgrids – highlight the need to understand the nature of dynamical interactions in power networks.

In response to these fundamental changes in energy systems, we aim to present a set of tools that enable frequency-selective modeling, simulation, and control. Regardless of whether the primary power distribution occurs at DC or low AC frequency (e.g., 60/50 or 400 Hz), the switching mode of operation present in the system gives rise to multiple frequency components in signals under study. The modeling concepts described here provide mechanisms to quantify dynamics of harmonics in power networks by developing a unified analysis framework for power electronic converters, electric drives, and other power system components. Dynamics of interest for analysis and control are often those of deviations from periodic behavior of voltages and currents. A precise characterization of such transients is of primary interest, particularly for emerging control and protection strategies that utilize signal processing and micro-controllers.

The conventional analysis of power systems relies on simplifications gained by approximating certain quantities (e.g., voltages and currents) as sinusoids ("phasors"), possibly with "slowly varying" magnitude and phase (often defined in an implicit, heuristic way). This "sinusoidal quasi-steady-state" approximation is widely used to study electro-mechanical dynamics such as transient stability of bulk power systems. There are, however, cases reported in the power system literature when this type of approximation leads to incorrect conclusions about stability [4, 5]. For faster phenomena such as electromagnetic transients (EMT), the primary tools are time-domain simulations that take no advantage of the particular nominal analytical form of the variables of interest. Time-domain simulations are not only a significant computational burden, but they also offer little insight into problem sensitivities to design quantities and almost no basis for control design. This type of fundamental analytical problem is not mitigated by improvements in computational technology. The goal of this book is to explore the stratum between these two approaches and to devise *dynamic multi-harmonic* phasor models that are analytically tractable, yet able to characterize the time variations of faster phenomena with a substantial decrease in analytical complexity and the required computational effort.

Energy processing systems are characterized by a very wide range of time scales of interest. These vary from 10^{-9} seconds, as in so-called blanking times of power electronic switches, to 10^3 seconds (i.e., hourly dynamics) in large electric power systems. Typical models at each time scale are given by both nonlinear and non-autonomous differential equations (the conceptual extension to differential-algebraic equations is possible), viz.,

$$\frac{d}{dt}x = f(x,t) \tag{\#}$$

where "x" is the state (often corresponding to energy storage elements), and the independent sources cause the time variation in (#). A general analysis framework for such systems is very involved [6], and we are interested in possible simplifications due to the particular nature of the independent sources in (#). The corresponding terms are often dominantly sinusoidal as, for example, in the case of equipment connected to a utility grid. We are interested in autonomous (time-invariant) *approximate* models in terms of transformed coordinates

$$\frac{d}{dt}X = F(X) \tag{\S a}$$

from which a time-domain approximation is derived as

$$\hat{x}(t) = \Phi(t)\,X(\varepsilon t) \qquad (\S b)$$

where $\Phi(t)$ is a known time-dependent matrix and X varies slowly (as indicated by its dependence on the "slow" time εt [7]). Quantities in these new coordinates reduce in steady-state to well-known phasor representations of the fundamental and its harmonics, all of which are sinusoidal waveforms with constant magnitude and phase. The hope is that the decomposition ($\S a$)–($\S b$) will simplify analysis and speed up simulations; the remaining issue, of course, is the accuracy of the approximate model. We aim to show that the fidelity of the approximate model is surprisingly good in energy processing systems. Many such systems spend long times near (approximately) steady-state operation, which is then interrupted by infrequent, but very important transients.

Various forms of frequency selective analysis have deep roots in power engineering in the form of phasor-based dynamical models, and the advantage of the analytical approach described here is its systematic derivation of phasor dynamics. The idea of deriving dynamical models for Fourier coefficients goes back to classical averaging theory. We focus on the development of related formalisms that are applicable to power networks containing switching subsystems. The dynamical equations for Fourier coefficients are often nonlinear, and their analytical usefulness stems from the availability of families of approximations that are based on physical and numerical insights offered by the underlying frequency decomposition.

Our approach is situated at the intersection of signal processing, control, and energy engineering. It recasts concepts from signal processing by deriving parametric descriptions that are uniquely suited for control tasks and for which power engineering provides a physical understanding that is crucial in model validation and simplification.

Control Engineering

Issues related to frequency-domain modeling of nonlinear systems and model-order reduction have been discussed at some length in the literature (e.g., [8] and references therein). Our methodology is tailored to energy processing networks; theoretical error bounds for application to general switched circuits are given in [9]. The structural stability of periodic orbits for power networks with switches has been a topic of considerable theoretical interest. The issue is far from settled, as modeling details have a major impact on the existence of such orbits. Promising results have been obtained with tools from bifurcation theory [10]. Our emphasis is somewhat different, as we aim at precisely characterizing dynamics around (demonstrated or assumed) limit cycles. The issues related to numerical simulations of dynamical systems via harmonic balancing have been addressed, for example, in [11].

An important goal of our analytical development is to further the use of physical and mathematical structures that are associated with energy storage, flow, and dissipation. The motivation for theoretical development of an energy-based control design methodology as a unifying design framework comes specifically from large, complex, and inherently nonlinear systems. Standard design techniques tend to fail when facing large model uncertainty, as may result from compounded modeling errors in a large number of sub-units in a complex structure. This applies particularly to linear robust control methods that, almost invariably, rely on small-gain uncertainty characteristics such as H_∞ [12]. On the other hand, basic properties of energy flow and dissipation are often relatively simple, even in complex structures, and readily lend themselves as a foundation for effective control design. In particular, this approach offers a way to piece together many nonlinear, but mostly local, descriptions in a manner that avoids the explosive growth of analytical and computational complexity.

In electric drives and polyphase power electronic converters, it is common to extract harmonic components for feedback control purposes. This is typically done via reference-frame (position) transformations followed by integral action in new coordinates, where quantities of interest are DC in balanced steady state. In unbalanced m-phase systems, it is common to consider m decoupled variables (the so-called sequences obtained from the Fortescue transformation, originally introduced before World War I). For mildly unbalanced systems, one of these components (typically the "positive" sequence) is the key quantity. Repetitive control is a technique that offers a practical and simple solution to the tracking or rejection of periodic signals. It has been used successfully in various power electronic applications; connections with dynamic phasors can be observed when comparing the elegant implementation in [13] with [14].

Power Engineering

Power quality problems have traditionally been treated from a steady-state point of view, in both deterministic and statistical formulations [15–18]. Our approach adds a temporal dimension to such studies, while reducing to the conventional framework in (periodic) steady state. Dynamical quantities such as time-varying harmonics (phasors) can in principle be measured or estimated in real time; for lower frequencies (multiples of 60/50 Hz), this technology has already matured [19]. While our premise is that a fundamental period is a bedrock of an AC system, many processes and events in energy conversion are inherently multiscale. The key roles of models are to provide predictions and interpretability. With the increasing emphasis on data, statistical concepts are useful at various scales, for example in providing estimates of the uncertainty of model predictions [20].

Modeling efforts that we describe provide a unifying framework for reexamination of the two most important coordinate transformations in power engineering: (1) the Blondel-Park transform (which addresses spatial non-homogeneity in electro-mechanical systems), and (2) symmetrical components (which replace an unbalanced polyphase system with a collection of balanced systems). We postpone the details for later chapters.

Dynamic phasor models have a number of useful features: (1) these large-signal models are time-invariant, allowing a tremendous simplification of analysis; (2) inputs and consequently states tend to vary slowly compared to the driving frequencies, resulting in faster simulations; (3) dynamic phasors achieve "simultaneous demodulation" as all variables are constant ("DC") in a steady state; (4) dynamic phasors are very effective in revealing dynamical couplings between various quantities; (5) both refinements and simplifications are possible for various accuracy requirements; (6) dynamic phasor models of particular components of energy processing systems are modular (i.e., involve only the components being modeled, and replace just these), and fully compatible with representations based on static phasors, used for components that are modeled in less detail (e.g., electrically distant portions).

Organization of This Book

The book has nine chapters, organized into three parts. Part I consists of five chapters dedicated to the theoretical foundations of dynamic phasors and some of their generalizations:

- Chapter 1 introduces the one-sided Fourier series representation of periodic finite-power polyphase waveforms with a given period. The coefficients of such a representation, which are known as *polyphase rms phasors*, provide a cost-efficient characterization of the time-domain waveform. Various derived quantities, such as the harmonic-based symmetrical sequence components, can be readily determined from a waveform's set of phasors. In particular, voltage and current phasors provide compact expressions for real and apparent power, as well as the rms voltage and rms current values. The process of evaluating phasors from acquired waveform samples is briefly examined, with a special focus on efficient computational schemes that use a sparse subset of samples to determine selected dominant harmonic phasors.
- In Chap. 2, we extend concepts from the preceding chapter to non-periodic waveforms, such as encountered in power systems in the presence of transients. This is achieved by introducing a sliding-window *Hilbert space* interpretation for cycle-long waveform segments. The resulting *time-dependent* (a.k.a. "dynamic") phasors provide a cost-efficient alternative to time-domain modeling of circuits and devices in terms of differential equations. In fact, dynamic phasors satisfy the same (linear, time-invariant) voltage-current circuit equations as the time-domain waveforms from which these phasors are derived. This unique attribute of dynamic phasors is exploited in Part II of this book to construct efficient phasor-domain models for power electronic devices, electric machines, and power systems.

- Chapter 3 examines in detail several schemes for efficient estimation of selected dynamic phasors from a (possibly sparse) subset of waveform samples. In particular, the *sub-cycle* approach uses the smallest subset of uniformly-spaced samples to estimate phasors of interest. Phasors – and phasor-like quantities such as the analytic signal – can also be used to determine the fundamental frequency of a periodic waveform, and to track variations of this frequency that occur in the presence of transients. In general, phasors and their various generalizations provide a *time-frequency representation* (TFR) for any given waveform of interest.
- Generalized phasors, such as the Short-Time Fourier Transform (STFT), the S-Transform, the Wavelet Transform, and the Wigner-Ville Distribution, are the subject of Chaps. 4 and 5. The added flexibility offered by such time-frequency representations facilitates controllable tradeoff between implementation cost, frequency-resolution, and time-resolution. In particular, such TFRs can be "tuned" to highlight a specific power system transient component of interest. Constant-bandwidth transforms (i.e., the STFT) are examined in Chap. 4, while Chap. 5 is devoted to variable-bandwidth (primarily constant-Q) transforms.

Part II of this book consists of two chapters dedicated to modeling and control applications of dynamic phasors:

- Chapter 6 explores modeling applications in power electronics, the domain where the concept of dynamic phasors originated. Standard pulse-width modulated (PWM) converters are addressed first, as they form the backbone of many grid-connected and autonomous applications. The following section looks into applications where the switching frequency is not significantly higher than the natural response frequencies of passive components in the circuit, necessitating a coordinated analysis of multiple frequency components. The modeling task is made more challenging when some of the power switches are semi-controllable, as in the case of thyristors, whose turn-off is governed by the rest of the system; this class of circuits is widespread because of low cost and control simplicity. The last section in this chapter illustrates the control options available in (nonlinear) dynamic phasor models in a utility system application: the thyristor-controlled series capacitor (TCSC).
- Chapter 7 addresses two broad groups of applications in energy processing systems. The first involves electric machines. While electric machines, especially AC, have been a part of electrical technology since its inception, the use of power electronic converters as means of efficiently controlling machines is of much more recent vintage, and the system-level interactions in the frequency domain need to be studied carefully. The second group of applications is exemplified by a system-level model of a generator during unbalanced faults; some other power system use cases of dynamic phasors are mentioned at the conclusion of this chapter.

Part III of this book consists of two chapters dedicated to applications of dynamic phasors to characterize power flow in electric energy processing systems:

- Chapter 8 presents several steady-state decompositions of apparent power into time-invariant components, starting with a 5-component decomposition which is subsequently refined into one consisting of 11 components. In addition to the active (= real) power and Budeanu's "reactive power," which are shared by all decompositions, the remaining components can be used to quantify the impact of higher harmonics and phase imbalance on power quality and, in particular, on the power factor. Moreover, each such apparent power decomposition induces a matching decomposition of the current waveform into a sum of mutually-orthogonal components. Construction of this induced current decomposition can be accomplished in either one of two distinct ways: (a) orthogonal projection on a set of voltage-derived subspaces in the Hilbert space of periodic polyphase waveforms, or (b) grade-splitting of an apparent power multivector within a suitably constructed Geometric Algebra.
- In Chap. 9, we use dynamic phasors, both full-cycle and sub-cycle, to extend the various time-invariant decompositions from the preceding chapter to the case where steady-state operation is disrupted by the occurrence of transients. The resulting *time-dependent* apparent power and its components can provide an accurate indication of the onset and duration of a transient, in addition to their role as power quality metrics during steady-state operation. Furthermore, decompositions constructed from sub-cycle phasors offer a flexible cost-performance tradeoff, both in steady-state and in the presence of transients, that spans the range from full-cycle-based metrics to purely-instantaneous power quality metrics.

The book concludes with 12 appendices that parallel the book's order of presentation, summarizing background material, technical details, and data for test scenarios. While some of this material is available elsewhere, we believe that the convenience and unified notation justify its inclusion in the book.

Potential Uses of This Book

The rate of change in electric energy consumption and generation – as well as its growing societal importance – is unprecedented and calls for a re-examination of engineering tools. We hope that this book will find its way to practitioners and researchers in the energy industry who are tasked with creatively solving numerous emerging problems. Engineering education is evolving in parallel, and we hope that the book can be used as a core reference for courses in power quality and in modeling of energy processing systems (power electronics, power systems, and electric drives). It can serve as a reference for classes on smart grids, microgrids, grid-connected power electronics, and distributed energy resources. Finally, it can serve as a source of use cases and application problems for courses in signal processing, estimation, and control.

Acknowledgments

The roots of this book go back more than three decades, and its growth has been supported by many individuals, funding agencies, and organizations. It is a pleasant task to thank those who have helped build up and systematize the foundations for the material presented in this book. Foremost on this list are discussions and collaborations with Seth Sanders, George Verghese, Paolo Mattavelli, and Bernie Lesieutre during the early years of developing the theory of dynamic phasors – which is presented in Part I of this book – and exploring its applications to modeling and control, which are covered in Part II. Our subsequent work on applying dynamic phasors to characterize power flows and dynamic power quality – which we introduce in Part III of this book – was influenced by discussions with Leszek Czarnecki and Alex Emanuel.

The funding provided by the Office of Naval Research (under programs directed by Al Tucker, Terry Ericsen and H. Scott Coombe) and by the National Science Foundation, in support of our research on dynamic phasors, has been instrumental in developing many of the new results presented in this book.

We wish to extend our appreciation to our colleagues who provided insightful suggestions through reviews of this book. These include George Verghese, Christopher Demarco, Paulo Ribeiro, Seth Sanders, Paolo Mattavelli, and Julius Smith, as well as a number of anonymous reviewers.

Several graduate students, postdoctoral fellows, and visiting scientists at Northeastern and Tufts universities have participated in research that contributed to this book. These include Gerardo Escobar, Pedja Stefanov, Andrija Sarić, Timur Aydin, Milun Perišić, Margret Ragnarsdottir, and Afsaneh Ghanavati. In particular, Chaps. 8 and 9 contain several results obtained in the course of Afsaneh's graduate work; she also generated the synthetic-short-transient dataset (Sect. L.2.1) and assisted in preparing several of the figures.

We also wish to thank the colleagues who graciously provided real-life data sets used as examples throughout this book:

- The adjustable-speed drive dataset (Sect. L.1.7) was provided by Sergio Ceballos from the University of Seville, Spain.
- The two Hydro-Ottawa datasets (Sects. L.1.8 and L.2.3) were provided by Branislav Djokić from NRC Canada.
- The wind-farm dataset (Sect. L.2.2) was provided by Moinul Islam, the first author of [303].
- The paper-mill dataset (Sect. L.2.4) was provided by Adel Nasiri, the senior author of [302].

Mark Transtrum provided much needed help with editing Encapsulated PostScript (eps) files for several figures.

Finally, we wish to express our thanks to Joe Chow and David Hill, the editors of the Springer book series on Power Electronics and Power Systems, for the expeditious handling of our manuscript, and to Michael McCabe and his team at Springer Nature Publishing for carefully supervising the production of the final book.

Part I

Theory of Dynamic Phasors

"What leads and drags the world
are not the machines but the ideas."

Victor Hugo

Chapter 1 Background

The dominant waveforms in electric energy processing systems are engineered to be periodic because of the physics and materials of loads and sources, and because of the need to exercise regular control actions at appropriate time scales. In utility and many other applications, sinusoidal AC waveforms are prevalent, due in part to historical reasons (e.g., the functionality, simplicity, and reliability offered by transformers and AC generators). While many sources (and often voltage sources) are designed to be nominally sinusoidal, the nonlinearities in energy systems, ranging from device geometries to materials to switching, tend to introduce harmonics in other key waveforms, such as currents. However, energy processing networks are engineered structures so they are designed to have desirable properties such as being convergent systems. This particular property guarantees that periodic inputs will yield periodic steady-state responses throughout the system, as is, for example, commonly observed in utility power systems.

Periodic waveforms have traditionally been represented by the celebrated Fourier series. The coefficients of such a representation are often called *phasors* because they provide information about both the amplitude and the phase of each harmonic component, and the phase difference is a key quantity in high-voltage transmission. In contrast to this "frequency-domain" or "phasor-domain" representation, one can also interpret periodic waveforms with a fixed pre-determined period T as elements in a suitably defined Hilbert space. This allows alternative (time-domain) representations in terms of any one of the many choices of orthogonal bases associated with the Hilbert space of *finite-rms* periodic waveforms, i.e., periodic waveforms that are square-integrable on any period-long interval. One such basis is provided by complex harmonics: using it results in the Fourier series, which forms a bridge between time-domain and frequency-domain representations.

The idea of representing waveforms of interest in energy conversion as points in function spaces occurred remarkably early. It was pioneered in the 1930s by S. Fryze [27], who was a faculty colleague of S. Banach at the Polytechnic University

H. Lev-Ari, A. M. Stanković, *Dynamic Phasors in Energy Processing Systems*, Power Electronics and Power Systems, https://doi.org/10.1007/978-3-031-50020-6_2

of Lwów in Poland (now Lviv in Ukraine), and by W. Quade [28] from the Technical University of Karlsruhe in Germany.

In fact, one can readily extend both the notion of Fourier series and the Hilbert space interpretation of periodic finite-rms waveforms to polyphase systems [29]. In the m-phase case, the Hilbert space elements are $1 \times m$ (i.e., row) vectors of periodic finite-rms waveforms, and the Fourier coefficients (a.k.a. polyphase phasors) are $1 \times m$ vectors of complex numbers, as described in Sect. 1.2. Polyphase phasors can be used to define symmetrical sequence components for each individual harmonic, as discussed in Sect. 1.3.

Digital evaluation of phasors involves sampling an acquired power system waveform at a suitably high rate, known as the "Nyquist rate," and then evaluating a discretized version of the Fourier coefficients. Such discrete-time Fourier coefficients can be efficiently determined via the celebrated *fast Fourier transform* (FFT) algorithm. However, when one is interested in selected dominant phasors, a direct *discrete Fourier transform* (DFT) evaluation may be the more efficient choice (see the discussion in Sects. 1.4 and 1.5). A more flexible alternative is provided by the least-squares approach, which is introduced in Sect. 3.2. In particular, accurate estimates of selected phasors can be determined via the least-squares approach from a *sparse subset of uniformly-spaced samples*, which results in a significant reduction in computational cost, as explained in Sect. 1.5.

1.1 Fourier-Series Representation of Periodic Waveforms

All time-domain waveforms $x(t)$ considered in this chapter are periodic, with a fixed fundamental period T; we also define the fundamental (angular) frequency as $\omega_0 = {}^{2\pi}/_{T}$. We restrict our discussion to *finite-rms* T-periodic waveforms, i.e., we assume that[1]

$$X_{rms}^2 \overset{\text{def}}{=} \frac{1}{T} \int_T \left| x(t) \right|^2 dt \quad < \quad \infty \tag{1.1}$$

where the notation $\int_T$ indicates integration over any contiguous interval of length T. The set of all such waveforms, all sharing the same given period length T, forms a *Hilbert space*: see (1.3) for additional details and Appendix J for an introduction to the theory of Hilbert spaces. In this book we are interested in energy processing systems, where T is typically fixed in advance to ${}^1/_{60}$ sec (or ${}^1/_{50}$ sec). Also, such real-life waveforms are always bounded, which guarantees their finite-rms property.

[1] In the Signals and Systems literature, finite-rms periodic waveforms provide a prime example of so-called finite-power waveforms.

Every finite-rms T-periodic waveform can be represented by a summable infinite two-sided Fourier series, viz.,

$$x(t) = \sum_{k=-\infty}^{\infty} \langle x\rangle_k \, e^{jk\omega_0 t} \tag{1.2a}$$

where

$$\langle x\rangle_k \stackrel{\text{def}}{=} \frac{1}{T}\int_T x(s)\, e^{-jk\omega_0 s}\, ds \tag{1.2b}$$

are known as the *Fourier coefficients* associated with the waveform $x(\cdot)$. In the context of our Hilbert space framework, *summability* means that the rms residual $\left[\frac{1}{T}\int_T \left| x(t) - \sum_{-N_1}^{N_2} \langle x\rangle_k\, e^{jk\omega_0 t}\right|^2 dt\right]^{1/2}$ converges to zero[2] as $N_1 \to \infty$ and $N_2 \to \infty$ (independently). Under suitable conditions, the Fourier-series expression (1.2a) may also hold pointwise: a rigorous introduction to summability of Fourier series can be found in [21, 22]. Several useful properties of Fourier series are listed in [23–25]. The key property is linearity of the waveform-to-phasor map (1.2b): in Chap. 2 we use this property to associate a linear time-invariant filter with a sliding-window version of this map.

Most time-domain power system waveforms of interest are real-valued, representing physical quantities such as voltages and currents. The Fourier coefficients of a real-valued waveform $x(\cdot)$ exhibit *conjugate symmetry*, namely, $\langle x\rangle_{-k} = \langle x\rangle_k^*$, where the superscript * denotes complex conjugation. This useful property makes it possible to replace the two-sided Fourier series (1.2a) by a more compact one-sided equivalent, as shown below. The representation of physical quantities by complex waveforms is relatively rare in power system modeling and analysis, though complex-variable models are used, for example, in describing geometrically non-salient electric machines [26]. Consequently, we restrict our discussion in the remainder of this chapter, and elsewhere in this book, to *real-valued waveforms*.

Thus we can view the waveform $x(\cdot)$ as an element in the *Hilbert space* of all real-valued, finite-rms, T-periodic waveforms (i.e., periodic waveforms with a given pre-determined period T), with an inner-product defined by

$$\langle\, x(\cdot)\,,\, y(\cdot)\,\rangle \stackrel{\text{def}}{=} \frac{1}{T}\int_T x(s)\, y(s)\, ds \tag{1.3}$$

so that $X_{rms}^2 = \langle\, x(\cdot)\,,\, x(\cdot)\,\rangle$. Every waveform belonging to this Hilbert space can be represented by a summable *one-sided Fourier series*, i.e.,

[2] See also the discussion in Sect. J.4.

$$x(t) = X_0 + \sum_{k=1}^{\infty} \sqrt{2}\,\Re\left\{X_k\, e^{jk\omega_0 t}\right\} \tag{1.4a}$$

where $\Re\{\ \}$ denotes the real part of a complex number, and X_k is the so-called *one-sided rms phasor*, defined as a scaled version of the Fourier coefficient $\langle x \rangle_k$ associated with the k-th harmonic of the waveform $x(\cdot)$, viz.,

$$X_k = \begin{cases} \langle x \rangle_0 & k = 0 \\ \sqrt{2}\,\langle x \rangle_k & k \geq 1 \end{cases}$$

or, equivalently,

$$X_k = a_k\,\langle x \rangle_k \qquad \text{where} \quad a_k = \begin{cases} 1 & k = 0 \\ \sqrt{2} & k \geq 1 \end{cases} \tag{1.4b}$$

This means that

$$X_k = \begin{cases} \dfrac{1}{T}\displaystyle\int_T x(s)\,ds & k = 0 \\[2ex] \dfrac{\sqrt{2}}{T}\displaystyle\int_T x(s)\,e^{-jk\omega_0 s}\,ds & k \geq 1 \end{cases} \tag{1.4c}$$

This representation relies on the fact that the (two-sided) Fourier coefficients $\langle x \rangle_k$ of a real-valued waveform are conjugate-symmetric, which allows us to replace the two-sided Fourier series (1.2a) with the one-sided version (1.4). Readers with interest in complex-valued models should be able to translate the one-sided formulation presented here into its equivalent two-sided version, as needed.

The corresponding (one-sided) Parseval identity is

$$\langle\, x(\cdot)\,,\, y(\cdot)\,\rangle = X_0\,Y_0 + \sum_{k=1}^{\infty} \Re\left\{\, X_k\,Y_k^* \right\} \tag{1.5}$$

where $\{Y_k\,;\ 0 \leq k < \infty\}$ are the one-sided rms phasors associated with the waveform $y(\cdot)$. The Parseval identity (1.5) implies that the phasor sequence $\{X_k\,;\ 0 \leq k < \infty\}$ is *square-summable*, since $\sum_{k=0}^{\infty} |X_k|^2 = \langle\, x(\cdot)\,,\, x(\cdot)\,\rangle = \frac{1}{T}\int_T x^2(t)\,dt \equiv X_{rms}^2 < \infty$. This identity establishes an isometry (= a one-to-one correspondence that preserves the inner product) between the Hilbert space of finite-rms T-periodic waveforms and the Hilbert space of square-summable rms phasor sequences associated with these waveforms. Notice that our Hilbert space consists of *real-valued* waveforms and employs real scalars, while the corresponding rms phasors are complex.

In Part II of this book we will encounter phasor representations of state-space models. In particular, we will represent the state-vector by a (column) vector of phasors. In Part III we will use (row) vectors of phasors to represent polyphase voltage and current waveforms. In general, we can represent any $n \times m$ *matrix* of real-valued, finite-rms T-periodic waveforms, say $x(t) = \left[x^{(i,j)}(t)\right]_{i=1:n,\, j=1:m}$, by a sequence of $n \times m$ matrix rms phasors $\{X_k\,;\ 0 \le k < \infty\}$, where $X_k = \left[X_k^{(i,j)}\right]_{i=1:n,\, j=1:m}$, and $\{X_k^{(i,j)};\ 0 \le k < \infty\}$ is the phasor sequence associated with the scalar waveform $x^{(i,j)}(\cdot)$ via the Fourier-series relation (1.4).

1.2 Phasor Representation of Periodic Polyphase Waveforms

In Part III we will use rms phasors to describe power flow in polyphase power systems. This motivates us to introduce here a Hilbert space whose elements are *finite-rms periodic polyphase waveforms.* To be specific, we will consider m-phase systems, i.e., systems with $m + 1$ conductors ("wires") in which the first m are referenced either to a common ground or to the $(m + 1)$-th ("neutral") conductor. Then we can define the m-dimensional voltage and current *row vectors*

$$
\begin{aligned}
v(t) &\stackrel{\text{def}}{=} \left[\, v_1(t)\ \ v_2(t)\ \ \ldots\ \ v_m(t) \,\right] \\
i(t) &\stackrel{\text{def}}{=} \left[\, i_1(t)\ \ i_2(t)\ \ \ldots\ \ i_m(t) \,\right]
\end{aligned}
\tag{1.6}
$$

where all currents have reference directions "toward" the load (see Fig. 1.1).

As before, we restrict our discussion to finite-rms T-periodic waveforms, i.e., we assume that the polyphase voltage waveform $v(t)$ satisfies the constraint

$$
V_{rms}^2 \stackrel{\text{def}}{=} \frac{1}{T} \int_T v(t)\, v^{\top}(t)\, dt \ < \ \infty \tag{1.7}
$$

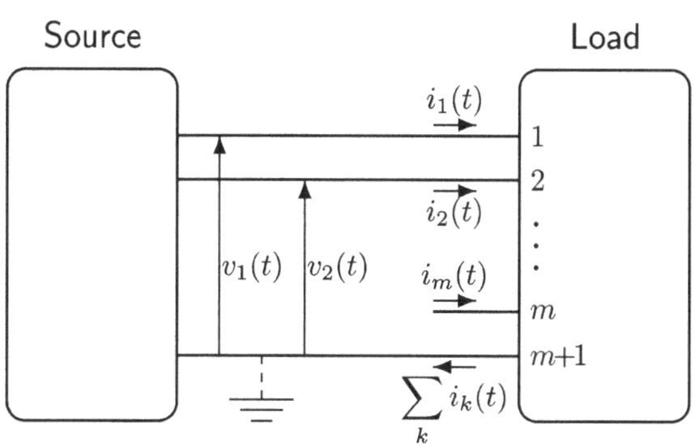

Fig. 1.1 Circuit schematic of a polyphase system.

(and similarly for $i(t)$), where the superscript $^{\top}$ denotes transposition. Thus we can view $v(t)$ and $i(t)$, the row vectors representing polyphase load voltage and current (nonlinear load, multi-harmonic source), as elements in the Hilbert space of all real-valued, m-phase, finite-rms T-periodic waveforms, with the inner-product defined by

$$\langle\, x(\cdot)\,,\; y(\cdot)\,\rangle \;\stackrel{\text{def}}{=}\; \frac{1}{T}\int_T x(s)\, y^{\top}(s)\, ds \tag{1.8a}$$

The corresponding (one-sided) *polyphase* Parseval identity is

$$\langle\, x(\cdot)\,,\; y(\cdot)\,\rangle \;=\; X_0\, Y_0^{\top} \;+\; \sum_{k=1}^{\infty} \Re\left\{\, X_k\, Y_k^{H} \right\} \tag{1.8b}$$

where the superscript ^{H} denotes conjugate transposition, and X_k and Y_k are the $1 \times m$ phasors associated via (1.4) with the polyphase waveforms $x(\cdot)$ and $y(\cdot)$, respectively. For instance, in this terminology the rms value of the polyphase voltage $v(t)$ is expressed as $V_{rms} = \|\, v(\cdot)\| \stackrel{\text{def}}{=} \sqrt{\langle\, v(\cdot)\,,\; v(\cdot)\,\rangle}$, and the average (real) power delivered to the load is $P = \langle\, v(\cdot)\,,\; i(\cdot)\,\rangle$ (see also [29]).

1.2.1 Polyphase Fourier Series

The constraint (1.7) implies that the finite-rms property is also satisfied by every component of the polyphase waveforms $v(t) = \left[\, v_1(t)\;\; v_2(t)\;\; \ldots\;\; v_m(t)\,\right]$, so that each individual $v_i(t)$ can be represented by a summable Fourier series. Consequently, the polyphase waveform $v(t)$ can be represented by a summable *polyphase* Fourier series, viz.,

$$v(t) = V_0 + \sum_{k=1}^{\infty} \sqrt{2}\, \Re\left\{ V_k\, e^{jk\omega_0 t} \right\} \tag{1.9a}$$

where $V_k = \left[\, V_k^{(1)}\;\; V_k^{(2)}\;\; \ldots\;\; V_k^{(m)}\,\right]$ is the polyphase rms phasor (= vector of scaled Fourier coefficients) describing the k-th harmonic of a polyphase voltage waveform $v(t)$. Consequently, recall (1.4),

$$V_k \;=\; \begin{cases} \dfrac{1}{T}\displaystyle\int_T v(s)\, ds & k = 0 \\[2ex] \dfrac{\sqrt{2}}{T}\displaystyle\int_T v(s)\, e^{-jk\omega_0 s}\, ds & k \geq 1 \end{cases} \tag{1.9b}$$

and similarly for I_k, the polyphase rms phasors associated with the current waveform $i(t)$.

It will be convenient to rewrite the Parseval identity in matrix-vector notation, by introducing the (row vector) voltage and current *phasor arrays*, viz.,

$$\mathbf{V} \overset{\text{def}}{=} \left[V_0 \; V_1 \; V_2 \; \ldots \; \right] \quad , \quad \mathbf{I} \overset{\text{def}}{=} \left[I_0 \; I_1 \; I_2 \; \ldots \; \right] \tag{1.10a}$$

We can now express the rms voltage and current, as well as the active power P, either in terms of waveforms or in terms of phasor arrays, viz.,

$$V_{rms}^2 \overset{\text{def}}{=} \| v(\cdot) \|^2 = \|\mathbf{V}\|^2 \quad , \quad I_{rms}^2 \overset{\text{def}}{=} \| i(\cdot) \|^2 = \| \mathbf{I} \|^2$$
$$P \overset{\text{def}}{=} \langle \, v(\cdot) \, , \, i(\cdot) \, \rangle = \Re \left\{ \mathbf{V} \, \mathbf{I}^H \right\} \tag{1.10b}$$

Notice how our decision to represent polyphase waveforms by *row vectors* of functions (rather than column vectors) simplifies notation in expressions like (1.10) and (1.8).

1.2.2 Harmonic-Phase Representation

The Fourier-series representation

$$x(t) = X_0 + \sum_{k=1}^{\infty} \sqrt{2} \, \Re \left\{ X_k e^{jk\omega_0 t} \right\} \tag{1.11a}$$

of a finite-rms periodic polyphase waveform $x(\cdot)$ can be further refined by observing that

$$X_k = \left[X_k^{(1)} \; X_k^{(2)} \; \ldots \; X_k^{(m)} \right] \equiv \sum_p X_k^{(p)} \, \eta_p$$

where η_p is a $1 \times m$ unit vector, viz.,

$$\eta_p \overset{\text{def}}{=} [\underbrace{0 \ldots 0 \; 1}_{p} \; 0 \ldots 0] \quad , \quad 1 \leq p \leq m \tag{1.11b}$$

This results in a representation of the polyphase waveform $x(t)$ in terms of its individual (k-th harmonic/p-th phase) components $x_k^{(p)}(t)$, viz.,

$$x(t) = \sum_{p=1}^{m} \underbrace{X_0^{(p)} \, \eta_p}_{x_0^{(p)}(t)} + \sum_{k=1}^{\infty} \sum_{p=1}^{m} \underbrace{\sqrt{2} \, \Re \left\{ X_k^{(p)} e^{jk\omega_0 t} \right\} \eta_p}_{x_k^{(p)}(t)} \tag{1.11c}$$

Notice that $x_k^{(p)}(\cdot)$ is a polyphase waveform and thus an element in the Hilbert space of polyphase waveforms that we introduced in the preceding section.

Furthermore, these k-th harmonic/p-th phase components are all mutually-orthogonal because of the orthogonality of distinct harmonics and phases. A direct calculation shows that

$$\begin{aligned}\langle\, x_k^{(p)}(\cdot)\,,\; y_\ell^{(q)}(\cdot)\,\rangle \;&=\; \frac{1}{T}\int_T \sqrt{2}\,\Re\left\{X_k^{(p)} e^{jk\omega_0 s}\right\}\left(\eta_p\;\eta_q^\top\right)\sqrt{2}\,\Re\left\{Y_\ell^{(q)} e^{j\ell\omega_0 s}\right\}\,ds \\ &=\; \delta_{p,q}\left[\frac{1}{T}\int_T\left(\sqrt{2}\,\Re\left\{X_k^{(p)} e^{jk\omega_0 s}\right\}\right)\right. \\ &\qquad \left.\times\left(\sqrt{2}\,\Re\left\{Y_\ell^{(q)} e^{j\ell\omega_0 s}\right\}\right)\,ds\right]\end{aligned}$$

where

$$\delta_{p,q} \;\stackrel{\text{def}}{=}\; \begin{cases} 1 & p = q \\ 0 & p \neq q \end{cases} \tag{1.12a}$$

We conclude that, for any two finite-rms periodic polyphase waveforms $x(\cdot)$ and $y(\cdot)$ of the same period T,

$$\langle\, x_k^{(p)}(\cdot)\,,\; y_\ell^{(q)}(\cdot)\,\rangle \;=\; \delta_{k,\ell}\;\delta_{p,q}\;\;\Re\left\{X_k^{(p)}\left[Y_k^{(p)}\right]^*\right\} \tag{1.12b}$$

This means that the (k, p)-subspace (i.e., the subspace of all k-th harmonic/p-th phase components) is orthogonal to the (ℓ, q)-subspace, except when both $k = \ell$ and $p = q$. In view of the decomposition (1.11c), our Hilbert space is an *orthogonal direct sum*[3] of all (k, p)-subspaces. Finally, we observe that each individual (k, p)-subspace is two-dimensional: every waveform in this subspace can be expressed, for instance, as a linear combination of $(\cos k\omega_0 t)\eta_p$ and $(\sin k\omega_0 t)\eta_p$.

1.3 Symmetrical Sequence Components

Symmetrical sequence components are a standard way to describe unbalanced operation and faults in polyphase systems [30, Ch. 8]. The symmetrical sequence components of a periodic three-phase waveform

$$x(t) \;=\; X_0 + \sum_{k=1}^{\infty}\sqrt{2}\,\Re\left\{X_k\, e^{jk\omega_0 t}\right\} \tag{1.13a}$$

[3] See Sect. J.5 for a definition of this concept.

are obtained by applying the unitary transformation

$$\begin{bmatrix} X_{k,0} & X_{k,+} & X_{k,-} \end{bmatrix} \stackrel{\text{def}}{=} X_k \, \mathcal{M}_{sc} \tag{1.13b}$$

to each of the (three-phase) phasors $\{X_k\,;\ k = 0, 1, 2, \ldots\}$. Here the subscript "0" denotes the zero-sequence component, while the subscripts "+" and "−" denote the positive and negative sequence components, respectively.

In general, the square $(m \times m)$ matrix $\mathcal{M}_{sc}$ is a normalized m-point DFT matrix: in three-phase systems it is

$$\mathcal{M}_{sc} = \frac{1}{\sqrt{3}} \begin{pmatrix} 1 & 1 & 1 \\ 1 & \alpha & \alpha^2 \\ 1 & \alpha^2 & \alpha \end{pmatrix} \quad , \quad \alpha \stackrel{\text{def}}{=} e^{\,j2\pi/3} = -\frac{1}{2} + j\frac{\sqrt{3}}{2} \tag{1.13c}$$

where we note that $\alpha^2 = \alpha^*$. It can be verified that the matrix $\mathcal{M}_{sc}$ is unitary, namely $\mathcal{M}_{sc}\,\mathcal{M}_{sc}^H = I$ (or, equivalently, $\mathcal{M}_{sc}^{-1} = \mathcal{M}_{sc}^H$), where the superscript ^{H} denotes complex conjugate transposition (a.k.a. Hermitian transpose). In other words, $\mathcal{M}_{sc}$ is a norm-preserving mapping from the phase-domain to the symmetrical-component (SC) domain, so that

$$\left\|X_k\right\|^2 = \left|X_{k,0}\right|^2 + \left|X_{k,+}\right|^2 + \left|X_{k,-}\right|^2 \tag{1.14a}$$

which in turn implies preservation of (rms) magnitude, i.e.,

$$X_{rms}^2 \equiv \sum_{k=0}^{\infty} \left\|X_k\right\|^2 = \underbrace{\sum_{k=0}^{\infty} \left|X_{k,0}\right|^2}_{X_{0,rms}^2} + \underbrace{\sum_{k=0}^{\infty} \left|X_{k,+}\right|^2}_{X_{+,rms}^2} + \underbrace{\sum_{k=0}^{\infty} \left|X_{k,-}\right|^2}_{X_{-,rms}^2} \tag{1.14b}$$

where $X_{rms}^2 = \frac{1}{T} \int_T x^2(s)\, ds$. As commonly encountered in transforms, scaling factors other than $1/\sqrt{3}$ can be used in the definition of the matrix $\mathcal{M}_{sc}$. However, changing the scaling of $\mathcal{M}_{sc}$ necessitates adjustments in the inverse transform, as well as in the norm identities (1.14).

Example 1.1 (Two-Frequency Waveform): Consider the case of a periodic (i.e., steady-state) balanced three-phase waveform that contains the first two harmonics, viz.,

$$\begin{aligned} x(t) = {} & A\Big[\cos(\omega_0 t) \quad \cos\big(\omega_0 t - \tfrac{2\pi}{3}\big) \quad \cos\big(\omega_0 t - \tfrac{4\pi}{3}\big)\Big] \\ & + B\Big[\sin(2\omega_0 t) \quad \sin[2(\omega_0 t - \tfrac{2\pi}{3})] \quad \sin[2(\omega_0 t - \tfrac{4\pi}{3})]\Big] \\ = {} & \sqrt{2}\,\Re\big\{X_1\, e^{j\omega_0 t}\big\} + \sqrt{2}\,\Re\big\{X_2\, e^{j2\omega_0 t}\big\} \end{aligned}$$

where (notice that $\alpha^2 = \alpha^{-1} = \alpha^*$)

$$X_1 = \frac{A}{\sqrt{2}}\begin{bmatrix} 1 & \alpha^* & \alpha \end{bmatrix} \quad , \quad X_2 = \frac{-jB}{\sqrt{2}}\begin{bmatrix} 1 & \alpha & \alpha^* \end{bmatrix}$$

Consequently,

$$X_1 \mathcal{M}_{sc} = A\begin{bmatrix} 0 & \sqrt{\frac{3}{2}} & 0 \end{bmatrix} \quad , \quad X_2 \mathcal{M}_{sc} = B\begin{bmatrix} 0 & 0 & -j\sqrt{\frac{3}{2}} \end{bmatrix}$$

so that in this case the only nonzero symmetrical components are $X_{1,+} = \sqrt{\frac{3}{2}}\,A$ and $X_{2,-} = -j\sqrt{\frac{3}{2}}\,B$. Note that this agrees with the practical notions that harmonics of the order $3k+1$ (where k is non-negative integer) belong to the positive sequence, harmonics of the order $3k+2$ belong to the negative sequence, and harmonics of the order $3k$ belong to the zero sequence [31]. ■

Next we can express the quantity $\mathbf{V}\,\mathbf{I}^H$, used to determine the steady-state real power P (recall (1.10)) in terms of the voltage and current symmetrical components, viz.,

$$\mathbf{V}\,\mathbf{I}^H \equiv \sum_{k=0}^{\infty} V_k\, I_k^H = \sum_{k=0}^{\infty} \begin{bmatrix} V_{k,0} & V_{k,+} & V_{k,-} \end{bmatrix}\begin{bmatrix} I_{k,0} & I_{k,+} & I_{k,-} \end{bmatrix}^H$$

where we used the property $\mathcal{M}_{sc}\,\mathcal{M}_{sc}^H = I$. Splitting this expression into contributions from individual sequence components, we obtain an additive decomposition for P, viz.,

$$P = P_0 + P_+ + P_- \tag{1.15a}$$

where

$$P_+ = \sum_{k=0}^{\infty} \Re\left\{V_{k,+}\, I_{k,+}^H\right\} \quad , \quad P_- = \sum_{k=0}^{\infty} \Re\left\{V_{k,-}\, I_{k,-}^H\right\} \tag{1.15b}$$

and

$$P_0 = \sum_{k=0}^{\infty} \Re\left\{V_{k,0}\, I_{k,0}^H\right\} \tag{1.15c}$$

These symmetrical sequence components of P can also be determined directly in the time-domain from the symmetrical sequence decomposition of voltage and current waveforms: see (1.18) below.

In general, the mapping (1.13b) defines a one-to-one correspondence between the three elements of the complex phasor X_k and the three (complex-valued) symmetrical components $[\, X_{k,0} \quad X_{k,+} \quad X_{k,-}\,]$. However, in the special case $k = 0$, the phasor X_0 is *real-valued,* so that the corresponding zero-sequence component $X_{0,0}$ is real-valued as well, because it is the (scaled) sum of the elements of X_0. Furthermore, the corresponding positive-sequence and negative-sequence components are *conjugates* of each other, viz.,

$$\left[\, X_{0,+} \quad X_{0,-} \,\right] = X_0 \begin{pmatrix} 1 & 1 \\ \alpha & \alpha^* \\ \alpha^* & \alpha \end{pmatrix} \frac{1}{\sqrt{3}}$$

where we used the fact that $\alpha^2 = \alpha^*$. This means that, for $k = 0$, the mapping (1.13b) reduces to a more compact, all real-valued, one-to-one correspondence, viz.,

$$X_0 \iff \left[\, X_{0,0} \quad \Re\{X_{0,+}\} \quad \Im\{X_{0,+}\} \,\right]$$

between the three elements of the real-valued X_0 and the three real-valued "independent components" $X_{0,0}$, $\Re\{X_{0,+}\}$, and $\Im\{X_{0,+}\}$. Since $\|X_0\|^2 \neq |X_{0,0}|^2 + |\Re\{X_{0,+}\}|^2 + |\Im\{X_{0,+}\}|^2$, this form of the one-to-one correspondence relation lacks the norm-preservation property (1.14). However, we observe that

$$\left\|X_0\right\|^2 = \left|X_{0,0}\right|^2 + \left|X_{0,+}\right|^2 + \left|X_{0,-}\right|^2 = \left|X_{0,0}\right|^2 + \left|\sqrt{2}\,\Re\{X_{0,+}\}\right|^2 + \left|\sqrt{2}\,\Im\{X_{0,+}\}\right|^2$$

where we used the property $X_{0,-} = X^*_{0,+}$. This implies that a norm-preserving correspondence can be obtained by appropriate scaling of the real and imaginary parts of $X_{0,+}$: the explicit form of this unitary correspondence is

$$\left[\, X_{0,0} \quad \sqrt{2}\,\Re\{X_{0,+}\} \quad \sqrt{2}\,\Im\{X_{0,+}\} \,\right] = X_0\, \Theta_0 \tag{1.16a}$$

where

$$\Theta_0 = \frac{1}{\sqrt{3}} \begin{pmatrix} 1 & \sqrt{2} & 0 \\ 1 & -\frac{\sqrt{2}}{2} & \sqrt{\frac{3}{2}} \\ 1 & -\frac{\sqrt{2}}{2} & -\sqrt{\frac{3}{2}} \end{pmatrix} \tag{1.16b}$$

is a real-valued unitary matrix (=orthogonal matrix), namely, it has the property $\Theta_0\, \Theta_0^\top = I$.

Finally, we observe that a three-phase finite-rms periodic waveform $x(t)$ can also be represented in terms of the three (finite-rms) *scalar* periodic waveforms

$$
\begin{aligned}
x_0(t) &\stackrel{\text{def}}{=} X_{0,0} + \sum_{k=1}^{\infty} \sqrt{2}\,\Re\left\{X_{k,0}\, e^{jk\omega_0 t}\right\} \\
x_+(t) &\stackrel{\text{def}}{=} X_{0,+} + \sum_{k=1}^{\infty} \sqrt{2}\,\Re\left\{X_{k,+}\, e^{jk\omega_0 t}\right\} \\
x_-(t) &\stackrel{\text{def}}{=} X_{0,-} + \sum_{k=1}^{\infty} \sqrt{2}\,\Re\left\{X_{k,-}\, e^{jk\omega_0 t}\right\}
\end{aligned}
\tag{1.17}
$$

which are associated with the *scalar* symmetrical component sets $\{X_{k,0}\,;\ k \geq 0\}$, $\{X_{k,+}\,;\ k \geq 0\}$, and $\{X_{k,-}\,;\ k \geq 0\}$, defined via (1.13). These three scalar waveforms are mutually orthogonal, viz.,

$$
\frac{1}{T}\int_T x_+(s)\, x_-(s)\, ds \;=\; 0 \;=\; \frac{1}{T}\int_T x_\pm(s)\, x_0(s)\, ds
$$

Consequently, we conclude (again) that $X_{rms}^2 = X_{0,rms}^2 + X_{+,rms}^2 + X_{-,rms}^2$, where $X_{0,rms}^2 = \frac{1}{T}\int_T x_0^2(s)\, ds$, and similarly for the positive and negative sequence components.

Thus, any (finite-rms) three-phase voltage waveform $v(t)$ can be fully represented by the mutually-orthogonal scalar waveforms $\{v_0(t), v_+(t), v_-(t)\}$, defined via (1.17) and, similarly, for the corresponding three-phase current waveform $i(t)$. It follows that the three power components defined in (1.15) in terms of *symmetrical component phasors* can be equivalently expressed in the *waveform domain* as

$$
P_0 = \langle\, v_0(\cdot)\,,\, i_0(\cdot)\,\rangle\;,\quad P_+ = \langle\, v_+(\cdot)\,,\, i_+(\cdot)\,\rangle\;,\quad P_- = \langle\, v_-(\cdot)\,,\, i_-(\cdot)\,\rangle \tag{1.18}
$$

by virtue of the Parseval relation between phasors and (finite-rms) waveforms.

1.4 Evaluation of Phasors from Waveform Samples

Digital evaluation of the harmonic phasors X_k associated with a periodic waveform $x(\cdot)$ requires, in a real-life scenario, (a) acquisition of the waveform $x(\cdot)$ by a suitable sensor, (b) discrete-time sampling, and (c) analog-to-digital (A/D) conversion. The resulting discrete-time sequence of samples is no longer perfectly periodic as a result of various (non-periodic) perturbations, such as sensor noise, aliasing of residual higher frequency harmonics, and A/D quantization noise – as well as possible presence of transients and inter-harmonics.

Nevertheless, in most practical cases the phasors X_k can still be *reliably estimated* from acquired noisy waveform samples: this challenge is analyzed in detail in Chap. 3. Here we limit our discussion to a brief overview of two distinct approaches to phasor estimation:

- A phasor extraction technique based on an idealized formulation, in which we assume that (a) the periodic waveform of interest is perfectly band-limited and (b) its uniformly-spaced samples can be acquired without any perturbations.
- A least-squares-based technique for phasor estimation from noisy waveform samples, which is a special case of the general solution provided in Sect. 3.2 of Chap. 3.

We show that both techniques *generate the same expression* for mapping waveform samples to estimated phasors: this expression coincides with a DFT/FFT of the sequence of N waveform samples acquired from a single period of $x(\cdot)$. It may come as a surprise that a phasor estimate obtained via an idealized formulation can still be used in a real-life scenario and even provide reliable suppression of the various perturbations associated with sensor-acquired waveform samples.

1.4.1 Idealized Formulation

When the waveform $x(\cdot)$ is perfectly band-limited and noise-free, we need to evaluate only a finite number of phasors, because $X_k = 0$ for all $k > K$ for some K. In this idealized case, we can obtain the exact values of $\{X_k \,;\; 0 \le k \le K\}$ by using samples of the continuous-time waveform $x(\cdot)$: this is a direct consequence of Nyquist's theorem, when the constraint $N \ge 2K$ (a.k.a. "Nyquist condition") is satisfied.

Instead of relying on Nyquist theorem arguments, we use here a direct approach, which results in a system of linear equations for $\{X_k \,;\; 0 \le k \le K\}$. For simplicity of presentation we focus in the remainder of this chapter on scalar (single-phase) waveforms. *A noise-free perfectly band-limited* single-phase periodic waveform $x(\cdot)$ can be represented by a *finite Fourier series*, viz.,

$$x(t) \;=\; X_0 + \sum_{k=1}^{K} \sqrt{2}\, \Re \left\{ X_k \, e^{jk\omega_0 t} \right\} \tag{1.19}$$

A finite set of N equally-spaced samples of this waveform, covering a single cycle, is obtained by choosing the sampling interval as $\frac{T}{N}$, so that

$$x\left(t_0 + n\frac{T}{N}\right) \;=\; X_0 + \sum_{k=1}^{K} \sqrt{2}\, \Re \left\{ X_k \, e^{jk\omega_0 t_0} e^{j(2\pi/N)kn} \right\} \quad , \qquad 0 \le n \le N-1$$

This is a set of N linear equations in the $2K+1$ *real-valued* unknowns

$$\{\, X_0,\; \Re X_1,\; \Im X_1,\; \Re X_2,\; \Im X_2, \ldots,\; \Re X_K,\; \Im X_K \,\}$$

A unique solution is obtained when we set $N = 2K + 1$ (which also satisfies the Nyquist condition), so that

$$\sqrt{2}\, x\left(t_0 + n\frac{T}{N}\right) = \sqrt{2}\, X_0 + \sum_{k=1}^{K} X_k\, e^{jk\omega_0 t_0} e^{j(2\pi/N)kn} + \sum_{k=1}^{K} X_k^*\, e^{-jk\omega_0 t_0} e^{-j(2\pi/N)kn}$$

which can be rewritten in the more compact form

$$\sqrt{2}\; x\left(t_0 + n\frac{T}{N}\right) = \sum_{k=0}^{N-1} \breve{X}_k\, e^{j(2\pi/N)kn} \quad , \qquad 0 \le n \le N-1 \tag{1.20a}$$

where

$$\breve{X}_k \;\stackrel{\text{def}}{=}\; \begin{cases} \sqrt{2}\, X_0 & k = 0 \\ X_k\, e^{jk\omega_0 t_0} & 1 \le k \le K \\ \left(\breve{X}_{N-k}\right)^* & K+1 \le k \le N-1 \end{cases} \tag{1.20b}$$

and we used the fact that $e^{-j(2\pi/N)kn} = e^{j(2\pi/N)(N-k)n}$.

We recognize (1.20a) as the N-point *inverse DFT relation* between the real-valued samples $\{x(t_0),\ x(t_0 + \frac{T}{N}),\ x(t_0 + 2\frac{T}{N}), \ldots,\ x(t_0 + (N-1)\frac{T}{N})\}$ and the (scaled) complex phasors $\{\breve{X}_0, \breve{X}_1, \breve{X}_2, \ldots, \breve{X}_{N-1}\}$. Inverting this relation produces a *forward DFT relation* (see, e.g., [32, 33]), viz.,

$$\breve{X}_k = \frac{\sqrt{2}}{N} \sum_{n=0}^{N-1} x\left(t_0 + n\frac{T}{N}\right) e^{-j(2\pi/N)kn} \quad , \qquad 0 \le k \le N-1 \tag{1.21}$$

The *Fast Fourier Transform* (FFT) [32, 33] provides an efficient computational scheme for evaluating $\{\breve{X}_k\,;\ 0 \le k \le N-1\}$ at a cost of $\frac{N}{2}\log_2 N$ complex multiplications.

The original phasors $\{X_0,\ X_1,\ X_2,\ \ldots,\ X_K\}$ can now be recovered via (1.20b), namely,

$$X_k = \begin{cases} \frac{1}{\sqrt{2}}\, \breve{X}_0 & k = 0 \\ e^{-jk\omega_0 t_0}\, \breve{X}_k & 1 \le k \le K \end{cases}$$

Notice that the corresponding explicit expression for X_k, viz.,

$$X_k = \begin{cases} \frac{1}{\sqrt{2}} \breve{X}_0 \equiv \frac{1}{N} \sum_{n=0}^{N-1} x\left(t_0 + n\frac{T}{N}\right) & k = 0 \\ e^{-jk\omega_0 t_0} \breve{X}_k \equiv \frac{\sqrt{2}}{N} \sum_{n=0}^{N-1} x\left(t_0 + n\frac{T}{N}\right) e^{-jk\omega_0\left(t_0 + n\frac{T}{N}\right)} & 1 \leq k \leq K \end{cases}$$

looks like a discretized approximation of the continuous-time phasor definition (1.4), applied to the cycle-long interval $t_0 \leq t < t_0 + T$. Thus, the final expression for X_k in the idealized case is

$$X_k = \frac{a_k}{N} \sum_{n=0}^{N-1} x\left(t_0 + n\frac{T}{N}\right) e^{-jk\omega_0\left(t_0 + n\frac{T}{N}\right)} \tag{1.22}$$

where the scaling coefficient a_k is as defined in (1.4b), namely $a_k = \sqrt{2}$ for $k \geq 1$, while $a_0 = 1$. The preceding analysis demonstrates that when the waveform $x(t)$ is perfectly band-limited and noise-free, as we assumed in (1.19), this discrete-time expression generates the *exact value* of the phasor X_k.

1.4.2 Real-Life Scenario

In reality, no sensor-acquired waveform is perfectly band-limited or noise-free, which means that the idealized Fourier series representation (1.19) is replaced, in a real-life scenario, by

$$x(t) = X_0 + \sum_{k=1}^{K} \sqrt{2}\, \Re\left\{X_k\, e^{jk\omega_0 t}\right\} + \text{perturbation} \tag{1.23}$$

The perturbation component of $x(\cdot)$ consists of (a) sensor measurement noise, (b) higher frequency harmonics (above the K-th harmonic) of the original periodic waveform, and (c) various non-harmonic components, such as inter-harmonics and transients. The sampled version of $x(\cdot)$ contains additional perturbations caused by sampling – such as folding of higher frequency components to lower frequencies, known as *aliasing* – as well as quantization of the sensor-acquired waveform, caused by analog-to-digital conversion. This means that exact recovery of the phasors X_k from samples of the waveform $x(\cdot)$, as described in Sect. 1.4.1, is no longer possible in a real-life scenario. Instead, one needs to construct a *reliable estimate* $\widehat{X}_k$ of the phasor X_k: this challenge is analyzed in detail in Chap. 3.

The quality of such an estimate depends on the strength of the perturbation component in (1.23). In particular, the least-squares estimate that we present in Sect. 3.2 is a linear function of the waveform samples, and thus the level of error (i.e., $\widehat{X}_k - X_k$) associated with this estimate is proportional to the relative strength of the perturbation component in (1.23).

Real-life sampling of a continuous-time waveform consists, actually, of three separate steps [32, 33]: (a) low-pass filtering by an *anti-aliasing filter*, (b) sampling at the uniformly-spaced time-instants $t_n = nT_s$, and (c) A/D conversion. Since the analog anti-aliasing filter cannot have a very narrow transition band, it is customary to set the sampling rate to be well above twice the highest frequency of interest, a practice known as "oversampling" [32, 33]. For instance, typical sampling rates of modern *Phasor Measurement Units* (PMUs) range from 128 samples/cycle to 512 samples/cycle [34–37], which results in $128 \le N \le 512$. Control-oriented considerations may dictate using even higher sampling rates [38, 39].

Anti-aliasing filtering results in suppressing the high-frequency components of the waveform, thereby reducing the effect of aliasing, as well as the level and bandwidth of additive noise. Notice that the cut-off frequency of the anti-aliasing filter determines the value of K in (1.23) and thus controls the magnitude of the higher frequency harmonic component of the perturbation – the only perturbation component whose level can be controlled at will. The K-value defined by this selection should ensure that the higher frequency harmonic component is only a minor contributor to the overall perturbation in the resulting waveform samples.

Section 3.2 provides a comprehensive least-squares solution to the problem of phasor estimation from polyphase waveform samples. It allows (a) using M out of the available N samples within an arbitrary time interval and (b) estimating a subset of L phasors (not necessarily contiguous). These design parameters must satisfy the constraint $2L \le M \le N$.

However, here we need to consider only the simplest version of this solution, with $M = N$, i.e., using all available N waveform samples within a single cycle. Notice that because of the presence of a non-periodic perturbation, the resulting estimate of X_k will be (slightly) dependent on the choice of the cycle-long interval of waveform samples used to form this estimate, say $t_1 - T < t \le t_1$. Under these simplifying assumptions, the set of waveform samples used to construct a least-squares estimate $\widehat{X}_k$ is $\left\{x(t_1 - i\frac{T}{N});\ 0 \le i \le N-1\right\}$, so that t_1 indicates the location of the last sample in the chosen time interval. This means that the sampling interval is $T_s = \frac{T}{N}$, and the resulting phasor estimate – derived from the general expression (3.18a) – is

$$\widehat{X}_k = \frac{\sqrt{2}}{N} \sum_{i=0}^{N-1} x\left(t_1 - i\frac{T}{N}\right) e^{-jk\omega_0\left(t_1 - i\frac{T}{N}\right)}$$

This samples-to-phasor mapping can be equivalently expressed in terms of the location of the first sample in the chosen time interval, i.e., $t_0 \stackrel{\text{def}}{=} t_1 - (N-1)\frac{T}{N}$,

so that

$$\widehat{X}_k = \frac{\sqrt{2}}{N} \sum_{n=0}^{N-1} x\left(t_0 + n\frac{T}{N}\right) e^{-jk\omega_0\left(t_0+n\frac{T}{N}\right)} \tag{1.24}$$

We observe that this expression coincides with the "exact recovery" solution (1.22) we obtained from the idealized formulation in Sect. 1.4.1. In other words, the direct DFT/FFT estimate (1.22) applies in both the idealized case and a real-life scenario: in the latter, it provides reliable suppression of the various perturbations associated with sensor-acquired waveform samples.

1.5 Sparse Evaluation of Dominant Harmonics

The concept of approximating a given scalar waveform with a small number of sinusoids is at the heart of so-called *sparse FFT* techniques (see, e.g., [40,41]). Such algorithms use a small subset of the available waveform samples to evaluate the Fourier transform of the given waveform only at the dominant peaks. When applied to general (non-periodic) waveforms, the main computational effort is directed at identifying which samples to use and which (waveform-specific) frequency components to evaluate.

In contrast, the dominant frequencies of steady-state power system waveforms are always multiples of the fundamental harmonic frequency and are thus *known in advance.* Moreover, a steady-state power system waveform has only a relatively small number of non-negligible phasors: long-term records of power system voltages and currents tell us that the most common frequencies in steady-state operation, in addition to the fundamental (50 Hz or 60 Hz), are the 5-th, 7-th, and 11-th harmonics and, occasionally, also the 13-th, 17-th, and 19-th harmonics.[4] In other words, $L \ll N$, where L is the number of non-negligible phasors and N is the number of samples/cycle.

The best L-harmonic approximation to the waveform $x(\cdot)$ is obtained by using the L phasors X_k with the largest norm. To establish this claim we consider the optimization problem

$$\min_{\mathcal{S}} \min_{\{\widehat{X}_k\,;\,k\in\mathcal{S}\}} \frac{1}{T}\int_T \left[x(t) - \widehat{x}(t)\right]^2 dt \tag{1.25a}$$

where the approximating waveform $\widehat{x}(\cdot)$ consists of L harmonics, viz.,

$$\widehat{x}(t) = \sum_{k\in\mathcal{S}} \sqrt{2}\,\Re\left\{\widehat{X}_k\, e^{jk\omega_0 t}\right\} \tag{1.25b}$$

[4] This observation holds even in the presence of most transients, as we demonstrate in Sect. 3.3.

and $\mathcal{S} = \{k_\ell\,;\ 1 \le \ell \le L\}$ denotes a selected set of L harmonic indices. In view of the Parseval identity (1.9c), this is equivalent to the phase-domain optimization problem

$$\min_{\mathcal{S}} \min_{\{\widehat{X}_k\,;\,k\in\mathcal{S}\}} \left\{ \sum_{k\in\mathcal{S}} \left|X_k - \widehat{X}_k\right|^2 + \sum_{k\notin\mathcal{S}} \left|X_k\right|^2 \right\} \tag{1.26}$$

Thus, the optimal choice for any prescribed index set $\mathcal{S}$ is to set $\widehat{X}_k = X_k$ for all $k \in \mathcal{S}$. The resulting approximation error is $\sum_{k\notin\mathcal{S}}\left|X_k\right|^2$, which, in turn, is minimized by selecting $\mathcal{S}$ to include the L harmonics with the largest phasor magnitudes $\left|X_k\right|$.

When L is sufficiently small, the computational cost of directly evaluating L dominant phasors via (1.21), which is $2LN$ real multiplications, is smaller than $\frac{N}{2}\log_2 N$, the cost of an FFT-based evaluation of a full set of N phasors. To be specific, this is the case when $L < \log_2 N$: for instance, we can evaluate nine dominant phasors via (1.21) using $N = 512$ samples/cycle for the same cost as a 512-point FFT. Thus, to achieve a meaningful reduction in cost, we are motivated to consider reducing the number of waveform samples used: this can be achieved by using a sparse subset of waveform samples.

Efficient computation of selected phasors is especially important in the context of *dynamic phasors* (see Chap. 2) because of the need to repeat this process for every position of a sliding window, namely for every new waveform sample. Recall that the overall set of waveform samples is usually densely spaced, so as to allow fast tracking and/or detection of transient phenomena. These observations lead us to introduce in Sect. 3.2 a least-squares approach for the estimation of L selected phasors from a *sparse subset* of uniformly-spaced waveform samples. In general, given a set of N samples per cycle, we can use a subset of M of these waveform samples, where $2L \le M \le N$. This means that the integer-valued design parameter M captures the excess degrees of freedom in the least-squares problem setting (1.26). In particular, this conceptual framework allows reliable estimation of the fundamental phasor: this involves setting $L = 1$ and using all available excess degrees of freedom, namely, $M = N$ (see also [42]).

On the other extreme, choosing $M = 2L$ results in the lowest computational cost. The resulting *sub-cycle phasor estimation scheme*, which we introduce in Sect. 3.3, uses the smallest possible number of samples to estimate a few dominant dynamic phasors. For instance, in the special case when $x(\cdot)$ consists of a finite number of harmonics (and no DC component), say $x(t) = \sqrt{2}\sum_{k\in\mathcal{S}} \Re\{X_k\, e^{jk\omega_0 t}\}$, where $\mathcal{S} = \{k_\ell\,;\ 1 \le \ell \le L\,,\ k_\ell \ne 0\}$, we can set up a system of linear equations using $2L$ samples, namely, a sparse equivalent of (1.22). The (true) phasors X_k can then be recovered without error by solving this set of equations.

Chapter 2 Dynamic Fourier Phasors

Chapter 1 of this book presented a conceptual framework for phasor-based representation of *periodic* finite-rms polyphase waveforms. In the current chapter, we extend this framework to *non-periodic* polyphase waveforms occurring as deviations from *periodic* steady-state operation with period T. This is achieved by modifying the original Hilbert space of waveforms, which was defined over the fixed interval $(0 \quad T]$, to a time-dependent Hilbert space of polyphase waveforms defined over the sliding interval $(t - T \quad t]$, where "t" denotes a time-instant of interest.

We show in Sect. 2.1 that all well-known attributes of the classical phasor-based representation are still valid in the non-periodic setting. However, the resulting *dynamic Fourier phasors,* as well as all metrics derived from them, become functions of the time variable "t." In particular, the cycle-averaged rms voltage, rms current, and real power are time-variant quantities. The same conceptual framework can also be used to generate dynamic (i.e., time-variant) symmetrical sequence components, as detailed in Sect. 2.3. In Chap. 9 of this book, we explore the key role played by dynamic phasors in constructing advanced power flow metrics that provide detailed information about power quality both in steady-state operation and in the presence of transients.

The dynamic behavior of each such time-variant phasor can be described in terms of a time-invariant differential equation, as demonstrated in Sect. 2.2. In particular, every dynamic voltage–current relation (i.e., a circuit equation) that can be mapped into a linear state-space model with constant coefficients induces a similar relation between the dynamic phasors associated with these voltage and current waveforms. This key property of dynamic phasors is explored in Sect. 2.4. It turns out, however, that mapping initial conditions from the original time-domain circuit equation into its phasor-domain counterpart is a highly non-trivial task. We introduce in Sect. 2.5 a complete characterization of all admissible choices of initial conditions in the phasor domain and highlight a few particularly useful schemes.

H. Lev-Ari, A. M. Stanković, *Dynamic Phasors in Energy Processing Systems*,
Power Electronics and Power Systems, https://doi.org/10.1007/978-3-031-50020-6_3

The discussion in Sect. 2.4 relies on a filter-bank interpretation of the map from a waveform to its dynamic phasors. This observation motivates a variety of generalizations, which are examined in Chaps. 4 and 5. The resulting schemes generate decompositions of the original time-domain waveform into a collection of frequency subband components. Since the bandwidth of each individual component is significantly smaller than that of the original input waveform, the subband sampling rate used in simulation can be much lower than the rate needed for a faithful representation of the time-domain input waveform. The same observation applies, of course, also to the Fourier-based dynamic phasors presented in the current chapter.

Dynamic phasors have a variety of applications in monitoring power flow in transmission and distribution grids, as well as in modeling and control of electric drives and power electronic converters. In particular, dynamic phasors are highly useful for modeling of unbalanced polyphase systems that include power converters (e.g., rectifiers and inverters) and electric machines. In such analyses, we can vary the number of phasors at different frequencies to address a particular problem. In ac motor examples presented in Chap. 7, all frequencies are harmonically related. A more general case occurs, for example, when mechanical torque load has angular dependence. In another example, we can include a dynamic phasor at the PWM frequency and some of its multiples, in addition to the fundamental and harmonics of the ac supply. This hierarchical nature of the dynamic phasor approach is valuable when models of a system are needed with varying levels of detail.

Phasor dynamic models are typically developed from time-domain descriptions (differential equations) using a procedure that we denote as *generalized averaging* (see Sect. 2.1). In the case of nonlinear equations, a key element in the modeling process is the development of approximations to the right-hand side of the time-domain equations at a particular frequency. These approximations are based on the describing function method [43, 44] and typically problem-dependent; some general error bounds are provided in [9, 44]. The concept of dynamic phasors was introduced in [45] and applied to series resonant and switched mode dc–dc converters. Extensions to multi-frequency averaging that takes into account interactions between harmonics were presented in [46].

Dynamic phasor models are nonlinear in general and represent large-signal descriptions. As such, they are different from models expressed in terms of spiral vectors [47], which rely on system linearity to propagate the signals in the model. Spiral vectors in a way echo the classical approach to differential equations that requires modeling a system and then providing an explicit formula for the solution. Our approach is more in line with a control-theoretic perspective, based on the qualitative theory of differential equations, that studies equilibria, stability, and effects of parametric changes without requiring explicit formulas for the solution. Dynamic phasors have connections with the concept of space vectors that is extensively used in modeling of electric machines; we explore those connections in Chap. 7.

2.1 Dynamic Fourier-Series Representation

We now turn to introduce an explicitly time-variant Hilbert space framework for real-valued (scalar) waveforms with finite local power, viz.,

$$\frac{1}{T}\int_{t-T}^{t} x^2(s)\, ds \; < \; \infty \qquad \text{for all } t \tag{2.1a}$$

For a fixed time-instant "t," our space consists of finite waveform segments, and we define the local (= sliding-window) inner-product in this space as

$$\langle\, x(\cdot)\,,\; y(\cdot)\,\rangle\,(t) \;\stackrel{\text{def}}{=}\; \frac{1}{T}\int_{t-T}^{t} x(s)\, y(s)\, ds \tag{2.1b}$$

The sliding-window Fourier-series representation is based on the concept of dynamic phasors. This means that a *one-sided Fourier series* can be associated with the real-valued finite waveform segment $\{x(s)\,;\; t-T \;<\; s \;\le\; t\}$ where T is the duration of a cycle (or the period in a steady-state), $\omega_0 \stackrel{\text{def}}{=} 2\pi/T$, and we consider "$t$" as a parameter, viz.,

$$x(s) = X_0(t) \;+\; \sqrt{2}\sum_{k=1}^{\infty} \Re\left\{X_k(t)\, e^{jk\omega_0 s}\right\}\;, \qquad t-T \;<\; s \;\le t \tag{2.2a}$$

The rms dynamic phasor $X_k(t)$ is determined via a sliding-window operation, viz.,

$$X_k(t) \;=\; \begin{cases} \langle\, x\,\rangle_0\,(t) & k = 0 \\ \sqrt{2}\;\langle\, x\,\rangle_k\,(t) & k \ge 1 \end{cases} \tag{2.2b}$$

where

$$\langle\, x\,\rangle_k\,(t) \;\stackrel{\text{def}}{=}\; \frac{1}{T}\int_{t-T}^{t} x(s)\, e^{-jk\omega_0 s}\, ds \qquad , \qquad -\infty < k < \infty \tag{2.2c}$$

is known as a *generalized average* [45]. The dynamic phasor $X_k(t)$ is a function of the window location parameter t, except when $x(\cdot)$ is a periodic waveform with period T: in this case, $X_k(t)$ is independent of t and coincides with the standard (rms, one-sided) Fourier phasor. We notice that the Fourier-series representation (2.2a) is "non-causal" in the sense that $x(s)$ for $s < t$ is represented in terms of $X_k(t)$. It is also non-unique, in the sense that the same $x(s)$ can be associated with $X_k(t)$ for multiple values of t. For these reasons, we restrict our attention in the sequel to $s = t$, so that (2.2a) reduces to a causal and unique representation, viz.,

$$x(t) = X_0(t) + \sqrt{2} \sum_{k=1}^{\infty} \Re \left\{ X_k(t)\, e^{jk\omega_0 t} \right\} \tag{2.2d}$$

The Parseval identity for the Fourier series (2.2) can be written in the form (recall (1.9c))

$$\langle\, x(\cdot)\,,\; y(\cdot)\,\rangle\,(t) \;=\; X_0(t)\, Y_0(t) \;+\; \sum_{k=1}^{\infty} \Re \left\{\, X_k(t)\, Y_k^*(t) \right\} \tag{2.3}$$

where we used our definition (2.1b). The Parseval identity defines an isometry between the local Hilbert space of waveforms on the t-dependent, sliding-window, interval $(t - T \quad t]$, and the Hilbert space of time-dependent phasor sequences associated with these waveforms. This means that all local inner-products can be evaluated via appropriate dynamic phasors.

The dynamic Fourier-series relationship extends naturally to *matrices of waveforms*. In general, we can represent any matrix of real-valued, finite-rms waveforms, say $x(t) = \left[x^{(i,j)}(t) \right]_{i=1:n,\, j=1:m}$ (for any n, m), in terms of the dynamic matrix rms phasors $X_k(t) = \left[X_k^{(i,j)}(t) \right]_{i=1:n,\, j=1:m}$, where $\{X_k^{(i,j)}(\cdot)\,;\; 0 \le k < \infty\}$ is the dynamic rms phasor sequence associated with the scalar waveform $x^{(i,j)}(\cdot)$ via the waveform-to-phasor map (2.2 b,c).

A case of particular interest (in Part III) is that of *polyphase waveforms:* in this case, the waveforms $x(t)$ and $y(t)$ are $1 \times m$ matrices (i.e., row vectors), and

$$\begin{aligned} \langle\, x(\cdot)\,,\; y(\cdot)\,\rangle\,(t) &\overset{\text{def}}{=} \frac{1}{T} \int_{t-T}^{t} x(s)\, y^{\top}(s)\, ds \\ &= X_0(t)\, Y_0^{\top}(t) + \sum_{k=1}^{\infty} \Re \left\{ X_k(t)\, Y_k^H(t) \right\} \;=\; \Re \left\{ \mathbf{X}(t)\, \mathbf{Y}^H(t) \right\} \end{aligned} \tag{2.4a}$$

where the superscript $^{\top}$ denotes transposition, the superscript ^{H} denotes conjugate transposition, and we use the dynamic (time-dependent) *phasor arrays*

$$\begin{aligned} \mathbf{X}(t) &\overset{\text{def}}{=} \left[\, X_0(t) \quad X_1(t) \quad X_2(t) \quad \ldots \,\right] \\ \mathbf{Y}(t) &\overset{\text{def}}{=} \left[\, Y_0(t) \quad Y_1(t) \quad Y_2(t) \quad \ldots \,\right] \end{aligned} \tag{2.4b}$$

The polyphase Parseval identity (2.4) implies that all local inner-products can be evaluated via appropriate dynamic phasors.

In particular, we can express the dynamic rms voltage and current, as well as the dynamic active power $P(t)$, either in terms of (polyphase) waveforms or in terms of phasor arrays, viz.,

$$V_{rms}^2(t) \stackrel{\text{def}}{=} \| v(\cdot)\|^2(t) = \|\mathbf{V}(t)\|^2 \quad , \quad I_{rms}^2(t) \stackrel{\text{def}}{=} \| i(\cdot)\|^2(t) = \| \mathbf{I}(t) \|^2$$

$$P(t) \stackrel{\text{def}}{=} \langle\, v(\cdot)\,,\, i(\cdot)\,\rangle(t) = \Re\left\{\mathbf{V}(t)\,\mathbf{I}^H(t)\right\} \tag{2.5}$$

where $\mathbf{V}(t)$ (resp., $\mathbf{I}(t)$) is the dynamic phasor array associated with the (non-periodic) polyphase waveform $v(\cdot)$ (resp., $i(\cdot)$).

Example 2.1 (Piecewise-constant Fourier series): Consider the multi-harmonic waveform

$$x(t) = \begin{cases} \sqrt{2}\displaystyle\sum_{k=1}^{\infty} \Re\left\{A_k\, e^{jk\omega_0 t}\right\} & t \geq 0 \\ \sqrt{2}\displaystyle\sum_{k=1}^{\infty} \Re\left\{B_k\, e^{jk\omega_0 t}\right\} & t < 0 \end{cases} \tag{2.6}$$

which exhibits a jump in its (piecewise-constant) Fourier coefficients at $t = 0$. Clearly, $X_k(t)$ remains constant in steady-state, namely

$$X_k(t) = \begin{cases} A_k & t \geq T \\ B_k & t \leq 0 \end{cases} \qquad \text{for } k \geq 1 \tag{2.7}$$

while $X_0(t) = 0$ outside the interval $0 \leq t \leq T$. In particular, if $A_k = B_k$ (no jump), the dynamic phasor $X_k(t)$ is time-invariant, namely $X_k(t) = A_k$ and $X_0(t) = 0$ for all t.

However, in general, $X_k(t)$ is truly time-variant in the transition interval $0 \leq t \leq T$. Indeed, in this interval, for $k \geq 1$,

$$\begin{aligned} X_k(t) &\stackrel{\text{def}}{=} \frac{\sqrt{2}}{T}\int_{t-T}^{t} x(s)\, e^{-jk\omega_0 s}\, ds \\ &= \frac{1}{T}\left\{\int_{t-T}^{0}\left(\sum_{m=1}^{\infty} 2\,\Re\left\{B_m\, e^{jm\omega_0 s}\right\}\right) e^{-jk\omega_0 s}\, ds \right. \\ &\qquad \left. + \int_{0}^{t}\left(\sum_{m=1}^{\infty} 2\,\Re\left\{A_m\, e^{jm\omega_0 s}\right\}\right) e^{-jk\omega_0 s}\, ds\right\} \end{aligned}$$

$$= \left[A_k(t/T) + B_k(1 - t/T)\right] + \sum_{\substack{m\geq 1\\ m\neq k}} \left(A_m - B_m\right) \frac{e^{j(m-k)\omega_0 t} - 1}{2\pi j(m-k)}$$

$$+ \sum_{m\geq 1} \left(A_m^* - B_m^*\right) \frac{1 - e^{j(m+k)\omega_0 t}}{2\pi j(m+k)}$$

or, equivalently,

$$X_k(t) = A_k - \left(A_k - B_k\right)\left(1 - t/T\right) \tag{2.8}$$

$$+ \left\{ \sum_{\substack{m\geq 1\\ m\neq k}} \left(A_m - B_m\right) \frac{\sin\frac{(k-m)\,\omega_0 t}{2}}{(k-m)\pi}\; e^{-j\frac{(k-m)\omega_0 t}{2}} + \sum_{m\geq 1} \left(A_m^* - B_m^*\right) \frac{\sin\frac{(k+m)\,\omega_0 t}{2}}{(k+m)\pi}\; e^{-j\frac{(k+m)\omega_0 t}{2}} \right\}$$

This expression consists of two parts: the first part is a weighted average, gradually changing from B_k (at $t = 0$) to A_k (at $t = T$), while the second part is oscillatory, vanishing at the end points $t = 0$ and $t = T$. In particular, if $A_m = B_m$ for all m (no jump), the second part vanishes completely, so that $X_k(t) = A_k$ for all t. It is important to notice that $X_k(t) - A_k$ is made of contributions from the coefficient jumps $A_m - B_m$ *for all m values*, i.e., we have a "cross-harmonic" effect.

The cross-harmonic effect also generates a non-vanishing "DC dynamic phasor" $X_0(t)$ in the transition interval $0 \leq t \leq T$, even though $A_0 = 0 = B_0$. Indeed,

$$X_0(t) \overset{\text{def}}{=} \frac{1}{T}\int_{t-T}^{t} x(s)\, e^{-jk\omega_0 s}\, ds$$

$$= \frac{1}{\sqrt{2}} \left\{ \sum_{m=1}^{\infty} \left(A_m - B_m\right) \frac{e^{jm\omega_0 t} - 1}{2\pi jm} + \sum_{m=1}^{\infty} \left(A_m^* - B_m^*\right) \frac{1 - e^{jm\omega_0 t}}{2\pi jm} \right\}$$

which we can rewrite as

$$X_0(t) = \sqrt{2} \sum_{m=1}^{\infty} \Re\left\{\left(A_m - B_m\right) e^{j\frac{m\omega_0 t}{2}}\right\} \frac{\sin\frac{m\,\omega_0 t}{2}}{m\pi} \tag{2.9}$$

In summary, a jump in the amplitude of even a single harmonic results in a transient in all harmonics.

The dynamic phasors $X_k(t)$ decay with increasing k, a property shared with all Fourier-series expansions of finite-local-power waveforms. This phenomenon is clearly evident, for instance, in the special case when $x(t)$ consists of a single harmonic, so that $A_k = 0 = B_k$ for all $k \neq 1$. Then, for $0 \leq t \leq T$,

$$X_1(t) = A_1 - \left(A_1 - B_1\right)\left(1 - t/T\right) + \left[\left(A_1 - B_1\right)^*\right] \frac{\sin\omega_0 t}{2\pi}\; e^{-j\omega_0 t} \tag{2.10a}$$

while

$$X_0(t) = \sqrt{2}\,\Re\left\{\left(A_1 - B_1\right)e^{j\frac{\omega_0 t}{2}}\right\}\frac{\sin\frac{\omega_0 t}{2}}{\pi} \tag{2.10b}$$

and, for $k \geq 2$,

$$X_k(t) = \left(A_1 - B_1\right)\frac{\sin\frac{(k-1)\,\omega_0 t}{2}}{(k-1)\pi}\,e^{-j\frac{(k-1)\omega_0 t}{2}} + \left[\left(A_1 - B_1\right)^*\right]\frac{\sin\frac{(k+1)\,\omega_0 t}{2}}{(k+1)\pi}\,e^{-j\frac{(k+1)\omega_0 t}{2}} \tag{2.10c}$$

Thus, in this special case, $|X_k(t)|$ decays as $1/k$ at each time-instant within the transition interval.

In summary, we observe again that a step change in the input waveform results in a cycle-long transition interval $0 \leq t \leq T$, in which the dynamic phasors exhibit a cross-harmonic effect. Since the oscillatory component of each $X_k(t)$ is limited to a cycle-long interval, its frequency bandwidth is, approximately, $\frac{1}{T} \equiv f_0$. ■

2.2 Dynamics of $X_k(t)$

As observed in Sect. 2.1, the dynamic Fourier-series relationship (2.2) extends naturally to *matrices of waveforms*. In general, we can represent any matrix $x(t)$ of real-valued, finite-rms waveforms, in terms of the dynamic matrix rms phasors $X_k(t)$ associated with the matrix waveform $x(\cdot)$ via the waveform-to-phasor map (2.2 b,c). It will be convenient to rewrite (2.2 b,c) in the form

$$X_k(t) = \frac{a_k}{T}\int_0^T x(t-\tau)\,e^{-jk\omega_0(t-\tau)}\,d\tau \tag{2.11a}$$

where

$$a_k \overset{\text{def}}{=} \begin{cases} 1 & k = 0 \\ \sqrt{2} & k \geq 1 \end{cases} \tag{2.11b}$$

Consequently,

$$\begin{aligned} \frac{d}{dt}X_k(t) &= \frac{a_k}{T}\int_0^T \frac{d}{dt}\left\{x(t-\tau)\,e^{-jk\omega_0(t-\tau)}\right\}d\tau \\ &= \frac{a_k}{T}\int_0^T \dot{x}(t-\tau)\,e^{-jk\omega_0(t-\tau)}\,d\tau \\ &+ \frac{a_k}{T}\int_0^T x(t-\tau)\left(-jk\omega_0\right)e^{-jk\omega_0(t-\tau)}\,d\tau \end{aligned}$$

so that

$$\frac{d}{dt}\, X_k(t) \;=\; a_k \left\langle \frac{d}{dt}\, x \right\rangle_k (t) \;-\; jk\omega_0\, X_k(t) \tag{2.12}$$

In deriving this relation, we assumed that $x(\cdot)$ is known in the cycle-long interval $[t - T \quad t]$. Thus, if the waveform $x(t)$ is known for every $-\infty < t < \infty$, the dynamic phasor $X_k(t)$ can be defined via (2.11) for every t, which implies that the differential equation (2.12) also holds for every t.

However, in general, waveforms are known only over one-sided intervals. Thus a more realistic assumption would be that $x(t)$ is known only for $t \geq t_0$, for some given t_0. Without loss of generality, we can set $t_0 = 0$: in this case, the dynamic phasor $X_k(t)$ can be evaluated via (2.11) only for $t \geq T$: consequently, the differential equation (2.12) holds in the same interval. However, when we have information about the dynamical evolution of $x(t)$ – e.g., a state-space model description – it may be possible to extend the region of definition of the differential equation (2.12) to the interval $t \geq 0$ (see Sect. 2.5 for a detailed discussion).

Example 2.2 (RL circuit): Consider a single-phase RL circuit, described for $t \geq 0$ by the differential equation

$$L\,\frac{di}{dt} + R\,i(t) = v(t) \tag{2.13a}$$

and driven by the voltage

$$v(t) = \sqrt{2}\,\Re\left\{V_1\, e^{j\omega_0 t}\right\} \quad , \qquad t \geq 0 \tag{2.13b}$$

where $\omega_0 = \frac{2\pi}{T}$. Nothing is known (or assumed) about the RL circuit for $t < 0$. However, we will assume a known initial value $i(0)$ for the current.

The differential equation (2.13) admits the solution (for all $t \geq 0$)

$$i(t) = i_{ss}(t) + i_{tr}(t) \tag{2.14a}$$

with the steady-state and transient components of $i(t)$ given by

$$i_{ss}(t) = \sqrt{2}\;\Re\left\{\frac{V_1}{R(1+j\zeta)}\; e^{j\omega_0 t}\right\} \quad , \qquad i_{tr}(t) = \Big[i(0) - i_{ss}(0)\Big]\, e^{-\alpha t} \tag{2.14b}$$

Here $\alpha \stackrel{\text{def}}{=} R/L$, $\zeta \stackrel{\text{def}}{=} \omega_0/\alpha = 2\pi/\alpha T$, and $i(0)$ is the (arbitrary) initial value for the current.

Since the steady-state component $i_{ss}(t)$ is periodic, it consists of a single Fourier harmonic with a time-invariant amplitude and phase. As a result, for $t \geq T$,

$$\frac{a_k}{T}\int_{t-T}^{t} i_{ss}(s)e^{-jk\omega_0 s}ds = \begin{cases} \frac{V_1}{R(1+j\zeta)} & k=1 \\ 0 & \text{otherwise} \end{cases}$$

where we recall that $a_k = \sqrt{2}$ for $k \geq 1$, and $a_0 = 1$. On the other hand, the transient component $i_{tr}(t)$ contributes to all harmonics, viz.,

$$\xi_k(t) \stackrel{\text{def}}{=} \frac{a_k}{T}\int_{t-T}^{t} i_{tr}(s)e^{-jk\omega_0 s}ds = a_k\, i_{tr}(0)\,\frac{e^{\alpha T}-1}{\alpha T + jk2\pi}\, e^{-(\alpha+jk\omega_0)t} \tag{2.15a}$$

which corresponds to a decaying sinusoid with frequency $k\omega_0$. In summary, the dynamic (rms, one-sided) Fourier phasors associated with $i(t)$ are (for $t \geq T$)

$$I_k(t) = \begin{cases} \xi_0(t) & k=0 \\ \frac{V_1}{R(1+j\zeta)} + \xi_1(t) & k=1 \\ \xi_k(t) & k \geq 2 \end{cases} \tag{2.15b}$$

Notice that the transient part of $I_k(t)$ exists for all k, even though the steady-state part of the same dynamic phasor vanishes for $|k| \neq 1$. Also notice that $|\xi_k(t)|^2$ is proportional to $e^{-2\alpha t}/(1+k^2\zeta^2)$, decaying with time and with increasing harmonic index. This means that we need only a small number of higher harmonics to capture the transient in $i(t)$. In fact, if we choose $R = 0.1\,\Omega$, $L = 1\,\text{mH}$, $f = 60\,\text{Hz}$, then the only significant harmonics are the zero-th and the first one.

We now apply the generalized averaging operation $\langle\cdot\rangle_k$ to the differential equation (2.13a), resulting in

$$L\, a_k \left\langle \frac{d}{dt} i \right\rangle_k (t) + R\, I_k(t) = V_k(t)$$

Combining this result with the identity (2.12), we conclude that, for $t \geq T$, the current phasor $I_k(t)$ satisfies the equation

$$\begin{aligned} \frac{d}{dt} I_k(t) &= a_k \left\langle \frac{d}{dt} i \right\rangle_k (t) - jk\omega_0\, I_k(t) \\ &= -\left(\alpha + jk\omega_0\right) I_k(t) + \frac{1}{L} V_k(t) \end{aligned} \tag{2.16a}$$

where we recall that $\alpha \stackrel{\text{def}}{=} R/L$, and

$$V_k(t) = \begin{cases} 0 & k=0 \\ V_1 & k=1 \\ 0 & k \geq 2 \end{cases} \qquad \text{for } t \geq T \tag{2.16b}$$

We can now propagate this differential equation for $t \geq T$, using the initial conditions

$$I_k(T) = \begin{cases} \xi_0(T) & k = 0 \\ \frac{V_1}{R(1+j\zeta)} + \xi_1(T) & k = 1 \\ \xi_k(T) & k \geq 2 \end{cases} \quad (2.16c)$$

where

$$\xi_k(T) \;=\; a_k\, i_{tr}(0)\, \frac{e^{\alpha T} - 1}{\alpha T + jk2\pi}\, e^{-\alpha T} \quad (2.16d)$$

A direct calculation confirms that the solution of (2.16) coincides with (2.15b) for $t \geq T$.

Notice that we can, in fact, *define*

$$\widehat{\xi}_k(0) \stackrel{\text{def}}{=} a_k\, i_{tr}(0)\, \frac{e^{\alpha T} - 1}{\alpha T + jk2\pi} \quad (2.17a)$$

and use it to initialize the differential equation (2.16a) at $t = 0$ (rather than at $t = T$). To be specific, we propose to use the expression (2.16b) for $V_k(t)$ for all $t \geq 0$ and replace (2.16c) by a new set of initial conditions at $t = 0$, viz.,

$$I_k(0) = \begin{cases} \widehat{\xi}_0(0) & k = 0 \\ \frac{V_1}{R(1+j\zeta)} + \widehat{\xi}_1(0) & k = 1 \\ \widehat{\xi}_k(0) & k \geq 2 \end{cases} \quad (2.17b)$$

Again, a direct calculation confirms that the solution of the phasor differential equation (2.16a), using the initial conditions (2.17), coincides with (2.15b) for all $t \geq 0$. These observations motivate us to introduce in Sect. 2.5 a systematic approach for initialization of dynamic phasor equations. ■

Again, we observe that the waveform-to-phasor map (2.2b,c) and the dynamic relation (2.12) can be used (without modification) with matrix waveforms of arbitrary dimensions. In the following section, we consider the dynamics of phasors associated with row vector waveforms, while in Sects. 2.4 and 2.5 we examine the evolution of dynamic phasors associated with column-vector waveforms that satisfy a linear state-space equation.

2.3 Dynamic Symmetrical Sequence Components

The symmetrical sequence components of a *non-periodic* 3-phase waveform $x(t)$ are determined via a *dynamic* version of the unitary transformation used in Sect. 1.3 for the periodic case, namely

$$\left[\, X_{k,0}(t) \quad X_{k,+}(t) \quad X_{k,-}(t) \,\right] \stackrel{\text{def}}{=} X_k(t)\, \mathcal{M}_{sc} \tag{2.18a}$$

where $X_k(t)$ are the dynamic phasors associated with the waveform $x(\cdot)$ via (2.2 b,c), and the 3×3 unitary matrix $\mathcal{M}_{sc}$ is defined in (1.13).[1] Recall the subscript "0" denotes the zero-sequence component, while the subscripts "+" and "−" denote the positive and negative sequence components, respectively.

The symmetrical sequence components can also be expressed directly in terms of the 3-phase waveform $x(t)$. Multiplying (2.2c) on the right by the unitary matrix $\mathcal{M}_{sc}$, and using (2.18a), results in

$$\left[\, X_{k,0}(t) \quad X_{k,+}(t) \quad X_{k,-}(t) \,\right] \equiv X_k(t)\, \mathcal{M}_{sc} =$$

$$= \begin{cases} \dfrac{1}{T} \displaystyle\int_{t-T}^{t} \left[\, x(s)\, \mathcal{M}_{sc} \,\right] ds & k = 0 \\[2ex] \dfrac{\sqrt{2}}{T} \displaystyle\int_{t-T}^{t} \left[\, x(s)\, \mathcal{M}_{sc} \,\right] e^{-jk\omega_0 s}\, ds & k \geq 1 \end{cases} \tag{2.18b}$$

Among the salient features of this definition are compatibility with conventional symmetrical components in a periodic steady-state: when the waveform $x(t)$ is periodic, the associated dynamic phasors $X_k(t)$ are constant (i.e., time-invariant), and the corresponding symmetrical sequence components coincide with the standard definition from polyphase circuit theory, which was presented in Sect. 1.3.

Our definition of dynamic symmetrical components differs from the notion of instantaneous symmetrical components introduced by Lyon [48] in one very important aspect – the equation (2.18b) includes integration over a period of the fundamental waveform, and this is absent in [48]. This difference has important consequences – while the (time-varying) transformation used in instantaneous symmetrical components proved useful in certain problems, it does not change the time-varying nature of the model in phase (a-b-c) coordinates *during transients*. On the other hand, the presence of the integral term in (2.18b) will allow us to develop time-invariant models with an adjustable degree of accuracy, as the key mathematical objects of interest (namely the dynamic symmetrical components) will have transient dynamics that can be derived from the original model.

---

[1] For $k = 0$, use instead a dynamic version of the simpler, all real-valued, relation (1.16).

Symmetrical sequence components inherit, via (2.18a), the dynamic characteristics of their parent phasors. Thus, we observe that the dynamic phasor equation (2.12), viz.,

$$\frac{d}{dt} X_k(t) = a_k \left\langle \frac{d}{dt} x \right\rangle_k (t) - jk\omega_0 X_k(t)$$

induces a similar dynamic equation for the symmetrical sequence components of $X_k(t)$, namely,

$$\frac{d}{dt} \left[X_{k,0}(t) \quad X_{k,+}(t) \quad X_{k,-}(t) \right] = \tag{2.19}$$
$$\left[a_k \left\langle \frac{d}{dt} x \right\rangle_k (t) \right] \mathcal{M}_{sc} - jk\omega_0 \left[X_{k,0}(t) \quad X_{k,+}(t) \quad X_{k,-}(t) \right]$$

where the time-dependent generalized averaging operation $\langle \cdot \rangle_k (t)$ is defined in (2.2c). Notice that this dynamic equation couples all three symmetrical sequence components: in contrast, the dynamic phasor equation (2.12) involves no coupling between the three phases.

Example 2.3 (3-phase RL circuit): Consider the case of a possibly coupled three-phase inductor, described by a symmetric inductance matrix L and supplied by (possibly unbalanced) voltages at a fixed frequency ω_0, through a three-phase resistor R (= a diagonal 3×3 matrix). The corresponding circuit equation is

$$\left[v_a(t) \; v_a(t) \; v_c(t) \right] = \left[i_a(t) \; i_a(t) \; i_c(t) \right] R + \left(\frac{d}{dt} \left[i_a(t) \; i_a(t) \; i_c(t) \right] \right) L$$

or, in compact notation,

$$v(t) = i(t) R + \left(\frac{d}{dt} i(t) \right) L \tag{2.20}$$

where $i(t) = \left[i_a(t) \quad i_a(t) \quad i_c(t) \right]$ and $v(t) = \left[v_a(t) \quad v_a(t) \quad v_c(t) \right] = \sqrt{2}\, \Re \left\{ V_1 e^{j\omega_0 t} \right\}$. Applying the time-variant generalized averaging operation $\langle \cdot \rangle_k (t)$, we obtain a differential equation for the k-th harmonic dynamic phasors, viz.,

$$V_k(t) = I_k(t) R + \left(a_k \left\langle \frac{d}{dt} i \right\rangle_k (t) \right) L \ , \qquad k \geq 0$$

Notice that in this example $V_1(t) = V_1$ (for $t \geq 0$), and $V_k(t) = 0$ for all $k \neq 1$.

Next, multiplying this equation on the right by $\mathcal{M}_{sc}$ results in conversion from the phase domain to the symmetrical component (SC) domain, viz.,

$$\begin{aligned}\left[\, V_{k,0}(t) \;\; V_{k,+}(t) \;\; V_{k,-}(t) \,\right] &= \left[\, I_{k,0}(t) \;\; I_{k,+}(t) \;\; I_{k,-}(t) \,\right]\left(\mathcal{M}_{sc}^{-1} R\, \mathcal{M}_{sc}\right) \\ &\quad + \left[a_k \left\langle \frac{d}{dt} i \right\rangle_k (t) \right] \mathcal{M}_{sc}\left(\mathcal{M}_{sc}^{-1} L\, \mathcal{M}_{sc}\right)\end{aligned}$$

Finally, applying (2.19) produces a dynamic equation relating the (k-th harmonic) voltage and current symmetrical sequence components, viz.,

$$\left[\, V_{k,0}(t) \;\; V_{k,+}(t) \;\; V_{k,-}(t) \,\right] = \left[\, I_{k,0}(t) \;\; I_{k,+}(t) \;\; I_{k,-}(t) \,\right]\left(\mathcal{M}_{sc}^{-1} R\, \mathcal{M}_{sc}\right) \tag{2.21}$$

$$+ \left(\frac{d}{dt} \left[\, I_{k,0}(t) \;\; I_{k,+}(t) \;\; I_{k,-}(t) \,\right] + jk\omega_0 \left[\, I_{k,0}(t) \;\; I_{k,+}(t) \;\; I_{k,-}(t) \,\right] \right) \left(\mathcal{M}_{sc}^{-1} L\, \mathcal{M}_{sc}\right)$$

In steady-state, (2.21) reduces to the standard result from polyphase circuit theory (e.g., [49, p. 875]), as the time derivative on the right-hand side vanishes. ■

2.4 Linear State-Space Models

The differential equation (2.12) can be combined with any linear state-space model, say,

$$\frac{d}{dt} x(t) = A\, x(t) + B\, u(t) \quad , \qquad t \geq 0 \tag{2.22a}$$

to produce the phase-domain dynamical relation

$$\frac{d}{dt} X_k(t) = A\, X_k(t) - jk\omega_0\, X_k(t) + B\, U_k(t) \tag{2.22b}$$

where $x(\cdot)$ is an $n \times 1$ vector, and $u(\cdot)$ is a $p \times 1$ vector. We assume that $\Re\{\lambda_i(A)\} < 0$, i.e., the real part of each eigenvalue of the matrix A is negative. Consequently, we have $\Re\{\lambda_i(A - jk\omega_0 I)\} < 0$ for all k, so that both (2.22a) and (2.22b) describe stable systems. This also implies that the matrix $(A - jk\omega_0 I)$ is invertible for every integer k.

An alternative derivation of the key dynamic relation (2.22b) can be obtained by relying on a *linear time-invariant* (LTI) filtering interpretation of the phasor definition (2.2 b,c). Starting with the equivalent phasor expression (2.11), we observe that

$$\mathcal{X}_k(t) \stackrel{\text{def}}{=} X_k(t)\, e^{jk\omega_0 t} = \frac{a_k}{T} \int_0^T x(t-\tau)\, e^{jk\omega_0 \tau}\, d\tau = \int_{-\infty}^{\infty} x(t-\tau)\, h_k(\tau)\, d\tau$$

$$= h_k(t) \circledast x(t) \tag{2.23a}$$

where $\circledast$ denotes convolution, $\omega_0 = {}^{2\pi}/_T$, and $h_k(t)$ is the impulse response of a causal *band-pass filter*, viz.,

$$h_k(t) \stackrel{\text{def}}{=} \begin{cases} \dfrac{a_k}{T}\, e^{jk\omega_0 t} & 0 \le t \le T \\ 0 & \text{else} \end{cases} \tag{2.23b}$$

This means that the *passband phasor* $\mathcal{X}_k(t)$ is obtained by passing the waveform $x(t)$ through a *causal linear time-invariant filter* (Fig. 2.1), whose impulse response $h_k(t)$ is given by (2.23b).

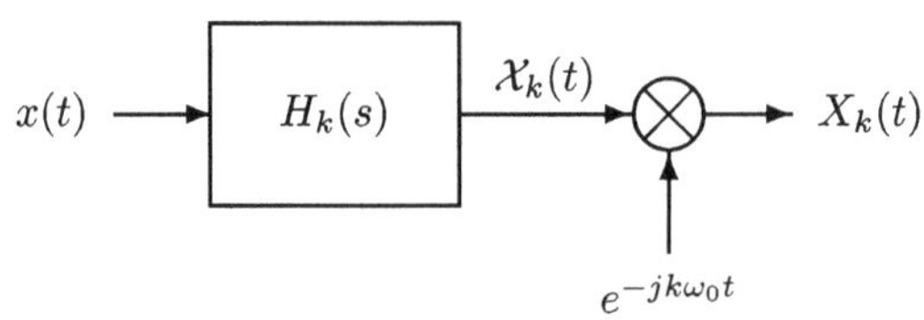

Fig. 2.1 Linear filtering interpretation of Fourier phasors.

The frequency response of this filter is centered at $\omega = k\omega_0$, namely $H_k(s) = a_k\, H_0(s - jk\omega_0)$, where

$$H_0(s) = \frac{1 - e^{-sT}}{sT} = \int_0^{\infty} h_0(t)\, e^{-st}\, dt$$

is a low-pass filter with a 3 dB bandwidth of ${}^{0.44}/_T$ Hz (Fig. 2.2). The corresponding frequency response is

$$H_0(j\omega) = e^{-j\omega T/2}\, \frac{\sin \omega T/2}{\omega T/2} \tag{2.24a}$$

Notice that this filter has a *linear phase response*, resulting in a constant group delay, viz.,

$$\text{group-delay} \stackrel{\text{def}}{=} -\frac{d}{d\omega}\left\{\arg\, H_0(j\omega)\right\} = \frac{T}{2} \tag{2.24b}$$

Certain techniques for *estimation* of dynamic phasors from a limited number of waveform samples can achieve a shorter group delay: examples are presented in Sects. 3.2 and 3.3.

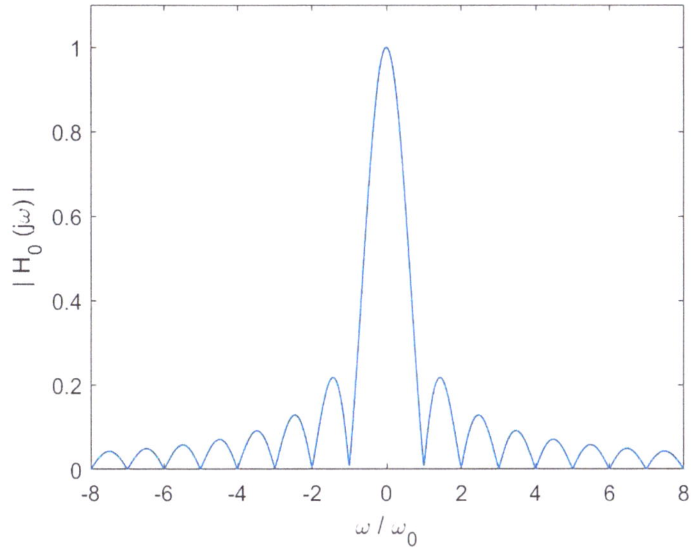

Fig. 2.2 Magnitude response of $H_0(s)$.

The frequency spectrum of the passband phasor $\mathcal{X}_k(t)$ is centered at $\omega = k\omega_0$, and its 3 dB bandwidth is $0.44/T \approx 1/2T \equiv f_0/2$ Hz. Multiplying $\mathcal{X}_k(t)$ by $e^{-jk\omega_0 t}$ generates the *baseband dynamic phasor* $X_k(t)$: this operation, known as "demodulation," shifts the frequency spectrum down by $k\omega_0$. Thus the frequency spectrum of $X_k(t)$ is centered at the origin, and its bandwidth is approximately $f_0/2$. We notice that the passband phasor $\mathcal{X}_k(t)$ is completely determined by the input waveform $x(t)$ and the transfer function $H_k(s)$. In contrast, the baseband phasor $X_k(t)$ is determined only up to a constant phase-shift because the demodulating waveform $e^{-jk\omega_0 t}$ depends on the choice made for the origin of the time axis "t." In other words, the actual (i.e., implemented) demodulating waveform is $e^{-jk\omega_0(t-t_0)}$, where the value of t_0 can be chosen at will: this results in a constant phase-shift of $k\omega_0 t_0$.

When the waveform $x(t)$ is generated by a linear state-space model, such as (2.22a), the overall transfer function from the input $u(t)$ to the passband phasor $\mathcal{X}_k(t)$ is a cascade connection of the input-to-state map $G(s) \stackrel{\text{def}}{=} (sI - A)^{-1}B$ and the waveform-to-phasor map $H_k(s)$. Since single-input/single-output linear time-invariant systems commute, we can flip the order of these two operations: applying $H_k(s)$ first, followed by $G(s)$, we obtain a linear filtering relation between $\mathcal{U}_k(t)$, the k-th passband phasor associated with the waveform $u(t)$, and $\mathcal{X}_k(t)$ (Fig. 2.3).

In other words, passband phasors satisfy the same dynamic linear model as the original waveforms, viz.,

$$\frac{d}{dt}\mathcal{X}_k(t) = A\,\mathcal{X}_k(t) + B\,\mathcal{U}_k(t) \tag{2.25a}$$

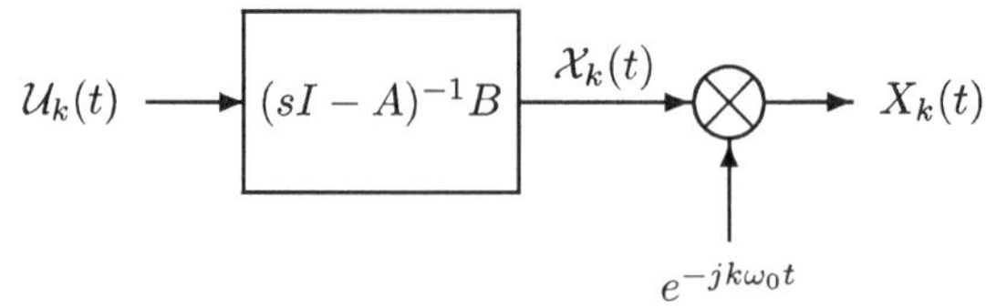

Fig. 2.3 Inheritance property of linear state-space models.

Next, we can use the relations $\mathcal{X}_k(t) = X_k(t)\, e^{jk\omega_0 t}$ and $\mathcal{U}_k(t) = U_k(t)\, e^{jk\omega_0 t}$ to obtain an equivalent dynamic equation for the k-th *baseband phasor*, viz.,

$$\frac{d}{dt} X_k(t) = A\, X_k(t) - jk\omega_0 X_k(t) + B\, U_k(t) \tag{2.25b}$$

which is identical to (2.22b).

Notice that the inheritance property shown in (2.25) is completely independent of the transfer function $H_k(s)$. The choice (2.23) produces the standard (Fourier) dynamic phasors of (2.2): however, we have the flexibility of defining *generalized* dynamic phasors $\{X_k(t)\}$ via the linear filtering relation of Fig. 2.1 with any desired choice of the filters $\{H_k(s)\}$. This observation motivates our interest in phasor estimation schemes, which are examined in Chap. 3, and non-Fourier dynamic phasors, which are discussed in Chap. 4.

The state-space equations (2.25) allow efficient propagation of the dynamic phasors $\{X_k(t)\}$ for selected k-values of interest: this is an attractive alternative to repeated evaluation of $X_k(t)$, *for every time-instant* "t," directly from the original definition (2.2c). Notice, however, that this approach relies on the assumption that the input phasors $\{U_k(t)\}$ are either readily available or can be computed with minimal effort: this is usually the case in power system applications, where the input waveform $u(\cdot)$ is generated by a mechanical or electronic source and thus is *periodic for all* $t \geq 0$ with a known period T, namely

$$u(t) = U_0 + \sqrt{2} \sum_{k=1}^{\infty} \Re\left\{U_k\, e^{jk\omega_0 t}\right\} \quad , \quad \omega_0 = \frac{2\pi}{T} \tag{2.26}$$

Consequently, $U_k(t) = U_k$ for all $t \geq T$ (and all k), where the constant phasor U_k can be determined *from a single cycle segment* of the waveform $u(\cdot)$ via the original definition (2.2c). We present in Sect. 2.5 an initialization approach that extends the range of constancy of the dynamic phasor $U_k(t)$ to all $t \geq 0$, by relying on the (periodic) nature of the expression (2.26).

2.5 Ersatz-Past (EP) Initialization

We have observed in the previous section that a waveform $x(t)$ generated by a linear state-space model gives rise to passband dynamic phasors $\{\mathcal{X}_k(t)\}$ that satisfy the same model, as is evident from (2.25a). However, the initial conditions for this differential equation cannot be uniquely determined from the initial conditions of the original state-space model for the waveform $x(t)$. This is so because the linear state-space equation

$$\frac{d}{dt}\,x(t) \;=\; A\,x(t) + B\,u(t) \quad , \qquad \text{for } t \geq 0 \quad , \qquad \text{with } x(0) = x_0 \tag{2.27}$$

holds only for non-negative t-values, so that neither $u(t)$ nor $x(t)$ are known for $t < 0$. Consequently, $U_k(t)$ is unknown (i.e., undefined) for $t < T$, which makes it impossible to propagate the dynamic phasor equation (2.25) in the interval $0 \leq t < T$. Similarly, $X_k(t)$ is also undefined for $t < T$. We shall use the terms "input phasor gap" and "state phasor gap" to refer to the undefined phasors sets $\{U_k(t)\,;\; 0 \leq t < T\}$ and $\{X_k(t)\,;\; 0 \leq t < T\}$, respectively.

One way to resolve this difficulty is by selecting a convenient extension of the input waveform $u(t)$ to the time interval $-T \leq t < 0$, and use it to propagate the state-space equation (2.27) for $t \geq -T$: this makes it possible to define $X_k(t)$ and $U_k(t)$ and propagate (2.25), for $t \geq 0$. There are infinitely many possible choices for such *Ersatz-Past* (EP) input history $\{u^{\mathsf{EP}}(t)\,;\; -T \leq t < 0\}$: each choice generates a corresponding EP state trajectory $\{x^{\mathsf{EP}}(t)\,;\; -T \leq t < 0\}$. Since the EP interval $-T \leq t < 0$ is exactly one-cycle-long, we can uniquely represent each choice of the EP input history via a Fourier series or, equivalently, via the associated set of (rms) Fourier phasors $\{U_k^{\mathsf{EP}}\}$, viz.,

$$u^{\mathsf{EP}}(t) \;=\; U_0^{\mathsf{EP}} + \sqrt{2}\,\sum_{k=1}^{\infty} \Re\left\{U_k^{\mathsf{EP}}\, e^{jk\omega_0 t}\right\} \quad , \qquad -T \leq t < 0 \tag{2.28}$$

We conclude that the resulting EP state trajectory $\{x^{\mathsf{EP}}(t)\,;\;\; -T \leq t < 0\}$ is completely determined by the phasors $\{U_k^{\mathsf{EP}}\,;\;\; k \geq 0\}$ and the initial condition $x_{00} \stackrel{\text{def}}{=} x^{\mathsf{EP}}(-T)$.

Due to causality of the state-space equation (2.27), the only constraint we need to impose on the chosen EP input history (and the chosen x_{00}) is

$$\lim_{t\downarrow 0}\, x(t) \;\stackrel{\text{def}}{=}\; x_0 \;=\; \lim_{t\uparrow 0}\, x^{\mathsf{EP}}(t) \tag{2.29}$$

When this constraint is satisfied, the choice of the EP input history impacts the state trajectory only in the interval $-T \leq t \leq 0$.

Our next step is to determine the input phasor gap $\{U_k(t);\ 0 \le t < T\}$ via (2.2c), namely,

$$U_k(t) = \frac{a_k}{T}\left[\int_{t-T}^{0} u^{\mathsf{EP}}(s)\, e^{-jk\omega_0 s}\, ds + \int_{0}^{t} u(s)\, e^{-jk\omega_0 s}\, ds\right] \tag{2.30}$$

where $0 \le t < T$, and $a_k = \sqrt{2}$ for all $k \ne 0$, while $a_0 = 1$.

We can now use the results of Appendix A (with $t_0 = -T$ and $t_1 = 0$) to express the EP state trajectory as a sum of a steady-state (periodic) component and a transient component, i.e., $x^{\mathsf{EP}}(t) = x_{ss}^{\mathsf{EP}}(t) + x_{tr}^{\mathsf{EP}}(t)$. The steady-state component $x_{ss}^{\mathsf{EP}}(t)$ is completely determined by the EP input $\{u^{\mathsf{EP}}(t);\ -T \le t < 0\}$: thus, it depends only on the phasors $\{U_k^{\mathsf{EP}};\ k \ge 0\}$, but not on x_{00}. In particular, $x_{ss}^{\mathsf{EP}}(0) = -A^{-1} B\, U_0^{\mathsf{EP}} - \sqrt{2}\, \sum_{k=1}^{\infty} \Re\left\{\left(A - jk\omega_0 I\right)^{-1} B\, U_k^{\mathsf{EP}}\right\}$. The transient component $x_{tr}^{\mathsf{EP}}(t)$ is given by the expression $x_{tr}^{\mathsf{EP}}(t) = e^{At} x_{tr}^{\mathsf{EP}}(0)$, where (recall (A.4a))

$$x_{tr}^{\mathsf{EP}}(0) = e^{AT} x_{tr}^{\mathsf{EP}}(-T) \equiv e^{AT}\left[x_{00} - x_{ss}^{\mathsf{EP}}(0)\right] \tag{2.31}$$

and we used the fact that $x_{ss}^{\mathsf{EP}}(-T) = x_{ss}^{\mathsf{EP}}(0)$, due to the periodicity of $x_{ss}^{\mathsf{EP}}(\cdot)$. Since this relation defines a one-to-one correspondence between x_{00} and $x_{tr}^{\mathsf{EP}}(0)$, we can use $x_{tr}^{\mathsf{EP}}(0)$ as a design parameter that determines the EP initial condition x_{00} via (2.31), for any given choice of the EP phasors $\{U_k^{\mathsf{EP}};\ k \ge 0\}$.

Similarly, the EP terminal condition (2.29) translates into a constraint on the set of EP design parameters, namely (recall (A.4b)),

$$\begin{aligned} x_0 &= x_{tr}^{\mathsf{EP}}(0) + x_{ss}^{\mathsf{EP}}(0) \\ &= x_{tr}^{\mathsf{EP}}(0) - A^{-1} B\, U_0^{\mathsf{EP}} - \sqrt{2} \sum_{k=1}^{\infty} \Re\left\{\left(A - jk\omega_0 I\right)^{-1} B\, U_k^{\mathsf{EP}}\right\} \end{aligned} \tag{2.32}$$

Finally, setting $t_0 = -T$ and $t_1 = 0$ in (A.8) produces an explicit expression for the initial condition $X_k(0)$ of the state dynamic phasor, viz.,

$$X_k(0) = \left(A - jk\omega_0 I\right)^{-1} \left\{ a_k\, \frac{I - e^{-AT}}{T}\, x_{tr}^{\mathsf{EP}}(0) - B\, U_k^{\mathsf{EP}} \right\} \tag{2.33}$$

which shows explicitly the dependence of $X_k(0)$ on the design parameters $x_{tr}^{\mathsf{EP}}(0)$ and $\{U_k^{\mathsf{EP}};\ k \ge 0\}$.

In summary, given any choice of the design parameters $\{U_k^{\mathsf{EP}};\ k \ge 0\}$, we can determine the input phasor gap via (2.30), and the state-vector initial conditions $\{X_k(0);\ k \ge 0\}$ via (2.32) and (2.33). This information provides the starting point for propagating the state-equation (2.25b) for all $t \ge 0$. In the special case when $u(t)$ is perfectly known in advance, we can optimize the selection of the EP phasors $\{U_k^{\mathsf{EP}}\}$: in particular, we may wish to control the smoothness of transition between $\{u^{\mathsf{EP}}(t)\,;\ -T \le t < 0\}$ and $\{u(t)\,;\ t \ge 0\}$, and/or the fluctuation of the input phasor gap $\{U_k(t)\,;\ 0 \le t < T\}$. Alternatively, we may opt to control the smoothness of transition between $\{x^{\mathsf{EP}}(t)\,;\ -T \le t < 0\}$ and $\{x(t)\,;\ t \ge 0\}$, and/or the fluctuation of the state phasor gap $\{X_k(t)\,;\ 0 \le t < T\}$. Such considerations are especially important in applications, such as *hybrid simulation* [50, 51], that repeatedly switch between time-domain (i.e., electromagnetic transient) simulation and phasor-domain simulation. The need for reducing fluctuation is even more acute in modeling/simulation of systems with nonlinear components: making a judicious EP choice can slow variation of the state variables $\{X_k(t)\}$ within the gap interval $0 \le t \le T$ [52].

Maximally-Smooth EP Initialization

This relatively simple initialization scheme relies on the assumption that the input waveform $u(t)$ is periodic with a known Fourier-series representation, viz.,

$$u(t) \;=\; U_0 + \sqrt{2}\sum_{k=1}^{\infty} \Re\left\{U_k\, e^{jk\omega_0 t}\right\}\,, \qquad t \ge 0 \tag{2.34a}$$

This allows us to define the EP input segment $\{u^{\mathsf{EP}}(t)\,;\ -T \le t < 0\}$ as a *periodic extension* of $u(t)$, viz.,

$$u^{\mathsf{EP}}(t) \;\stackrel{\text{def}}{=}\; u(t+T) \qquad \text{for} \quad -T \le t < 0$$

so that

$$u^{\mathsf{EP}}(t) \;=\; U_0 + \sqrt{2}\sum_{k=1}^{\infty} \Re\left\{U_k\, e^{jk\omega_0 t}\right\}\,, \qquad -T \le t < 0 \tag{2.34b}$$

or, equivalently, $U_k^{\mathsf{EP}} = U_k$. This periodic extension implies that $U_k(t) = U_k$ for all $t \ge 0$ (and all k), so that there is no need to evaluate $U_k(t)$ via (2.30). Moreover, the transition from $\{u^{\mathsf{EP}}(t)\,;\ -T \le t < 0\}$ to $\{u(t)\,;\ t \ge 0\}$ is maximally-smooth, namely,

$$\lim_{t\uparrow 0}\left(\frac{d}{dt}\right)^m u^{\mathsf{EP}}(t) \;=\; \lim_{t\downarrow 0}\left(\frac{d}{dt}\right)^m u(t) \qquad \text{for all } m \ge 0 \tag{2.35a}$$

The same property also holds for the state trajectory, viz.,

$$\lim_{t\uparrow 0}\left(\frac{d}{dt}\right)^m x^{\mathsf{EP}}(t) \;=\; \lim_{t\downarrow 0}\left(\frac{d}{dt}\right)^m x(t) \qquad \text{for all } m \ge 0 \tag{2.35b}$$

Indeed, since $X_k^{\mathsf{EP}} = X_k$, the steady-state component of the state-vector satisfies (2.35a), viz.,

$$\lim_{t\uparrow 0} \left(\frac{d}{dt}\right)^m x_{ss}^{\mathsf{EP}}(t) = \lim_{t\downarrow 0} \left(\frac{d}{dt}\right)^m x_{ss}(t) \qquad \text{for all } m \geq 0$$

In particular, $x_{ss}^{\mathsf{EP}}(0) = x_{ss}(0)$, so that (recall (2.32)) $x_{tr}^{\mathsf{EP}}(0) = x_{tr}(0)$. Now, since $x_{tr}^{\mathsf{EP}}(t) = e^{At} x_{tr}^{\mathsf{EP}}(0)$ for $-T \leq t < 0$, while $x_{tr}(t) = e^{At} x_{tr}(0)$ for $t \geq 0$, it follows that the transition (at $t = 0$) of the transient component is also maximally-smooth, which completes the proof of (2.35b). Moreover, we observe that since the $x(t)$ waveform transition (at $t = T$) is maximally-smooth, the resulting dynamic phasors $\{X_k(t)\}$ also exhibit a maximally-smooth transition.

We can now use (2.32) to determine $x_{tr}^{\mathsf{EP}}(0)$, namely

$$x_{tr}^{\mathsf{EP}}(0) = x_0 + A^{-1} B\, U_0 + \sqrt{2} \sum_{k=1}^{\infty} \Re\left\{\left(A - jk\omega_0 I\right)^{-1} B\, U_k\right\} \tag{2.36a}$$

and insert it into (2.33) to obtain the initial condition for $X_k(0)$, viz.,

$$X_k(0) = \left(A - jk\omega_0 I\right)^{-1} \left\{ a_k \frac{I - e^{-AT}}{T}\, x_{tr}^{\mathsf{EP}}(0) - B\, U_k \right\} \tag{2.36b}$$

The state phasor trajectory $\{X_k(t)\,;\; t \geq 0\}$ can now be determined by propagating the differential equation (2.25) for $t \geq 0$, using (2.36b) as the initial condition, and the time-invariant input phasors $U_k(t) = U_k$ for all $t \geq 0$.

Example 2.4 (RL circuit: maximally-smooth EP initialization): The RL circuit from Example 2.2 can be interpreted as a single-state model, viz.,

$$\frac{d}{dt}\, i(t) = -\alpha\, i(t) + \tfrac{1}{L}\, v(t) \;, \qquad t \geq 0 \tag{2.37}$$

where $\alpha \overset{\text{def}}{=} \frac{R}{L}$, and the initial condition is $i(0)$. Mapping into the state-space model notation (2.27), we observe that $x(t) \equiv i(t)$ (= inductor current), $u(t) \equiv v(t)$ (= driving voltage), $A = -\alpha = -\frac{R}{L}$, and $B = \frac{1}{L}$. Thus, the corresponding phasor dynamic relation is

$$\frac{d}{dt}\, I_k(t) = -\left(\alpha + jk\omega_0\right) I_k(t) + \tfrac{1}{L}\, V_k(t) \tag{2.38}$$

In order to propagate this differential equation for $t \geq 0$ we need to know the voltage phasor gap segment $\{\, V_k(t)\,;\; 0 \leq t < T \,\}$ and the initial dynamic current phasor $I_k(0)$.

Since $v(t) = \sqrt{2}\,\Re\left\{V_1\, e^{j\omega_0 t}\right\}$ for $t \geq 0$, we can apply the maximally-smooth EP initialization approach and use the expression (2.36) to determine the initial condition $I_k(0)$. Extending the voltage $v(t)$ periodically into the EP interval results in

$$v^{\mathsf{EP}}(t) = \sqrt{2}\,\Re\left\{V_1\, e^{j\omega_0 t}\right\}\,, \qquad -T \leq t < 0$$

so that $V_k(t) = V_k$ for all $t \geq 0$, with $V_k = 0$ for all $k \neq 1$. Consequently, the solution of (2.38) is $I_k(t) = I_k + \xi_k(t)$, where $\xi_k(t) = e^{-(\alpha + jk\omega_0)t}\,\xi_k(0)$, and (recall (A.2b)) $I_k = \frac{1}{L(\alpha+jk\omega_0)}\,V_k$ or, more explicitly,

$$I_k = \begin{cases} \frac{V_1/R}{1+j\zeta} & k = 1 \\ 0 & \text{otherwise} \end{cases} \tag{2.39a}$$

where $\zeta \stackrel{\text{def}}{=} \frac{\omega_0 L}{R} \equiv \frac{\omega_0}{\alpha}$.

According to (2.36), the initial condition for propagating the phasor differential equation (2.38) is

$$I_k(0) = \frac{1}{\alpha + jk\omega_0}\left\{\frac{1}{L}\,V_k + a_k\,\frac{e^{\alpha T} - 1}{T}\,i_{tr}^{\mathsf{EP}}(0)\right\}$$

or, equivalently, $I_k(0) = I_k + \xi_k(0)$, where

$$\xi_k(0) = a_k\,\frac{e^{\alpha T} - 1}{\alpha T + jk2\pi}\,i_{tr}^{\mathsf{EP}}(0) \tag{2.39b}$$

and

$$i_{tr}^{\mathsf{EP}}(0) = i(0) - \sqrt{2}\,\Re\left\{\frac{V_1/R}{1 + j\zeta}\right\} \equiv i(0) - \sqrt{2}\,\Re\left\{I_1\right\} \tag{2.39c}$$

which reproduces the expressions (2.14) of Example 2.2 for $i_{ss}(0)$ and $i_{tr}(0)$. Also, a direct calculation confirms that the solution of (2.38) coincides with (2.15b) for $t \geq 0$.

In summary, the maximally-smooth EP initialization approach provides a rigorous validation for the "ad-hoc" choice of initial conditions proposed in (2.17). ■

The maximally-smooth EP initialization scheme can be extended, in principle, to any system whose input is given by a known closed-form expression, say,

$$u(\cdot) = \left\{\, f_\theta(t)\,;\ t \geq 0\right\}$$

where θ denotes a set of known parameters. This allows us to extend $u(\cdot)$ to obtain the EP input segment $\{u^{\mathsf{EP}}(t)\,;\ -T \le t < 0\}$ via the relation

$$u^{\mathsf{EP}}(t) \overset{\text{def}}{=} f_\theta(t) \qquad \text{for} \quad -T \le t < 0 \tag{2.40}$$

and evaluate the corresponding phasor gap segment $\left\{ U_k(t)\,;\ 0 \le t < T \right\}$, so that now the dynamic input phasor $U_k(t)$ is known for all $t \ge 0$ (and all k). Notice that $U_k(t)$ is no longer constant, since $f_\theta(t)$ is not necessarily periodic.

Optimized EP Initialization

In this EP initialization approach, we aim to control variation of the state phasors $\{X_k(t)\}$ within the gap interval $0 \le t < T$. To achieve this objective, we reduce the number of smoothness constraints and use the resulting slack to optimize a cost function of interest. We still rely on explicit knowledge of the Fourier-series representation of the periodic input waveform $u(t)$, viz.,

$$u(t) = \sqrt{2} \sum_{k=1}^{K} \Re\left\{U_k\, e^{jk\omega_0 t}\right\}, \qquad t \ge 0 \tag{2.41a}$$

where we have assumed that $u(\cdot)$ has no DC component, and its harmonic content is limited to the finite range $1 \le k \le K$. In this approach, we allow for $U_k^{\mathsf{EP}} \neq U_k$, so that

$$u^{\mathsf{EP}}(t) = \sqrt{2} \sum_{k=1}^{K} \Re\left\{U_k^{\mathsf{EP}}\, e^{jk\omega_0 t}\right\} \qquad \text{for} \quad -T \le t < 0 \tag{2.41b}$$

We can now use the expressions (2.41a,b) to evaluate the time-variant input phasor segment $\{U_k(t)\,;\ 0 \le t < T\}$ via (2.30), namely,

$$U_k(t) = \frac{a_k}{T}\left[\int_{t-T}^{0} u^{\mathsf{EP}}(s)\, e^{-jk\omega_0 s}\, ds + \int_0^t u(s)\, e^{-jk\omega_0 s}\, ds\right] \tag{2.42}$$

Due to the jump from U_k^{EP} to U_k at $t = 0$, the resulting $\{U_k(t)\}$ gap phasors coincide with those evaluated in Example 2.1, with A_k replaced by U_k, and B_k replaced by U_k^{EP} (see Example 2.5 for further details).

Each choice of $\{U_k^{\mathsf{EP}}\}$ determines a specific trajectory for the gap phasor segments $\{U_k(t)\,;\ 0 \le t < T\}$ and $\{X_k(t)\,;\ 0 \le t < T\}$. We have the option to control the variation of these gap phasors by appropriately selecting the $2K$ real-valued vector parameters $\{\Re\{U_k^{\mathsf{EP}}\}\,;\ 1 \le k \le K\}$ and $\{\Im\{U_k^{\mathsf{EP}}\}\,;\ 1 \le k \le K\}$. For instance, we can impose the smoothness constraints

$$\lim_{t\uparrow 0}\left(\frac{d}{dt}\right)^m u^{\mathsf{EP}}(t) \;=\; u^{(m)}(0) \;\stackrel{\text{def}}{=}\; \lim_{t\downarrow 0}\left(\frac{d}{dt}\right)^m u(t) \qquad 0 \le m \le r-1$$

which translate to a set of constraints on the EP phasors $\{U_k^{\mathsf{EP}}\}$, viz.,

$$\sqrt{2}\sum_{k=1}^{K}\Re\left\{(jk\omega_0)^m\, U_k^{\mathsf{EP}}\right\} \;=\; \sqrt{2}\sum_{k=1}^{\infty}\Re\left\{(jk\omega_0)^m\, U_k\right\}\;, \qquad 0 \le m \le r-1 \tag{2.43}$$

This set of constraints defines a set of rp real-coefficient equations in $2Kp$ real unknowns (the real and imaginary parts of the $1\times p$ phasors U_k^{EP}), where p denotes the number of phases. In general, a (non-unique) solution can exist only if $2Kp \ge rp$ or, equivalently,

$$2K \ge r \tag{2.44}$$

When $2K > r$, the number of real-valued design parameters exceeds the number of constraints, so we can use the excess $2K - r$ degrees of freedom to minimize any cost function of interest. Once the phasors $\{U_k^{\mathsf{EP}}\;;\; 1 \le k \le K\}$ have been determined, the value of $x_{tr}^{\mathsf{EP}}(0)$ is obtained via the boundary constraint (2.32).

In the following example, we use a single degree of freedom to control the variation of the state phasor gap.

Example 2.5 (RL circuit: optimized EP initialization): We now consider applying optimized EP initialization to the RL circuit of Example 2.4. In order to facilitate a comparison with Example 2.4, we consider here the case $K = 1$, namely $V_k^{\mathsf{EP}} = 0$ for all $k \ne 1$. We will use a single smoothness constraint (so $r = 1$), namely

$$\sqrt{2}\,\Re\{V_1^{\mathsf{EP}}\} \;=\; v(0) \;\equiv\; \sqrt{2}\,\Re\{V_1\}$$

or, equivalently, $\Re\{V_1^{\mathsf{EP}}\} = \Re\{V_1\}$. Thus the initial condition for propagation of the phasor differential equation (2.38) is (recall (2.33))

$$I_k(0) \;=\; \frac{1}{\alpha + jk\omega_0}\left\{\frac{1}{L}\,V_k^{\mathsf{EP}} + a_k\,\frac{e^{\alpha T}-1}{T}\,i_{tr}^{\mathsf{EP}}(0)\right\}$$

where (recall (2.32))

$$i_{tr}^{\mathsf{EP}}(0) \;=\; i(0) - \sqrt{2}\,\Re\left\{\frac{V_1^{\mathsf{EP}}/R}{1+j\zeta}\right\}$$

and

$$V_k^{\mathsf{EP}} = \begin{cases} \Re\{V_1\} + j\eta & k = 1 \\ 0 & k \ne 1 \end{cases}$$

The real-valued unconstrained variable $\eta \equiv \Im\{V_1^{\mathsf{EP}}\}$ can now be used to control the behavior of the state phasor gap $\{I_k(t)\,;\; 0 \leq t < T\}$.

First, we need to determine the input phasor gap $\{V_k(t)\,;\; 0 \leq t < T\}$ from the combined waveform $\{v^{\mathsf{EP}}(t)\,;\; -T \leq t < 0\} \bigcup \{v(t)\,;\; 0 \leq t < T\}$ via (2.30). Since $v(t) = \sqrt{2}\,\Re\{V_1\, e^{j\omega_0 t}\}$ and $v^{\mathsf{EP}}(t) = \sqrt{2}\,\Re\{V_1^{\mathsf{EP}}\, e^{j\omega_0 t}\}$, the corresponding $\{V_k(t)\}$ phasors are given, within the interval $0 \leq t \leq T$, by the expressions (2.10) from Example 2.1, with $A_1 = V_1$ and $B_1 = V_1^{\mathsf{EP}}$, viz.,

$$
\begin{aligned}
V_0(t) &= \sqrt{2}\,\Re\left\{\left(V_1 - V_1^{\mathsf{EP}}\right) e^{j\frac{\omega_0 t}{2}}\right\} \frac{\sin\frac{\omega_0 t}{2}}{\pi} \\
V_1(t) &= V_1 - \left(V_1 - V_1^{\mathsf{EP}}\right)\left(1 - {}^{t}/_{T}\right) + \left(V_1 - V_1^{\mathsf{EP}}\right)^* \frac{\sin\omega_0 t}{2\pi}\; e^{-j\omega_0 t}
\end{aligned}
$$

and, for $k \geq 2$,

$$
V_k(t) = \left(V_1 - V_1^{\mathsf{EP}}\right) \frac{\sin\frac{(k-1)\,\omega_0 t}{2}}{(k-1)\pi}\; e^{-j\frac{(k-1)\omega_0 t}{2}} + \left(V_1 - V_1^{\mathsf{EP}}\right)^* \frac{\sin\frac{(k+1)\,\omega_0 t}{2}}{(k+1)\pi}\; e^{-j\frac{(k+1)\omega_0 t}{2}}
$$

Notice that for $t \uparrow T$ these expressions yield $V_k(T) = 0$ for all $k \neq 1$, and $V_1(T) = V_1$.

The phasor differential equation (2.38) can now be propagated (numerically) to evaluate the state phasors $\{I_k(t)\}$ within the interval $0 \leq t \leq T$, using the expressions presented above for the initial conditions $I_k(0)$ and the input phasors $V_k(t)$. The resulting phasor magnitude trajectory $\{|I_1(t)|\,;\; 0 \leq t \leq T\}$ is compared in Fig. 2.4 with the maximally-smooth version from Example 2.4, where we set

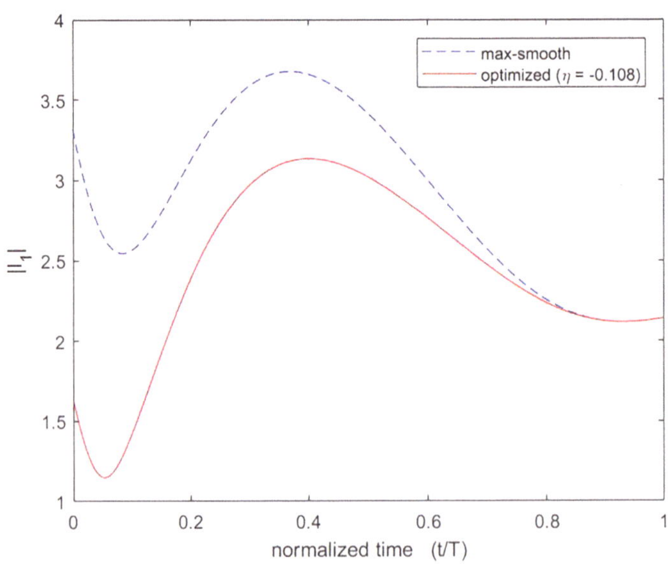

Fig. 2.4 State phasor magnitude trajectory: fundamental harmonic.

$R = 0.2\,\Omega$, $L = 1\,\text{mH}$, $V_1 = 1\,\text{V}$, and $i(0) = 0.5\,\text{A}$. Both trajectories converge to the same terminal value, but the one corresponding to optimized EP initialization exhibits "equiripple" fluctuation, in the sense that η is chosen to minimize the magnitude deviation $\max_{0 \le t \le T} \| \, |I_1(t)| - |I_1(T)| \, \|$. ■

We now observe that the family of EP initialization schemes obtained by varying $\eta \equiv \Im\{V_1^{\mathsf{EP}}\}$ in the preceding example contains, as a special case, the maximally-smooth EP initialization. The latter corresponds to setting $\eta = \Im\{V_1\}$, which results in $V_1^{\mathsf{EP}} = V_1$, and matches the choice made in the maximally-smooth EP approach. The following theorem generalizes this observation to the case when the input waveform consists of a finite number of harmonics, and the number of smoothness constraints matches the number of design parameters.

Theorem 2.1 (EP equivalence): *When the periodic input waveform is given by the expression*

$$u(t) = \sqrt{2} \sum_{k=1}^{K} \Re\left\{U_k e^{jk\omega_0 t}\right\} \quad , \qquad t \ge 0 \tag{2.45}$$

and the number of smoothness constraints matches the number of design parameters, namely $r = 2K$ in (2.43), the optimized EP initialization scheme reduces to maximally-smooth EP initialization.

Proof: If we set $r = 2K$ in (2.43), the number of constraints matches the number of design parameters. In this case we can rewrite the constraints (2.43) as two sets of $K \times K$ linear equations, involving only real-valued variables, viz.,

$$\frac{1}{\sqrt{2}}\, u^{(m)}(0) = \begin{cases} (-1)^{m/2}\, \omega_0^m \sum_{k=1}^{K} k^m\, \Re\{U_k^{\mathsf{EP}}\} & m = 0, 2, 4, \ldots, 2K-2 \\[2ex] (-1)^{(m+1)/2}\, \omega_0^m \sum_{k=1}^{K} k^m\, \Im\{U_k^{\mathsf{EP}}\} & m = 1, 3, 5, \ldots, 2K-1 \end{cases}$$

These equations establish a one-to-one correspondence between the set of smoothness constraints $\{u^{(m)}(0)\,;\ 0 \le m \le 2K-1\}$, and the set of EP phasors $\{U_k^{\mathsf{EP}}\,;\ 1 \le k \le K\}$. This one-to-one relation becomes even more evident when we rewrite these equations in matrix form, viz.,

$$\mathcal{A}_{even} \begin{pmatrix} \Re\{U_1^{\mathsf{EP}}\} \\ \Re\{U_2^{\mathsf{EP}}\} \\ \vdots \\ \Re\{U_K^{\mathsf{EP}}\} \end{pmatrix} = \frac{1}{\sqrt{2}} \begin{pmatrix} u(0) \\ -u^{(2)}(0)/\omega_0^2 \\ \vdots \\ (-1)^{K-1}\, u^{(2K-2)}(0)/\omega_0^{2K-2} \end{pmatrix} \tag{2.46a}$$

$$\mathcal{A}_{odd}\begin{pmatrix}\Im\{U_1^{\mathsf{EP}}\}\\ \Im\{U_2^{\mathsf{EP}}\}\\ \vdots\\ \Im\{U_K^{\mathsf{EP}}\}\end{pmatrix} = -\frac{1}{\sqrt{2}}\begin{pmatrix}u^{(1)}(0)/\omega_0\\ -u^{(3)}(0)/\omega_0^3\\ \vdots\\ (-1)^{K-1}u^{(2K-1)}(0)/\omega_0^{2K-1}\end{pmatrix}$$

where

$$\mathcal{A}_{even} \stackrel{\text{def}}{=} \begin{pmatrix}1 & 1 & 1 & \dots & 1\\ 1 & 2^2 & 3^2 & \dots & K^2\\ \vdots & \vdots & \vdots & & \vdots\\ 1 & 2^{2K-2} & 3^{2K-2} & \dots & K^{2K-2}\end{pmatrix}$$

$$\mathcal{A}_{odd} \stackrel{\text{def}}{=} \mathcal{A}_{even} \bullet \mathsf{diag}\{1, 2, 3, \dots, K\} \tag{2.46b}$$

Since $\mathcal{A}_{even}$ is a *Vandermonde matrix* with distinct columns, it is non-singular, which implies that $\mathcal{A}_{odd}$ is non-singular as well. Consequently, the resulting solution $\{U_k^{\mathsf{EP}};\ 1 \le k \le K\}$ is completely determined by the equation (2.46). Moreover, it coincides with the one obtained via the maximally-smooth EP initialization approach. Indeed, comparing (2.41b) with (2.45), we conclude that

$$\mathcal{A}_{even}\begin{pmatrix}\Re\{U_1^{\mathsf{EP}}\}\\ \Re\{U_2^{\mathsf{EP}}\}\\ \vdots\\ \Re\{U_K^{\mathsf{EP}}\}\end{pmatrix} = \mathcal{A}_{even}\begin{pmatrix}\Re\{U_1\}\\ \Re\{U_2\}\\ \vdots\\ \Re\{U_K\}\end{pmatrix}$$

Since $\mathcal{A}_{even}$ is always non-singular, it follows that $\Re\{U_K^{\mathsf{EP}}\} = \Re\{U_K\}$ for $k = 1, 2, \dots, K,$ and similarly for the imaginary parts. The final conclusion is that $U_k^{\mathsf{EP}} = U_k$ for $k = 1, 2, \dots, K,$ which is precisely the maximally-smooth EP initialization associated with the periodic input $u(t) = \sqrt{2}\sum_{k=1}^{K} \Re\left\{U_k e^{jk\omega_0 t}\right\}$.

Pre-windowed EP Initialization

When we have no prior information about $\{u(t);\ t \ge 0\}$, e.g., when it is periodic with unknown $\{U_k\}$ or contains non-periodic components, we have no way to optimize the choice of $\{U_k^{\mathsf{EP}}\}$. In this case, it makes sense to reduce the number of degrees of freedom to the minimum required to satisfy the constraint (2.32). The simplest choice of this kind is to set $U_k^{\mathsf{EP}} = 0$ for all k, which results in $u^{\mathsf{EP}}(t) = 0$ and justifies the designation "pre-windowed." Notice that in this case the transition at $t = 0$ from $u^{\mathsf{EP}}(t)$ to $u(t)$ is, in general, discontinuous and non-smooth, i.e.,

$$\lim_{t\uparrow 0} \left(\frac{d}{dt}\right)^m u^{\mathsf{EP}}(t) \neq \lim_{t\downarrow 0} \left(\frac{d}{dt}\right)^m u(t) \qquad \text{for all } m \geq 0$$

Nevertheless, the transition at $t = 0$ from $x^{\mathsf{EP}}(t)$ to $x(t)$ is always continuous, by virtue of the state-continuity constraint (2.29).

The pre-windowed choice reduces the constraint (2.32) to $x_0 = x_{tr}^{\mathsf{EP}}(0)$, so that the state phasor initialization expression (2.33) simplifies to

$$X_k(0) = a_k \left(A - jk\omega_0 I\right)^{-1} \frac{I - e^{-AT}}{T} x_0 \tag{2.47}$$

Notice that in this case $x_{ss}^{\mathsf{EP}}(t) = 0$, so that $x^{\mathsf{EP}}(t) = x_{tr}^{\mathsf{EP}}(t) = e^{At} x_0$. The input phasors $\{U_k(t);\ 0 \leq t < T\}$ must be evaluated via the expression (2.30), which in the pre-windowed case reduces to

$$U_k(t) = \frac{a_k}{T} \int_0^t u(s) e^{-jk\omega_0 s} ds \ , \qquad 0 \leq t < T \tag{2.48}$$

because now $u^{\mathsf{EP}}(t) = 0$. Notice that this integral can be evaluated only when the (previously unknown) input $u(t)$ becomes available in real time.

Example 2.6 (RL circuit: pre-windowed EP initialization): We now demonstrate applying pre-windowed EP initialization to the RL circuit of Example 2.4. We set $v^{\mathsf{EP}}(t) = 0$ (equivalently, $V_k^{\mathsf{EP}} = 0$ for all k). Thus the initial condition for propagation of the phasor differential equation (2.38) is (according to (2.47))

$$I_k(0) = a_k \frac{e^{\alpha T} - 1}{\alpha T + jk2\pi} i(0)$$

The phasor differential equation (2.38) can now be propagated for $0 \leq t \leq T$, using the initial conditions $I_k(0)$ and the input phasors $V_k(t)$. The latter must be evaluated in real time as the (previously unknown) voltage $v(t)$ becomes available, using the expression (2.48).

In order to facilitate a comparison with Examples 2.4 and 2.5, we now assume that the input used in our numerical simulation for $t \geq 0$ will, in fact, be $v(t) = \sqrt{2}\ \Re\{V_1\, e^{j\omega_0 t}\}$. Since we have set $v^{\mathsf{EP}}(t) = 0$, the corresponding gap phasors $\{V_k(t);\ 0 \leq t < T\}$ are now given by the expressions (2.10) from Example 2.1, with $A_1 = V_1$ and $B_1 = V_1^{\mathsf{EP}} = 0$, viz.,

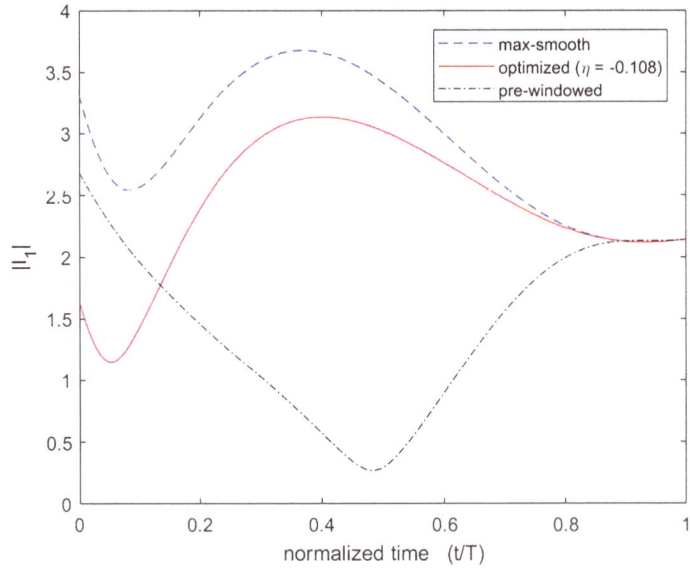

Fig. 2.5 State phasor magnitude trajectory: fundamental harmonic.

$$V_k(t) = \begin{cases} \sqrt{2}\,\Re\left\{V_1\, e^{j\frac{\omega_0 t}{2}}\right\} \dfrac{\sin\frac{\omega_0 t}{2}}{\pi} & k = 0 \\ V_1\,\dfrac{t}{T} + V_1^*\,\dfrac{\sin\omega_0 t}{2\pi}\; e^{-j\omega_0 t} & k = 1 \\ V_1\,\dfrac{\sin\frac{(k-1)\,\omega_0 t}{2}}{(k-1)\pi}\; e^{-j\frac{(k-1)\omega_0 t}{2}} + V_1^*\,\dfrac{\sin\frac{(k+1)\,\omega_0 t}{2}}{(k+1)\pi}\; e^{-j\frac{(k+1)\omega_0 t}{2}} & k \geq 2 \end{cases}$$

Notice that for $t \uparrow T$ these expressions yield $V_k(T) = 0$ for all $k \neq 1$, and $V_1(T) = V_1$, which is consistent with the terminal values obtained in the maximally-smooth and the optimized EP initialization schemes (i.e., Examples 2.4 and 2.5).

Using these expressions for $\{V_k(t)\}$, we can now evaluate the state phasors $\{I_k(t)\}$ by propagating (numerically) the phasor differential equation (2.38). The resulting $|I_1(t)|$ phasor magnitude is compared in Fig. 2.5 with the maximally-smooth version from Example 2.4, and the optimized version from Example 2.5, where again we set $R = 0.2\,\Omega$, $L = 1\,\text{mH}$, $V_1 = 1\,\text{V}$, and $i(0) = 0.5\,\text{A}$. All three trajectories converge to the same terminal value (at $t = T$), but again we observe that the trajectory corresponding to optimized EP initialization minimizes the fluctuation of $|I_1(t)|$ around $|I_1(T)|$. ■

Chapter 3 Efficient Estimation of Dynamic Phasors

The well-known DFT/FFT is currently the method of choice for digital evaluation of phasors [42]. Since the waveform sampling rate of modern grid monitoring devices is in the range of 128–512 samples/cycle (see, e.g., [34–37]), the application of a DFT to a data record consisting of one (or more) full cycle generates a very large number of higher harmonic phasors, in addition to the fundamental. However, only a few of these phasors may be of interest in most applications, while the remaining ones are quite negligible and thus of little interest. Moreover, half of the phasors generated by the DFT are completely redundant, since they represent negative frequencies, and thus are complex conjugates of the other half.

In general, any digital scheme for the evaluation of phasors, including the DFT/FFT, generates an approximation, because it must rely on a finite number of waveform samples, rather than on the continuous-time integral expression (2.2c). This means that digital phasor evaluation necessarily becomes *phasor estimation,* with some attendant estimation error [53]. In this chapter we review several established schemes for efficient estimation of dynamic phasors.

A large variety of power system applications require only a few dominant-harmonic phasors and thus could potentially benefit from an alternative to DFT-based phasor estimation. Such applications can be divided into three broad categories:

- **Category 1 applications** require a highly accurate estimate of only the fundamental phasor, generated at a low refresh rate (once per cycle or less). Examples include (a) static state estimation, (b) frequency monitoring, and (c) parameter identification for validation and calibration. Phasor measurement unit (PMU) devices, which use waveform samples extracted from a multi-cycle-long block, were designed specifically to meet the needs of this category of applications [54–57]. Alternatively, one can use a least-squares formulation of the phasor estimation problem or one of its variants, such as Fourier–Taylor, Kalman–Taylor,

H. Lev-Ari, A. M. Stanković, *Dynamic Phasors in Energy Processing Systems*,
Power Electronics and Power Systems, https://doi.org/10.1007/978-3-031-50020-6_4

Fourier–Kalman–Taylor, or Modified Kalman–Taylor to improve estimation accuracy [58–61].

- **Category 2 applications** require a moderately accurate estimate of only the fundamental phasor but can benefit from a higher refresh rate and lower latency [54, 62]. Examples include (a) computer-based protection, (b) fault detection and location, (c) monitoring transmission system stability, and (d) control schemes designed to prevent inter-area oscillations.
- **Category 3 applications** require moderate accuracy estimation of a few dominant phasors [63, 64]. Examples include (a) monitoring distribution system stability, (b) integration and monitoring of renewable sources, (c) load monitoring and characterization, (d) control schemes designed to prevent sub-synchronous oscillation, and (e) monitoring, detection, and classification of distribution system power quality events.

Since PMUs cannot provide low latency and a high refresh rate that is comparable to the waveform sampling rate, applications that require such features can be implemented only in dedicated high-speed grid monitoring devices (a.k.a. continuous point-on-wave monitors) [64]. A high refresh rate can be achieved by using a real-time sample-by-sample processing approach, in which phasor estimates are reevaluated as soon as a new waveform sample becomes available. In addition, low latency is achieved when the map from waveform samples to a phasor estimate involves only the most recent samples, located within a single cycle or a fraction of a cycle.

We provide in this chapter a formulation that can accommodate all three application categories via an appropriate choice of design parameters. We consider both processing options: (a) block processing schemes which result in a low refresh rate, and (b) non-decimated (= sample-by-sample) processing schemes which result in a refresh rate, which equals the waveform sampling rate.

When an accurate model of the system or device in question is available, a model-based Kalman filter can provide a significant reduction in the impact of measurement noise at the expense of increased computational cost. In this approach we must include all dominant phasors in the estimation process: omitting non-negligible phasors increases the estimation error of the Kalman filter. We provide a detailed analysis of this approach in Sect. 3.1.

When such an accurate model is not available (e.g., a large power grid), the method of choice is least-squares estimation [61]. We formulate in Sect. 3.2 a polyphase linear least-squares (PLLS) problem with several controllable design parameters: (a) the choice of harmonic phasors to be estimated, (b) the number of waveform samples used to determine the desired phasor estimates, (c) the time intervals between these waveform samples, and (d) the refresh rate.

Efficient estimation of dynamic phasors can also be viewed as a structured sparse implementation of the Fourier transform. When analyzing long signal records, schemes known as the *sparse Fourier transform* (SFT) compute only the dominant frequency components from a small (sub-Nyquist) randomized subset of the input data [41]. In contrast, both the model-based Kalman filter and the PLLS scheme use predetermined regularly-spaced waveform samples. This conceptual simplification is made possible by relying on prior knowledge of the nominal fundamental frequency and most likely dominant harmonics of power system waveforms.

Restricting the number of waveform samples used at each (discrete) time-instant to determine a few relevant phasor estimates also results in a noticeable cost reduction, as compared with the FFT or even the SFT. This difference in computational cost is of particular significance in implementations that use sliding-window sample-by-sample processing to achieve a high refresh rate of the resulting phasor estimates, which is required in all Category 2 applications, as well as many of Category 3 applications.

Sub-cycle phasor estimation, which we discuss in Sect. 3.3, is a very special case of the non-decimated PLLS approach. In addition to a high refresh rate, it achieves low latency by using only the most recent waveform samples, located within half a cycle or less. Sub-cycle phasors also exhibit a (modified) Parseval property, which facilitates the construction of advanced power flow metrics, as detailed in Sect. 9.3. In addition, the low computational cost of the sub-cycle scheme is achieved by reducing the number of input samples to the minimum needed for ensuring a unique solution to the PLLS problem.

Finally, Sect. 3.4 examines an alternative approach to the construction of dynamic phasors, based on the notion of *analytic signal*, and some of its challenges and applications. In particular, we compare three schemes for estimation of the fundamental frequency based on (a) analytic signal approach, (b) full-cycle least-squares fundamental phasor estimate, and (c) sub-cycle fundamental phasor estimate. Our comparison indicates that all three techniques produce comparable frequency estimates, both in steady-state operation and in the presence of transients. However, the cost of implementing a sub-cycle-based scheme is significantly lower than that of the other two techniques.

3.1 Kalman Filter Estimation of Dynamic Phasors

When a waveform is acquired with additive noise, the dynamic phasors associated with this waveform will be noisy as well. The level of phase-domain noise can be reduced by evaluating the generalized averaging expression (2.2c) over a longer interval, consisting of several full cycles: when a phasor $X_k(t)$ is evaluated by generalized averaging over multiple cycles of a noisy version of the waveform $x(\cdot)$, the resulting signal-to-noise ratio (SNR) is proportional to the length of the averaging interval. While this technique works well in steady-state operation, any transients contained in the waveform of interest, such as initialization transients, will be strongly attenuated by multi-cycle averaging.

Alternatively, one can rely on system model information to obtain estimates of the dynamic phasors of interest that exhibit a gradually diminishing level of noise. When a waveform $x(\cdot)$ satisfies a linear, constant-coefficient state-space model with a known (periodic) input $u(\cdot)$, the impact of additive measurement noise can be reduced by using a Kalman filter: this results in phasor estimates $\widehat{X}_k(t)$ that exploit the growing history $\{x(s)\,;\ 0 \leq s \leq t\}$ of the waveform $x(\cdot)$ to achieve a gradual decline in the level of the noise-induced error component of $\widehat{X}_k(t)$.

We recall that the waveform-domain state-equation $\frac{d}{dt}\,x(t) = A\,x(t) + B\,u(t)$ gives rise to the baseband phase-domain equation (2.25b) or, equivalently, its passband phasor version (2.25a), viz.,

$$\frac{d}{dt}\,\mathcal{X}_k(t) = A\,\mathcal{X}_k(t) + B\,\mathcal{U}_k(t) \tag{3.1}$$

where $\mathcal{X}_k(t)$ is an $n \times 1$ vector and $\mathcal{U}_k(t)$ is a $p \times 1$ vector. Notice that, in order to streamline our presentation, we use here the *passband version* of all dynamic phasors.

Next, we intend to use the (one-sided) Fourier-series representation (2.2d) as our measurement equation. However, in order to limit the number of state variables, we must restrict the choices of the harmonic index k in the state-equation (3.1) to non-negligible harmonics. This means that the set of harmonics we use in our state-space model is some finite set $\boldsymbol{\mathcal{S}} \stackrel{\text{def}}{=} \{k_1, k_2, \ldots, k_L\}$, so that the corresponding (noisy) measurement equation is given by the expression $x(t) = \sum_{k \in \boldsymbol{\mathcal{S}}} a_k\,\Re\left\{X_k(t)\,e^{jk\omega_0 t}\right\} + \eta(t)$ (recall that $a_k = \sqrt{2}$ for all $k \geq 1$, and $a_0 = 1$) or, equivalently, in passband-version form, viz.,

$$x(t) = \sum_{k \in \boldsymbol{\mathcal{S}}} a_k\,\Re\left\{\mathcal{X}_k(t)\right\} + \eta(t) \tag{3.2}$$

The additive "noise signal" $\eta(\cdot)$ in this equation represents the combined effect of both sensor measurement noise *and* the modeling error caused by ignoring some (weaker) harmonics.

The measurement equation (3.2) is linear in $\Re\{\mathcal{X}_k(t)\} = \frac{1}{2}\{\mathcal{X}_k(t) + \mathcal{X}_k^*(t)\}$: this type of relation is known as *widely-linear* [65]. One way to take into account this generalized linearity relation is by augmenting the set of state variables to include both $\mathcal{X}_k(t)$ and its conjugate $\mathcal{X}_k^*(t)$, for every $k \in \boldsymbol{\mathcal{S}}$. The corresponding generalization of the Kalman filter, which is known as the *Widely-Linear Complex Kalman Filter* (WLCKF) [66, 67], can be used to obtain estimates of the phasors $\{\mathcal{X}_k(t)\,,\ k \in \boldsymbol{\mathcal{S}}\}$ whose quality improves with time.

3.1.1 A Dual-Channel Phase-Domain Linear State-Space Model

The WLCKF does not fully exploit the special structure of the linear state-space model (3.1)–(3.2), which involves *real-valued* measurements and state-equation parameters. A simpler alternative is to use the real and imaginary parts of the passband phasor $\mathcal{X}_k(t)$ as state variables, resulting in a purely real-valued state-equation:[1] we observe that (3.1) maps into a pair of decoupled equations for the real

[1] This is known as a "dual-channel" or "bi-variate" approach [66, 67].

and imaginary parts of $\mathcal{X}_k(t)$, viz.,

$$\frac{d}{dt}\begin{pmatrix}\Re\{\mathcal{X}_k(t)\}\\ \Im\{\mathcal{X}_k(t)\}\end{pmatrix} = \begin{pmatrix}A & 0\\ 0 & A\end{pmatrix}\begin{pmatrix}\Re\{\mathcal{X}_k(t)\}\\ \Im\{\mathcal{X}_k(t)\}\end{pmatrix} + \begin{pmatrix}B & 0\\ 0 & B\end{pmatrix}\begin{pmatrix}\Re\{\mathcal{U}_k(t)\}\\ \Im\{\mathcal{U}_k(t)\}\end{pmatrix} \tag{3.3}$$

for each $k \in \boldsymbol{\mathcal{S}}$. Consequently, both the complete state-vector and input vector consist of multiple blocks of size $n \times 1$ each, namely,

$$\mathbf{X}(t) \stackrel{\text{def}}{=} \begin{pmatrix} a_{k_1}\Re\{\mathcal{X}_{k_1}(t)\}\\ a_{k_1}\Im\{\mathcal{X}_{k_1}(t)\}\\ a_{k_2}\Re\{\mathcal{X}_{k_2}(t)\}\\ a_{k_2}\Im\{\mathcal{X}_{k_2}(t)\}\\ \vdots\\ a_{k_L}\Re\{\mathcal{X}_{k_L}(t)\}\\ a_{k_L}\Im\{\mathcal{X}_{k_L}(t)\} \end{pmatrix}, \quad \mathbf{U}(t) \stackrel{\text{def}}{=} \begin{pmatrix} a_{k_1}\Re\{\mathcal{U}_{k_1}(t)\}\\ a_{k_1}\Im\{\mathcal{U}_{k_1}(t)\}\\ a_{k_2}\Re\{\mathcal{U}_{k_2}(t)\}\\ a_{k_2}\Im\{\mathcal{U}_{k_2}(t)\}\\ \vdots\\ a_{k_L}\Re\{\mathcal{U}_{k_L}(t)\}\\ a_{k_L}\Im\{\mathcal{U}_{k_L}(t)\} \end{pmatrix} \tag{3.4}$$

In general, the number of such blocks is $N = 2L$, as is evident from the above definition. However, when $\boldsymbol{\mathcal{S}}$ contains the 0-th harmonic, the corresponding state and input phasors are real-valued, i.e., $\Im\{\mathcal{X}_0(t)\} = 0$ and $\Im\{\mathcal{U}_0(t)\} = 0$, so that the number of non-vanishing blocks in both $\mathbf{X}(t)$ and $\mathbf{U}(t)$ is reduced to $N = 2L - 1$.

Thus, the final form of the phase-domain linear state-space model is

$$\frac{d}{dt}\mathbf{X}(t) = \boldsymbol{\mathcal{A}}\,\mathbf{X}(t) + \boldsymbol{\mathcal{B}}\,\mathbf{U}(t) \tag{3.5a}$$

where

$$\boldsymbol{\mathcal{A}} = I_N \otimes A \quad , \quad \boldsymbol{\mathcal{B}} = I_N \otimes B \tag{3.5b}$$

Here $\otimes$ denotes the Kronecker product,[2] and I_N is the $N \times N$ identity matrix, where N is the number of blocks in the state-vector $\mathbf{X}(t)$ (i.e., either $2L$ or $2L-1$). Notice that the measurement equation (3.2) now becomes

$$\begin{aligned} x(t) &= \begin{bmatrix} I_n & 0 & I_n & 0 & \ldots & I_n & 0 \end{bmatrix}\mathbf{X}(t) + v(t)\\ &= \left(\begin{bmatrix} 1 & 0 & 1 & 0 & \ldots & 1 & 0 \end{bmatrix} \otimes I_n\right)\mathbf{X}(t) + v(t) \end{aligned}$$

so that the resulting state-space model is not observable [69]: in particular, we cannot estimate $\Im\{\mathcal{X}_k(t)\}$ for any $k \in \boldsymbol{\mathcal{S}}$.

[2] A detailed discussion of the Kronecker product and its properties can be found in [68].

One possible remedy is to increase the number of $x(\cdot)$ measurements used to update the state estimate in each time-instant. This means we exploit the expression (2.2a) for $0 \leq \Delta < T$, viz.,

$$x(t-\Delta) = X_0(t) + \sqrt{2} \sum_{k=1}^{\infty} \Re\left\{X_k(t)\, e^{jk\omega_0(t-\Delta)}\right\}$$

Expressing this relation in terms of passband phasors, and using only the set of harmonics defined by $\mathcal{S}$, results in the identity

$$\begin{aligned} x(t-\Delta) &= \sum_{k \in \mathcal{S}} a_k\, \Re\left\{\mathcal{X}_k(t)\, e^{-jk\omega_0\Delta}\right\} + \eta(t-\Delta) \\ &= \sum_{k \in \mathcal{S}} a_k \Big[\big(\cos k\omega_0\Delta\big)\, \Re\big\{\mathcal{X}_k(t)\big\} + \big(\sin k\omega_0\Delta\big)\, \Im\big\{\mathcal{X}_k(t)\big\}\Big] + \eta(t-\Delta) \\ &= \sum_{k \in \mathcal{S}} \Big[\big(\cos k\omega_0\Delta\big) I_n \quad \big(\sin k\omega_0\Delta\big) I_n\Big] \begin{pmatrix} a_k\, \Re\{\mathcal{X}_k(t)\} \\ a_k\, \Im\{\mathcal{X}_k(t)\} \end{pmatrix} + \eta(t-\Delta) \end{aligned}$$

where $\eta(\cdot)$ represents the combined effect of measurement noise and modeling error. We can now form an extended measurement vector by acquiring multiple measurements of the waveform $x(\cdot)$ within the interval $(t-T \quad t]$, viz.,

$$\mathbf{y}(t) = \begin{pmatrix} x(t) \\ x(t-h) \\ \vdots \\ x(t-(M-1)h) \end{pmatrix} = \mathcal{C}\,\mathbf{X}(t) + \begin{pmatrix} \eta(t) \\ \eta(t-h) \\ \vdots \\ \eta(t-(M-1)h) \end{pmatrix} \tag{3.6a}$$

for some $h \leq T/M$, where $\mathcal{C} = C \otimes I_n$, and

$$C = \begin{pmatrix} 1 & 0 & \dots & 1 & 0 \\ \cos k_1\varphi & \sin k_1\varphi & \dots & \cos k_L\varphi & \sin k_L\varphi \\ \cos 2k_1\varphi & \sin 2k_1\varphi & \dots & \cos 2k_L\varphi & \sin 2k_L\varphi \\ \vdots & \vdots & & \vdots & \vdots \\ \cos(M-1)k_1\varphi & \sin(M-1)k_1\varphi & \dots & \cos(M-1)k_L\varphi & \sin(M-1)k_L\varphi \end{pmatrix} \tag{3.6b}$$

with $\varphi \overset{\text{def}}{=} \omega_0 h \equiv 2\pi \frac{h}{T} \leq \frac{2\pi}{M}$. We can assume, without loss of generality, that $k_1 < k_2 < \ldots < k_L$. When $k_1 = 0$, the second column of the matrix C consists entirely of zero elements and should be removed. Consequently, the number of non-vanishing columns of C always matches the number of $n \times 1$ blocks of the phase-domain state-vector $\mathbf{X}(t)$: it is $2L$ when $k_1 > 0$ but reduces to $2L-1$ when $k_1 = 0$.

In summary, the complete dual-channel phase-domain linear state-space model is given by (3.5) and (3.6). Applying the standard (linear) Kalman filter to this model generates estimates $\widehat{\mathcal{X}}_k(t)$ of the passband phasors $\{\mathcal{X}_k(t)\,,\ k\in\boldsymbol{\mathcal{S}}\}$ from noisy measurements of the waveform $x(t)$. We demonstrate in Sect. 3.1.2 that the observability of this model, and hence the quality of the resulting phasor estimates, depends primarily on the singular value profile of the matrix C. Finally, estimates $\widehat{X}_k(t)$ of the *baseband phasors* can be obtained, as needed, from their *passband versions* $\widehat{\mathcal{X}}_k(t)$ via the relation $\widehat{X}_k(t)=\widehat{\mathcal{X}}_k(t)e^{-jk\omega_0 t}$ or, equivalently, via the dual-channel relation

$$\begin{pmatrix}\Re\{\widehat{X}_k(t)\}\\ \Im\{\widehat{X}_k(t)\}\end{pmatrix} = \left[\begin{pmatrix}\cos k\omega_0 t & \sin k\omega_0 t\\ -\sin k\omega_0 t & \cos k\omega_0 t\end{pmatrix}\otimes I_n\right]\begin{pmatrix}\Re\{\widehat{\mathcal{X}}_k(t)\}\\ \Im\{\widehat{\mathcal{X}}_k(t)\}\end{pmatrix} \tag{3.7}$$

which involves only real quantities.

Example 3.1 (RL circuit: phasor estimation): The single-phase RL circuit from Example 2.2 can be interpreted as a single-state model, viz.,

$$\frac{d}{dt}\,i(t) \;=\; -\,\alpha\, i(t) + \tfrac{1}{L_0}\,v(t)\;, \qquad t\geq 0 \tag{3.8}$$

where[3] $\alpha \overset{\text{def}}{=} \frac{R}{L_0}$, and the initial condition is $i(0)$. Mapping into the notation of the state-space model (2.22), we observe that $x(t)\equiv i(t)$ (=inductor current), $u(t)\equiv v(t)$ (=driving voltage), $A=-\alpha=-\frac{R}{L_0}$, and $B=\frac{1}{L_0}$. According to the analysis in Example 2.2, the magnitude of the current phasors $I_k(t)$ decays with increasing k (recall (2.15b)). It makes sense to set in this case $\boldsymbol{\mathcal{S}}=\{0,1,2,\ldots,L-1\}$, so that every harmonic up to the $(L-1)$-th will be estimated.

Since the source voltage in this case is $v(t)=\sqrt{2}\,\Re\left\{V_1\,e^{j\omega_0 t}\right\}$ (for $t\geq 0$), there is only one non-zero voltage phasor (for $t\geq T$), i.e., $V_1(t)=V_1$, and $V_k(t)=0$ for all $k\neq 1$. Thus, the complete (passband) phase-domain state and input vectors are

$$\mathbf{I}(t)\overset{\text{def}}{=}\begin{pmatrix} I_0(t)\\ \sqrt{2}\,\Re\{\mathcal{I}_1(t)\}\\ \sqrt{2}\,\Im\{\mathcal{I}_1(t)\}\\ \vdots\\ \sqrt{2}\,\Re\{\mathcal{I}_{L-1}(t)\}\\ \sqrt{2}\,\Im\{\mathcal{I}_{L-1}(t)\}\end{pmatrix}\,,\qquad \mathbf{V}(t)\overset{\text{def}}{=}\begin{pmatrix}0\\ \sqrt{2}\,\Re\{V_1e^{j\omega_0 t}\}\\ \sqrt{2}\,\Im\{V_1e^{j\omega_0 t}\}\\ 0\\ \vdots\\ 0\end{pmatrix} \tag{3.9}$$

[3] We use here the notation L_0 for the circuit inductance in order to distinguish it from the number of harmonics in the set $\boldsymbol{\mathcal{S}}$.

Notice that the length of these vectors is $N = 2L - 1$ and that the baseband and passband phasors for the 0-th harmonic coincide, i.e., $I_0(t) = \mathcal{I}_0(t)$. Also,

$$\mathcal{A} = I_N \otimes A = -\alpha I_N \quad , \quad \mathcal{B} = I_N \otimes B = \frac{1}{L_0} I_N$$

and $\mathcal{C} = C \otimes I_1 = C$, so that the phase-domain state-space model (3.5)–(3.6) reduces to

$$\begin{aligned} \frac{d}{dt}\mathbf{I}(t) &= -\alpha\,\mathbf{I}(t) + \frac{1}{L_0}\mathbf{V}(t) \\ \mathbf{y}(t) &\equiv \begin{pmatrix} i(t) \\ i(t-h) \\ \vdots \\ i(t-(M-1)h) \end{pmatrix} = C\,\mathbf{I}(t) + \begin{pmatrix} \eta(t) \\ \eta(t-h) \\ \vdots \\ \eta(t-(M-1)h) \end{pmatrix} \end{aligned} \tag{3.10}$$

This example is analyzed in [69] using the parameter values

$$\omega_0 = 120\pi \; rad/sec\,, \quad V_1 = 10 \; Volt\,, \quad R = 1\,\Omega\,, \quad L_0 = 100 \; mH,$$

and a measurement signal-to-noise ratio of 21 dB. We observe that, with these parameter values, $\left|\frac{\mathcal{I}_2(t)}{I_0(t)}\right| = \frac{1}{53.3}$ because (recall (2.15))

$$\left|\frac{\mathcal{I}_k(t)}{I_0(t)}\right| = \left|\frac{I_k(t)}{I_0(t)}\right| = \frac{1}{\sqrt{1 + k^2\left(\frac{\omega_0 L_0}{R}\right)^2}} \quad , \quad \text{where} \quad \frac{\omega_0 L_0}{R} = 12\pi$$

This means that the (noise-free) phasors $\mathcal{I}_k(t)$ are negligible, for all $k \geq 2$, in comparison with $I_0(t)$, so that it makes sense to set $L = 2$ in (3.9), viz.,

$$\mathbf{I}(t) = \begin{pmatrix} I_0(t) \\ \sqrt{2}\,\Re\{\mathcal{I}_1(t)\} \\ \sqrt{2}\,\Im\{\mathcal{I}_1(t)\} \end{pmatrix} \quad , \quad \mathbf{V}(t) = \begin{pmatrix} 0 \\ \sqrt{2}\,\Re\{V_1 e^{j\omega_0 t}\} \\ \sqrt{2}\,\Im\{V_1 e^{j\omega_0 t}\} \end{pmatrix} \tag{3.11}$$

The superiority of the Kalman-based phasor estimates, as compared with the standard (sliding-window) phasors, is demonstrated by Fig. 3.1. The Kalman-based estimates coincide (for $t \geq 2T$, or so) with the true (noise-free) $I_k(t)$, as given by (2.15). In contrast, the sliding-window DFT-based phasors exhibit a significant error for all t.

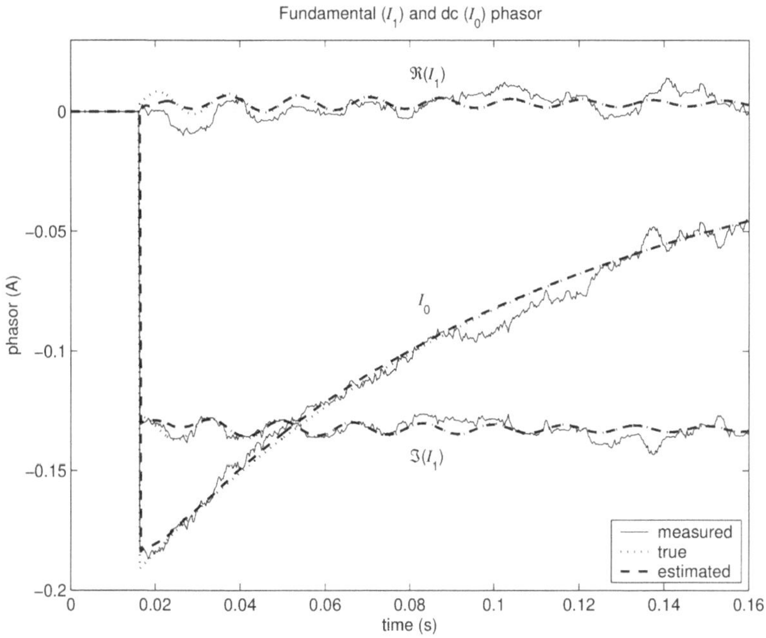

Fig. 3.1 Comparison of Kalman-based phasor estimates (dashed line) with standard DFT-based phasors (solid line) evaluated in a noisy environment. The dotted line shows the true (noise-free) phasor.

■

3.1.2 Observability of the Dual-Channel Phase-Domain Model

The observability of the dual-channel phase-domain model (3.5)–(3.6) depends entirely on the conditioning of the matrix C, as demonstrated by the following theorem.

Theorem 3.1 (Observability of phase-domain model): *The state-space model (3.5)–(3.6) is observable if, and only if, the Hermitian matrix $C^{\top}C$ is positive-definite.*

Proof: The observability Gramian for this model is

$$\mathcal{O} \stackrel{\text{def}}{=} \sum_{r=0}^{\infty} (\mathcal{C}\mathcal{A}^r)^{\top}(\mathcal{C}\mathcal{A}^r) \tag{3.12}$$

The system is observable if, and only if, $\mathcal{O}$ is positive-definite [70]. However, since $\mathcal{A} = I_N \otimes A$ and $\mathcal{C} = C \otimes I_n$, it follows from properties of the Kronecker product [68] that

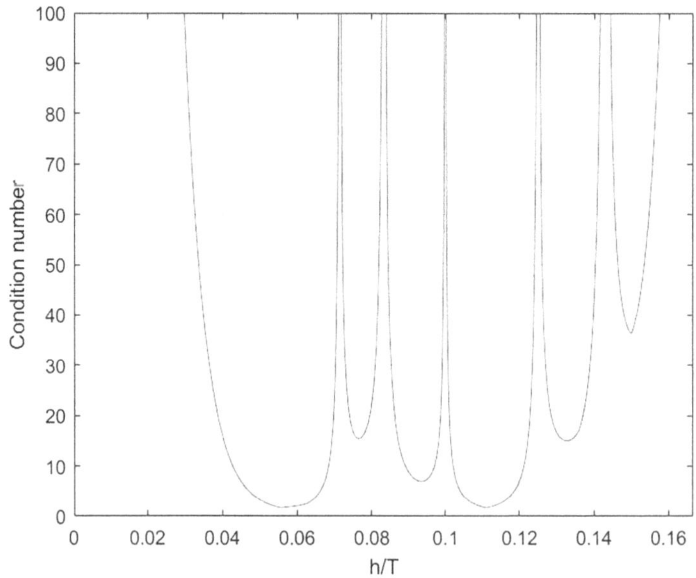

Fig. 3.2 Condition number of the C-matrix, for $\mathcal{S} = \{1, 5, 7\}$.

$$
\begin{aligned}
(\mathcal{C}\mathcal{A}^r)^\top(\mathcal{C}\mathcal{A}^r) &= (\mathcal{A}^r)^\top\mathcal{C}^\top\mathcal{C}\mathcal{A}^r = (I_N \otimes A^r)^\top(C \otimes I_n)^\top(C \otimes I_n)(I_N \otimes A^r) \\
&= C^\top C \otimes (A^r)^\top A^r
\end{aligned}
$$

and, consequently,

$$
\mathcal{O} = C^\top C \otimes \sum_{r=0}^{\infty}(A^r)^\top A^r
$$

Notice that $\sum_{r=0}^{\infty}(A^r)^\top A^r \equiv I_n + \sum_{r=1}^{\infty}(A^r)^\top A^r$ is positive-definite regardless of A, so that $\operatorname{rank}\mathcal{O} = n \cdot \operatorname{rank}(C^\top C)$: it follows that $\mathcal{O}$ is positive-definite if, and only if, $C^\top C$ is positive-definite. This completes the proof of the theorem. ■

Corollary: *A necessary (but not sufficient) condition for* $C^\top C > 0$ *is that the number of rows of the matrix* C *equals (or exceeds) the number of its columns, i.e.,* $M \geq N$. ■

The singular value profile (and hence also the rank) of the matrix C depends on the set $\mathcal{S}$ of selected harmonic indices, as well as the value of the parameter

$\varphi \equiv 2\pi\frac{h}{T}$. For instance, the phase-domain state-space model (3.5)–(3.6), with $\mathcal{S} = \{1, 5, 7\}$, is *not observable* for several distinct choices of h in the range $0 < h < \frac{T}{M}$, as is evident from plotting the *condition number* [4] of the matrix C as a function of $\frac{h}{T}$ (Fig. 3.2). We conclude that the choice of the inter-measurement shift h should be guided by the conditioning of the resulting C-matrix: a judicious choice of h will yield a well-conditioned C-matrix, resulting in more accurate phasor estimates.

3.1.3 Estimation of Symmetrical Sequence Components

The estimation of dynamic symmetrical components is a central task in numerous power system applications. In protection, it is a means to detect and classify faults [71, 72]; in control and power quality-related tasks, it is important for proper functioning of components and subsystems [73]. It is also a key part in simulation environments that combine detailed representations of some portions of the systems ("the study part") with reduced representations for other less important portions ("neighboring systems") [74,75].

The phase-domain state-space model (3.5)–(3.6) can also be used to obtain Kalman filter-based estimates of the *symmetrical sequence components* associated with elements of the state-vector $\mathbf{X}(t)$. This involves using a suitable *similarity transformation*, such as the one used (implicitly) in Example 2.3 to replace the voltage and current dynamic phasors of a three-phase RL circuit by their symmetrical component (a.k.a. *SC-domain*) equivalents $\{V_{k,0}(t)\,, V_{k,+}(t)\,, V_{k,-}(t)\}$ and $\{I_{k,0}(t)\,, I_{k,+}(t)\,, I_{k,-}(t)\}$, respectively.

An explicit derivation of this similarity transformation is provided in Appendix K: we show that, in general, $\mathbf{X}(t) = \mathcal{T}\mathbf{X}^{(sc)}(t)$, where $\mathbf{X}^{(sc)}(t)$ consists of SC-domain equivalents of elements of the phase-domain state-vector $\mathbf{X}(t)$, and $\mathcal{T}$ is a block-diagonal matrix, viz.,

$$\mathcal{T} = \begin{pmatrix} \mathcal{T}_{k_1} & & \mathrm{O} \\ & \mathcal{T}_{k_2} & \\ \mathrm{O} & & \ddots \\ & & & \mathcal{T}_{k_L} \end{pmatrix} \tag{3.13}$$

We present in the Appendix an explicit expression for the block elements $\mathcal{T}_k$ of this matrix and use it to show that $\mathcal{T}$ is orthogonal, i.e., $\mathcal{T}^\top\mathcal{T} = I$. Applying the

[4] The condition number of a matrix is the ratio between its largest and smallest singular values. Hence, the condition number of a singular matrix is infinite.

linear transformation $\mathbf{X}(t) = \mathcal{T}\mathbf{X}^{(sc)}(t)$ to the phase-domain state-equation, (3.5) converts it into an SC-domain equivalent, namely,

$$\frac{d}{dt}\mathbf{X}^{(sc)}(t) = \left(\mathcal{T}^{-1}\mathcal{A}\mathcal{T}\right)\mathbf{X}^{(sc)}(t) + \left(\mathcal{T}^{-1}\mathcal{B}\mathcal{T}\right)\mathbf{U}^{(sc)}(t) \tag{3.14a}$$

where we also replaced $\mathbf{U}(t)$ by its SC-domain version $\mathbf{U}^{(sc)}(t) \stackrel{\text{def}}{=} \mathcal{T}^{-1}\mathbf{U}(t)$. Similarly, the phase-domain measurement equation (3.6) is also converted into an SC-domain equivalent, viz.,

$$\mathbf{y}(t) = \begin{pmatrix} x(t) \\ x(t-h) \\ \vdots \\ x(t-(M-1)h) \end{pmatrix} = \left(\mathcal{C}\mathcal{T}\right)\mathbf{X}^{(sc)}(t) + \begin{pmatrix} \eta(t) \\ \eta(t-h) \\ \vdots \\ \eta(t-(M-1)h) \end{pmatrix} \tag{3.14b}$$

where we recall that $\mathcal{C} = C \otimes I_n$, with C given by (3.6).

The simplest SC-domain linear state-space model is obtained when the time-domain state-vector $x(\cdot)$ consists of a *single three-phase waveform*, as illustrated in the following example.

Example 3.2 (SC-domain state-space model for three-phase RL circuit): Consider the three-phase RL circuit from Example 2.3. Transposing the differential equation (2.20) and multiplying it on the left by L^{-1} results in the state-equation

$$\frac{d}{dt}x(t) = A\,x(t) + B\,u(t)$$

where $x(t) = i^{\top}(t)$, $u(t) = v^{\top}(t) = \sqrt{2}\,\Re\left\{V_1^{\top}e^{j\omega_0 t}\right\}$, $A = -L^{-1}R$, and $B = L^{-1}$ (recall that, in this example, the matrix R is diagonal, and the matrix L is symmetric). Thus the number of state variables is $n = 3$ and, in the language of Appendix K, the number of three-phase waveforms of interest is $p = 1$.

The dual-channel phase-domain version of this state-equation is (3.5), with the phase-domain state-vector $\mathbf{X}(t)$ and input vector $\mathbf{U}(t)$ given by (3.4). The elements of these vectors, i.e., the (passband) state-phasors and input phasors, are given by

$$\mathcal{X}_k(t) = \mathcal{I}_k^{\top}(t)$$

$$\mathcal{U}_k(t) = \begin{cases} \left(L^{-1}V_1^{\top}\right)e^{j\omega_0 t} & k = 1 \\ 0 & \text{else} \end{cases}$$

In order to convert this phase-domain state-space model into an equivalent SC-domain version, we need to employ the state transformation $\mathbf{X}(t) = \mathcal{T}\mathbf{X}^{(sc)}(t)$, where the matrix $\mathcal{T}$ is given by (3.13).

The resulting SC-domain model is still given by (3.14). However, since $p = 1$ and $n = 3$, the expressions (K.8)–(K.9) for $\mathcal{T}_k$ reduce to

$$\mathcal{T}_k = \begin{cases} \Theta_0 & k = 0 \\ \Theta_{sc} & k \neq 0 \end{cases} \tag{3.15}$$

because in this case $\mathcal{P}_1 = I_6$. In other words, the state-vector transformation $\mathbf{X}(t) = \mathcal{T}\mathbf{X}^{(sc)}(t)$ reduces, for each non-zero harmonic, to

$$\begin{pmatrix} \Re\{\mathcal{X}_k(t)\} \\ \Im\{\mathcal{X}_k(t)\} \end{pmatrix} \equiv \begin{pmatrix} \Re\{\mathcal{I}_k^\top(t)\} \\ \Im\{\mathcal{I}_k^\top(t)\} \end{pmatrix} = \Theta_{sc} \begin{pmatrix} \Re\{\mathcal{I}_k^{(sc)}(t)\} \\ \Im\{\mathcal{I}_k^{(sc)}(t)\} \end{pmatrix}$$

where

$$\Theta_{sc} = \begin{pmatrix} \Re\{\mathcal{M}_{sc}\} & \Im\{\mathcal{M}_{sc}\} \\ -\Im\{\mathcal{M}_{sc}\} & \Re\{\mathcal{M}_{sc}\} \end{pmatrix}, \quad \mathcal{I}_k^{(sc)}(t) \stackrel{\text{def}}{=} \begin{pmatrix} I_{k,0}(t) \\ I_{k,+}(t) \\ I_{k,-}(t) \end{pmatrix} e^{jk\omega_0 t}$$

and we used the fact that $x(t) = i^\top(t)$, so that $\mathcal{X}_k(t) \equiv \mathcal{I}_k^\top(t)$ for all $k \in \mathcal{S}$.

The transformation used for $k = 0$ in this example still coincides with (K.3), namely,

$$X_0(t) \equiv I_0^\top(t) = \Theta_0 \underbrace{\begin{pmatrix} I_{0,0}(t) \\ \sqrt{2}\,\Re\{I_{0,+}(t)\} \\ \sqrt{2}\,\Im\{I_{0,+}(t)\} \end{pmatrix}}_{I_0^{(sc)}(t)}$$

with

$$\Theta_0 = \frac{1}{\sqrt{3}} \begin{pmatrix} 1 & \sqrt{2} & 0 \\ 1 & -\frac{\sqrt{2}}{2} & \sqrt{\frac{3}{2}} \\ 1 & -\frac{\sqrt{2}}{2} & -\sqrt{\frac{3}{2}} \end{pmatrix}$$

This completes the characterization, for this example, of the phase-domain/SC-domain transformation $\mathbf{X}(t) = \mathcal{T}\mathbf{X}^{(sc)}(t)$.

Finally, we notice that the measurement equation for this example is given by the expression (3.14b), with $\mathcal{C} = C \otimes I_3$. In addition, we recall that the transformation matrix $\mathcal{T}$ used in (3.14b) reduces to a simpler expression, as described by (3.15). ■

Next, we can apply a Kalman filter to the SC-domain state-space model (3.14) to generate an estimate $\widehat{\mathbf{X}}^{(sc)}(t)$ of the SC-domain state-vector from noisy measurements of the time-domain state $x(t)$. The quality of the resulting estimate depends on the conditioning of the SC-domain observability Gramian (recall the definition (3.12))

$$\begin{aligned}
\mathcal{O}^{(sc)} &= \sum_{r=0}^{\infty} \left[(\mathcal{C}\mathcal{T})(\mathcal{T}^{-1}\mathcal{A}\mathcal{T})^r\right]^\top \left[(\mathcal{C}\mathcal{T})(\mathcal{T}^{-1}\mathcal{A}\mathcal{T})^r\right] \\
&= \sum_{r=0}^{\infty} \left[(\mathcal{C}\mathcal{T})(\mathcal{T}^{-1}\mathcal{A}^r\mathcal{T})\right]^\top \left[(\mathcal{C}\mathcal{T})(\mathcal{T}^{-1}\mathcal{A}^r\mathcal{T})\right] \\
&= \sum_{r=0}^{\infty} \mathcal{T}^\top (\mathcal{C}\mathcal{A}^r)^\top (\mathcal{C}\mathcal{A}^r)\, \mathcal{T} \\
&= \mathcal{T}^\top \mathcal{O}\, \mathcal{T}
\end{aligned}$$

where $\mathcal{O}$ is the observability Gramian of the phase-domain linear state-space model (3.5)–(3.6), as defined in (3.12). Since the matrix $\mathcal{T}$ is orthogonal, the eigenvalues of $\mathcal{O}^{(sc)}$ coincide with those of $\mathcal{O}$, so that both Gramians have the same condition number. This implies that the quality of SC-domain estimates is the same as that of phase-domain estimates.

Finally, we can obtain estimates of individual symmetrical components of interest by converting the passband SC-domain state-vector estimate $\widehat{\mathbf{X}}^{(sc)}(t)$ into its *baseband form* via the transformation (3.7). Notice that this transformation is not needed in applications that require only *magnitudes* of symmetrical sequence components.

3.2 Least-Squares Estimation of Dynamic Phasors

In this section we formulate a polyphase linear least-squares (PLLS) problem with several adjustable design parameters: (a) the choice of harmonic phasors to be estimated, (b) the number of waveform samples used to determine the desired phasor estimates, and (c) the lengths of time intervals between these waveform samples.

In order to estimate only the dominant harmonics of an acquired polyphase waveform $x(t)$, we can use a "signal+perturbation" model, namely

$$x(t) = \widehat{x}(t) + \text{perturbation}$$

where

$$\widehat{x}(t) \stackrel{\text{def}}{=} \sum_{k \in \boldsymbol{S}} \sqrt{2}\, \Re\left\{ \widehat{X}_k \, e^{jk\omega_0 t} \right\} \tag{3.16a}$$

Here $\boldsymbol{S} = \{k_\ell \, ; \; 1 \le \ell \le L\}$ is a set of selected harmonic indices, and ω_0 is the *a-priori selected* nominal steady-state fundamental frequency. The difference $x(t) - \widehat{x}(t)$ (a.k.a. "error") consists, in general, of the excluded (non-dominant) harmonics, various non-harmonic components (e.g., inter-harmonics and transients), measurement noise, and perturbations caused by sampling and quantization of the continuous-time waveform $x(t)$.

Notice that the set of dominant harmonics included in (3.16a) does not include a DC component. Our choice to exclude this component from the "estimated waveform" $\widehat{x}(t)$ is motivated by the observation that power system waveforms rarely contain a DC component: even when such a component is present, its magnitude does not justify including it among the dominant harmonics. When needed, the formulation (3.16a) can be extended to include a DC component $\widehat{X}_0$, and the results presented in this section, as well as in Appendix G, can be straightforwardly modified to accommodate such an extension.

A least-squares approach is commonly used to suppress the impact of such perturbations on the resulting phasor estimates. This means that $\widehat{X}_k$ are obtained by solving a *polyphase linear least-squares* (PLLS) problem, viz.,

$$\min_{\{\widehat{X}_k \, ; \, k \in \boldsymbol{S}\}} \sum_{i=0}^{M-1} \left\| x(t - \Delta_i) - \widehat{x}(t - \Delta_i) \right\|^2 \tag{3.16b}$$

where $\boldsymbol{\Delta} = \left\{ \Delta_i \, ; \; 0 \le i \le M-1 \right\}$ is a set of (distinct, non-negative) inter-sample shifts, with $\Delta_0 = 0$. We can assume, without loss of generality, that

$$0 = \Delta_0 < \Delta_1 < \Delta_2 < \; \ldots < \Delta_{M-1} \tag{3.16c}$$

and

$$k_1 < k_2 < \; \ldots < k_L \tag{3.16d}$$

We also assume that the fundamental harmonic is always included in $\boldsymbol{S}$: typically, $k_1 = 1$. Notice that the solution of this least-squares problem is a function of the sliding-window set of polyphase samples $\{x(t), x(t - \Delta_1), \ldots, x(t - \Delta_{M-1})\}$ and thus depends, in general, on the time-instant "t." We will use the notation $\widehat{X}_k(t)$ to highlight the time dependence of estimated phasors.

The PLLS formulation (3.16) covers a wide range of possible implementations, ranging from the low-rate estimation of the fundamental harmonic phasor, as done in PMUs, to the fast detection of transients via the sub-cycle approach [78], which is needed in system-level protection applications. The former involves solving the PLLS once every 1–3 cycles, using $L = 1$ and a high value of M (see Sect. 3.2.2). The latter generates a new PLLS solution at every (discrete) time-instant, using a moderate value of L and the maximally-sparse choice $M = 2L$ (see Sect. 3.3).

3.2.1 PLLS Solution: A Linear Waveform-to-Phasor Map

The least-squares formulation (3.16) contains the operator $\Re\{\cdot\}$, which makes it somewhat different from a standard linear least-squares problem. The solution to this type of optimization problem is derived in Appendix G: as shown in Sect. G.1, the solution involves an $M \times 2L$ matrix $\mathcal{W}$, viz.,

$$\mathcal{W} \overset{\text{def}}{=} \begin{bmatrix} \boldsymbol{\Psi} & \boldsymbol{\Psi}^* \end{bmatrix} \quad , \qquad \boldsymbol{\Psi} \overset{\text{def}}{=} \begin{bmatrix} \psi_{k_1} & \psi_{k_2} & \cdots & \psi_{k_L} \end{bmatrix} \tag{3.17a}$$

where the asterisk (*) denotes complex conjugation and, for each $k \in \mathcal{S}$,

$$\psi_k \overset{\text{def}}{=} \frac{1}{\sqrt{M}} \begin{bmatrix} 1 & e^{-jk\omega_0 \Delta_1} & \ldots & e^{-jk\omega_0 \Delta_{M-1}} \end{bmatrix}^{\mathsf{T}} \tag{3.17b}$$

We will assume from now on that $M \geq 2L$, which is a necessary (but not sufficient) condition for (3.16) to have a unique solution.

When the matrix $\mathcal{W}$ has *full column rank*, the (unique) optimizing solution of the polyphase least-squares problem (3.16) is, according to Theorem G.1,

$$\begin{pmatrix} \widehat{X}_{k_1}(t)\, e^{jk_1\omega_0 t} \\ \widehat{X}_{k_2}(t)\, e^{jk_2\omega_0 t} \\ \vdots \\ \widehat{X}_{k_L}(t)\, e^{jk_L\omega_0 t} \end{pmatrix} = \sqrt{\frac{2}{M}} \begin{bmatrix} \mathbf{I}_L & \mathbf{0}_L \end{bmatrix} \left(\mathcal{W}^H \mathcal{W}\right)^{-1} \mathcal{W}^H \begin{pmatrix} x(t) \\ x(t-\Delta_1) \\ \vdots \\ x(t-\Delta_{M-1}) \end{pmatrix} \tag{3.18a}$$

where $\mathbf{I}_L$ is an $L \times L$ identity matrix, and $\mathbf{0}_L$ is an $L \times L$ matrix consisting entirely of zero elements. We will provide in the sequel necessary and sufficient conditions for the matrix $\mathcal{W}$ to have full column rank, which is a prerequisite for the invertibility of $\mathcal{W}^H \mathcal{W}$.

When the waveform-to-phasor map is implemented in digital hardware, evaluation of the phasor estimates $\{\widehat{X}_k(t)\, ; \; k \in \mathcal{S}\}$ occurs only at the discrete time-instants $t = nT_s$, where n is an integer and T_s is the sampling interval. The cost of evaluating the estimated m-phase phasors for a single discrete time-instant is $\mathcal{O}(LMm)$, where we assume that the time-invariant $L \times M$ matrix $\sqrt{\frac{2}{M}} \begin{bmatrix} \mathbf{I}_L & \mathbf{0}_L \end{bmatrix} \left(\mathcal{W}^H \mathcal{W}\right)^{-1} \mathcal{W}^H$ has been evaluated and stored in advance.

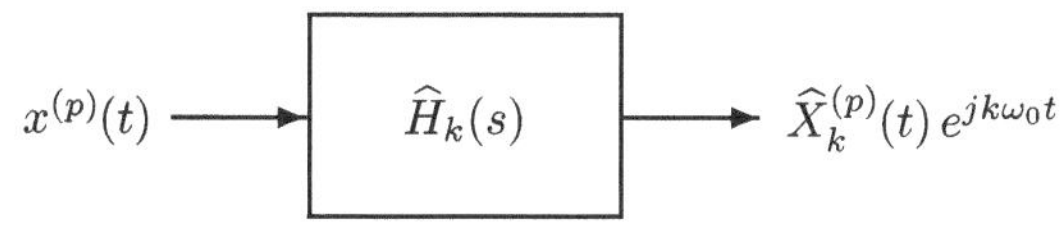

Fig. 3.3 Single-input/single-output filtering interpretation of the waveform-to-phasor map for a given $k \in \boldsymbol{\mathcal{S}}$: here $x^{(p)}(t)$ denotes the p-th phase of the polyphase waveform $x(t)$ and similarly for $\widehat{X}_k^{(p)}(t)$.

Notice that the waveform-to-phasor map (3.18a) is *linear and time-invariant, with no coupling between phases.* For each individual phase it acts as a filter-bank, consisting of L single-input/single-output filters $\{\widehat{H}_k(s)\,;\ k \in \boldsymbol{\mathcal{S}}\}$. Each $\widehat{H}_k(s)$ maps a single element of the $m \times 1$ vector $x(t)$ of input waveforms into the corresponding estimated k-th dynamic phasor (Fig. 3.3).

The same observations can be directly demonstrated by applying the Laplace transform to (3.18a), which generates an explicit expression for the transfer function of this filter-bank, viz.,

$$\begin{pmatrix} \widehat{H}_{k_1}(s) \\ \widehat{H}_{k_2}(s) \\ \vdots \\ \widehat{H}_{k_L}(s) \end{pmatrix} = \sqrt{\frac{2}{M}} \begin{bmatrix} \mathbf{I}_L & \mathbf{0}_L \end{bmatrix} \left(\mathcal{W}^H \mathcal{W}\right)^{-1} \mathcal{W}^H \begin{pmatrix} 1 \\ e^{-s\Delta_1} \\ \vdots \\ e^{-s\Delta_{M-1}} \end{pmatrix} \tag{3.18b}$$

The transfer function $\widehat{H}_k(s)$ determines the steady-state and transient properties of the estimated phasor $\widehat{X}_k(t)$. In particular, $\widehat{H}_k(j\omega)$ has a *steady-state harmonics-blocking property,* as described in detail in Theorem G.2, viz.,

$$\begin{aligned} \widehat{H}_k(j\ell\omega_0) &= \sqrt{2}\,\delta_{k,\ell} \\ \widehat{H}_k(-j\ell\omega_0) &= 0 \end{aligned} \tag{3.19}$$

for every $k \in \boldsymbol{\mathcal{S}}$ and $\ell \in \boldsymbol{\mathcal{S}}$. This property implies that the estimated phasors $\widehat{X}_k(t)$ are error-free when the input of the filter $\widehat{H}_k(s)$ is a periodic waveform whose harmonic content is contained in $\boldsymbol{\mathcal{S}}$: when $x(t) = \sum_{\ell \in \boldsymbol{\mathcal{S}}} \sqrt{2}\, \Re\{X_\ell\, e^{j\ell\omega_0 t}\}$, the corresponding (steady-state) estimated phasor is simply $\widehat{X}_k(t) = X_k$ for all t, regardless of the selected inter-sample shift values $\{\Delta_i\}$. However, when a periodic waveform $x(t)$ contains *spurious harmonics* (i.e., harmonics excluded from $\boldsymbol{\mathcal{S}}$), the estimated phasor $\widehat{X}_k(t)$ will exhibit fluctuations around X_k [78, 79], viz.,

$$\widehat{X}_k(t) = X_k + \frac{\sqrt{2}}{2} \sum_{\ell \notin \mathcal{S}} \Big[\widehat{H}_k(j\ell\,\omega_0)\, X_\ell\, e^{j(\ell-k)\,\omega_0 t} + \widehat{H}_k(-j\ell\,\omega_0)\, X_\ell^*\, e^{-j(\ell+k)\,\omega_0 t} \Big] \tag{3.20a}$$

which results in the mean-square expression

$$\frac{1}{T}\int_0^T \| \widehat{X}_k(t) - X_k \|^2 dt = \sum_{\ell \notin \mathcal{S}} \frac{\left[\, | \widehat{H}_k(j\ell\,\omega_0) |^2 + | \widehat{H}_k(-j\ell\,\omega_0) |^2 \right]}{2} \, \|X_\ell\|^2 \tag{3.20b}$$

Thus the level of fluctuation depends on the selection of $\mathcal{S}$, the magnitudes $\|X_\ell\|$ of the spurious harmonics of $x(t)$, and the frequency response $\widehat{H}_k(j\omega)$.

For digital processing purposes, we use samples of the polyphase waveform $x(t)$, generated with a sampling interval $T_s = \frac{T}{N}$, and we set $t = nT_s \equiv n\frac{T}{N}$ in the waveform-to-phasor map (3.18a). The resulting refresh rate for the estimated phasors is equal to the waveform sampling rate, i.e., an updated estimate is generated every T_s seconds, which means we generate a *sliding-window* phasor estimate. In contrast, today's PMUs generate a block-based phasor estimate, using non-overlapping blocks of waveform samples.

3.2.2 PLLS Estimation of the Fundamental Phasor

A variety of power system applications require estimation of only the fundamental phasor, as already pointed out in the introduction to this chapter. The corresponding *simplified PLLS problem* is obtained by setting $\mathcal{S} = \{1\}$, which implies that $L = 1$. For digital processing purposes, we will assume that the evaluation of the waveform-to-phasor map (3.18a) is preceded by an analog-to-digital conversion, which generates N samples per cycle of the continuous-time waveform $x(\cdot)$.

Example 3.3 (Full-cycle estimation of the fundamental phasor): Consider the problem of generating the fundamental polyphase phasor estimate $\widehat{X}_1(t)$ from N waveform samples that are uniformly-spaced within the cycle-long interval $(t - T \quad t]$. This means that we set $\mathcal{S} = \{1\}$, $M = N$, and

$$\Delta_i = i\,\frac{T}{M} \quad , \qquad 0 \le i \le M-1,$$

in the PLLS problem (3.16), where T denotes the length of a single cycle. In this case the matrix $\boldsymbol{\Psi}$ consists of a single column ψ_1, where

$$\psi_1 \stackrel{\text{def}}{=} \frac{1}{\sqrt{M}} \begin{bmatrix} 1 & \alpha & \alpha^2 & \ldots & \alpha^{(M-1)} \end{bmatrix}^{\top}$$

and $\alpha \stackrel{\text{def}}{=} e^{-j\omega_0\Delta_1} = e^{-j2\pi/M}$. The two columns of the resulting matrix $\mathcal{W}$ are mutually-orthogonal, viz.,

$$[\psi_1^*]^H\,\psi_1 \equiv \psi_1^{\top}\,\psi_1 = \frac{1}{M}\sum_{i=0}^{M-1}\alpha^{2i} = \frac{1}{M}\cdot\frac{1-\alpha^{2M}}{1-\alpha} = 0$$

because $\alpha \neq 1$ and $\alpha^{2M} = e^{-j4\pi} = 1$. Consequently, $\mathcal{W}^H\mathcal{W}$ is a 2×2 identity matrix, and

$$\begin{bmatrix} \mathbf{I}_L & \mathbf{0}_L \end{bmatrix} \left(\mathcal{W}^H \mathcal{W}\right)^{-1} \mathcal{W}^H = \begin{bmatrix} 1 & 0 \end{bmatrix} \mathcal{W}^H = \psi_1^H$$

where we used the fact that $L = 1$. We conclude that

$$\widehat{X}_1(t)\, e^{j\omega_0 t} = \sqrt{\frac{2}{M}}\; \psi_1^H \begin{pmatrix} x(t) \\ x(t-\Delta_1) \\ \vdots \\ x(t-\Delta_{M-1}) \end{pmatrix} = \frac{\sqrt{2}}{M} \sum_{r=0}^{M-1} (\alpha^*)^r\, x\big(t - r\tfrac{T}{M}\big)$$

so that

$$\widehat{X}_1(t)\, e^{j\omega_0 t} = \frac{\sqrt{2}}{M} \sum_{r=0}^{M-1} e^{j\left(2\pi/M\right)r}\, x\left(t - r\tfrac{T}{M}\right) \tag{3.21a}$$

The filter associated with this linear map is, of course,

$$\widehat{H}_1(s) = \frac{\sqrt{2}}{M} \sum_{r=0}^{M-1} e^{j\left(2\pi/M\right)r}\, e^{-s(T/M)r} = \frac{\sqrt{2}}{M}\; \frac{1 - e^{-j2\pi\left(\frac{s}{j\omega_0}-1\right)}}{1 - e^{-j(2\pi/M)\left(\frac{s}{j\omega_0}-1\right)}} \tag{3.21b}$$

where we used the fact that $T = \frac{2\pi}{\omega_0}$. The corresponding frequency response, viz.,

$$\widehat{H}_1(j\omega) = \sqrt{2}\, e^{-j\left(\frac{\omega}{\omega_0}-1\right)(M-1)(\pi/M)}\; \frac{\sin(\frac{\omega}{\omega_0} - 1)\pi}{M\, \sin(\frac{\omega}{\omega_0} - 1)\frac{\pi}{M}} \tag{3.21c}$$

has a magnitude response of $\sqrt{2}$ at $\omega = \omega_0$ and nulls at the other harmonic frequencies (Fig. 3.4). Notice that, according to (3.21), the frequency response is periodic in ω with a period of $M\omega_0$.

Recall that, for digital processing purposes, we use samples of the polyphase waveform $x(t)$, generated with a sampling interval $T_s = \frac{T}{N}$, and we set $t = nT_s \equiv n\frac{T}{N}$ in the waveform-to-phasor map (3.21). The resulting discrete-time expression is (recall that $M = N$ in this example)

$$\widehat{X}_1(nT_s)\, e^{j\omega_0 n T_s} = \frac{\sqrt{2}}{N} \sum_{r=0}^{N-1} e^{j\left(2\pi/N\right)r}\, x\left(n\tfrac{T}{N} - r\tfrac{T}{N}\right)$$

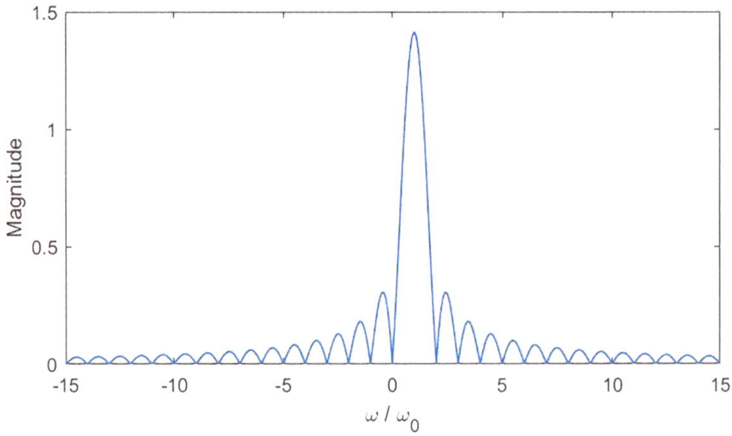

Fig. 3.4 Magnitude response of the full-cycle (=DFT) estimation filter $\widehat{H}_1(j\omega)$.

Since $\omega_0 \, n \, T_s = \omega_0 \, \frac{T}{N} \, n = \frac{2\pi}{N} \, n$, it follows that

$$\begin{aligned}\widehat{X}_1(nT_s) &= \frac{\sqrt{2}}{N} \sum_{r=0}^{N-1} x\Big([n-r]\tfrac{T}{N}\Big) \, e^{-j\left(2\pi/N\right)(n-r)} \\ &= \frac{\sqrt{2}}{N} \sum_{\nu=n-N+1}^{n} x\Big(\nu\tfrac{T}{N}\Big) \, e^{-j\left(2\pi/N\right)\nu}\end{aligned}$$

which matches exactly the fundamental phasor obtained by applying the DFT to the full-cycle set of samples $\{x(\nu T_s)\,;\ n-N+1 \le \nu \le n\}$: see, for instance, the DFT expression (1.21), for the special case $k = 1$ and $t_0 = (n - N + 1)\frac{T}{N}$.

■

Example 3.4 (Half-cycle estimation of the fundamental phasor): Next, consider estimating the same fundamental phasor $\widehat{X}_1(t)$ from $\frac{N}{2}$ polyphase waveform samples that are uniformly-spaced within the *half-cycle-long* interval $(t - \frac{T}{2} \;\; t]$, where T denotes the length of a single cycle. This means that we set $\mathcal{S} = \{1\}$, $M = \frac{N}{2}$, and

$$\Delta_i \; = \; i \, \frac{T}{N} \; \equiv \; i \, \frac{T}{2M} \qquad , \qquad 0 \le i \le M-1$$

in the least-squares problem (3.16). In this case, the matrix $\boldsymbol{\Psi}$ still consists of a single column ψ_1, which is half as long as the one in Example 3.3, viz.,

$$\psi_1 \;\stackrel{\text{def}}{=}\; \frac{1}{\sqrt{M}} \begin{bmatrix} 1 & \alpha & \alpha^2 & \dots & \alpha^{(M-1)} \end{bmatrix}^{\mathsf{T}}$$

and $\alpha \stackrel{\text{def}}{=} e^{-j\omega_0 \Delta_1} = e^{-j2\pi/N} \equiv e^{-j\pi/M}$, because now $M = \frac{N}{2}$. Again, $\psi_1 \perp \psi_1^*$, viz.,

$$[\psi_1^*]^H \psi_1 \equiv \psi_1^\top \psi_1 = \frac{1}{M} \sum_{i=0}^{M-1} \alpha^{2i} = \frac{1}{M} \cdot \frac{1-\alpha^{2M}}{1-\alpha} = 0$$

because $\alpha \neq 1$ and $\alpha^{2M} = e^{-j2\pi} = 1$. Consequently, (3.18a) reduces to

$$\widehat{X}_1(t)\, e^{j\omega_0 t} = \sqrt{\frac{2}{M}}\, \psi_1^H \begin{pmatrix} x(t) \\ x(t-\Delta_1) \\ \vdots \\ x(t-\Delta_{M-1}) \end{pmatrix} = \frac{\sqrt{2}}{M} \sum_{r=0}^{M-1} (\alpha^*)^r\, x\left(t - r\frac{T}{2M}\right)$$

which means that the waveform-to-phasor map in this case is

$$\widehat{X}_1(t)\, e^{j\omega_0 t} = \frac{\sqrt{2}}{M} \sum_{r=0}^{M-1} e^{j(\pi/M)r}\, x\left(t - r\frac{T}{2M}\right) \tag{3.22a}$$

and the filter associated with this linear map is

$$\widehat{H}_1(s) = \frac{\sqrt{2}}{M} \sum_{r=0}^{M-1} e^{j(\pi/M)r}\, e^{-s(T/2M)r} = \frac{\sqrt{2}}{M} \frac{1 - e^{-j\pi(\frac{s}{j\omega_0}-1)}}{1 - e^{-j(\pi/M)(\frac{s}{j\omega_0}-1)}} \tag{3.22b}$$

The corresponding frequency response, viz.,

$$\widehat{H}_1(j\omega) = \sqrt{2}\, e^{-j(\frac{\omega}{\omega_0}-1)(M-1)(\pi/2M)} \frac{\sin(\frac{\omega}{\omega_0}-1)\frac{\pi}{2}}{M \sin(\frac{\omega}{\omega_0}-1)\frac{\pi}{2M}} \tag{3.22c}$$

has a magnitude response of $\sqrt{2}$ at $\omega = \omega_0$ but has nulls only at *odd harmonic* frequencies (Fig. 3.5). Notice that the even harmonics are not suppressed: in particular, the DC gain is relatively high, so that an external front-end filter has to be used to suppress the exponentially-decaying "DC offset" component that may be generated by fault currents in a power system. The same observations were made, for instance, in [62, Sect. 5.3.3] and [42, Sects. 2.4 and 2.6]. Also notice that, according to (3.22), the frequency response is periodic in ω with a period of $2M\omega_0$.

■

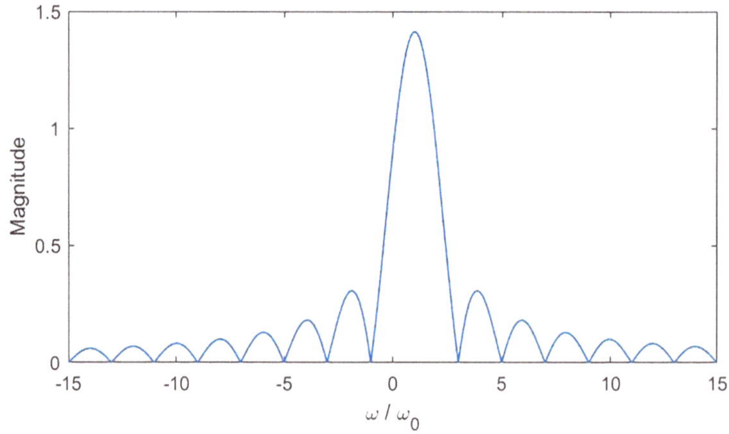

Fig. 3.5 Magnitude response of the half-cycle estimation filter $\widehat{H}_1(j\omega)$.

3.2.3 Fractional-Cycle Estimation from Uniformly-Shifted Samples

Motivated by Examples 3.3 and 3.4, we now consider a more structured version of the least-squares estimation problem (3.16), which uses a sparse set of *uniformly-shifted* waveform samples, namely,

$$\Delta_i = i\,\Delta \quad , \qquad i = 0, 1, 2, \ \ldots\ , M - 1 \tag{3.23a}$$

The inter-sample spacing Δ can be chosen at will. In particular, when the input waveform is sampled at a rate of N samples per cycle, and the sampled-data sequence $x(n\frac{T}{N})$ serves as input for a digital signal processing implementation of this estimation scheme, the inter-sample shift can be chosen as any integral multiple of the waveform sampling interval $T_s = \frac{T}{N}$. Also, while the value of N used in modern grid monitoring devices is typically in the range of $128 - 512$ (see, e.g., [35–37]), the fractional least-squares estimation scheme uses, at any given time-instant, a subset of the available waveform samples, acquired from a fraction of a single cycle, so that $2L \leq M \leq N$, where L is the number of harmonics used in the estimation scheme (= the number of elements in the harmonic index set $\boldsymbol{S}$). In the sequel we will focus mainly on *sparse estimation schemes* where $M \ll N$, with L limited to a few dominant harmonics.

We define the *fractionality* of a uniformly-shifted estimation scheme as the normalized length of the sliding interval from which samples are acquired, viz.,

$$\text{Fractionality} \stackrel{\text{def}}{=} \frac{M\Delta}{T} = M\,\xi \tag{3.23b}$$

where $\xi \overset{\text{def}}{=} \frac{\Delta}{T}$ is a *normalized inter-sample shift*, and T is the length of a single steady-state waveform cycle. Since fractionality impacts the group-delay associated with the waveform-to-phasor map (3.18a), we choose to impose the constraint

$$0 < \text{Fractionality} \leq 1 \qquad \text{or, equivalently,} \qquad 0 < \xi \leq \frac{1}{M} \tag{3.23c}$$

This choice limits the selection of input samples for this map to the sliding interval $[t - M\xi T \quad t]$, which is contained within the *single-cycle* interval $[t - T \quad t]$.

Incorporating the uniform-shift property into the waveform-to-phasor map (3.18a) endows the matrix $\mathcal{W}$ with a *Vandermonde structure*, namely,

$$\psi_k = \frac{1}{\sqrt{M}} \begin{bmatrix} 1 & \alpha^k & \alpha^{2k} & \dots & \alpha^{(M-1)k} \end{bmatrix}^{\mathsf{T}}, \qquad \text{for every } k \in \mathcal{S} \tag{3.24a}$$

where

$$\alpha \overset{\text{def}}{=} e^{-j\omega_0 \Delta} = e^{-j2\pi\xi} \quad , \qquad 0 < \xi \leq \frac{1}{M} \tag{3.24b}$$

This is similar to the ψ_1-structure observed in Examples 3.3 and 3.4: however, here the value of α is adjustable, and we allow the inclusion of multiple (dominant) harmonics in $\mathcal{S}$. Notice that $0 < 2\pi\xi \leq \frac{2\pi}{M}$ according to (3.24b), which implies that $\alpha \neq 1$.

The $M \times 2L$ Vandermonde matrix $\mathcal{W} = \begin{bmatrix} \Psi & \Psi^* \end{bmatrix}$ has full column rank (recall that $M \geq 2L$) if, and only if, its columns are all *distinct*. To be specific, this means that for every $k \in \mathcal{S}$ and every $\ell \in \mathcal{S}$: (i) $\psi_k \neq \psi_\ell$ (when $k \neq \ell$) and (ii) $\psi_k \neq \psi_\ell^*$ (including the case $k = \ell$). In view of (3.24), these constraints are equivalent to

$$\left.\begin{array}{l} \text{(i) } \alpha^{k-\ell} \neq 1 \ \ (\text{when } k \neq \ell) \\ \text{(ii) } \alpha^{k+\ell} \neq 1 \end{array}\right\} \quad \text{for every } k \in \mathcal{S}\,,\ \ell \in \mathcal{S} \tag{3.25a}$$

where we used the fact that $\alpha^* = \alpha^{-1}$.

In digital implementations of the fractional-cycle uniformly-shifted estimation scheme, both the nominal cycle length T and the inter-sample shift Δ are integral multiples of the sampling interval T_s. Consequently, the normalized inter-sample shift $\xi = \frac{\Delta}{T}$ is always a *rational number*, i.e., $\xi = {}^{m}/_{D}$, where m and D are positive co-prime integers. In turn, this implies that the sequence $\alpha^k = e^{-jk2\pi\xi} \equiv e^{-jk2\pi m/D}$ is periodic in k with period D, viz.,

$$\alpha^k = \alpha^{k+D} = \alpha^{\{k \bmod D\}} \qquad \text{for every} k \in \mathcal{S}$$

where $k \bmod D$ denotes the remainder of dividing k by D. To highlight the fact that the period D always depends on the value of ξ, we will use in the sequel the explicit notation $D(\xi)$.

In summary, for rational values of ξ, the constraint (3.25a) induces an equivalent constraint on the harmonic indices k and ℓ, viz.,

$$\left.\begin{array}{l}\text{(i)}\ \ (k-\ell) \bmod D(\xi) \neq 0 \ \ (\text{when } k \neq \ell) \\[2ex] \text{(ii)}\ \ (k+\ell) \bmod D(\xi) \neq 0\end{array}\right\} \text{ for every } k \in \boldsymbol{\mathcal{S}}\,,\ \ell \in \boldsymbol{\mathcal{S}} \tag{3.25b}$$

which limits the harmonics that can be included in the index set $\boldsymbol{\mathcal{S}}$.

One simple way of satisfying the full-rank constraint (3.25b) is to limit the indices included in $\boldsymbol{\mathcal{S}}$ to lie within a single period of the sequence $\alpha^k \equiv e^{-jk2\pi m/D}$, say $0 \leq k \leq D(\xi) - 1$. With this special choice, the first part of (3.25b) is automatically satisfied, while the second part can be replaced by the *non-complementarity* restriction

$$k + \ell \neq D(\xi) \quad \text{for every } k \in \boldsymbol{\mathcal{S}}\,,\ \ell \in \boldsymbol{\mathcal{S}}$$

Since typical power system (voltage and current) waveforms contain primarily odd harmonics, it makes sense to choose $\boldsymbol{\mathcal{S}}$ as a subset (of size L) of the odd-harmonic indices in the range $\left[0 \quad D(\xi)-1\right]$, subject to the non-complementarity constraint. This simplified characterization of the full rank condition is used in the construction of sub-cycle phasor estimates (see Sect. 3.3, as well as the analysis in Sect. G.6).

The introduction of uniform shifts (i.e., $\Delta_i = i\Delta$) simplifies somewhat the transfer function expression (3.18b), i.e.,

$$\begin{pmatrix} \widehat{H}_{k_1}(s) \\ \widehat{H}_{k_2}(s) \\ \vdots \\ \widehat{H}_{k_L}(s) \end{pmatrix} = \sqrt{\frac{2}{M}}\begin{bmatrix} \mathbf{I}_L & \mathbf{0}_L \end{bmatrix} \left(\mathcal{W}^H \mathcal{W}\right)^{-1} \mathcal{W}^H \begin{pmatrix} 1 \\ e^{-s\Delta} \\ \vdots \\ e^{-s(M-1)\Delta} \end{pmatrix} \tag{3.26}$$

The resulting frequency responses $\{\widehat{H}_k(j\omega)\,;\ k \in \boldsymbol{\mathcal{S}}\}$ are all periodic in ω with period ω_0/ξ : this is so because

$$e^{-s\Delta_i}\Big|_{s=j\omega} = \left(e^{-j\left(\frac{\omega}{\omega_0}\right)2\pi\xi}\right)^i \quad , \qquad 0 \leq i \leq M-1$$

is periodic, for every value of i, with the same period ω_0/ξ. Notice that each transfer function $\widehat{H}_k(s)$ is now a *polynomial of degree* $M - 1$ in the complex variable $z = e^{-s\Delta}$, with $2L-1$ of its zeros explicitly specified by the Harmonics-Blocking Theorem. In the special case when $M = 2L$, these observations will make it possible to obtain a simplified expression for the frequency response $\widehat{H}_k(j\omega)$ for every $k \in \mathcal{S}$ (see (3.31) in Sect. 3.3).

3.2.4 Column-Orthogonality of $\mathcal{W}$

The impact of finite-precision computation, as well as other perturbations, on the accuracy of least-squares phasor estimates obtained via the waveform-to-phasor map (3.18a) depends, primarily, on the condition number of the matrix $\mathcal{W}$ [76, Chap. 9]. In the uniformly-shifted version of this least-squares problem, the condition number of $\mathcal{W}$ is controlled by the harmonic index set $\mathcal{S}$ and the parameters M and ξ. An illustration of this relation is provided by Fig. 3.6: the ideal condition number of unity is achieved when $M\xi = \frac{1}{2}$ or $M\xi = 1$, while reduction of fractionality below $M\xi = 0.3$ results in a steep increase of $\mathrm{cond}(\mathcal{W})$.

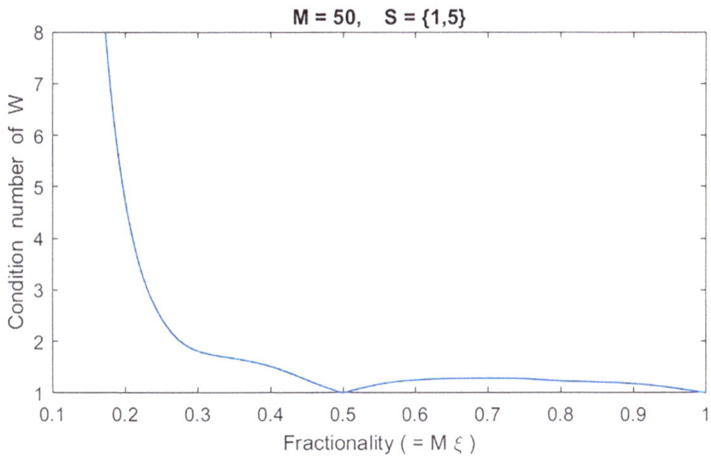

Fig. 3.6 Condition number of the matrix $\mathcal{W}$ for $\mathcal{S} = \{1, 5\}$ and $M = 50$.

The key to achieving a unity condition number is *column-orthogonality* of the matrix $\mathcal{W} = \left[\Psi\ \Psi^*\right]$. In addition to improving numerical robustness, column-orthogonality results in a simple expression for the frequency response $\widehat{H}_k(j\omega)$ (see (3.28b) below), as well as an explicit characterization of all the nulls of this frequency response. Moreover, in the special case $M = 2L$, which we examine in detail in Sect. 3.3, orthogonality of the matrix $\mathcal{W}$ results in a Parseval-type relation between instantaneous power (= product of voltage and current) and estimated dynamic phasors (as detailed in Sect. 9.3.2), as well as a reasonably accurate low-cost estimate of the cycle-averaged real power in an electric power system, in both steady-state operation and in the presence of common transients (see Sects. 9.4.1–9.4.2).

The elements of the matrix $\mathcal{W} = \left[\Psi\ \Psi^*\right]$ are determined by the normalized inter-sample shift ξ, because $\alpha = e^{-j2\pi\xi}$ (recall (3.24)). Thus, column-orthogonality can be achieved only for carefully selected values of this control parameter. For instance, since $\alpha \neq 1$ due to the fractionality constraint $0 < \xi \leq \frac{1}{M}$, it follows that

$$[\psi_1^*]^H \psi_1 \equiv \psi_1^\top \psi_1 = \frac{1}{M}\sum_{i=0}^{M-1} \alpha^{2i} = \frac{1}{M} \cdot \frac{1-\alpha^{2M}}{1-\alpha^2}$$

which means that $\psi_1 \perp \psi_1^*$ if, and only if, $\alpha^{2M} = 1 = e^{-j4\pi\xi M}$. This can happen only if $2\xi M$ is an integer or, equivalently, if ξ is an integral multiple of $\frac{1}{2M}$. Since the columns of the matrix $\mathcal{W}$ always include ψ_1 and ψ_1^*, this result serves as a *necessary condition* for the orthogonality of $\mathcal{W}$. Taking into account the fractionality constraint $0 < \xi \leq \frac{1}{M}$, we conclude that there are precisely two choices of ξ which satisfy this necessary condition:

- Full-cycle sampling, resulting from the choice $\xi = \frac{1}{M}$. In this case the samples used as input to the least-squares estimation scheme are acquired from the sliding interval $[t - T \;\; t]$.
- Half-cycle sampling, resulting from the choice $\xi = \frac{1}{2M}$. In this case the samples used as input to the least-squares estimation scheme are acquired from the sliding interval $[t - \frac{T}{2} \;\; t]$.

These are, of course, exactly the choices used in Examples 3.3 and 3.4. However, in contrast to the setting of these two examples, here we are concerned with estimating several dominant harmonic phasors, not just the fundamental.

Theorem G.3 (in Sect. G.3) provides a complete characterization of column-orthogonality in terms of the value of ξ and the content of the harmonic index set $\mathcal{S}$. In particular, the theorem states that, when $\mathcal{W}$ has full column rank,

- The full-cycle choice $\xi = \frac{1}{M}$ results in mutual-orthogonality of all the columns of $\mathcal{W}$, regardless of the content of the harmonic index set $\mathcal{S}$.
- The half-cycle choice $\xi = \frac{1}{2M}$ ensures column-orthogonality of $\mathcal{W}$ when the index set $\mathcal{S}$ consists entirely of odd harmonics.

These serve as a ξ-dependent *sufficient condition* (when $\xi \leq \frac{1}{M}$) for the column-orthogonality of $\mathcal{W}$. Notice that, according to this result, the columns of $\mathcal{W}$ can never be mutually-orthogonal when $0 < \xi < \frac{1}{2M}$ or $\frac{1}{2M} < \xi < \frac{1}{M}$.

When the orthogonality property holds, the waveform-to-phasor map (3.18a) reduces to the simple uniformly-shifted form

$$\widehat{X}_k(t)\, e^{jk\omega_0 t} = \sqrt{\frac{2}{M}}\; \psi_k^H \begin{pmatrix} x(t) \\ x(t-\xi T) \\ \vdots \\ x(t-(M-1)\xi T) \end{pmatrix} = \frac{\sqrt{2}}{M} \sum_{r=0}^{M-1} (\alpha^*)^{kr}\, x\big(t - r(\xi T)\big)$$

for every $k \in \mathcal{S}$, so that

$$\widehat{X}_k(t)\, e^{jk\omega_0 t} = \frac{\sqrt{2}}{M} \sum_{r=0}^{M-1} e^{j(2\pi\xi)kr}\, x\big(t - r(\xi\, T)\big) \quad , \qquad k \in \mathcal{S} \tag{3.27}$$

Notice that setting $k = 1$ and $\xi = \frac{1}{M}$ recovers the full-cycle expression (3.21a) from Example 3.3, while the choice $\xi = \frac{1}{2M}$ recovers the half-cycle expression (3.22a) from Example 3.4.

Following the same steps as in Examples 3.3 and 3.4, we find that the filter associated with this linear waveform-to-phasor map is

$$\widehat{H}_k(s) = \frac{\sqrt{2}}{M} \sum_{r=0}^{M-1} e^{j(2\pi\xi)kr}\, e^{-s(\xi\, T)r} = \frac{\sqrt{2}}{M}\, \frac{1 - e^{-j2\pi\xi M(\frac{s}{j\omega_0} - k)}}{1 - e^{-j2\pi\xi(\frac{s}{j\omega_0} - k)}}$$

so that, in fact,

$$\widehat{H}_k(s) = \begin{cases} \dfrac{\sqrt{2}}{M}\, \dfrac{1 - e^{-j2\pi\,(\frac{s}{j\omega_0} - k)}}{1 - e^{-j(2\pi/M)\,(\frac{s}{j\omega_0} - k)}} & \xi = \frac{1}{M} \\[2ex] \dfrac{\sqrt{2}}{M}\, \dfrac{1 - e^{-j\pi\,(\frac{s}{j\omega_0} - k)}}{1 - e^{-j(\pi/M)\,(\frac{s}{j\omega_0} - k)}} & \xi = \frac{1}{2M} \end{cases} \tag{3.28a}$$

The corresponding frequency response is

$$\widehat{H}_k(j\omega) = \begin{cases} \sqrt{2}\, \dfrac{\sin(\frac{\omega}{\omega_0} - k)\pi}{M\, \sin(\frac{\omega}{\omega_0} - k)\frac{\pi}{M}}\, e^{-j(\frac{\omega}{\omega_0} - k)(\frac{M-1}{M})\pi} & \xi = \frac{1}{M} \\[2ex] \sqrt{2}\, \dfrac{\sin(\frac{\omega}{\omega_0} - k)\frac{\pi}{2}}{M\, \sin(\frac{\omega}{\omega_0} - k)\frac{\pi}{2M}}\, e^{-j(\frac{\omega}{\omega_0} - k)(\frac{M-1}{M})\frac{\pi}{2}} & \xi = \frac{1}{2M} \end{cases} \tag{3.28b}$$

Notice that $\widehat{H}_k(j\omega)$ is a frequency-shifted version of $\widehat{H}_1(j\omega)$, i.e., $\widehat{H}_k(j\omega) = \widehat{H}_1\big(j\omega - (k-1)\omega_0\big)$. Also, for $k = 1$, these frequency response expressions coincide with (3.21c) and (3.22c), respectively. In particular, the pattern of nulls is identical to the one shown in Fig. 3.4 (for the full-cycle case) and Fig. 3.5 (for the half-cycle case).

The relation between fractionality and column-orthogonality underscores the basic dichotomy in selecting the value of the control parameter ξ. On the one hand, choosing $\xi = \frac{1}{M}$ or $\xi = \frac{1}{2M}$ usually results in lower steady-state estimation error. This is so because, for most odd r values, $\big|H_k(jr\omega_0)\big| = 0$ (see, e.g., Figs. 3.4 and 3.5), leading to a significant reduction in steady-state fluctuations of sub-cycle phasors, caused by leakage of spurious harmonics (recall (3.20)). On the other hand, selecting a ξ-value in the range $\frac{1}{4M} < \xi < \frac{1}{2M}$ is desirable in applications where achieving a faster transient response is the primary objective, and a moderate steady-state error can be tolerated, as demonstrated in Sect. 9.4.3.

Finally, we observe that (3.28b) exhibits a *linear phase response,* resulting in a *constant group-delay,* which is proportional to $\xi(M-1)$ in both cases. To be specific,

$$\text{group-delay} \stackrel{\text{def}}{=} -\frac{d}{d\omega}\arg\{\widehat{H}_k(j\omega)\} = \xi\,(M-1)\,\tfrac{T}{2}$$

$$= \begin{cases} \frac{M-1}{M}\cdot\frac{T}{2} & \xi = \frac{1}{M} \\ \frac{M-1}{2M}\cdot\frac{T}{2} & \xi = \frac{1}{2M} \end{cases} \tag{3.29}$$

In other words, the group-delay of the full-cycle scheme is (almost) a half-cycle, while using the half-cycle scheme results in a 50% reduction of group-delay.

3.3 Sub-Cycle Dynamic Phasors

The sub-cycle approach to estimation of dominant phasors is a cost-efficient variant of the uniformly-shifted fractional least-squares scheme that we introduced in Sect. 3.2. It is obtained by imposing the following added constraints:

- **Restricted harmonics:** In addition to the fundamental, the index set $\mathcal{S}$ contains only dominant odd harmonics, such as the 5th, 7th, etc.
- **Maximal sparsity** ($M = 2L$): This can be interpreted as using the smallest number of sparsely-located waveform samples (and thus, requiring the lowest computational cost) for a given number of dominant harmonics or, alternatively, allowing the largest number of estimated phasors for a given number of samples.
- **Low fractionality** ($\xi \le \frac{1}{2M}$): This results in reduced group-delay, which is essential for a fast response to fault-induced transients.
- **Non-decimated digital implementation:** The waveform-to-phasor map is implemented in discrete time as a sliding-window operation, resulting in the highest possible output rate, equal to the waveform sampling rate. This facilitates a fast response to the onset of a transient.

In contrast, the phasor estimation scheme used in PMUs involves radically different parameter choices, namely, (a) $L = 1$ and $M \gg L$, (b) a much higher fractionality, using blocks of 1–3 cycles with no overlap, and (c) a decimated time scale, yielding a single phasor output per block. The resulting estimation scheme ensures high accuracy of the estimated fundamental phasor in steady-state operation, albeit at the expense of poor representation of transient phenomena.

The $M \times 2L$ matrix $\mathcal{W}$ is square, as a result of the maximal sparsity constraint. When the full column rank condition (3.25b) is satisfied, the matrix $\mathcal{W}$ is invertible, and the waveform-to-phasor map (3.18a) reduces to

$$\begin{pmatrix} \widehat{X}_{k_1}(t)\, e^{jk_1\omega_0 t} \\ \widehat{X}_{k_2}(t)\, e^{jk_2\omega_0 t} \\ \vdots \\ \widehat{X}_{k_L}(t)\, e^{jk_L\omega_0 t} \end{pmatrix} = \frac{1}{\sqrt{L}} \begin{bmatrix} \mathbf{I}_L & \mathbf{0}_L \end{bmatrix} \mathcal{W}^{-1} \begin{pmatrix} x(t) \\ x(t-\Delta) \\ \vdots \\ x(t-(M-1)\Delta) \end{pmatrix} \tag{3.30a}$$

where $\Delta \equiv \xi T$ is the inter-sample shift, $\mathbf{I}_L$ is an $L \times L$ identity matrix, and $\mathbf{0}_L$ is an $L \times L$ matrix consisting entirely of zero elements. Again, we observe that the cost of evaluating the set of estimated m-phase phasors is $\mathcal{O}(LMm)$, where we assume that the time-invariant $L \times M$ matrix $\frac{1}{\sqrt{L}}\begin{bmatrix} \mathbf{I}_L & \mathbf{0}_L \end{bmatrix}\mathcal{W}^{-1}$ has been evaluated and stored in advance.

The invertible relation (3.30a) defines a *one-to-one correspondence* between the set of sub-cycle phasor estimates $\{\widehat{X}_{k_\ell}(t)\,,\ 1 \le \ell \le L\}$ and the set of waveform samples $\{x(t - i\Delta)\,,\ 0 \le i \le M-1\}$. The positive integer L (or, equivalently, M) serves as a design parameter that controls a fundamental trade-off: increasing L results in improved steady-state behavior (because of fewer spurious harmonics), at the expense of longer group-delay, a less agile transient response, and increased implementation cost. On the other extreme, setting $L = 1$ results in a very low cost sub-cycle scheme that has applications in frequency estimation (see Sect. 3.4) and in control of power-electronic devices (see Sect. 9.5).

As we observed in Sect. 3.2, the waveform-to-phasor map (3.30a) is, in fact, a linear time-invariant filter-bank, with a (polyphase) input $x(t)$ and L (polyphase) outputs, one for each harmonic index $k \in \mathcal{S}$ (Fig. 3.7). Applying the Laplace transform to (3.30a), we obtain a transfer-function characterization of this map, viz.,

$$\begin{pmatrix} \widehat{H}_{k_1}(s) \\ \widehat{H}_{k_2}(s) \\ \vdots \\ \widehat{H}_{k_L}(s) \end{pmatrix} = \frac{1}{\sqrt{L}} \begin{bmatrix} \mathbf{I}_L & \mathbf{0}_L \end{bmatrix} \mathcal{W}^{-1} \begin{pmatrix} 1 \\ e^{-s\Delta} \\ \vdots \\ e^{-s(M-1)\Delta} \end{pmatrix} \tag{3.30b}$$

which should be compared with (3.26). Notice that $\widehat{H}_k(s)$ is a single-input/single-output filter: each phase of the vector input $x(t)$ is mapped into its own scalar phasor (i.e., there is no coupling between phases).

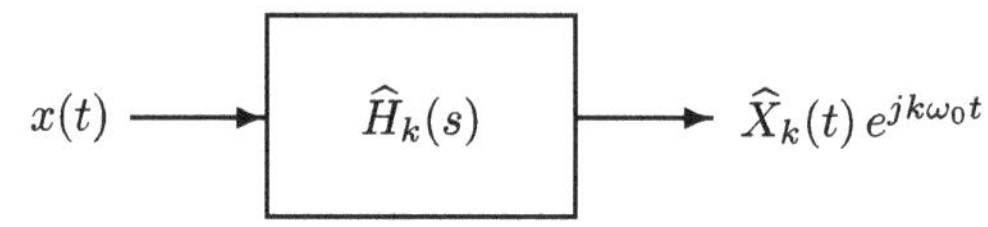

Fig. 3.7 Single-input/single-output linear filtering interpretation of the waveform-to-phasor map for a given $k \in \mathcal{S}$.

The sub-cycle scheme inherits the *harmonics-blocking* property (3.19), viz.,

$$\begin{aligned}\widehat{H}_k(j\ell\omega_0) &= \sqrt{2}\,\delta_{k,\ell}\\ \widehat{H}_k(-j\ell\omega_0) &= 0\end{aligned}$$

for every $k \in \mathcal{S}$ and $\ell \in \mathcal{S}$. Thus, when the input of $\widehat{H}_k(s)$ is a periodic waveform whose harmonic content is contained in $\mathcal{S}$ (i.e., $x(t) = \sum_{\ell \in \mathcal{S}} \sqrt{2}\,\Re\{X_\ell\, e^{j\ell\omega_0 t}\}$), the corresponding sub-cycle phasors are simply $\widehat{X}_k(t) = X_k$, regardless of the selected value for the inter-sample shift Δ. However, when $x(t)$ contains *spurious harmonics* (i.e., harmonics excluded from $\mathcal{S}$), the sub-cycle phasor $\widehat{X}_k(t)$ will exhibit fluctuations around the (steady-state) value X_k [78,79] (see also (3.20)).

We now observe that each transfer function $\widehat{H}_k(s)$ in (3.30b) is a polynomial of degree $M - 1 \equiv 2L - 1$ in the complex variable $z = e^{-s\Delta}$, with *all of its zeros* explicitly specified by the harmonics-blocking property. This means that we can express $\widehat{H}_k(s)$ explicitly in terms of its zeros, which also results in a simple expression for the frequency response $\widehat{H}_k(j\omega)$ for every $k \in \mathcal{S}$ [78], viz.,

$$\widehat{H}_k(j\omega) = \\ \sqrt{2}\left[\frac{\sin(\frac{\omega}{\omega_0}+k)\pi\xi}{\sin(2k)\pi\xi}\prod_{\ell\in\mathcal{S},\ell\neq k}\frac{\sin(\frac{\omega}{\omega_0}+\ell)\pi\xi}{\sin(k+\ell)\pi\xi}\cdot\frac{\sin(\frac{\omega}{\omega_0}-\ell)\pi\xi}{\sin(k-\ell)\pi\xi}\right]e^{-j(\frac{\omega}{\omega_0}-k)(M-1)\pi\xi} \tag{3.31}$$

where $\xi = \frac{\Delta}{T}$ is the normalized inter-sample shift parameter. This expression clearly satisfies the harmonics-blocking property, as is also evident, for instance, from Fig. 3.9. Notice that the magnitude response $|\widehat{H}_k(j\omega)|$ is a periodic function of ω, with a period of ω_0/ξ.

Certain combinations of $\mathcal{S}$ and ξ result in a vanishing denominator term in the expression (3.31) for $\widehat{H}_k(j\omega)$. This occurs when the square matrix $\mathcal{W}$ is rank-deficient, so that the waveform-to-phasor map of (3.18a) cannot be defined. According to Sect. 3.2.3, a necessary and sufficient condition for avoiding this singularity is

$$\left.\begin{aligned}&\text{(i) } (k-\ell) \bmod D(\xi) \neq 0 \ \text{(when } k \neq \ell\text{)}\\ &\text{(ii) } (k+\ell) \bmod D(\xi) \neq 0\end{aligned}\right\}\ \text{for every } k \in \mathcal{S},\ \ell \in \mathcal{S} \tag{3.32}$$

which limits the harmonics that can be included in the index set $\mathcal{S}$ (see also the discussion in Sect. G.2). One simple way of satisfying this constraint is to limit the indices included in $\mathcal{S}$ to lie the range $1 \leq k \leq D(\xi) - 1$. With this special choice, part (i) of the constraint is automatically satisfied, while part (ii) can be replaced by the *non-complementarity* restriction

$$k + \ell \neq D(\xi) \quad \text{for every } k \in \mathcal{S}\,,\ \ell \in \mathcal{S} \tag{3.33}$$

Valid choices of the index set $\mathcal{S}$, for $L = 2, 3, 4, 5$, are provided in Tables G.1 and G.2 of Sect. G.6.

Since $\mathcal{W}$ is now a square matrix, column-orthogonality is equivalent to row-orthogonality, namely

$$\mathcal{W}^H \mathcal{W} = \mathbf{I}_M = \mathcal{W}\mathcal{W}^H$$

The discussion in Sect. 3.2.4 implies that when $M = 2L$, a sufficient condition for orthogonality of the matrix $\mathcal{W}$, subject to the sub-cycle fractionality constraint $\xi \leq \frac{1}{2M}$, is that (a) $\xi = \frac{1}{2M}$ and (b) the index set $\mathcal{S}$ consists entirely of odd harmonics (see also Theorem G.3 in Appendix G). An orthogonal $\mathcal{W}$ satisfies the identity $\mathcal{W}^{-1} = \mathcal{W}^H$, so that (3.30a) reduces to

$$\widehat{X}_k(t)\, e^{jk\omega_0 t} = \frac{1}{\sqrt{L}}\, \psi_k^H \begin{pmatrix} x(t) \\ x(t-\Delta) \\ \vdots \\ x(t-(M-1)\Delta) \end{pmatrix} \tag{3.34}$$

where (compare with (3.27))

$$\psi_k^H = \frac{1}{\sqrt{M}} \begin{bmatrix} 1 & e^{jk\pi/M} & \ldots & e^{jk(M-1)\pi/M} \end{bmatrix} \tag{3.35}$$

Similarly, the frequency response expression (3.28b) still holds, which means that, for $\xi = \frac{1}{2M}$,

$$\widehat{H}_k(j\omega) = \sqrt{2}\, \frac{\sin(\frac{\omega}{\omega_0} - k)\,\frac{\pi}{2}}{M \sin(\frac{\omega}{\omega_0} - k)\,\frac{\pi}{2M}}\; e^{-j(\frac{\omega}{\omega_0} - k)\,(\frac{M-1}{M})\frac{\pi}{2}} \tag{3.36}$$

The special choice $\xi = \frac{1}{2M} \stackrel{\text{def}}{=} \xi_{orth}(M)$ produces the most accurate results for periodic waveforms, as we demonstrate below.

3.3.1 Steady-State Behavior of Sub-Cycle Dynamic Phasors

To illustrate the steady-state performance of the uniformly-shifted sub-cycle scheme, consider a periodic waveform $x(t)$ (= the "phase a" current from the RLC load dataset in Sect. L.1.5), consisting of the fundamental, 7-th and 11-th harmonics, viz.,

$$x(t) = \sqrt{2}\,\Re\left\{X_1\,e^{j\omega_0 t} + X_7\,e^{j7\omega_0 t} + X_{11}\,e^{j11\omega_0 t}\right\}$$

The full-cycle scheme recovers exactly the original phasors, namely $X_k(t) = X_k$ for all k (recall, e.g., the special case $A_k = B_k$ in Example 2.1). In particular, the dynamic phasor $X_1(t)$ coincides with X_1 for all $t \geq T$, following a cycle-long initialization transient (Fig. 3.8, blue trace). We obtain the sub-cycle dynamic phasor $\widehat{X}_1(t)$ by passing the input waveform through the waveform-to-phasor filter $\widehat{H}_1(j\omega)$, which is determined by applying the choices $\boldsymbol{S} = \{1, 5\}$ and $\xi = \frac{1}{8}$ to the expression (3.31). Since this filter has nulls at $\frac{\omega}{\omega_0} = -11, -1, 7, 11$ (see, e.g., Fig. 3.9), the only harmonics at its output in this case will be the fundamental and the (−7)-th. Consequently, if we set $X_7 = 0$, the sub-cycle $\widehat{X}_1(t)$ (Fig. 3.8, black trace) coincides with the full-cycle $X_1(t)$ (Fig. 3.8, blue trace), except in the initialization interval. In contrast, when we restore the 7-th harmonic, the sub-cycle $\widehat{X}_1(t)$ exhibits fluctuation (Fig. 3.8, green trace), which is caused by a non-vanishing value of $\widehat{H}_1(-j7\omega_0)$. Notice, however, that the average value of $\widehat{X}_1(t)$ always coincides with the value of the input phasor X_1.

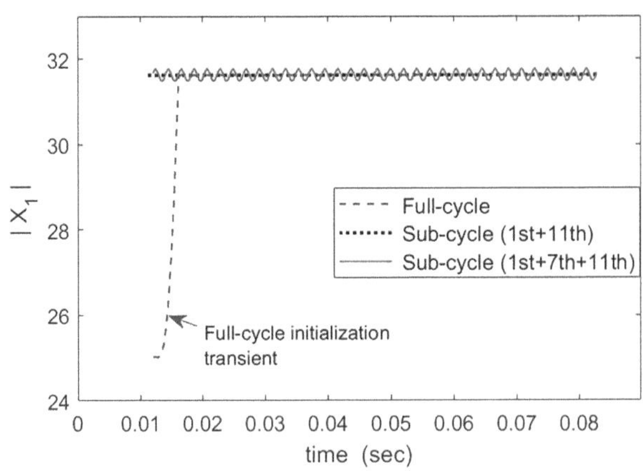

Fig. 3.8 Sub-cycle response to a periodic input ("phase a" current from the RLC load dataset in Sect. L.1.5): the full-cycle dynamic phasor $X_1(t)$ and two versions of the sub-cycle dynamic phasor $\widehat{X}_1(t)$, obtained with $\boldsymbol{S} = \{1, 5\}$ and $\xi = \frac{1}{8}$.

When the input waveform $x(t)$ is periodic (i.e., $x(t) = \sum_r \sqrt{2}\,\Re\{X_r\,e^{jr\omega_0 t}\}$), the mean-square value of the (steady-state) fluctuating component $\widehat{X}_k(t) - X_k$ is

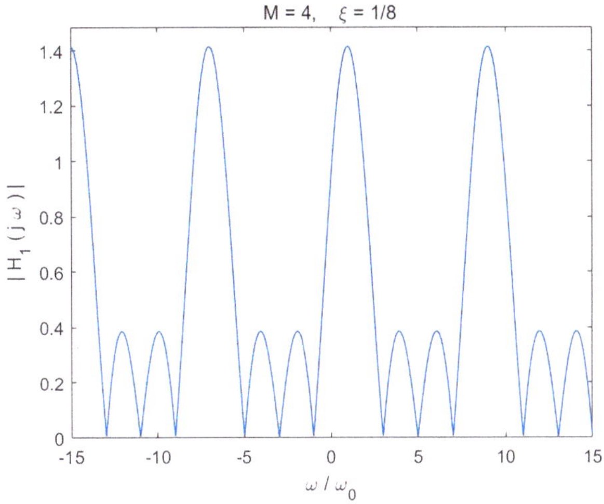

Fig. 3.9 Magnitude response of $\widehat{H}_1(s)$: $\boldsymbol{S} = \{1, 5\}$ and $\xi = 1/8$.

$$\frac{1}{T}\int_0^T |\widehat{X}_k(t) - X_k|^2 dt = \sum_{r \notin \mathcal{S}} \frac{[\,|\widehat{H}_k(jr\,\omega_0)|^2 + |\widehat{H}_k(-jr\,\omega_0|^2]}{2} |X_r|^2 \tag{3.37a}$$

where we used the expression (3.20) for steady-state estimation error. According to (3.31), the magnitude response of the waveform-to-phasor map is

$$\left|\widehat{H}_k(jr\omega_0)\right| = \sqrt{2}\left|\frac{\sin(r+k)\pi\xi}{\sin(2k)\pi\xi}\right| \prod_{\ell \in \mathcal{S},\ \ell \neq k} \left|\frac{\sin(r+\ell)\pi\xi}{\sin(k+\ell)\pi\xi} \cdot \frac{\sin(r-\ell)\pi\xi}{\sin(k-\ell)\pi\xi}\right| \tag{3.37b}$$

which means that the level of steady-state fluctuation, for a given choice of the harmonic index set $\boldsymbol{S}$, depends on the value of ξ (the normalized inter-sample shift) and the magnitudes $\{|X_r|\,;\ r \notin \boldsymbol{S}\}$ of the spurious harmonics of $x(t)$.

The contribution of the r-th (spurious) harmonic to the level of steady-state fluctuation of $\widehat{X}_k(t)$ is controlled by the gain factor

$$\sqrt{\frac{[\,|\widehat{H}_k(jr\,\omega_0)|^2 + |\widehat{H}_k(-jr\,\omega_0|^2]}{2}}$$

which depends on ξ via the expression (3.37b). Plotting this gain as a function of ξ reveals a "plateau," namely a wide range of ξ-values for which the gain factor is less than unity, for all (odd) spurious harmonic indices (Fig. 3.10). This range of ξ-values includes the special choice $\xi_{orth}(M) \stackrel{\text{def}}{=} \frac{1}{2M}$ (indicated by a ◇ in Fig. 3.10), which plays a key role in the construction of power-quality metrics from sub-cycle phasors (see Sect. 9.4 for further details). For now, we observe that any choice of ξ within this "plateau" results in a relatively low level of steady-state $\widehat{X}_k(t)$-fluctuation (for a periodic waveform $x(\cdot)$), regardless of the spurious-harmonics content.

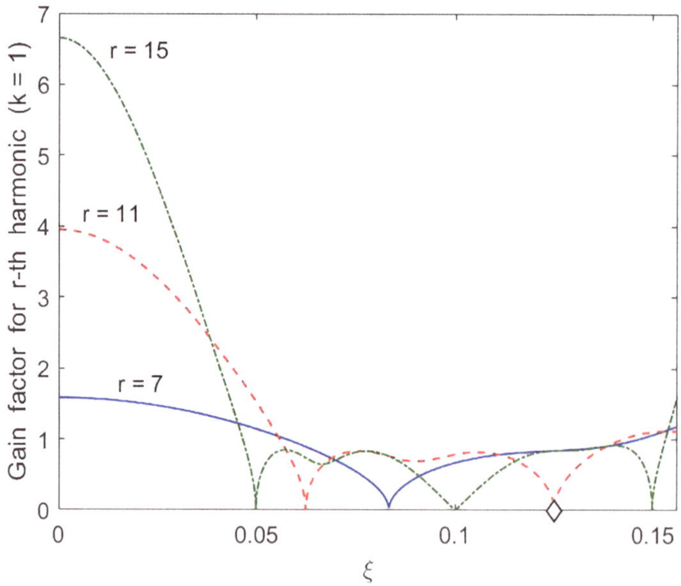

Fig. 3.10 RMS level of steady-state fluctuation of $\widehat{X}_1(t)$: gain factor for r-th harmonic ($r = 7, 11, 15$), as a function of ξ, for $\boldsymbol{S} = \{1, 5\}$.

3.3.2 Transient Response of Sub-Cycle Dynamic Phasors

The combined choice of the normalized shift ξ and the number of waveform samples used (M) controls the trade-off between *accuracy* (= steady-state error) and *agility* (= rate of transient response). Accuracy is the only concern in steady-state applications: Fig. 3.10 implies that the most accurate sub-cycle phasors for a periodic waveform are obtained when $\xi \approx \frac{1}{2M}$. In addition, increasing the value of M results in better accuracy, since fewer input harmonics are classified as "spurious": when $\xi = \xi_{orth}(M) \equiv \frac{1}{2M}$, the only spurious harmonics in $\widehat{X}_k(t)$ are those separated from k by a multiple of $2M$ (recall Fig. 3.9). The same observations apply to *in-band-transient* waveforms, i.e., waveforms that exhibit slow variation of their fundamental (and possibly higher) harmonic components: in this case, dynamic sub-cycle phasors constructed from a modest number of waveform samples can provide accurate estimates of the corresponding dynamic full-cycle phasors, but at a fraction of the cost (see Table 3.1). A detailed comparison of full-cycle and sub-cycle computational costs is provided in [80] (see also Table 9.1).

Table 3.1 Cost (per phase, per sample) comparison for full-cycle vs. sub-cycle (N = sampling rate, in samples/cycle)

	Full-cycle ($N = 128$)	Full-cycle ($N = 256$)	Sub-cycle ($M = 2$)	Sub-cycle ($M = 4$)	Sub-cycle ($M = 8$)
Cost of evaluating one phasor set	1792	4096	4	16	64

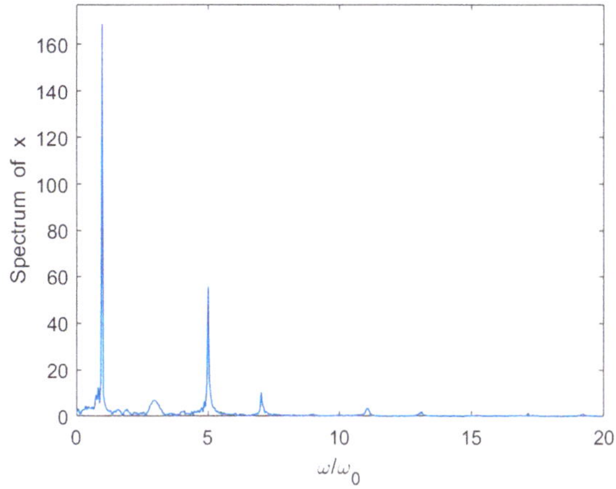

Fig. 3.11 Spectrum of "phase a" current from the paper-mill dataset.

To illustrate the capabilities of sub-cycle phasors for in-band transients, we consider a real-life dataset that exhibits a 5-cycle-long transient (see Sect. L.2.4 for a detailed description of this example, recorded at a paper-mill). We choose to focus here on a single waveform – the "phase a" current. Frequency analysis of this waveform reveals a significant 5-th harmonic (in addition to the fundamental) and traces of the 3-rd, 7-th, 11-th, and 13-th harmonics (Fig. 3.11). Several full-cycle phasors provide a clear depiction of the transient, and so do the corresponding sub-cycle phasors: Figs. 3.12–3.13 show the dynamic full-cycle and sub-cycle phasors for the 1-st, 3-rd, and 5-th harmonic, as well as the original waveform. The sub-cycle phasors were evaluated using $M = 8$, $\boldsymbol{\mathcal{S}} = \{1, 3, 5, 7\}$, and $\xi = \frac{1}{2M} = \frac{1}{16}$. In this example, selecting a smaller value for M, and either increasing or decreasing ξ, degrades the accuracy of all sub-cycle phasors $\{\widehat{X}_k(t)\,;\; k \in \boldsymbol{\mathcal{S}}\}$. We conclude that, for periodic (i.e., steady-state) or in-band-transient waveforms, good accuracy of sub-cycle phasors requires using somewhat higher values of M (e.g., $M = 8$) and the special choice $\xi = \xi_{orth}(M) \equiv \frac{1}{2M}$.

We observe that the level of the 3-rd harmonic phasor in the paper-mill example (both full-cycle and sub-cycle) increases by an order of magnitude during the occurrence of the transient. Clearly, the 3-rd harmonic phasor (and other triplen harmonics) cannot, and should not, be ignored in the presence of transients.[5] Such *transient-generated* metrics can provide an accurate indication of the occurrence of a transient and its duration.

This is even more evident in waveform examples that exhibit wide-band transients, typically in the high-frequency range. For instance, consider a real-life dataset that exhibits a 1-cycle-long capacitor-switching transient (see Sect. L.2.2 for a detailed description of this example, obtained from a wind-farm). Again

[5] Non-negligible triplen harmonics can also be generated in steady-state operation by resonance of distributed inverters in a power network [81].

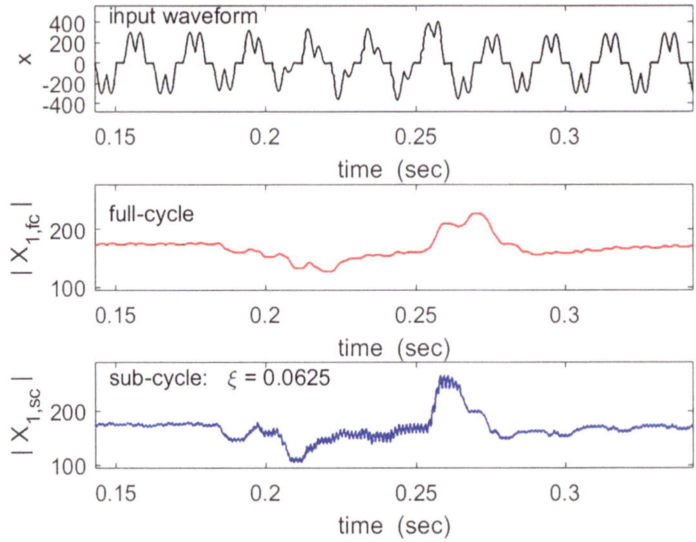

Fig. 3.12 Response to an in-band transient (paper-mill example): the input waveform $x(t)$ (top), the full-cycle dynamic phasor magnitude $|X_1(t)|$ (middle), and the sub-cycle dynamic phasor magnitude $|\widehat{X}_1(t)|$ obtained with $\boldsymbol{\mathcal{S}} = \{1, 3, 5, 7\}$ and $\xi = \frac{1}{16}$ (bottom).

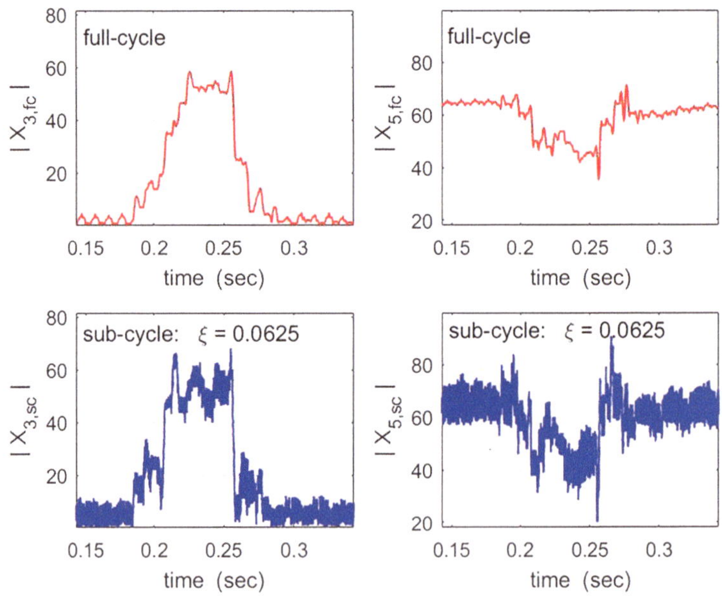

Fig. 3.13 Response to an in-band transient (paper-mill example): full-cycle and sub-cycle (with $\boldsymbol{\mathcal{S}} = \{1, 3, 5, 7\}$ and $\xi = \frac{1}{16}$) dynamic phasor magnitudes for the 3-rd and 5-th harmonics.

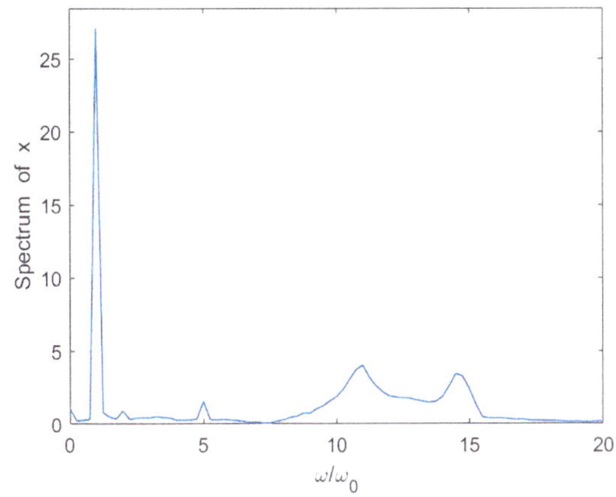

Fig. 3.14 Spectrum of "phase a" current from the wind-farm dataset.

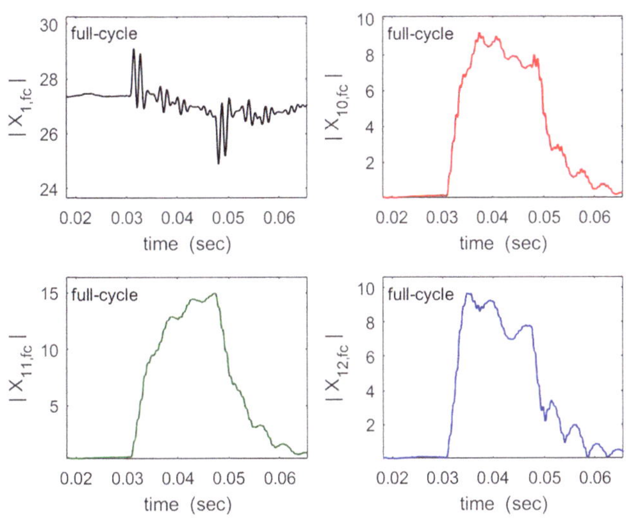

Fig. 3.15 Response to a wide-band transient (wind-farm example): full-cycle dynamic phasor magnitudes for the 1-st, 10-th, 11-th, and 12-th harmonics.

we consider here only the "phase a" current waveform. The frequency analysis of this waveform reveals a sizable continuous spectrum component in the range $8 \leq \frac{\omega}{\omega_0} \leq 16$, in addition to the fundamental and a trace of the 5-th harmonic (Fig. 3.14).

Examination of the complete set $\{X_k(t),\ 0 \leq k \leq 63\}$ of full-cycle dynamic phasors for the wind-farm example reveals multiple transient-generated harmonics. Figure 3.15 shows the most prominent of these (the 10-th, 11-th, and 12-th harmonics), alongside the fundamental. Notice that the transient is barely visible in the full-cycle dynamic phasor $X_1(t)$, due to the high-frequency nature of the capacitor-switching transient, and the narrow bandwidth (around $\omega = \omega_0$) of the waveform-to-phasor map used to generate this fundamental harmonic phasor. Again, we conclude that transient-generated metrics, such as $|X_3(t)|$ in the paper-mill example and $|X_{12}(t)|$ in the wind-farm example, can provide information about the occurrence and duration of a variety of transients.

In general, the transient behavior of a dynamic phasor (both sub-cycle and full-cycle) depends on the magnitude and phase response of the associated waveform-to-phasor map. However, we notice that, in the uniformly-shifted case, the phase response of the sub-cycle filter $\widehat{H}_k(s)$ is linear in ω, according to (3.31), so that the group-delay of $\widehat{H}_k(s)$ is frequency-independent and proportional to $M - 1$ and ξ, viz.,

$$\text{group-delay} \stackrel{\text{def}}{=} -\frac{d}{d\omega}\left\{\arg \widehat{H}_k(j\omega)\right\} = \xi \frac{(M-1)T}{2} \tag{3.38}$$

Thus the only impact of the phase response $\arg \widehat{H}_k(j\omega)$ is to impart a fixed delay to every frequency component of the input waveform $x(t)$. Similarly, we recall that the group-delay of the full-cycle waveform-to-phasor map is also frequency-independent and equal to $\frac{T}{2}$ (see (2.24b)), which imparts an even longer delay to the full-cycle phasors associated with $x(t)$.

The long trailing edge of the higher-order full-cycle phasors in Fig. 3.15 is caused primarily by *dispersion of the magnitude response*, which is due to the narrow-band nature of the full-cycle waveform-to-phasor map: compare the full-cycle magnitude response in Fig. 2.2 with the sub-cycle magnitude response in Fig. 3.9. We conclude that, in the presence of transients, we can no longer view the full-cycle phasor as a performance benchmark: its narrow-bandwidth nature results in a distorted transient response, especially in the case of wide-band transients. In fact, any LTI filter, including the sub-cycle waveform-to-phasor map (3.31), will cause some distortion of the transient. While increasing the bandwidth of such a filter may generate a more accurate depiction of the transient, it will inevitably degrade the accuracy of its steady-state response. This is an instance of the ubiquitous trade-off between accuracy and agility: one cannot simultaneously optimize both the steady-state accuracy and the transient response. This observation motivates us to adopt a two-criterion approach to optimized design of sub-cycle phasor schemes:

- Sub-cycle phasors should closely match their full-cycle counterparts in the steady-state interval that precedes the onset of a transient.
- Sub-cycle phasors should provide accurate information about the time of onset of a transient and its duration.

Applying this approach to the design of a sub-cycle scheme suitable for handling wide-band transients leads to specific choices of M and ξ, as we now demonstrate in the context of the wind-farm example.

The cost of evaluating the full-cycle phasor-set $\{X_k(t)\,;\ 0 \le k \le N/2 - 1\}$ for the wind-farm example is considerable, given the sampling rate of $N = 128$

samples/cycle. In addition, we are faced with the challenge of combining information from a large number of transient-generated phasor waveforms. In contrast, the cost of sub-cycle schemes is significantly lower, and there are fewer high-order phasors to consider. In our example, using $\boldsymbol{\mathcal{S}} = \{1, 5\}$ (i.e., $M = 4$) generates a single higher-order dynamic phasor (i.e., $\widehat{X}_5(t)$) at a cost that is two orders of magnitude lower than that of the full-cycle scheme (recall Table 3.1).

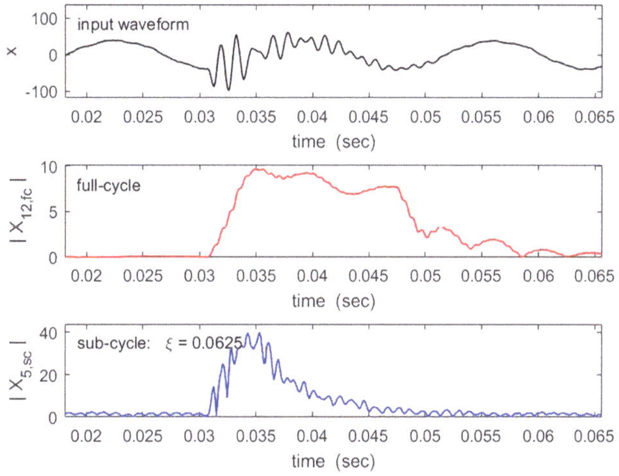

Fig. 3.16 Response to a wide-band transient (wind-farm example): the input waveform $x(t)$ (top), the full-cycle dynamic phasor magnitude $|X_{12}(t)|$ (middle), and the sub-cycle dynamic phasor magnitude $|\widehat{X}_5(t)|$ obtained with $\boldsymbol{\mathcal{S}} = \{1, 5\}$ and $\xi = \frac{1}{16}$ (bottom).

Figure 3.16 compares the sub-cycle phasor magnitude $|\widehat{X}_5(t)|$ with $|X_{12}(t)|$, the magnitude of the full-cycle dynamic phasor with the best depiction of the transient. Notice that the sub-cycle $|\widehat{X}_5(t)|$ provides a more accurate estimate of the duration of the transient than any of the full-cycle phasors: this is due to the long trailing edge of the transient in all full-cycle phasor waveforms. Also notice that the best sub-cycle transient response is achieved in this case by using a smaller number of waveform samples ($M = 4$) and a smaller value of normalized inter-sample shift ($\xi = \frac{1}{16} = \frac{1}{2}\,\xi_{orth}(4)$) than that used in the previous (=paper-mill) example. This is consistent with our earlier statement about the role of M and ξ in controlling the trade-off between accuracy and agility.

In summary, sub-cycle dynamic phasors constructed with a carefully-selected value of ξ (and a small M) provide the capability for efficient detection of wide-band transients and accurate determination of their duration and time of onset, while still offering a significant reduction in computational cost, as compared with full-cycle phasors. The analysis provided in this section suggests that in practical power system implementations it may be worthwhile to evaluate two (or more) sets of dynamic phasors, each one customized for a different operating regime. The overall computational burden of running several parallel sub-cycle schemes will still be significantly lower than that of a full-cycle calculation, while offering more accurate information for various categories of transients. Further discussion of customizing the choice of M, ξ for wide-band transient waveforms is provided in Sect. 9.4.3, in the context of power quality.

Finally, we recall that the linear filtering interpretation of sub-cycle dynamic phasors (see Fig. 3.7) relies on the waveform-to-phasor map $\widehat{H}_k(s)$ which, in turn, depends on the design parameters ξ and $\boldsymbol{\mathcal{S}}$. These observations motivate our interest in other possible choices of $\widehat{H}_k(s)$, leading to the introduction of *customized* (i.e., non-Fourier) dynamic phasors, which we discuss in Chaps. 4 and 5.

3.4 Beyond Phasors: Time–Frequency Analysis

The preceding sections of this chapter described several methods for estimation of dynamic phasors from an acquired, possibly non-periodic, waveform $x(\cdot)$, such as may occur in power systems in the presence of transients and faults. The estimated time-variant dynamic phasor $\widehat{X}_k(t)$ can be interpreted as the output of a *band-pass subband filter* $\widehat{H}_k(s)$ whose frequency response is centered at $f_{c,k} \stackrel{\text{def}}{=} kf_0$ (where $f_0 = \frac{\omega_0}{2\pi}$), as shown in Figs. 2.1, 3.3, and 3.7. Notice that all of the techniques discussed in Sects. 3.1–3.3 rely on reasonably accurate knowledge of the steady-state fundamental harmonic frequency ω_0. The combined set of harmonic frequencies and associated (rms) dynamic phasors, viz.,

$$\left\{ \left(\widehat{X}_k(t)\,,\, kf_0\right)\,;\ -\infty < t < \infty\,,\ 0 \le k < \infty \right\}$$

provides a *time–frequency representation* (TFR) of the acquired waveform $x(\cdot)$.

We discuss in Chaps. 4 and 5 a broader family of TFR techniques which do not rely on prior knowledge of the fundamental harmonic frequency f_0. These include, in particular, the short-time Fourier transform (STFT), the S-transform, and the continuous wavelet transform (CWT). In general, such TFR techniques allow for customized design of the subband filters $\widehat{H}_k(s)$, which means that one has (a) complete freedom in choosing the subband center frequencies $f_{c,k}$ and the frequency responses $\widehat{H}_k(j\omega)$, and (b) complete control of the frequency grid density. While it is common practice to use available prior information about the acquired waveform to guide the design of a TFR subband pattern, it is nevertheless entirely possible to generate a meaningful TFR without any reliance on such prior knowledge.

Chapter 4 is dedicated to examination of the STFT and some related TFR techniques, all of which involve constant-bandwidth subband patterns, while Chap. 5 presents the most commonly used variable-bandwidth TFR techniques. In contrast, the following three subsections provide a brief overview of a completely different approach to time–frequency analysis, which does not involve pre-determined frequency subbands or any prior information about the acquired waveform. This alternative TFR approach relies on the Hilbert transform and the notion of *analytic waveforms*.

3.4.1 Analytic Waveforms

The analytic version of a given real-valued waveform $\{x(t)\,;\ -\infty < t < \infty\}$ is the complex waveform

$$\widehat{x}(t) \stackrel{\text{def}}{=} x(t) + j\,\mathcal{H}\{x(t)\} \tag{3.39a}$$

where $\mathcal{H}\{x(t)\}$ is the well-known Hilbert transform [82–84] of the waveform $x(t)$. The Hilbert transform is a *non-causal* linear time-invariant filter with impulse response $h_{\text{Hilbert}}(t)$, i.e.,

$$\mathcal{H}\{x(t)\} \stackrel{\text{def}}{=} h_{\text{Hilbert}}(t) \circledast x(t) \tag{3.39b}$$

The frequency response of the Hilbert transform filter is

$$H_{\text{Hilbert}}(j\omega) = \begin{cases} -j & \omega > 0 \\ 0 & \omega = 0 \\ j & \omega < 0 \end{cases} \tag{3.39c}$$

Notice that the Fourier transform of the analytic waveform $\widehat{x}(t)$ vanishes for negative frequencies, viz.,

$$\widehat{X}(j\omega) = \int_{-\infty}^{\infty} \widehat{x}(t)\, e^{-j\omega t}\, dt = \begin{cases} 2X(j\omega) & \omega > 0 \\ X(0) & \omega = 0 \\ 0 & \omega < 0 \end{cases}$$

Since $x(t) = \Re\{\widehat{x}(t)\}$, we conclude that the relation $x(\cdot) \Longleftrightarrow \widehat{x}(t)$ is an invertible one-to-one correspondence (see also the discussion in [85]).

The notion of an analytic version of a given real-valued waveform was introduced by D. Gabor [86] in an attempt to provide a convenient characterization of nonlinear oscillation phenomena. It is often used to define time-dependent (a.k.a. "instantaneous") amplitude and frequency of a given waveform $x(\cdot)$ via the polar representation of its analytic version, viz.,

$$a(t) \stackrel{\text{def}}{=} \left|\widehat{x}(t)\right| \ , \qquad \omega(t) = \frac{d}{dt} \arg\{\widehat{x}(t)\} \tag{3.40}$$

Since $x(t) = \Re\{\widehat{x}(t)\} \equiv \Re\{a(t)e^{j\phi(t)}\}$, where $\phi(t) = \arg\{\widehat{x}(t)\}$, the analytic waveform $\widehat{x}(\cdot)$ is often perceived as an *instantaneous phasor* representing the real waveform $x(\cdot)$. In particular, if $x(t) = A\cos(\omega_0 t + \varphi_0)$, then $\widehat{x}(t) = \left(A\, e^{j\varphi_0}\right) e^{j\omega_0 t}$, so that

$$A = \left|\widehat{x}(t)\right| \quad , \qquad \omega_0 = \frac{d}{dt} \arg\{\widehat{x}(t)\}$$

which matches the standard notions of amplitude and frequency for a sinusoidal waveform. Analytic signals became a standard tool in the study of modulated carrier waveforms, characterization of band-pass communication channels, and generation of single-sideband AM signals (see,. e.g., [85, 87–89]).

It soon became evident that the analytic version of a real-valued waveform can provide a physically meaningful interpretation in terms of time-dependent amplitude and center frequency only if the input waveform is a "mono-component" signal [90, 91]. Although there is no consensus about the precise meaning of this term, most researchers tend to define a mono-component signal as a narrow-band waveform with a single dominant peak in the frequency domain, whose center frequency is much larger than its bandwidth[6] [91–93]. A classical example of a mono-component signal is an amplitude modulated (AM) waveform, viz.,

$$x(t) = \big[1 + s(t)\big] \cos \omega_c t$$

where $s(t)$ is known as the modulation signal: it has bounded magnitude (i.e., $|s(t)| < 1$) and is restricted to the frequency range $|\omega| \leq \omega_m$, i.e., the Fourier transform of $s(t)$ vanishes for $|\omega| > \omega_m$. Thus, when $\omega_m \ll \omega_c$ (or even just $\omega_m < \omega_c$), the corresponding analytic version of the AM waveform is

$$\widehat{x}(t) = \big[1 + s(t)\big]\, e^{j\omega_c t}$$

which can be viewed as a "generalized passband dynamic phasor" associated with the modulated waveform $x(t)$. The resulting time-dependent (a.k.a. "instantaneous") amplitude and frequency, namely,

$$\left|\widehat{x}(t)\right| = 1 + s(t) \quad , \qquad \frac{d}{dt} \arg\left\{\widehat{x}(t)\right\} = \omega_c$$

are both continuous functions of time, which coincide with the modulated amplitude and the carrier frequency, respectively.

However, it is easy to construct simple examples of waveforms whose instantaneous amplitude and frequency do not match the standard notions. One such example is the sinusoidally-modulated (a.k.a. "double-sideband") waveform

$$x(t) = 2 \cos \omega_m t \, \cos \omega_c t \equiv \cos\big[(\omega_c + \omega_m)t\big] + \cos\big[(\omega_c - \omega_m)t\big]$$

which has two equal-height peaks in the frequency domain, located at $\omega = \omega_c \pm \omega_m$ and, thus, cannot be considered a mono-component signal. The center frequency

[6] Some authors choose to impose additional constraints: see, e.g., [90].

of this waveform is, of course, ω_c. However, its instantaneous frequency has multiple Dirac-delta components, located at instants corresponding to phase-jumps of $\arg\{\widehat{x}(t)\}$, caused by sign changes of $\cos \omega_m t$. This counter-intuitive behavior of the instantaneous frequency is a direct consequence of the definition (3.40), as explained in detail in [94]. This example, and similar ones, clearly indicates the need for a comprehensive approach to "multi-component" waveforms.

3.4.2 Waveform Decomposition into Single-Mode Components

One direct approach to extract narrow-band single peak components from a given waveform is via the use of a filter-bank. This results in a representation of arbitrary waveforms in terms of their frequency subbands, as discussed in detail in Chaps. 4 and 5 of this book.

A different approach, which completely avoids subband filtering, is the Hilbert–Huang transform (HHT) [95], originally motivated by applications in geophysics, oceanography, and meteorology. This method uses a so-called Empirical Mode Decomposition (EMD) scheme to represent any given waveform as a sum of *intrinsic mode functions* (IMFs). The HHT constructs the intrinsic mode functions of a given waveform in such a way that the instantaneous frequency associated with each IMF is "consistent with the physics of the system studied" [95]. Once the decomposition into IMFs has been obtained, the analytic version of each IMF is used to define IMF-specific instantaneous amplitude and frequency: these are used as a time–frequency representation of the waveform of interest.

It should be pointed out that the original HHT technique cannot resolve modes whose frequencies are separated by less than an octave: an example of this unwanted behavior is illustrated in Figs. 25–28 of [95]. This shortcoming is of particular concern in power system applications in the presence of non-negligible harmonic distortion. For instance, when a waveform contains noticeable 5-th, 7-th, and 11-th harmonics in addition to the fundamental, the EMD combines all three higher harmonics into a single "composite IMF." The time-variant frequency associated with this IMF oscillates within the frequency range of the higher harmonics that are represented by this composite mode.

Subsequent research resulted in improvements to the HHT that make it possible to associate separate IMFs with closely-spaced higher frequency components and accurately recover their individual frequencies, see, e.g., [96, 97]. In particular, the recently introduced Variational Mode Decomposition (VMD) significantly outperforms EMD in terms of harmonic frequency detection and separation, while being more robust with respect to additive noise [98]. An example of the improvement achieved by using VMD instead of EMD is provided in the following subsection.

The Hilbert–Huang transform is, by its very nature, an offline technique, because of both the mode decomposition step (either EMD or VMD), which relies on the entire waveform record, and non-causality of the Hilbert transform. Thus, it cannot be used in applications, such as under-frequency load-shedding, that rely on real-time online phasor and frequency estimation.

3.4.3 Time–Frequency Representation from Analytic Signal Inputs

We examine in Chaps. 4 and 5 a variety of techniques for time–frequency representation (TFR) of *real-valued waveforms*. In general, one can replace the real input $x(t)$ of such TFR techniques by its analytic version $\widehat{x}(t)$. However, the consequences of such a replacement depend on the TFR technique in consideration.

TFR techniques that rely on filter-bank implementations, such as the short-time Fourier transform (STFT) and the wavelet transform, suffer from leakage caused by the sidelobes of their (non-ideal) subband filters. In other words, the output of each subband contains spurious contributions from frequency components outside the subband. This problem is especially noticeable when the impulse response of a subband filter is real-valued, for instance, when using a wavelet transform based on a real-valued mother wavelet function. Such subband filters have a conjugate-symmetric frequency response, which results in leakage from negative frequency components of the input waveform. *Analytic wavelets*, such as the Morse wavelet function, whose frequency response vanishes at negative frequencies, have been proposed as a remedy for this predicament. Alternatively, applying an analytic signal input to any subband-based TFR of choice completely eliminates the undesired leakage from negative frequency components and thus makes it possible to use arbitrary (non-analytic and/or real-valued) subband filters [99].

In contrast to subband-based TFRs, the benefit of using analytic inputs to quadratic TFRs such as the Wigner–Ville decomposition (WVD) is debatable. In

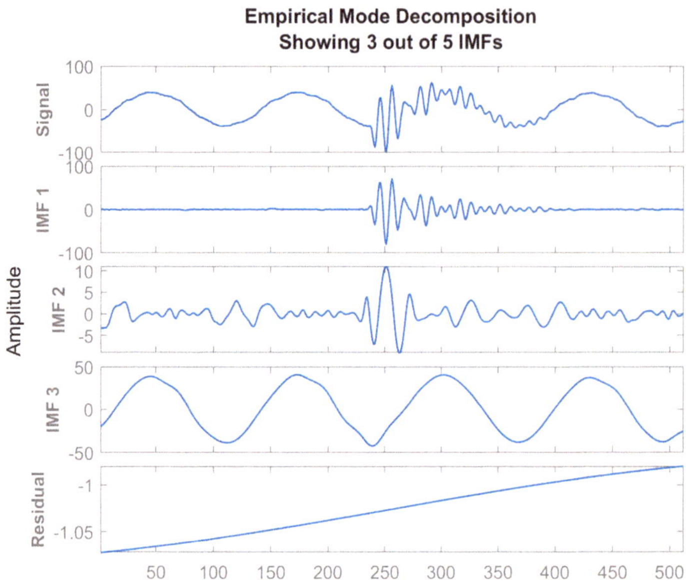

Fig. 3.17 EMD-generated IMFs for the wind-farm example.

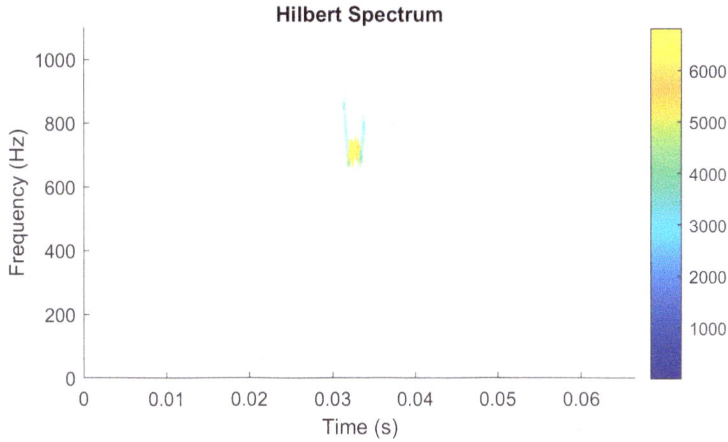

Fig. 3.18 Hilbert spectrum (2D) for the wind-farm example.

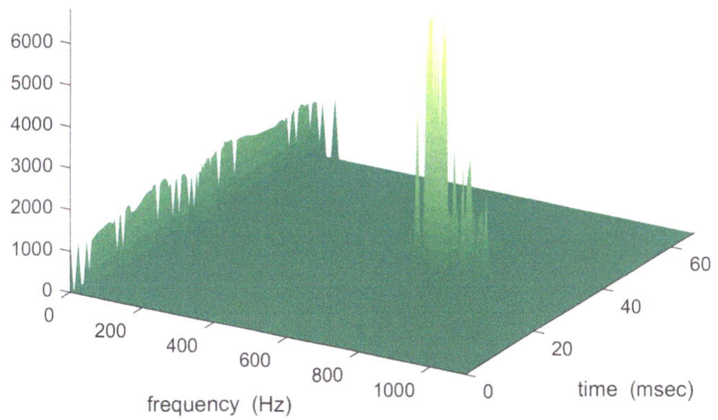

Fig. 3.19 EMD-based three-dimensional Hilbert spectrum for the wind-farm example.

general, the standard WVD has some strong interference around zero-frequency, caused by cross-terms involving positive and negative frequency components. Using an analytic signal input can reduce this effect, but this comes at the cost of smearing out sharp (pulse-like) transitions, as demonstrated in [100, Sect. 7.5].

The Hilbert–Huang transform (HHT) offers an alternative TFR, known as the *Hilbert spectrum*, which is based on using the EMD to generate IMFs (Figs. 3.17 and 3.18): these plots were generated via the `emd` and `hht` functions in MATLAB R2021a. The three-dimensional plot of the Hilbert spectrum (Fig. 3.19), generated by applying the `mesh` function to the output of the `hht` function, should be compared with the TFR generated by the short-time Fourier transform (Fig. 3.20). We notice that the STFT-based TFR shows an almost constant 60 Hz fundamental harmonic, as well as a wide-band transient with peaks at, approximately, 660 Hz and 870 Hz, and limited to the time interval of 20–50 ms.

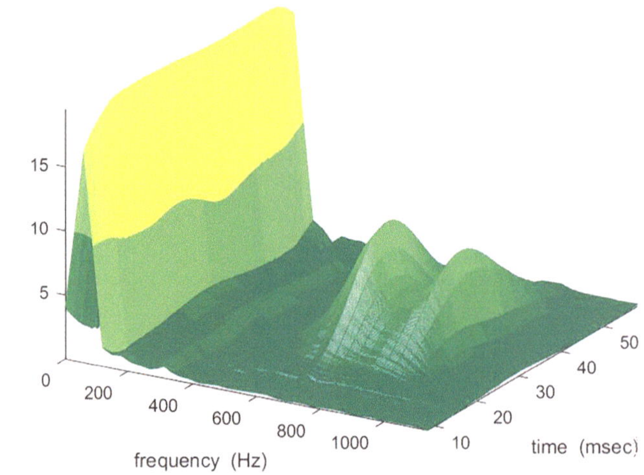

Fig. 3.20 Short-time Fourier transform (using 2-cycle-long Hamming window) for the wind-farm example.

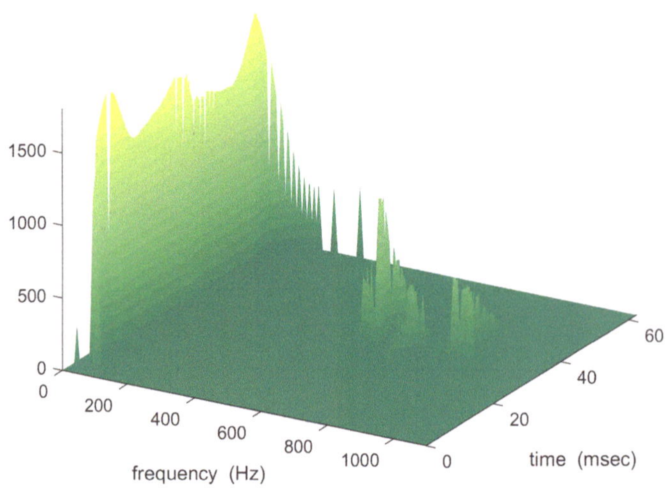

Fig. 3.21 VMD-based three-dimensional Hilbert spectrum for the wind-farm example.

In contrast, the information provided by the 3D Hilbert spectrum in Fig. 3.19 is much more ambiguous, especially in depicting the fundamental harmonic. This shortcoming of the EMD can be partially remedied by using the `vmd` function in MATLAB to construct a different set of intrinsic mode functions (IMFs). A three-dimensional Hilbert spectrum is obtained by using the `vmd` and `hht` functions in MATLAB, followed by the `mesh` function (Fig. 3.21). Notice the clearly defined fundamental harmonic (albeit with some amplitude variation) and the multiple (2 or 3) separate peaks in the 600–1000 Hz frequency range which roughly match the two transient-related ridges in Fig. 3.20.

3.4.4 Estimation of Fundamental Harmonic Frequency

Accurate estimation of the fundamental harmonic frequency f_0 of an acquired power system waveform $x(t)$ is a special case of time–frequency analysis, with a focus on a single frequency subband, centered at f_0. Numerous techniques have been proposed for achieving this goal (see, e.g., [42, 101]). Many among these can be described in terms of a three-stage process (Fig. 3.22): (a) extraction of a single-mode component of interest $x_\star(t)$, (b) generation of a corresponding time-dependent phase waveform $\phi_\star(t)$, and (c) translation of this phase waveform into a time-dependent estimate $\widehat{f_0}(t)$ of the fundamental harmonic frequency.

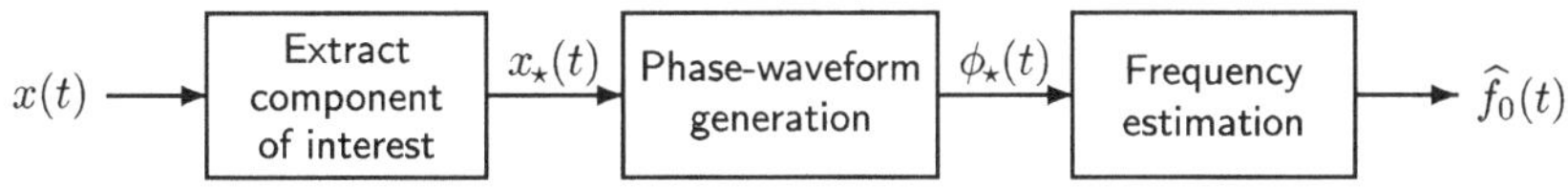

Fig. 3.22 Three-stage structure of frequency estimation.

Two commonly used methods for decomposing an acquired waveform into its single-mode components are (a) using a filter-bank with suitably-designed subbands or (b) using the empirical mode decomposition (EMD), or the variational mode decomposition (VMD). In particular, we observe that estimation of the fundamental frequency in power system applications involves only one single-mode component, which occupies a narrow frequency band around a known nominal frequency f_0. Extracting this component is easily achieved by a low-pass filter (or even a suitably designed band-pass filter) that suppresses all higher harmonics as well as wide-band transients. In contrast, using the EMD (or VMD) for the same purpose is much more costly and does not guarantee that the first IMF generated by this technique will correspond to the fundamental harmonic component.

Generation of a Phase Waveform

Once a correct single-mode component $x_\star(t)$ has been extracted, we can choose among a number of schemes that generate a time-dependent phase waveform for a given mono-component signal, such as:

- **Analytic signal approach:** Construct the analytic version of $x_\star(t)$, viz.,

$$\widehat{x}_\star(t) = x_\star(t) + j\mathcal{H}\{x_\star(t)\}$$

  and define the phase waveform as $\phi_\star(t) = \arg\{\widehat{x}_\star(t)\}$.
- **Dynamic phasor approach:** Construct a least-squares estimate $\widehat{\mathcal{X}}_1(t)$ of the passband fundamental phasor, such as (3.21a) or (3.22a). This is, in fact, the most frequently used approach, known as the *DFT method*, for extracting the fundamental frequency from acquired power systems waveforms [42]. An

interesting low-cost variant is to use a re-scaled 2-sample sub-cycle phasor estimate, viz.,

$$\widehat{\mathcal{X}}_1(t) = x_\star(t) + j\, x_\star(t - \tfrac{T}{4})$$

which is obtained by setting $M = 2$ and $\Delta = \frac{T}{4}$ in (3.30a) and dropping the constant scaling factor. Again, the desired phase waveform output is determined via the usual relation, namely, $\phi_\star(t) = \arg\{\widehat{\mathcal{X}}_1(t)\}$.

- **Clarke transform approach:** Construct (scaled) Clarke components $x_\alpha(t)$ and $x_\beta(t)$ of a given mono-component *three-phase waveform*, and then define $\phi_\star(t) = \arg\{x_\alpha(t) + j\, x_\beta(t)\}$. A version of this approach, intended for single-phase systems, is presented in [102].

Notice that all three approaches rely on combining a suitably constructed mono-component waveform with a $90°$-shifted version of itself.

Estimation of the Fundamental Frequency and Its Rate of Change

The two most frequently used approaches for frequency estimation from a sampled phase waveform are (a) direct numerical differentiation with smoothing and (b) least-squares polynomial fit [42, 101]. The first approach relies on a discrete-time implementation of the key relation (3.40), followed by suitable smoothing, i.e.,

$$\widehat{f_0}(nT_s) = \Gamma[n] \circledast \frac{\phi_\star(nT_s) - \phi_\star(nT_s - T_s)}{2\pi\, T_s} \tag{3.41}$$

where $\Gamma[n]$ can be chosen as the impulse response of any discrete-time low-pass filter with a unity DC gain. Since differentiation enhances the impact of higher harmonics, wide-band transients, and noise, smoothing of the "instantaneous frequency" estimate $\left[\phi_\star(nT_s) - \phi_\star(nT_s - T_s)\right]/2\pi\, T_s$ is required to obtain a reliable frequency estimate in power system applications.

The most common smoothing choice is uniform smoothing, which is achieved by setting

$$\Gamma[n] = \begin{cases} \frac{1}{L} & 0 \le n \le L-1 \\ 0 & \text{else} \end{cases} \tag{3.42}$$

where the positive integer L determines the bandwidth of the smoothing filter. Notice that the mapping (3.41) from phase waveform samples to a frequency estimate is, in fact, a discrete-time linear time-invariant FIR filter of length $L + 1$. Other choices of $\Gamma[n]$, designed to optimize specific performance criteria, have also been proposed: see, e.g., [101, 103, 104]. A similar approach can also be used to estimate the *rate of change of frequency* (ROCOF), which is defined as the derivative of $f_0(t)$, viz.,

$$\rho_0(t) \stackrel{\text{def}}{=} \frac{d}{dt} f_0(t) \tag{3.43}$$

ROCOF is often used as an indicator of frequency instability: ROCOF relays are used for fast load-shedding and to detect islanding [42, 105]. In the event of a sudden loss of generation, ROCOF relays can prevent frequency collapse by initiating corrective action much before generator under-frequency relays are tripped.

In contrast to smoothed differencing schemes, the *polynomial fit approach* makes it possible to calculate derivatives of the phase waveform without explicit numerical differentiation. This is achieved by constructing a continuous-time interpolation $\widehat{\phi}(t)$ of the phase waveform, based on a second-order Taylor series centered at $t = nT_s$ [42, 101], viz.,

$$\widehat{\phi}(t) = \varphi_0[n] + \varphi_1[n]\,(t - nT_s) + \varphi_2[n]\,(t - nT_s)^2 \tag{3.44}$$

The unknown coefficients $\{\varphi_0, \varphi_1, \varphi_2\}$ are determined by solving a least-squares approximation problem, i.e.,

$$\min_{\{\varphi_0,\varphi_1,\varphi_2\}} \sum_{i=0}^{M-1} \left[\phi_\star(nT_s - i\Delta) - \widehat{\phi}(nT_s - i\Delta) \right]^2 \tag{3.45a}$$

$$= \min_{\{\varphi_0,\varphi_1,\varphi_2\}} \left\| \begin{pmatrix} \phi_\star(nT_s) \\ \phi_\star(nT_s - \Delta) \\ \vdots \\ \phi_\star(nT_s - (M-1)\Delta) \end{pmatrix} - \mathcal{C}_\phi \begin{pmatrix} \varphi_0 \\ \varphi_1 \\ \varphi_2 \end{pmatrix} \right\|_2^2$$

where $\mathcal{C}_\phi$ is a constant (i.e., time-invariant) $M \times 3$ matrix that depends only on Δ and M, viz.,

$$\mathcal{C}_\phi = \begin{pmatrix} 1 & 0 & 0 \\ 1 & -\Delta & \Delta^2 \\ 1 & -2\Delta & 4\Delta^2 \\ \vdots & \vdots & \vdots \\ 1 & -(M-1)\Delta & (M-1)^2\Delta^2 \end{pmatrix} \tag{3.45b}$$

and $\Delta = DT_s$ for some positive integer D, known as a "decimation factor" [32, 33]. The resulting least-squares solution is

$$\varphi_r[n] = G_r \begin{pmatrix} \phi_\star(nT_s) \\ \phi_\star(nT_s - \Delta) \\ \vdots \\ \phi_\star(nT_s - (M-1)\Delta) \end{pmatrix}, \qquad r = 0, 1, 2 \tag{3.46a}$$

where the three row vectors $\{G_r\}$ are determined via the time-invariant expression

$$\begin{pmatrix} G_0 \\ G_1 \\ G_2 \end{pmatrix} = \left(C_\phi^\top C_\phi\right)^{-1} C_\phi^\top \tag{3.46b}$$

The corresponding frequency estimate is

$$\widehat{f_0}(nT_s) \stackrel{\text{def}}{=} \frac{1}{2\pi}\frac{d\widehat{\phi}}{dt}\bigg|_{t=nT_s} = \frac{1}{2\pi} G_1 \begin{pmatrix} \phi_\star(nT_s) \\ \phi_\star(nT_s-\Delta) \\ \vdots \\ \phi_\star(nT_s-(M-1)\Delta) \end{pmatrix} \tag{3.47a}$$

and the associated rate of change of frequency (ROCOF) is

$$\widehat{\rho_0}(nT_s) \stackrel{\text{def}}{=} \frac{1}{2\pi}\frac{d^2\widehat{\phi}}{dt^2}\bigg|_{t=nT_s} = \frac{1}{\pi} G_2 \begin{pmatrix} \phi_\star(nT_s) \\ \phi_\star(nT_s-\Delta) \\ \vdots \\ \phi_\star(nT_s-(M-1)\Delta) \end{pmatrix} \tag{3.47b}$$

Both expressions can be implemented as discrete-time finite-impulse-response (FIR) filters, whose time-invariant coefficients can be evaluated and stored in advance.

Notice that the length of the waveform window used to evaluate such frequency and ROCOF estimates (a.k.a. the "measuring window") is $(M-1)DT_s \approx MDT_s$ or, equivalently, a segment of MD waveform samples. This length should be comparable to the time constant of the sub-system that relies on the estimates (3.47). For instance, when used in support of under-frequency load-shedding relays, the recommended window length is 5–6 nominal cycles [42, Sect. 4.5]. In reality, commercially available ROCOF relays come in a much wider range of measuring window lengths, ranging from 2 cycles up to as high as 100 cycles [105]: however, in most cases the ROCOF measuring window is 2–6 cycles long.

In summary, *unsmoothed estimates* of "instantaneous frequency" are *highly unreliable indicators of performance* in power system applications. Suitably designed smoothing, either via low-pass filtering or by relying on a least-squares polynomial fit, can provide accurate estimates of the fundamental frequency and its rate of change.

Performance Comparison

We use two real-life datasets to compare the performance of three prospective techniques for generating an estimated phase waveform $\phi_\star(\cdot)$: (a) analytic signal approach, (b) full-cycle dynamic phasor approach, and (c) 2-sample sub-cycle dynamic phasor approach. When using direct numerical differentiation with a uniform 4-cycle smoothing scheme, all three approaches result in similar frequency trajectories: Fig. 3.23 presents a steady-state example, while Fig. 3.25 shows a transient case. The same observation holds when the differentiation with uniform smoothing scheme for frequency estimation is replaced by estimation via a least-squares polynomial fit, as described in (3.45)–(3.47): see Figs. 3.24 and 3.26.

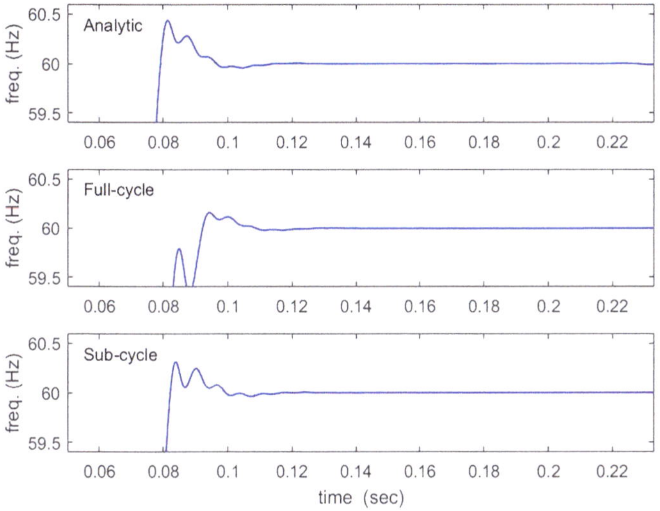

Fig. 3.23 Frequency estimates for prospective phase-generation approaches, using differentiation and 4-cycle smoothing (steady-state Hydro-Ottawa example).

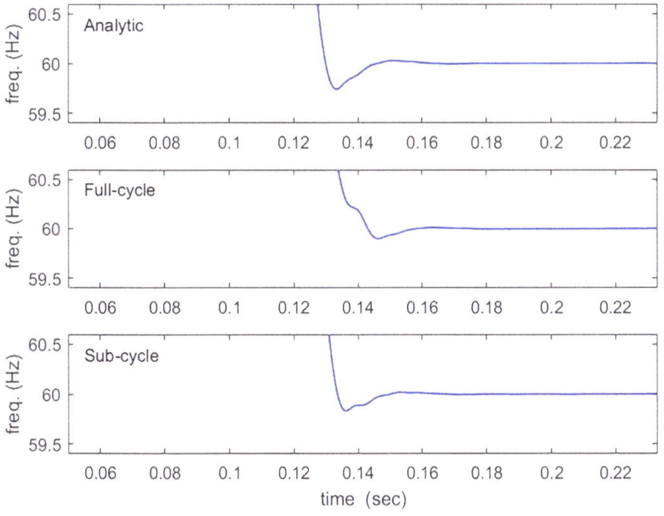

Fig. 3.24 Frequency estimates for prospective phase-generation approaches, using 7-cycle polynomial fit with $D = 1$ (steady-state Hydro-Ottawa example).

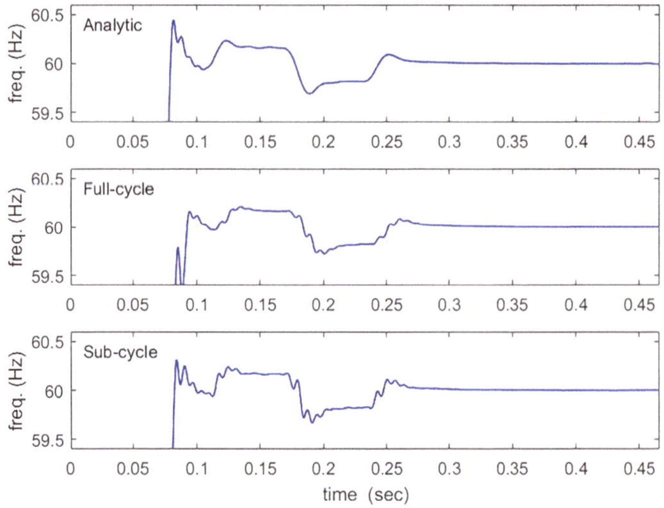

Fig. 3.25 Frequency estimates for prospective phase-generation approaches, using differentiation and 4-cycle smoothing (transient Hydro-Ottawa example).

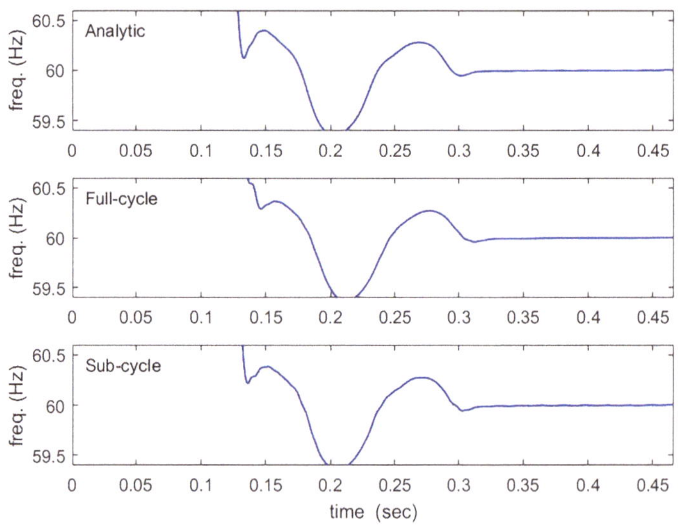

Fig. 3.26 Frequency estimates for prospective phase-generation approaches, using 7-cycle polynomial fit with $D = 1$ (transient Hydro-Ottawa example).

Additional information about the two real-life datasets used here can be found in Sects. L.1.8 and L.2.3 of Appendix L, respectively. All examples use the same discrete-time front-end filter (an eighth-order Butterworth low-pass filter with a cutoff frequency of 90 Hz), to suppress the impact of higher harmonics and wide-band transients.

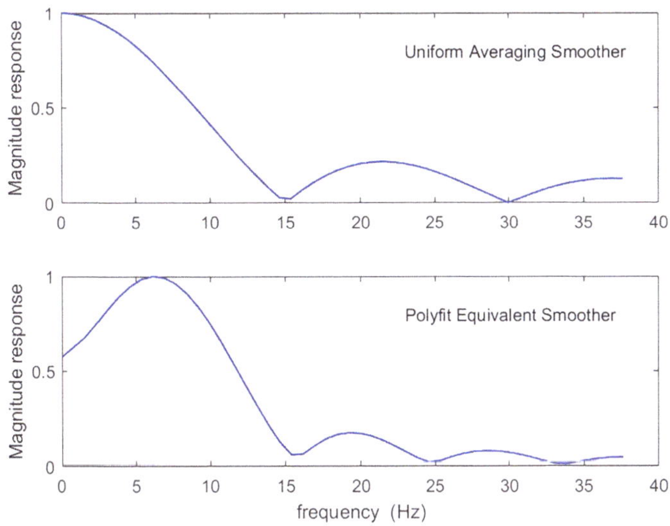

Fig. 3.27 Comparison of smoothing filter bandwidth: 4-cycle uniform averaging (top) vs. 7-cycle polynomial fit equivalent smoothing (bottom), both designed for a nominal frequency of 60 Hz.

In order to facilitate a fair comparison of the two proposed smoothing schemes (i.e., the third stage in Fig. 3.22), we adjust the measuring window of the polynomial fit scheme so that its equivalent smoothing bandwidth is comparable with that of the uniform averaging scheme, as illustrated in Fig. 3.27. The "polyfit equivalent smoother" is obtained by dividing the discrete-time transfer function $G_1[1\ z^{-1} \ldots z^{-(M-1)}]$, used in (3.47a) to map $\phi_\star(\cdot)$ into the frequency estimate $\widehat{f}_0(\cdot)$, by the differencing transfer function $1 - z^{-1}$, and scaling appropriately.

Examination of Figs. 3.23–3.24 reveals no significant difference in performance in steady-state operation between the two frequency estimation schemes under consideration (i.e., differencing with uniform averaging vs. a least-squares polynomial fit). On the other hand, Figs. 3.25–3.26 suggest that the polynomial fit approach may be somewhat inferior to uniform averaging in the presence of a transient. Finally, the value of the "decimation factor" D can be increased, which results in reduced implementation cost, at the expense of some degradation in smoothing performance.

Cost comparison

In view of the similarity in performance of the three prospective schemes for generating a frequency estimate, we conclude that the main difference between these approaches is their computational cost.

When the selected frequency estimation technique uses direct differencing followed by smoothing, as in (3.41), the 2-sample sub-cycle approach involves the lowest overall cost, i.e., $\mathcal{O}(2N_B + L)$ real multiplications per input waveform sample, where N_B is the order of the low-pass front-end filter, and L is the length of the discrete-time smoothing filter $\Gamma[n]$ in (3.41). The full-cycle approach requires $\mathcal{O}(2N_B + 2N + L)$ real multiplications per waveform sample, where N is the number of samples per cycle. However, when uniform averaging, as defined in (3.42), is used for smoothing the frequency estimate, the number of multiplications required for implementing (3.41) reduces to one (i.e., scaling by $\frac{F_s}{2\pi} \cdot \frac{1}{L}$). Moreover, inserting the uniform averaging choice (3.42) into the smoothed difference expression (3.41) results in a very simple expression, shown in (3.48e) below, which requires only one addition and one subtraction per waveform sample. The total computational cost for schemes that use uniform averaging is summarized in Table 3.2. For instance, $N_B = 8$ and $N = 128$ in the steady-state Hydro-Ottawa example (Figs. 3.23 and 3.24), so that the total cost per waveform sample of the 2-sample sub-cycle approach is 18 real multiplications, as compared with a significantly higher cost of 274 real multiplications for the full-cycle approach.

Table 3.2 Cost per sample comparison of prospective approaches to frequency estimation (with uniform averaging): single-phase waveform

	Front-end filter	Phasor evaluation	Smoothed difference	Total cost
Full-cycle approach	$2N_B+1$	$2N$	1	$\mathcal{O}(2N_B+2N)$
Sub-cycle approach	$2N_B+1$	0	1	$\mathcal{O}(2N_B)$
Analytic approach	$2N_B+1$	$4\log_2 N_x$	1	$\mathcal{O}(2N_B+4\log_2 N_x)$

The analytic signal approach requires $\mathcal{O}(2N_B + 4\log_2 N_x + L)$ real multiplications per waveform sample, where N_x is the number of samples in the entire signal record being processed. This reduces to $\mathcal{O}(2N_B + 4\log_2 N_x)$ when uniform averaging is used as the smoothing choice. In the steady-state Hydro-Ottawa example (Fig. 3.23) $N_x = 1789$ samples, which means that in this example the cost of the analytic approach (i.e., $2N_B + 4\log_2 N_x = 61$ real multiplications per waveform sample) is somewhat higher than that of the 2-sample sub-cycle approach. However, unlike the two alternative approaches, which can be implemented online to provide real-time frequency estimation, the analytic signal approach is a non-causal, offline technique. Still, this approach can be implemented online in block-fashion, with its block length providing a controllable trade-off between cost, steady-state error, and latency. Alternatively, the Hilbert transform can be *approximated* by a causal discrete-time linear time-invariant filter[32], making it possible to construct an online implementation of an approximate analytic signal-based frequency estimation scheme.

When the selected frequency estimation technique uses a least-squares polynomial fit, as in (3.45)-(3.47), the corresponding computational cost has the same expressions as smoothed differencing schemes, except that the smoothing window length L is replaced by M, which is the length of the G_r coefficient vectors in

(3.47). This means that all approaches will include an $\mathcal{O}(M)$ added cost term, which cannot be eliminated. For instance, using a 7-cycle polynomial fit window with the transient Hydro-Ottawa example adds a cost of $7 \times 128 = 896$ real multiplications per waveform sample. Clearly, the polynomial fit scheme, with $D = 1$, is much more expensive than the "differencing plus uniform averaging" scheme. Increasing the decimation factor D reduces the computational cost to $\mathcal{O}(M/D)$ but results in fluctuation of the frequency estimate $\widehat{f_0}(\cdot)$ due to increased leakage from higher harmonics and wide-band transients.

Finally, we should point out that estimation of the fundamental frequency is a special case of the more general problem of line spectrum estimation. There exist a large variety of (mostly offline) techniques for solving this general problem, including so-called high-resolution methods, such as Prony, Pisarenko, MUSIC, and ESPRIT [106, 107]. In particular, the power systems literature contains numerous publications that rely on the Prony method to estimate phasors and their frequency.

3.4.5 Fundamental Frequency Estimation in a Polyphase System

Using an acquired *three-phase waveform* to estimate the fundamental frequency offers an additional modality for smoothing the impact of higher harmonics and wide-band noise/transients.

The added information provided by a polyphase waveform can be exploited in several ways, including:

- **Across-phase averaging:** Obtain a separate frequency estimate for each phase, and then combine them into a single estimate, using the arithmetic average.
- **Positive-sequence input:** Evaluate the positive sequence component of the input waveform, and use it as input to one of the prospective frequency estimation schemes discussed above.

Here we will focus only on the case of three-phase waveforms and use uniform averaging exclusively as the smoothing operation. Both methods generate very similar frequency estimates, as is evident from Fig. 3.28. Notice that deviations of the estimated fundamental harmonic from the nominal frequency are much smaller in a three-phase-based estimate, as compared with the single-phase estimates shown in Fig. 3.25. Both of these two methods can be used with any of the three approaches used to generate Figs. 3.25 and 3.26. However, the computational cost of the "across-phase averaging" method is, in general, somewhat higher than that of the "positive-sequence input" method.

The preceding discussion suggests that the most efficient scheme for estimating the fundamental harmonic frequency from an acquired three-phase waveform is to apply the 2-sample sub-cycle dynamic phasor approach with uniform averaging to a positive-sequence input. The performance of this scheme is barely distinguishable from that obtained with the other two approaches (see Fig. 3.29), and its implementation cost is the lowest among all the schemes discussed in this subsection. And, as

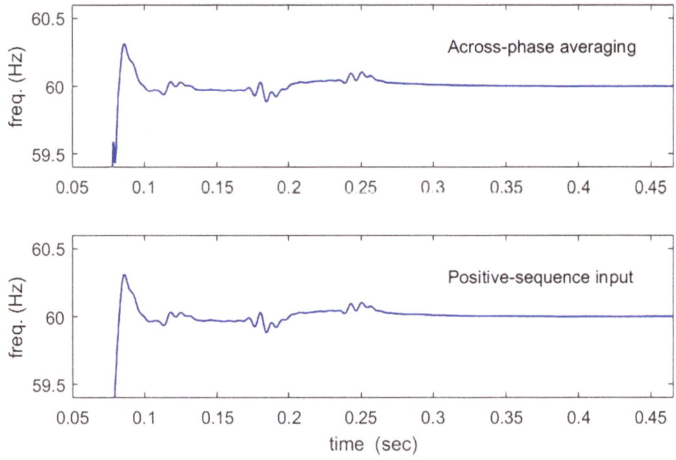

Fig. 3.28 Frequency estimates from a three-phase waveform (transient Hydro-Ottawa example), using the 2-sample sub-cycle approach, with 4-cycle uniform smoothing: (**a**) across-phase averaging (top) and (**b**) using positive-sequence input (bottom).

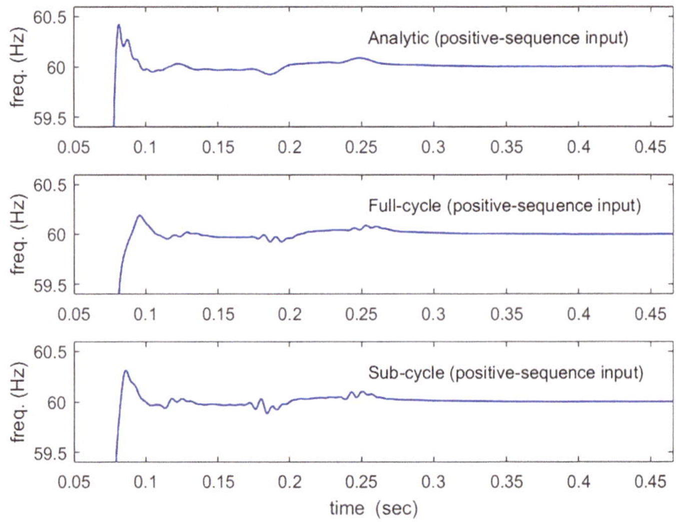

Fig. 3.29 Frequency estimates from a three-phase waveform (transient Hydro-Ottawa example), with 4-cycle uniform averaging and a positive-sequence input: (**a**) analytic signal approach (top), (**b**) full-cycle approach (middle), and (**c**) 2-sample sub-cycle approach (bottom).

demonstrated in Fig. 3.29, three-phase schemes (whether analytic signal, full-cycle, or sub-cycle) generate a more reliable frequency estimate than their single-phase counterparts, whose performance is illustrated in Fig. 3.26.

Since single-input/single-output linear time-invariant systems can be ordered at will, we can re-order the front-end operations in the diagram of Fig. 3.22 and apply the "three-phase to positive-sequence" map directly to the acquired three-phase waveform $x(t)$. The resulting (reduced) computational cost of a frequency estimation scheme that uses uniform averaging and a positive-sequence input is $\mathcal{O}(4N_B + 12\log_2 N_x)$ real multiplications per waveform sample for the analytic signal approach, $\mathcal{O}(4N_B + 4N)$ for the full-cycle approach, and $\mathcal{O}(4N_B)$ for the 2-sample sub-cycle approach. For instance, in the steady-state Hydro-Ottawa example, the full cost of a re-ordered three-phase 2-sample sub-cycle algorithm is only 39 real multiplications per waveform sample, as compared with 551 real multiplications required by its full-cycle counterpart, and 173 real multiplications for the (non-causal) analytic signal version.

We observe that the 2-sample sub-cycle three-phase frequency estimator is the most cost-efficient among the three approaches: it has by far the lowest computational cost, while providing essentially the same estimation quality as the full-cycle or analytic signal approaches (recall Fig. 3.29). We present here a complete listing of this algorithm, which highlights the relative simplicity of this approach, viz.,

- **Construct a 2-sample sub-cycle positive-sequence phasor,** i.e.,

$$\widehat{X}_{1,+}(t) = \underbrace{\left[x(t) - j\,x(t - \tfrac{T}{4})\right]}_{\text{three-phase sub-cycle phasor}} \begin{pmatrix} 1 \\ \alpha \\ \alpha^2 \end{pmatrix}$$

  where $\alpha = e^{j2\pi/3}$, according to (1.13c) and (2.18a), and we omitted the scaling factor $1/\sqrt{3}$, since we need only the phase of this complex phasor. This means we can evaluate first the scalar (= single-phase) waveform

$$\widetilde{x}(t) \overset{\text{def}}{=} x(t) \begin{pmatrix} 1 \\ \alpha \\ \alpha^2 \end{pmatrix} \tag{3.48a}$$

  and then construct the corresponding scalar 2-sample sub-cycle phasor

$$\widehat{X}_{1,+}(t) = \widetilde{x}(t) + j\,\widetilde{x}(t - \tfrac{T}{4}) \tag{3.48b}$$

  Interestingly, $\widetilde{x}(\cdot)$ is also a scaled version of the so-called *space vector,* a quantity most often used in analysis of polyphase electric machines (see Sect. 6.1.2).
- **Apply a low-pass filter** to generate a single-component version $\left[\widehat{X}_{1,+}(t)\right]_\star$ with attenuated high-frequency content, viz.,

$$\widehat{X}_{1,+}(t) \xrightarrow{\text{low-pass filter}} \left[\widehat{X}_{1,+}(t)\right]_\star \tag{3.48c}$$

- **Generate the corresponding phase waveform,** i.e.,

$$\phi_{\star}(t) \;=\; \arg\left\{\left[\widehat{X}_{1,+}(t)\right]_{\star}\right\} \tag{3.48d}$$

- **Combine differencing with uniform averaging,** according to (3.41)-(3.42), to obtain a frequency estimate $\widehat{f_0}(t)$. This results in the combined expression

$$\widehat{f_0}(nT_s) \;=\; \frac{F_s}{2\pi L}\left[\phi_{\star}(nT_s) - \phi_{\star}(nT_s - LT_s)\right] \tag{3.48e}$$

Notice that the overall computational cost of this scheme, i.e., $4N_B + 7 = \mathcal{O}(4N_B)$ real multiplications, is only a factor of two higher than that of a single-phase version of the same scheme (recall Table 3.2).

3.4.6 Least-Squares Phasor Estimation from Analytic Signal Input

We have observed in Sect. 3.4.3 that using the analytic version of an acquired waveform as input to a subband-based time–frequency representation (TFR) technique results in complete elimination of undesired leakage from negative frequency components. A similar observation applies to least-squares estimation of dynamic phasors.

We presented in Sect. 3.2 a comprehensive conceptual framework for linear least-squares estimation of dynamic phasors from a waveform of interest $x(t)$: the resulting polyphase linear least-squares (PLLS) phasor estimates were given by the vector–matrix expression (3.18a). In order to generate modified phasor estimates $\widehat{X}_k(t)$ that rely only on the positive frequency content of the waveform of interest, we need to replace the waveform $x(\cdot)$ in this expression by its *scaled* analytic version $\frac{1}{2}\widehat{x}(\cdot)$, namely (assuming that $M \geq 2L$),

$$\begin{pmatrix} \widehat{X}_{k_1}(t)\, e^{jk_1\omega_0 t} \\ \widehat{X}_{k_2}(t)\, e^{jk_2\omega_0 t} \\ \vdots \\ \widehat{X}_{k_L}(t)\, e^{jk_L\omega_0 t} \end{pmatrix} = \sqrt{\frac{2}{M}}\left[\mathbf{I}_L \;\; \mathbf{0}_L\right]\left(\mathcal{W}^H \mathcal{W}\right)^{-1}\mathcal{W}^H \begin{pmatrix} \frac{1}{2}\widehat{x}(t) \\ \frac{1}{2}\widehat{x}(t-\Delta_1) \\ \vdots \\ \frac{1}{2}\widehat{x}(t-\Delta_{M-1}) \end{pmatrix} \tag{3.49}$$

We will refer to this expression as a "PLLS with analytic input" phasor estimate.

In particular, when the waveform $x(\cdot)$ is periodic, i.e.,

$$x(t) \;=\; \sum_{\ell=1}^{\infty} \sqrt{2}\,\Re\{X_\ell\, e^{j\ell\omega_0 t}\}$$

then

$$\frac{1}{2}\,\widehat{x}(t) \;=\; \sum_{\ell=1}^{\infty} \frac{\sqrt{2}}{2}\; X_\ell\; e^{j\ell\omega_0 t}$$

and the resulting phasor estimates are (compare with (3.20a))

$$\widehat{X}_k(t) \;=\; X_k \;+\; \frac{\sqrt{2}}{2} \sum_{\ell \notin \mathcal{S}} \widehat{H}_k(j\ell\,\omega_0)\; X_\ell\, e^{j(\ell-k)\,\omega_0 t} \quad , \qquad k \in \mathcal{S}$$

The corresponding cycle-averaged error is (compare with (3.20b))

$$\frac{1}{T}\int_0^T \| \,\widehat{X}_k(t) - X_k\, \|^2 dt \;=\; \frac{1}{2} \sum_{\ell \notin \mathcal{S}} \| \,\widehat{H}_k(j\ell\,\omega_0)\, \|^2 \; \|X_\ell\|^2 \tag{3.50}$$

which is, clearly, smaller than the one in (3.20b).

An alternative approach is to change the formulation of the PLLS problem (3.16) itself, replacing *both* $x(\cdot)$ *and* $\widehat{x}(\cdot)$ by their analytic versions, as done for instance in [108]. This produces a different expression for the estimated phasors, viz.,

$$\begin{pmatrix} \widehat{X}_{k_1}(t)\, e^{jk_1\omega_0 t} \\ \widehat{X}_{k_2}(t)\, e^{jk_2\omega_0 t} \\ \vdots \\ \widehat{X}_{k_L}(t)\, e^{jk_L\omega_0 t} \end{pmatrix} \;=\; \frac{1}{\sqrt{2M}} \left(\boldsymbol{\Psi}^H\, \boldsymbol{\Psi}\right)^{-1} \boldsymbol{\Psi}^H \begin{pmatrix} \widehat{x}(t) \\ \widehat{x}(t-\Delta_1) \\ \vdots \\ \widehat{x}(t-\Delta_{M-1}) \end{pmatrix} \tag{3.51}$$

where $\boldsymbol{\Psi} \overset{\text{def}}{=} \left[\, \psi_{k_1} \;\; \psi_{k_2} \;\; \ldots \;\; \psi_{k_L} \,\right]$, exactly as defined in (3.17). However, in this case a necessary condition for the existence of a unique solution is $M \geq L$ (rather than $M \geq 2L$), so that the highest number of phasors we can estimate (denoted by L in (3.51)) is the same as the number of waveform samples used to form the estimate (denoted by M in (3.51)). In other words, this "fully-analytic PLLS" technique makes it possible to estimate twice as many phasors as the "PLLS with analytic input" version (3.49) or the original version that was presented in Sect. 3.2.

The level of estimation error of both techniques depends, of course, on the choices made for $\mathcal{S}$, M, and the shifts $\{\Delta_i\}$. For instance, when $\Delta_i = i\,\xi T$ and the input waveform is periodic, the "fully-analytic PLLS" estimate (3.51), constructed with $M = 2$, $\xi = 1/4$, and $\mathcal{S} = \{1, 3\}$, exhibits the same level of estimation error as the one shown in (3.50) for the "PLLS with analytic input" estimate (3.49), constructed with $M = 2$, $\xi = 1/4$, and $\mathcal{S} = \{1\}$.

We should point out, however, that many of the special properties of sub-cycle and full-cycle metrics, such as the Parseval property, are lost or modified when we rely on an analytic version of the input waveform. In particular, the connection between resulting voltage and current dynamic phasor estimates and some form of averaged instantaneous power is completely lost. This is potentially a serious drawback in a variety of applications ranging from characterization of transients to multi-scale system modeling [109, 110]. In particular, the power flow metrics we examine in Chaps. 8–9 rely heavily on the Parseval property of full-cycle and sub-cycle dynamic phasors.

Chapter 4 Beyond Phasors: Constant-Bandwidth Transforms

Full cycle dynamic phasors provide information about time variation in the harmonic content of any power system waveform $x(\cdot)$, as demonstrated in Chaps. 2 and 3. Assembling this information into a *spectrogram* generates a *time–frequency representation* (TFR) of the waveform $x(\cdot)$. For instance, the spectrogram constructed from full-cycle dynamic phasors of the transient Hydro-Ottawa dataset[1] exhibits significant power ridges at the 1st, 3rd, 5th, 7th, 9th, and 11th harmonics, which are all disrupted by a short transient in the $100–180\,\text{msec}$ interval (Fig. 4.1). This spectrogram also reveals an abrupt increase in the magnitude of even harmonics within the same interval.

The full-cycle waveform-to-phasor map $x(\cdot) \longrightarrow \{X_k(\cdot)\,;\ 0 \le k < \infty\}$, which was analyzed in Sect. 2.4, can be interpreted as a linear time-invariant filtering operation that generates the passband phasor $\mathcal{X}_k(t) = X_k(t)\, e^{jk\omega_0 t}$ by passing the waveform $x(\cdot)$ through a band-pass filter $H_k(s)$ (recall Fig. 2.1). A more detailed time–frequency representation can be obtained by using the well-known *short-time Fourier transform* (STFT), which extends the definition (2.23) of full-cycle dynamic phasors. The STFT produces a generalized dynamic phasor, viz.,

$$X(t\,;\,f_c)\, e^{j2\pi f_c t} \;=\; \underbrace{\left[\, h(t)\, e^{j2\pi f_c t} \,\right]}_{h(t\,;\,f_c)} \;\circledast\; x(t) \tag{4.1}$$

where $\circledast$ denotes linear convolution, and the frequency f_c can be chosen at will (see Sect. 4.1 for a detailed discussion). Selection of the *window function* $h(t)$ determines the level of detail (a.k.a. "time–frequency resolution") provided by the spectrogram $|X(t\,;\,f_c)|^2$. In particular, using a single-cycle long *rectangular window function* produces the standard (full-cycle) dynamic phasors $X_k(t)$,

[1] See Sect. L.2.3 of Appendix L for details about this example.

H. Lev-Ari, A. M. Stanković, *Dynamic Phasors in Energy Processing Systems*,
Power Electronics and Power Systems, https://doi.org/10.1007/978-3-031-50020-6_5

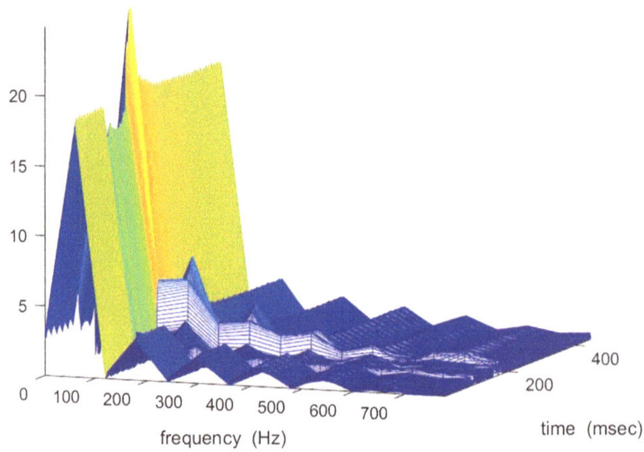

Fig. 4.1 Time–frequency representation of the transient Hydro-Ottawa dataset constructed from full-cycle dynamic phasors ($f_0 = 60\,\text{Hz}$).

i.e., in this case $X(t\,;\,f_c)\Big|_{f_c=kf_0} = \frac{1}{a_k}X_k(t)$, where a_k is a coefficient defined in (2.11). Digital evaluation of the STFT is examined in Sect. 4.2: it requires discretization of both the time variable "t" and the frequency variable "f_c," resulting in a *discretized STFT* (D-STFT), which can be efficiently implemented as a filter-bank, as detailed in Sect. 4.3.

The effective length of the window function $h(t)$ controls a trade-off between time-resolution and frequency-resolution: increasing window length improves frequency-resolution, while impairing time-resolution. Single-cycle long window functions are often used for accurate representation of harmonic frequency content, while multi-cycle windows are useful in representing sub-harmonic content. In contrast, accurate detection of transient onset and estimation of its duration typically require a short (i.e., sub-cycle long) window. In addition, the impact of spectral leakage can be minimized by employing classical window functions, such as Hamming, Kaiser, or Dolph-Chebyshev. Explicit guidelines for selecting window type and length are provided in Sect. 4.5.

IEEE Standard 1159-2019 provides a detailed taxonomy of transient types and a discussion of their causes [111]. The real-life transient examples that we use in this chapter, as well as in the rest of this book, include two frequently-encountered transient categories: *short rms variations* and *low-frequency oscillatory transients,* which we have labeled in the preceding chapters as "in-band transients" and "wide-band transients," respectively. Additional details about our examples can be found in Sect. L.2 of Appendix L.

It is essential at this point to draw a clear distinction between two very different categories of STFT applications: (a) those that involve only analysis and/or feature extraction, which rely solely on a time–frequency representation of the waveform of interest, and (b) applications that require *both analysis and synthesis*, allowing for perfect reconstruction of the original waveform from its time–frequency representation, which relies on invertibility of the STFT.

The first category of applications is very diverse, including spectrum display, parameter/feature extraction, detection, and pattern recognition (e.g., fault/event detection and classification in power systems). Such applications are very common in audio/speech analysis, biomedical signal processing, geophysics/oceanography, radar/sonar imaging, and quantum physics. Power system applications in this category include monitoring and analytics of power quality [112, Chap. 8], [113], protection [42, Chap. 9], [114], and event detection and classification [115–118].

The second category of applications is usually related to power system modeling, estimation, and control. Some modeling examples from this book include multi-frequency modeling of power electronic converters in Sects. 6.1 and 6.2. In our experience, the STFT approach is very effective in estimation [69] as well as control applications ranging from active filters [119] and distribution static synchronous compensators (D-Statcom) [120] to uninterruptible power supplies (UPS) [121]. The STFT approach is also well suited for tasks such as synchronization and control under the unbalanced and harmonic-rich operation of ac networks [122].

We now recall that perfect reconstruction of a waveform from its rms dynamic phasors via the Fourier-series representation (2.2d) relies on mutual orthogonality of the functions $\{e^{jk\omega_0 t}\ ;\ -\infty < k < \infty\}$ over the sliding interval $[t - T \quad t]$. Unfortunately, the continuous-parameter family $\{h(t)\, e^{j2\pi f_c t}\ ;\ -\infty < f_c < \infty\}$, which the STFT uses as a generalization of harmonically-related sinusoids, is not mutually orthogonal. Since this function set is not even linearly-independent, it cannot serve as a basis for the time-variant Hilbert space of finite-rms waveforms on the interval $[t - T \quad t]$. Thus, function space analysis of the STFT map (4.1) must rely on the *theory of frames,* which provides a conceptual foundation for waveform representation in terms of a *linearly-dependent* set of functions [141, 142]. A frame-based inversion formula for the STFT is provided in Sect. 4.1.5: a brief introduction to frame theory can be found in Sect. 4.5.2. Nevertheless, the orthogonal waveform decomposition approach is still the method of choice for construction of invertible "constant-Q" transforms, such as the discrete wavelet transform (DWT) and the discrete orthogonal S-transform (DOST): see Sect. 5.6 for further details.

The need for a frame-theoretic analysis of STFT inversion can be avoided by relying on the theory of *perfect reconstruction (PR) filter-banks:* it provides a more structured approach to construction of a digitally implemented inverse STFT, which we examine in Sect. 4.4. This theory interprets the STFT and its inverse in terms of a pair of non-decimated filter-banks: a single-input/multiple-output *analysis filter-bank* (AFB), which maps the sampled input waveform into its discrete-time subband components, and a multiple-input/single-output *synthesis filter-bank* (SFB), which is used to reconstruct the sampled input waveform from its subband components. In addition, the theory of PR filter-banks establishes, for a given AFB, an explicit characterization of all possible valid choices of the SFB.

The perfect reconstruction filter-bank approach also facilitates mapping desired properties of the STFT and its inverse into explicit design guidelines:

- **Frequency localization:** is achieved by splitting the effective frequency range of the input waveform into moderately overlapping subbands, one for each subband component of the input waveform.
- **Time-localization:** is facilitated by using AFB subband filters with relatively short group-delay, and avoiding down-sampling of the subband components.
- **Online (real-time) processing:** is made possible by avoiding down-sampling, and constraining both the AFB and the SFB to be causal. Causality of the AFB is needed in protection and online monitoring applications. Causality of both the AFB and SFB is essential in closed-loop system control applications.
- **Closed-loop processing:** requires manageable end-to-end (i.e., AFB+SFB) delay, in addition to relying on non-decimated processing. We present in Sect. 4.4 a special SFB choice, known as a *phasor-bank*, which can provide zero end-to-end delay.

A detailed discussion of these design considerations is provided in Sect. 4.5.

In addition to the time-domain expression (4.1), which is the starting point for *online implementation* of the STFT, we can also express the same map in terms of frequency-domain entities. We present in Sect. 4.1.3 an explicit expression for the STFT that involves only the frequency response of the window function $h(t)$ and the Fourier transform of the entire input waveform. Efficient digital *offline implementation* of the STFT is discussed in Sect. 4.6: it is obtained by discretizing the time and frequency variables, and it involves the FFT of the entire available sampled waveform record.

4.1 Short-time Fourier Transform (STFT)

The waveform-to-phasor map (2.2c) is a special case of the well-known *short-time Fourier transform* (STFT), which provides information about the time-varying frequency content of a given waveform $x(\cdot)$ within a specified (relatively narrow) frequency band. The STFT generates a *time–frequency representation* (TFR) of the waveform $x(\cdot)$, viz.,

$$X(t\,;\,f_c) \overset{\text{def}}{=} \int_{-\infty}^{\infty} h(t-\tau)\left[x(\tau)\,e^{-j2\pi f_c \tau}\right]d\tau \tag{4.2a}$$

where $h(\cdot)$ is a real-valued non-negative function, known as a *window*. A detailed presentation of the STFT can be found in [32, 33, 106]. We restrict most of our discussion in this chapter to *real-valued waveforms* $x(\cdot)$: this is useful, in particular, in power system applications, where the waveforms of interest represent physical quantities, such as voltage or current. Consequently, the STFT of a real-valued waveform $x(\cdot)$ is conjugate symmetric in the frequency variable f_c, namely,

$$X(t\,;\,-f_c) \;=\; X^*(t\,;\,f_c) \tag{4.2b}$$

This observation allows us to focus our discussion of TFRs on the frequency range $f_c \geq 0$. In addition to the time-domain definition (4.2a), the STFT can also be expressed in the frequency domain (see (4.12) below) in terms of the transfer function

$$H(s) \;=\; \int_{-\infty}^{\infty} h(t)\, e^{-st}\, dt \tag{4.2c}$$

associated with the window $h(t)$. In general, the frequency response of $H(s)$ may depend on the parameter f_c: this flexibility is used in Sect. 5.5 to introduce the S-transform, and in Sect. 5.2 to introduce so-called constant-Q transforms.

The STFT expression (4.2a) was introduced in the early 1960s by researchers from Bell Labs. It was used in [123] to define a short-time power spectrum, known as a *spectrogram.* A filtering interpretation of the STFT (see below) was used in [124, 125] to construct an efficient speech coding system, known as a vocoder, using a fixed window of finite length. Allowing explicit dependence of the window $h(t)$ on the modulating frequency f_c results in a generalization of the STFT, which was first suggested in [126] and analyzed in detail in [127]. A detailed discussion of such generalized transforms is provided in Chap. 5.

The purpose of the (relatively short) window $\Big\{h(t-\tau)\,;\;-\infty < \tau < \infty\Big\}$ is to suppress the waveform $x(\cdot)$ outside of a time interval around the time-instant "t," so that $X(t\,;\,f_c)$ can be interpreted as a "local spectrum" in the vicinity of this time-instant. Indeed, the TFR $X(t\,;\,f_c)$ is the Fourier transform of the windowed waveform $\Big\{h(t-\tau)\,x(\tau)\,;\;-\infty < \tau < \infty\Big\}$. For instance, the waveform-to-phasor map (2.2c) that was used to define rms dynamic phasors in Sect. 2.1 is obtained by making the special (one cycle long, rectangular, causal) window choice

$$h(t) = \begin{cases} \dfrac{\sqrt{2}}{T} & 0 \leq t \leq T \\[2ex] 0 & \text{else} \end{cases}$$

and setting $f_c = k f_0$ in (4.2a), where $f_0 = \omega_0/2\pi$ is the fundamental steady-state frequency, and $T = 1/f_0$. Most practical windows have finite length, which we will denote by T_h. For infinite-length windows, we will *define* $T_h \overset{\text{def}}{=} \Delta_t^{(PD\star 40)}$, where $\Delta_t^{(PD\star A)}$ denotes the effective (A dB) time-duration of the window function $h(t)$, as detailed in Sect. B.4.

The effective length of the window $h(t)$ controls the *time-resolution* of the TFR $X(t\,;\,f_c)$. Similarly, the *frequency-resolution* of the TFR $X(t\,;\,f_c)$ is determined by the effective bandwidth of the transfer function $H(s)$ that is associated with the window $h(t)$. A detailed overview of several window-length and window-bandwidth metrics is provided in Appendix B. The relation of these notions to time- and frequency-resolution is discussed in Sects.B.5–B.7. In particular, we assert in Sect. B.5 that a meaningful metric of frequency-resolution is provided by $\Delta_f^{(PD\star 10)}$, the 10 dB bandwidth of $H(j2\pi\nu)$. Similarly, time-resolution can be quantified by evaluating $\Delta_t^{(PD\star 20)}$, the effective (20 dB) time-duration of the window function $h(t)$, as detailed in Sect.B.7.

The window function can be chosen to have a short effective duration as well as small effective bandwidth, so that it provides good resolution in both time and frequency. The trade-off between window duration and bandwidth is similar for all (single parameter) window families and is always described by "constant product"

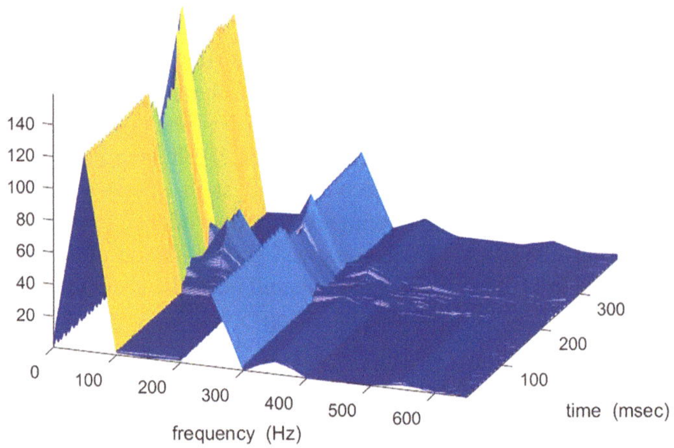

Fig. 4.2 Time–frequency representation of the paper-mill dataset, obtained with a single-cycle-long ($T = 20$ msec) rectangular window.

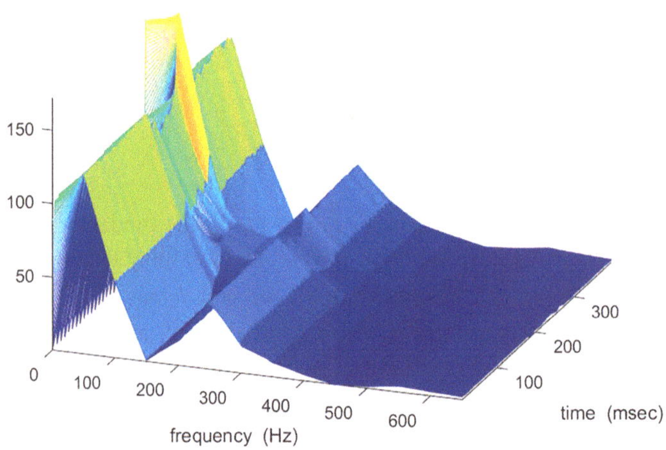

Fig. 4.3 Time–frequency representation of the paper-mill dataset, obtained with a single-cycle-long ($T = 20$ msec) Hamming window.

property, as detailed in Appendix B. The value of this constant depends only on (a) the metric(s) chosen to define the effective bandwidth and time-duration, and (b) the type of window.

Figures 4.2, 4.3, 4.4, and 4.5 illustrate some of the effects associated with a particular window choice. Windows such as Hamming, Hann, or Kaiser are used to reduce the impact of sidelobe leakage (which can be seen, for instance, in Fig. 4.4). However, such windows have a somewhat poorer frequency-resolution, as compared with the rectangular window, as is evident from Figs. 4.4–4.5. Section 4.5 provides a detailed set of window-design guidelines for the STFT.

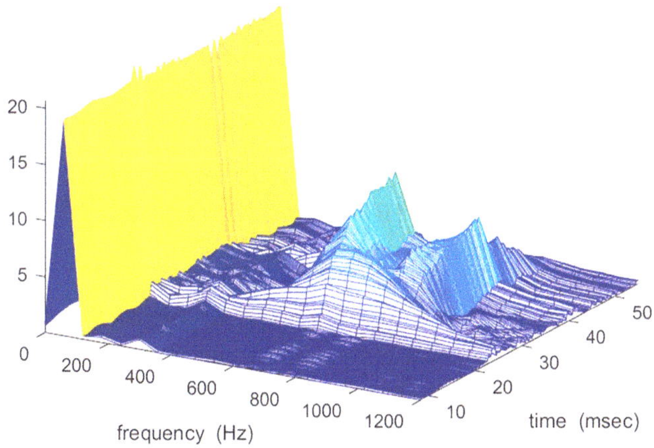

Fig. 4.4 Time–frequency representation of the wind-farm dataset, obtained with a single-cycle-long ($T = 16.7$ msec) rectangular window.

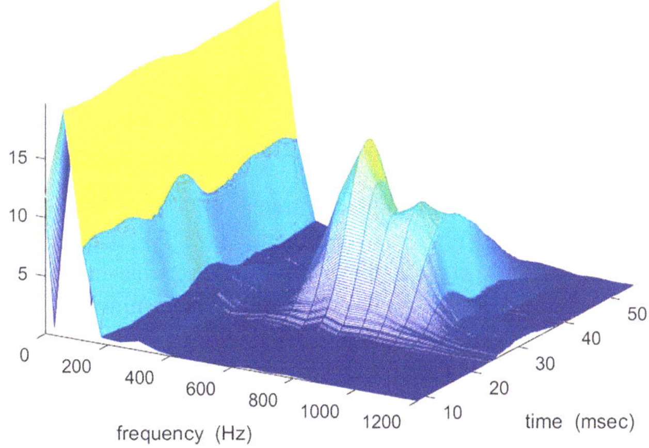

Fig. 4.5 Time–frequency representation of the wind-farm dataset, obtained with a single-cycle-long ($T = 16.7$ msec) Hamming window.

4.1.1 Dynamics of the TFR

The dynamic behavior of the TFR $X(t\,;\,f_c)$ is qualitatively similar to that of the dynamic phasors $X_k(t)$. In particular, an "STFT-version" of the differential equation (2.12) is easily obtained from (4.5), viz.,

$$\begin{aligned}\frac{d}{dt}\,X(t\,;\,f_c) &= \int_{-\infty}^{\infty} h(\tau)\,\frac{d}{dt}\left\{x(t-\tau)\,e^{-j2\pi f_c(t-\tau)}\right\}\,d\tau \\ &= \int_{-\infty}^{\infty} h(\tau)\,\dot{x}(t-\tau)\,e^{-j2\pi f_c(t-\tau)}\,d\tau \\ &\quad + \int_{-\infty}^{\infty} h(\tau)\,x(t-\tau)\left(-\,j2\pi f_c\right)e^{-j2\pi f_c(t-\tau)}\,d\tau\end{aligned}$$

so that

$$\frac{d}{dt}\,X(t\,;\,f_c) = \mathsf{TFR}\left\{\frac{d}{dt}\,x(t)\right\} - j2\pi f_c\,X(t\,;\,f_c) \tag{4.3}$$

where $\mathsf{TFR}\left\{\frac{d}{dt}\,x(t)\right\}$ is the short-time Fourier transform (STFT) of $\frac{d}{dt}\,x(t)$.

In addition, if $x(t)$ is described by a linear time-invariant state-space model, i.e., $\frac{d}{dt}\,x(t) = A\,x(t) + B\,u(t)$, then (compare with (2.22b))

$$\frac{d}{dt}\,X(t\,;\,f_c) = \left(A - j2\pi f_c\right)X(t\,;\,f_c) + B\,U(t\,;\,f_c) \tag{4.4}$$

where $U(t\,;\,f_c) = \mathsf{TFR}\left\{u(t)\right\}$ is the STFT of the input signal $u(t)$, and we used the linearity property of this transform.

4.1.2 Filtering Interpretation

We can interpret the constant-bandwidth transform (4.2a) as a combination of demodulation and filtering (Fig. 4.6): the output $X(t\,;\,f_c)$ captures the frequency content of the waveform $x(\cdot)$ in the vicinity of the time-instant t, within a frequency band centered at f_c, whose width is determined by the bandwidth of the transfer function $H(s)$. Some authors refer to $X(t\,;\,f_c)$ as a "time-varying spectrum" or a "local spectrum" (see, e.g., [128]). Additional insight into the meaning of such terminology is provided in Sect. 5.7, where we relate the squared magnitude $|X(t\,;\,f_c)|^2$ to the *Wigner–Ville distribution* (WVD) of the waveform $x(\cdot)$.

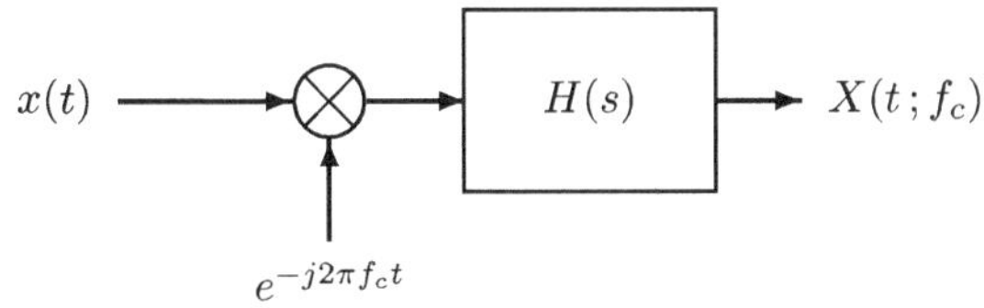

Fig. 4.6 Frequency-selective filtering interpretation of the STFT.

This interpretation can be made even more transparent by introducing a change of the integration variable in (4.2a), which allows us to rewrite this expression in the form

$$X(t\,;\,f_c) \;=\; \int_{-\infty}^{\infty} h(\tau)\,x(t-\tau)\,e^{-j2\pi f_c(t-\tau)}\,d\tau$$

or, equivalently,

$$X(t\,;\,f_c) \;=\; e^{-j2\pi f_c t}\int_{-\infty}^{\infty} \underbrace{\left[h(\tau)\,e^{j2\pi f_c \tau}\right]}_{h(t\,;\,f_c)}\,x(t-\tau)\,d\tau \tag{4.5}$$

The corresponding *band-pass* filtering interpretation is shown in Fig. 4.7. The band-pass filter $H\big(s-j2\pi f_c\big)=\int_{-\infty}^{\infty}\big[h(t)\,e^{j2\pi f_c t}\big]\,e^{-st}\,dt$ passes through frequencies in a band centered at f_c. The output of this band-pass filter is a "passband version" of the STFT, namely

$$\mathcal{X}(t\,;\,f_c) \;\stackrel{\text{def}}{=}\; X(t\,;\,f_c)\,e^{j2\pi f_c t} \;=\; h(t\,;\,f_c) \circledast x(t) \tag{4.6}$$

This output is then shifted to baseband as a result of multiplication by $e^{-j2\pi f_c t}$ (a.k.a. "demodulation" [32, 33]), which generates the STFT $X(t\,;\,f_c)$. Notice that $|\mathcal{X}(t\,;\,f_c)| = |X(t\,;\,f_c)|$.

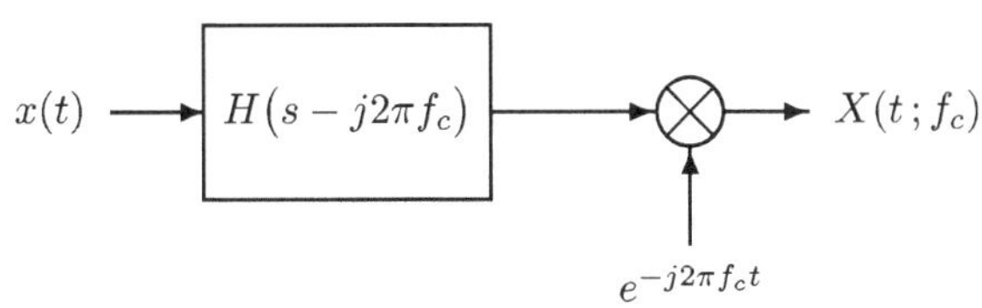

Fig. 4.7 Band-pass filtering interpretation of the STFT.

In order to ensure the validity of this frequency-selective filtering interpretation of the TFR $X(t\,;\,f_c)$ we need to impose the two following constraints on the transfer function $H(s)$:

- $H(s)$ is the transfer function of a low-pass filter: this guarantees that $X(t\,;\,f_c)$ represents the instantaneous frequency content of a segment of the waveform $x(\cdot)$ within a (small) contiguous frequency band that is centered at f_c.
- The frequency response of $H(s)$ has unity DC gain, viz.

$$H(0) \;=\; 1 \;=\; \int_{-\infty}^{\infty} h(t)\,dt \tag{4.7}$$

which guarantees that the magnitude of $X(t\,;\,f_c)$ correctly reflects the strength of the f_c-frequency component of $x(t)$.

The filtering interpretation of Figs. 4.6–4.7 provides a convenient starting point for constructing numerical implementations of the fixed-window STFT (see Sect. 4.2 for further details).

4.1.3 Relation to the Fourier Transform

One useful consequence of the "unity DC-gain" property (4.7) is a simple relation between the STFT and the continuous-time Fourier transform of the waveform $x(\cdot)$. Integrating (4.2a) with respect to the time variable "t" produces the identity

$$\int_{-\infty}^{\infty} X(t\,;\,f_c)\,dt \;=\; \int_{-\infty}^{\infty} x(\tau)\,e^{-j2\pi f_c\tau}\,d\tau \tag{4.8}$$

where we recognize the right-hand side expression as the Fourier transform of the waveform $x(\cdot)$, evaluated at the frequency f_c. We will use a convenient shorthand notation for the continuous-time Fourier transform, namely,

$$\mathcal{F}_{\omega}\Big\{x(\cdot)\Big\} \;\stackrel{\text{def}}{=}\; \int_{-\infty}^{\infty} x(t)\,e^{-j\omega t}\,dt \;\equiv\; X(j\omega) \tag{4.9}$$

so that (4.8) can be written compactly as

$$\int_{-\infty}^{\infty} X(t\,;\,f_c)\,dt \;=\; X(j2\pi f_c) \tag{4.10}$$

Since practically every waveform can be recovered from its Fourier transform, we also obtain an *inverse STFT* relation, viz.,

$$x(t) \;=\; \int_{-\infty}^{\infty} e^{j2\pi f_c t}\left[\int_{-\infty}^{\infty} X(\tau\,;\,f_c)\,d\tau\right] df_c \tag{4.11}$$

which recovers the waveform $x(t)$ from its TFR $X(t\,;\,f_c)$. Notice that a numerical implementation of this expression involves only a summation along the time dimension, followed by an inverse FFT.

4.1.4 Frequency-Domain Expression of the STFT

An equivalent *frequency-domain* expression for the STFT is obtained by mapping the baseband filtering interpretation of Fig. 4.6 into the frequency domain,[2] viz.,

$$\mathcal{F}_\omega\Big\{X(t\,;\,f_c)\Big\} \;=\; H(j\omega)\,\mathcal{F}_\omega\Big\{x(t)e^{-j2\pi f_c t}\Big\} \;=\; H(j\omega)\,X\big(j(\omega+2\pi f_c)\big)$$

Notice that here the Fourier transform $\mathcal{F}_\omega$ applies to the time variable t, mapping it into the frequency variable ω. Next, applying an inverse Fourier transform to this identity generates the explicit frequency-domain representation

$$X(t\,;\,f_c) \;=\; \int_{-\infty}^{\infty} H(j\omega)\,X\big(j(\omega+2\pi f_c)\big)\,e^{j\omega t}\,\frac{d\omega}{2\pi} \qquad (4.12)$$

Similarly, mapping the band-pass filtering interpretation of Fig. 4.7 into the frequency domain generates the relation

$$\mathcal{F}_\omega\Big\{\mathcal{X}(t\,;\,f_c)\Big\} \;=\; H\big(j(\omega-2\pi f_c)\big)\,X(j\omega)$$

so that the passband-TFR $\mathcal{X}(t\,;\,f_c)$ can be determined via the inverse Fourier transform, viz.,

$$\mathcal{X}(t\,;\,f_c) \;=\; \int_{-\infty}^{\infty} H\big(j(\omega-2\pi f_c)\big)\,X(j\omega)\,e^{j\omega t}\,\frac{d\omega}{2\pi} \qquad (4.13)$$

These frequency-domain expressions can only be used in offline implementations, since both require advance evaluation of the Fourier transform $X\big(j\omega\big)$ from the *entire waveform record.* Both (4.12) and (4.13) offer a convenient starting point for constructing offline numerical implementations of the STFT (see Sect. 4.6 for further details).

4.1.5 Inverting the STFT

The inverse (linear) transform from the TFR $X(t\,;\,f_c)$ back to the original waveform $x(t)$ is highly non-unique. A number of such *inverse STFT* formulas are

[2] Notice that setting $\omega = 0$ in this frequency-domain expression produces the identity (4.10).

listed in the literature. Perhaps the best known one [106, 128] involves a *synthesis window* $g(t)$, viz.,

$$x(t) = \int_{-\infty}^{\infty} e^{j2\pi f_c t} \left[\int_{-\infty}^{\infty} g(\tau - t)\, X(\tau\,;\, f_c)\, d\tau \right] df_c \tag{4.14a}$$

where $g(t)$ must satisfy the constraint

$$\int_{-\infty}^{\infty} g(\tau)\, h(\tau)\, d\tau = 1 \tag{4.14b}$$

For real-valued waveforms, this expression can be reduced to

$$x(t) = 2\Re \left\{ \int_{0}^{\infty} e^{j2\pi f_c t} \left[\int_{-\infty}^{\infty} g(\tau - t)\, X(\tau\,;\, f_c)\, d\tau \right] df_c \right\}$$

where we used the conjugate symmetry property of the TFR.

Notice that the inverse transform expression (4.11) corresponds to setting $g(t) = 1$ in (4.14), and relying on the unity DC-gain property $\int_{-\infty}^{\infty} h(\tau)\, d\tau = 1$ of the *analysis window* $h(t)$. Another simple choice that satisfies this constraint is obtained by choosing an analysis window with the property $h(0) = 1$, and setting $g(t) = \delta(t)$, which results in the inverse transform expression

$$x(t) = \int_{-\infty}^{\infty} e^{j2\pi f_c t}\, X(t\,;\, f_c)\, df_c$$

Alternative STFT inversion formulas have been introduced in [127, 129, 130].

The non-uniqueness of the inverse transform is due to the high level of redundancy in the STFT, as already pointed out in the introduction to this chapter. This redundancy can be somewhat reduced by considering only discrete values for the time and frequency variables of the TFR $X(t\,;\, f_c)$: this leads to the well-researched *Gabor Transform* [33, 106] (see also the discussion in Sect. 4.4). Similarly, the Fourier-series representation (2.2d) of an arbitrary finite-rms waveform $x(t)$ in terms of its rms dynamic phasors, i.e.,

$$x(t) = X_0(t) + \sqrt{2} \sum_{k=1}^{\infty} \Re \left\{ X_k(t)\, e^{jk\omega_0 t} \right\}$$

corresponds (up to scaling by $\sqrt{2}$) to a frequency- discretized version of the TFR, with $f_c = k f_0$, where $f_0 = 1/T \equiv \omega_0/2\pi$, and T is the duration of the averaging window, typically chosen as the length of one cycle in steady-state operation of a power system.

Continuous-time/continuous-frequency representations, such as the STFT, are not very useful for computational implementation. Instead, we opt to reformulate the problem of efficient inversion of the STFT via a discrete-time/discrete-frequency approach (see Sect. 4.4). To be specific, we use the theory of non-decimated perfect reconstruction (PR) filter-banks [131] to provide an exact (i.e., error-free) reconstruction of any discrete-time waveform from its discretized TFR.

4.2 Discrete Online Implementation: the D-STFT

Digital processing implementation of the STFT requires discretization of both the frequency variable "f_c" and the time variable "t." The resulting transform, which we name the D-STFT, can be interpreted as a discrete-time single-input/multiple-output filter-bank, namely an *analysis filter-bank* (AFB) [133]. Each channel of this filter-bank corresponds to a single subband, whose center frequency is a discretized value of f_c. In contrast to the common practice of down-sampling (a.k.a. decimation) of each subband component [131, 133], we opt here to maintain the highest available time-grid density in each subband, in order to achieve the best time-resolution and processing speed, as well as a fast response to onset of transients. A detailed analysis of the non-decimated AFB is provided in Sect. 4.3.

Discretizing Frequency

When used in practical applications, the continuous-time/continuous-frequency transform (4.2) can be evaluated only for a finite number of discrete frequencies, which are usually selected on a uniformly-spaced frequency grid, viz.,

$$f_{c,k} = k\frac{F_s}{L} \quad , \qquad 0 \le k \le K \tag{4.15a}$$

Here F_s is the intended time-sampling rate of the input waveform $x(t)$, the integer L controls the frequency-grid density, and

$$K \stackrel{\text{def}}{=} \mathsf{floor}\left(\frac{L}{2}\right) \tag{4.15b}$$

where $\mathsf{floor}(x)$ denotes the rounding down of x to the nearest integer less than or equal to x. Notice that we continue relying on the assumption that the input waveform $x(\cdot)$ is real-valued. Consequently, we need to be concerned only with the non-negative frequency range $0 \le f \le \frac{F_s}{2}$, which is covered by the discretized frequency grid $\{f_{c,k}\,;\ 0 \le k \le K\}$.

It will be convenient to use the passband version (4.6) of the continuous STFT as our starting point. The resulting *frequency-discretized version* is

$$\mathcal{X}_k(t) = h_k(t) \circledast x(t) \quad , \qquad 0 \le k \le K \tag{4.16a}$$

where we used (4.1), and introduced the frequency-discretized quantities

$$\mathcal{X}_k(t) \stackrel{\text{def}}{=} \mathcal{X}(t\,;f_{c,k}) \quad , \quad h_k(t) \stackrel{\text{def}}{=} h(t\,;f_{c,k}) \equiv h(t)\,e^{j2\pi f_{c,k}t} \tag{4.16b}$$

as well as (compare with Fig. 4.7),

$$H_k(s) \stackrel{\text{def}}{=} \mathcal{L}\left\{h_k(t)\right\} = H(s - j2\pi f_{c,k}) \tag{4.16c}$$

We can also introduce a frequency-discretized version of the passband-baseband relation (4.6), viz.,

$$X_k(t) \stackrel{\text{def}}{=} X(t\,;f_{c,k}) = \mathcal{X}_k(t)\,e^{-j2\pi f_{c,k}t} \tag{4.16d}$$

The frequency-discretized STFT (4.16) can be interpreted and implemented as a filter-bank, consisting of $K+1$ subbands. Each subband component $\mathcal{X}_k(t)$ is generated by a dedicated band-pass filter $H_k(s)$ (Fig. 4.8). The magnitude-response $|H_k(j\omega)|$ of this subband filter is centered at the frequency $f_{c,k} = k\frac{F_s}{L}$, but its bandwidth depends only on the choice of $h(t)$ and thus is independent of the subband index "k."

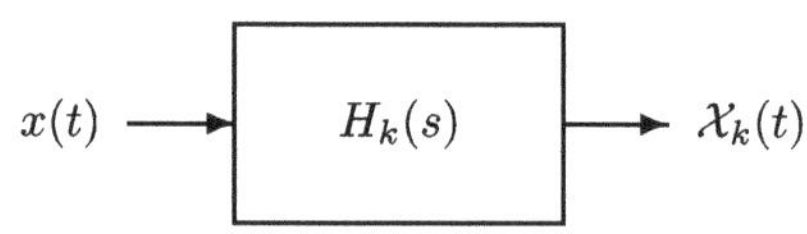

Fig. 4.8 Filtering interpretation of the frequency-discretized STFT.

Discretizing Time

Our starting point for discretization of the time variable "t" is a sampled version of the waveform of interest. Our approach is to preserve the filtering interpretation of the frequency-discretized STFT: this will lead us to a discrete-time filter-bank formulation. We continue to work with the passband version (shown in Fig. 4.8), which represents the relation (4.16), namely,

$$\mathcal{X}_k(t) = \int_{-\infty}^{\infty} \underbrace{\left[h(\tau)\,e^{j2\pi f_{c,k}\tau}\right]}_{h_k(\tau)} x(t-\tau)\,d\tau$$

A time-discretized version of this expression is obtained by setting $t = nT_s$ and approximating the integral by a sum, viz.,

$$\mathcal{X}_k(nT_s) \approx \sum_{m=-\infty}^{\infty} \mathcal{E}\Big[h(m\mathcal{E})\, e^{j2\pi f_{c,k} m\mathcal{E}}\Big]\, x(nT_s - m\mathcal{E})$$

where n is an integer, T_s is the time-sampling interval needed for a sufficiently accurate representation of the input waveform $x(t)$, and $\mathcal{E}$ is a discretization interval for the integration variable τ. Notice that this expression can be implemented only if $\mathcal{E}$ is an integral multiple of T_s. However, choosing $\mathcal{E}/T_s > 1$ results in reducing the accuracy of the integral-to-sum conversion. Thus, it makes sense to set $\mathcal{E} = T_s$ for the sake of enhanced accuracy, resulting in the discrete-time expression

$$\mathcal{X}_k(nT_s) \approx \sum_{m=-\infty}^{\infty} \underbrace{\Big[T_s\, h(mT_s)]\, e^{j2\pi f_{c,k} m T_s}\Big]}_{h_k[m]}\, x[n-m]$$

$$\equiv h_k[n] \circledast x[n] \tag{4.17a}$$

where we introduced the short-hand (discrete-time) notations

$$h_k[n] \stackrel{\text{def}}{=} T_s\, h_k(nT_s) = T_s\, h(nT_s)\, e^{j\left(2\pi/L\right)kn} \quad , \quad x[n] \stackrel{\text{def}}{=} x(nT_s) \tag{4.17b}$$

and used the explicit value of $f_{c,k}$ given in (4.15). Notice that

$$\sum_{m=-\infty}^{\infty} h_0[m] = T_s \sum_{m=-\infty}^{\infty} h(mT_s) \approx \int_{-\infty}^{\infty} h(t)\, dt = 1$$

which is the discrete equivalent of the unity DC-gain property (4.7).

The discrete TFR inherits the conjugate symmetry property of $\mathcal{X}(t\,;\,f_c)$ for real-valued input waveforms, namely, $\mathcal{X}_{-k}(nT_s) = \mathcal{X}_k^*(nT_s)$: this allows us to continue restricting our discussion to the one-sided frequency range $f_c \geq 0$. Moreover, notice that $\mathcal{X}_k(nT_s) = \mathcal{X}_{k+L}(nT_s)$: this periodicity property is a consequence of discretizing the time variable and *is not inherited* from the continuous-time $\mathcal{X}(t\,;\,f_c)$. Thus, $\mathcal{X}_k(nT_s)$ can provide a reasonably accurate representation of $\mathcal{X}(t\,;\,f_c)$ only within the *finite frequency range* $|f_c| \leq \frac{F_s}{2}$, which means that we need to consider only the one-sided subband range $0 \leq k \leq K$.

The baseband version of the discretized STFT expression (4.17) is, of course,

$$X_k(nT_s) = e^{-j\left(2\pi/L\right)kn}\, \mathcal{X}_k(nT_s) \approx e^{-j\left(2\pi/L\right)kn}\Big[h_k[n] \circledast x[n]\Big] \tag{4.18a}$$

which can also be written in the equivalent form (compare with Fig. 4.6)

$$X_k(nT_s) \approx \sum_{m=-\infty}^{\infty} h[n-m]\left(x[m]\, e^{-j(2\pi/L)km}\right), \quad 0 \le k \le K \tag{4.18b}$$

where

$$h[n] \overset{\text{def}}{=} T_s\, h(nT_s) \equiv h_0[n] \tag{4.18c}$$

is the discrete-time equivalent of the continuous-time window function $h(t)$. In addition, we observe that

$$\begin{aligned}\sum_{n=-\infty}^{\infty} X_k(nT_s) &\approx \sum_{m=-\infty}^{\infty} x[m]\, e^{-j(2\pi/L)km}\left[\sum_{n=-\infty}^{\infty} h[n-m]\right] \\ &\approx \underbrace{\sum_{m=-\infty}^{\infty} x[m]\, e^{-j(2\pi/L)km}}_{\text{DTFT of } x[\cdot]}\end{aligned} \tag{4.19}$$

which is the discretized equivalent of (4.10), relating the time-summation of $X_k(nT_s)$ to the discrete-time Fourier transform (DTFT) of the infinite sampled waveform record $\{x[m]\,;\ -\infty < m < \infty\}$, evaluated at the normalized frequency k/L.

4.3 Discrete-Time Filter-Bank Interpretation

The discrete STFT expression (4.17) describes a convolution operation (denoted by $\circledast$), viz.,

$$\mathcal{X}_k[n] \overset{\text{def}}{=} \mathcal{X}_k(nT_s) = h_k[n] \circledast x[n] \tag{4.20a}$$

where $h_k[n]$ and $x[n]$ are as defined in (4.17b). Thus, the mapping from a discrete-time input waveform $x[\cdot]$ to its D-STFT *passband output* $\mathcal{X}_k[n]$ can be interpreted as an analysis filter-bank (AFB), with one band-pass filter for each value of k, as shown in Fig. 4.9. The subband filter $H_k(z)$ is the transfer function associated with the discrete-time modulated window $h_k[n]$ of (4.17), namely,

$$h_k[n] \overset{\text{def}}{=} h[n]\, e^{j(2\pi/L)kn}, \quad 0 \le k \le K \tag{4.20b}$$

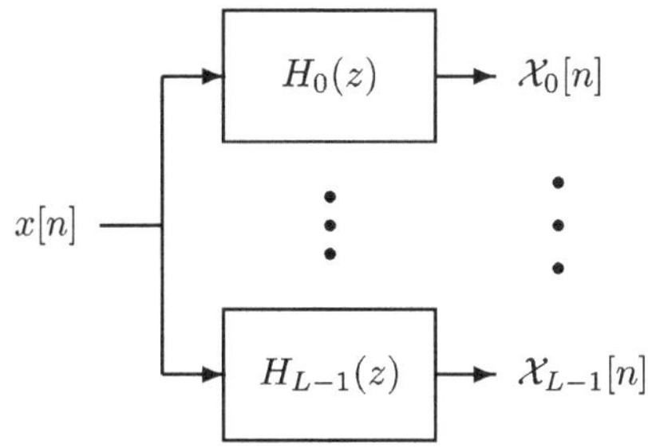

Fig. 4.9 Analysis filter-bank interpretation of the D-STFT.

so that

$$H_k(z) = H\left(e^{-j(2\pi/L)k}z\right) \quad , \quad H(z) \stackrel{\text{def}}{=} \sum_{n=-\infty}^{\infty} h[n]z^{-n} \tag{4.20c}$$

Selection of the *baseband generating filter* $H(z)$ controls time–frequency localization, spectral leakage and input-to-subband delay. Explicit guidelines for designing the window function $h[n]$ are examined in Sect. 4.5.

Notice that $H_k(z)$ is a band-pass filter, with a frequency response $H_k(e^{j\omega}) = H_k\left(e^{j[\omega-k(2\pi/L)]}\right)$. Thus our filter-bank interpretation, as shown in Fig. 4.9, focuses on the passband subband output $\mathcal{X}_k[n]$. The corresponding *baseband version* is

$$X_k[n] \stackrel{\text{def}}{=} X_k(nT_s) = \mathcal{X}_k[n]\,e^{-j(2\pi/L)kn} \quad , \quad 0 \le k \le L-1 \tag{4.20d}$$

Since $\left|\mathcal{X}_k[n]\right| = \left|X_k[n]\right|$, either one can be used to generate the discretized spectrogram $\left|\mathcal{X}_k[n]\right|^2 \equiv \left|X_k[n]\right|^2$.

We now observe that, under the assumption that both $x(t)$ and $h(t)$ are real-valued,

$$h_{L-k}[n] = h_{-k}[n] = h_k^*[n] \quad , \quad 1 \le k \le K$$

so that $H_{L-k}(z) = H_{-k}(z)$, as well as

$$\mathcal{X}_{L-k}[n] = \mathcal{X}_{-k}[n] = \mathcal{X}_k^*[n] \quad , \quad 1 \le k \le K$$

and similarly for the baseband waveforms $X_k[n]$. The practice of re-numbering subbands that correspond to a negative frequency index by adding L to their index is common in the digital signal processing literature: in particular, this is the standard nomenclature for FFT frequency bins [32, 33]. We conclude that the subbands numbered $K+1$ up to $L-1$ in Fig. 4.9 are redundant and need not be implemented in practice, when the input waveform $x(\cdot)$ is real-valued.

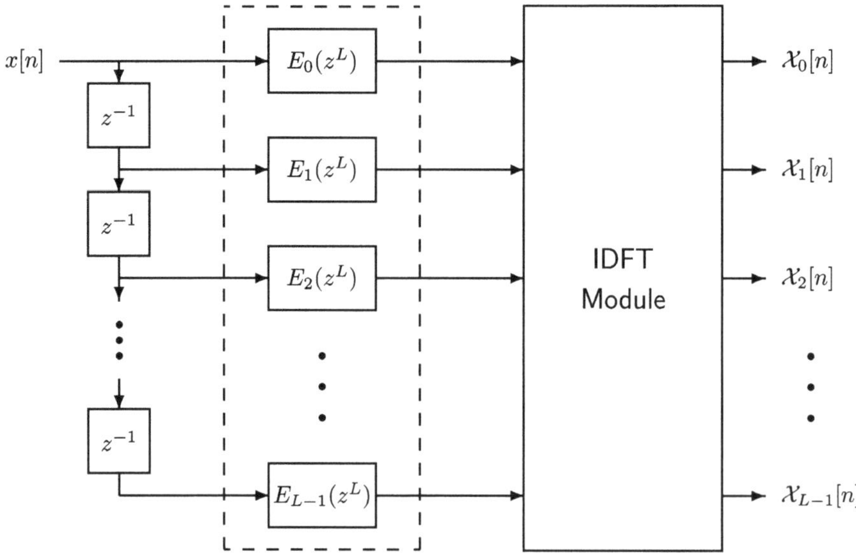

Fig. 4.10 Alternative implementation of the D-STFT.

Any set of discrete-time filters $\{H_k(z);\ 0 \le k \le L-1\}$ that satisfies the property (4.20c), namely $H_k(z) = H\big(e^{-j(2\pi/L)k}z\big)$, is known as a *uniform DFT filter-bank* [32, Sect. 11.10]. This terminology originates from the observation that such a filter-bank admits an alternative, more efficient, implementation that involves a DFT operation[3] (Fig. 4.10). The filtering module in this implementation consists of the *polyphase components* $\{E_i(z);\ 0 \le i \le L-1\}$ of the baseband generating filter $H(z) \equiv H_0(z)$, which are defined via the (Type 1) *polyphase decomposition*[4]

$$H(z) = E_0(z^L) + z^{-1}E_1(z^L) + z^{-2}E_2(z^L) + \ldots + z^{-(L-1)}E_{L-1}(z^L)$$

When $H(z)$ is a discretized window of finite length N_h, the length of each one of its polyphase components is upper-bounded by N_h/L, so that the implementation cost of the entire filtering module in Fig. 4.10 is the same as that of the single filter $H(z)$.

Common applications of uniform DFT filter-banks include coding, storage, and transmission of image and video signals, as well as the design of trans-multiplexers for multi-carrier communication systems. Such applications rely on the fact that down-sampling reduces the rate of the subband waveforms $\mathcal{X}_k[n]$, while still allowing perfect reconstruction. In contrast, our interest in non-decimated subband waveforms (a.k.a. generalized dynamic phasors) is motivated by the observation

[3] Actually, an inverse DFT.

[4] See, e.g., [32, Sect. 11.5.1] and/or [33, Sect. 6.6].

that down-sampling degrades time-resolution and processing speed and results in increased delay in power system monitoring, protection and (closed loop) control applications. Non-decimated filter-banks are also useful in applications that rely on the property of discrete-time shift-invariance. A comprehensive frame-theoretic analysis of partially-decimated and non-decimated PR filter-banks is provided in [131].

In conclusion, we observe that the filter-bank shown in Fig. 4.9 is a non-decimated *uniform DFT analysis filter-bank*, determined by the prototype analysis filter $H(z)$ and the number (L) of subband channels. We provide in Sect. 4.5 some guidelines regarding the choice of window type and the selection of the two design parameters N_h and L.

4.4 Causal Inversion of Non-Decimated D-STFT

As we explained in Sects. 4.2 and 4.3, we limit our discussion here entirely to the non-decimated causal case, which is the preferred choice in most power system monitoring and control applications. Causal (online) inversion of the non-decimated D-STFT can be achieved by using an appropriately designed *synthesis filter-bank* (Fig. 4.11), which generates a reconstructed waveform $\widehat{x}[n]$ from the (passband) subband components $\{\mathcal{X}_k[n];\ 0 \le k \le L-1\}$ or, equivalently, from their baseband versions $X_k[n] = \mathcal{X}_k[n]\, e^{-j(2\pi/L)\,kn}$. We recall that subbands in the index range $K+1 \le k \le L-1$ represent negative frequencies, because $H_{L-k}(z) = H_{-k}(z)$, and similarly for $G_k(z)$. Also, recall that $x[n]$ and $\mathcal{X}_k[n]$ are the sampled versions of $x(t)$ and $\mathcal{X}_k(t)$, respectively.

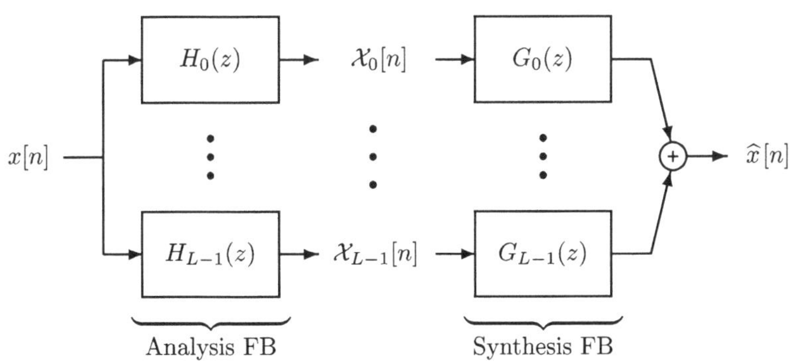

Fig. 4.11 Filter-bank interpretation of the analysis and synthesis D-STFT.

We say that *perfect reconstruction* is achieved when $\widehat{x}[n] = x[n-n_0]$, for some *reconstruction delay* $n_0 \ge 0$. Clearly, the perfect reconstruction (PR) condition, expressed in terms of the filter-bank components shown in Fig. 4.11, is

$$\sum_{k=0}^{L-1} G_k(z)\, H_k(z) \;=\; z^{-n_0} \quad , \qquad n_0 \ge 0 \tag{4.21}$$

4.4.1 Uniform DFT Filter-Banks

We now recall that the synthesis filters $H_k(z)$ used in constructing the D-STFT satisfy the constraint (4.20c), namely $H_k(z) = H\big(e^{-j(2\pi/L)\,k}z\big)$ for some discrete *analysis window* $H(z)$, so that the PR condition for reconstruction of the D-STFT is, in fact,

$$\sum_{k=0}^{L-1} G_k(z)\, H\big(e^{-j(2\pi/L)\,k}z\big) = z^{-n_0} \quad , \qquad n_0 \geq 0 \tag{4.22a}$$

We will assume, in accordance with the unity-gain constraint (4.7), that

$$H(z)\Big|_{z=1} = 1 = \sum_n h[n] \tag{4.22b}$$

This PR condition still admits infinitely many solutions, which is consistent with the observations made in Sect. 4.1.5 about the non-uniqueness of the inverse STFT transform. Here we shall focus on a number of specialized *causal constant-bandwidth solutions* to the PR condition (4.22a), with an emphasis on relevance for power system applications.

One well-known special solution is the uniform DFT synthesis filter-bank, which is defined by the constraint [32, 33]

$$G_k(z) = G\big(e^{-j(2\pi/L)\,k}z\big) \quad , \qquad 0 \leq k \leq L-1 \tag{4.23}$$

where $G(z)$ is some (yet to be determined) discrete *synthesis window*. In this case the PR condition (4.22a) reduces to

$$\sum_{k=0}^{L-1} F\big(e^{-j(2\pi/L)\,k}z\big) = z^{-n_0} \tag{4.24a}$$

where

$$F(z) \overset{\text{def}}{=} G(z)H(z) = \sum_{n=-\infty}^{\infty} f[n]\,z^{-n} \tag{4.24b}$$

Applying the inverse Z-transform and carrying out the summation with respect to the subband index "k" we obtain the time-domain version of this constraint, viz.,

$$\mathcal{S}_L[n]\; f[n] = \frac{1}{L}\,\delta[n-n_0] \tag{4.24c}$$

where $\delta[n]$ denotes the discrete-time (i.e., Kronecker) impulse, and

$$\mathcal{S}_L[n] \stackrel{\text{def}}{=} \frac{1}{L}\sum_{k=0}^{L-1} e^{-j(2\pi/L)\,kn} \equiv \sum_{i=-\infty}^{\infty} \delta[n-iL] \tag{4.24d}$$

is the discrete-time "sampling" (a.k.a. "picket fence") function. The PR equation (4.24c) admits a solution only when n_0 is an integer multiple of L. In particular, when $n_0 = 0$ this solution is

$$f[n] = \begin{cases} \frac{1}{L} & n = 0 \\ 0 & n = iL \;,\; i \neq 0 \end{cases}$$

which we recognize as the characterization of a Nyquist filter [133]. Once a particular Nyquist filter has been chosen, the synthesis filter-bank is uniquely determined, for any desired choice of the analysis window $H(z)$, via the relation $G(z) = \frac{F(z)}{H(z)}$. In particular, we notice that any FIR filter $F(z)$ of length less or equal to L is a (scaled) Nyquist filter. For instance, when $H(z)$ is a causal rectangular window of length L, then choosing $G(z) = \frac{1}{Lh[0]} \equiv 1$ results in $F(z) = H(z)$, which is indeed a Nyquist filter.

4.4.2 Phasor-Banks

Phasor-banks [134, 135] provide a useful modification of the uniform DFT characterization for the synthesis filter-bank (4.23). The synthesis part of a phasor-bank satisfies the constraint

$$G_k(z) = e^{-j(2\pi/L)kn_0}\, G\big(e^{-j(2\pi/L)k} z\big) \quad , \qquad 0 \le k \le L-1 \tag{4.25}$$

where n_0 is the desired reconstruction delay, while the analysis part is characterized by the (uniform DFT) constraint (4.20c). The resulting phasor-bank PR condition is a modified version of (4.24), viz.,

$$\sum_{k=0}^{L-1} e^{\,j(2\pi/L)kn_0}\, F\big(e^{-j(2\pi/L)k} z\big) = z^{-n_0} \tag{4.26a}$$

where $F(z) \stackrel{\text{def}}{=} G(z)H(z)$ is the same as in (4.24b). The time-domain version of this modified constraint is

$$\mathcal{S}_L[n-n_0]\, f[n] = \frac{1}{L}\,\delta[n-n_0] \tag{4.26b}$$

where $\delta[n]$ denotes the discrete-time (i.e., Kronecker) impulse, and $\mathcal{S}_L[n]$ is the discrete-time sampling function defined in (4.24d). The PR equation (4.26) always admits a solution, which is

$$f[n] = \begin{cases} \frac{1}{L} & n = n_0 \\ 0 & n = n_0 + iL \ , \ i \neq 0 \end{cases} \tag{4.26c}$$

In this case the transfer function $F(z)$ is a *delayed* Nyquist filter [133]. Although this solution to the PR equation holds for any value of n_0, we will restrict our subsequent discussion to the practical case when $0 \leq n_0 \leq L-1$.

Since we wish to avoid imposing constraints on the analysis window $H(z)$, the responsibility of satisfying the PR condition (4.26) rests entirely with our choice of the synthesis window $G(z)$. For instance, when we use a rectangular window of length L as the analysis window, i.e., $H(z) = \frac{1-z^{-L}}{L(1-z^{-1})}$, then $G(z) = 1$ is one possible choice of a synthesis window that satisfies the PR condition (4.26). In this special case

$$\widehat{x}[n] = \sum_{k=0}^{L-1} \mathcal{X}_k[n]\, e^{-j(2\pi/L)kn_0} \equiv \sum_{k=0}^{L-1} X_k[n]\, e^{j(2\pi/L)k(n-n_0)} = x[n-n_0] \tag{4.27}$$

which is a discrete-time version (with $0 \leq n_0 \leq L-1$) of the reconstruction expression (2.2a) for standard dynamic phasors.

More generally, when the analysis window $H(z)$ is any FIR filter of length L (or less), i.e., $H(z) = \sum_{n=0}^{L-1} h[n]\, z^{-n}$, the PR equation (4.26) admits solutions that are FIR of the same length, i.e., $G(z) = \sum_{n=0}^{L-1} g[n]\, z^{-n}$, resulting in a so-called *windowed-FIR phasor-bank* [134, 135]. In particular, one simple solution of this equation is

$$G(z) = \frac{1}{L\, h[n_0]} \tag{4.28}$$

In general, every choice (not necessarily FIR) of $H(z)$ has infinitely many matching choices of $G(z)$ that satisfy the phasor-bank PR condition (4.26). This added freedom of choice allows us to optimize the selection of the synthesis window $G(z)$ with respect to some meaningful metric of performance, such as sensitivity of the reconstruction step to perturbations in the subband phasors $X_k[n]$.

Optimally-Robust Causal Phasor-banks

Perturbations in phasor values used as input to the reconstruction step, viz.,

$$X_k[n] \implies X_k[n] + \Delta X_k[n]$$

are caused by imperfections in the analysis step, such as (a) noise in the acquired waveform, (b) quantization errors in waveform samples, (c) finite precision of discretized window function, (d) distortion caused by the communication channel between the originator of phasors (e.g., PMU) and the reconstruction processor, which may be located in a control center. In turn, these result in a perturbation of the reconstructed waveform, namely

$$\widehat{x}[n] \;=\; x[n-n_0] \;+\; \text{perturbation}$$

Given a multiplicity of valid choices for the reconstruction window $g[n]$, it makes sense to choose the one with lowest sensitivity. Notice that we are assuming here an unperturbed, infinite-precision, reconstruction step.

The subject of robust reconstruction is examined in [131] within the framework of the theory of frames [141, 142], which identifies the so-called canonical dual frame as the minimally-sensitive choice for the reconstruction step. Unfortunately, this choice results in an anti-causal synthesis filter-bank, which is of little use in online (real-time) implementations:[5] see also discussion in Sect. 4.5.2. Consequently, we need to re-examine from first principles the issue of reconstruction sensitivity, while imposing the constraints of *both PR and causality*, and using reconstruction delay as an *adjustable design parameter*.

The sensitivity of the reconstruction step to perturbations in the subband phasors $X_k[n]$ can be captured, for instance, via a *robustness index* for filter-banks that was introduced in [134], viz.,

$$\eta_{rob} \;\overset{\text{def}}{=}\; \sqrt{\frac{\left\|\widehat{x}[n]-x[n-n_0]\right\|_{rms}^2}{\left\|x[n]\right\|_{rms}^2}} \Bigg/ \sqrt{\frac{\sum_k \left\|\Delta X_k[n]\right\|_{rms}^2}{\sum_k \left\|X_k[n]\right\|_{rms}^2}} \tag{4.29a}$$

where

$$\left\|y[n]\right\|_{rms} \;\overset{\text{def}}{=}\; \lim_{N\to\infty} \sqrt{\frac{1}{N}\sum_{n=0}^{N-1} \left|y[n]\right|^2} \tag{4.29b}$$

A robust reconstruction step is indicated by a low value of η_{rob}.

We model the subband perturbations $\Delta X_k[n]$ as mutually-independent white noise processes, all with the same variance σ^2, which means that

$$\left\|\Delta \mathcal{X}_k[n]\right\|_{rms}^2 \;\equiv\; \left\|\Delta X_k[n]\right\|_{rms}^2 \;=\; \sigma^2$$

[5] Causality can be restored in some special cases by allowing non-adjustable reconstruction delay.

The resulting output perturbation level is

$$\| \widehat{x}[n]-x[n-n_0] \|_{rms}^2 = \sigma^2 \int_{-\pi}^{\pi} \sum_{k=0}^{L-1} \left|G_k(e^{j\omega})\right|^2 \frac{d\omega}{2\pi} = \sigma^2 L \int_{-\pi}^{\pi} \left|G(e^{j\omega})\right|^2 \frac{d\omega}{2\pi}$$

where we used the fact that $\int_{-\pi}^{\pi} \left|G_k(e^{j\omega})\right|^2 \frac{d\omega}{2\pi} = \int_{-\pi}^{\pi} \left|G(e^{j\omega})\right|^2 \frac{d\omega}{2\pi}$ for all k (recall (4.25)).

Notice that the norms $\left\|\Delta X_k[n]\right\|_{rms} \equiv \left\|\Delta \mathcal{X}_k[n]\right\|_{rms}$, as well as $\| x[n] \|_{rms}$ and $\left\|X_k[n]\right\|_{rms}$, are all independent of the synthesis window $G(z)$. Consequently, the robustness index η_{rob} is minimized by finding, among all causal synthesis windows $G(z)$ that satisfy the PR condition (4.26), one that minimizes the quadratic cost function

$$\int_{-\pi}^{\pi} \left|G(e^{j\omega})\right|^2 \frac{d\omega}{2\pi} = \sum_{n=0}^{\infty} |\, g[n] \,|^2 \stackrel{\text{def}}{=} \| \boldsymbol{g} \|^2$$

In summary, the optimally-robust (causal) synthesis window $G(z)$ is the solution to the linearly-constrained quadratic optimization problem [135]

$$\min_{\boldsymbol{g}} \; \| \boldsymbol{g} \|^2 \tag{4.30a}$$

subject to the constraint (= causal vector-matrix equivalent of (4.26c))

$$\mathcal{H} \underbrace{\begin{pmatrix} g[0] \\ g[1] \\ g[2] \\ \vdots \end{pmatrix}}_{\boldsymbol{g}} = \frac{1}{L} \begin{pmatrix} 1 \\ 0 \\ 0 \\ \vdots \end{pmatrix} \equiv \begin{pmatrix} f[n_0] \\ f[n_0+L] \\ f[n_0+2L] \\ \vdots \end{pmatrix} \tag{4.30b}$$

where, assuming that $H(z)$ is causal, and $0 \leq n_0 \leq L-1$,

$$\mathcal{H} = \begin{pmatrix} h[n_0] & \dots & h[0] & 0 & \dots & & & & & & & & \\ h[n_0+L] & \dots & h[L] & \dots & h[0] & 0 & \dots & & & & & & \\ h[n_0+2L] & \dots & \vdots & \dots & h[L] & \dots & h[0] & 0 & \dots & & & & \\ h[n_0+3L] & \dots & \vdots & \dots & \vdots & \dots & h[L] & \dots & h[0] & 0 & \dots \\ \vdots & & \vdots & & \vdots & & \vdots & & \vdots & & \end{pmatrix} \tag{4.30c}$$

We can also assume, without loss of generality, that $h[0] \neq 0$. Notice that this time-domain formulation relies explicitly on the causality of both the analysis impulse response $h[n]$ and the synthesis impulse response $g[n]$. Also notice that the constraint (4.30b) is an under-determined system of equations, admitting multiple solutions. Our objective is to find among these a solution $\boldsymbol{g}_{opt}^{(\boldsymbol{h},n_0)}$ with minimal Euclidean norm.

The minimum-norm solution of the under-determined set of equations (4.30b) is $\boldsymbol{g}_{opt}^{(\boldsymbol{h},n_0)} = \frac{1}{L}\,\mathcal{H}^{\dagger}\,\mathbf{e}_0$, where

$$\mathbf{e}_0 \stackrel{\text{def}}{=} [1\ 0\ \ldots\ 0]^{\top} \tag{4.31a}$$

and $\mathcal{H}^{\dagger}$ is the Moore-Penrose inverse of the matrix $\mathcal{H}$ [76, Chap. 7]: this is the causal counterpart of the *canonical dual* SFB described in [131]. Since $h[0] \neq 0$, the rows of $\mathcal{H}$ are linearly-independent, which implies that its generalized inverse is given by the expression $\mathcal{H}^{\dagger} = \mathcal{H}^{H}\left(\mathcal{H}\mathcal{H}^{H}\right)^{-1}$ [76, Exercise (7.24)], and thus

$$\boldsymbol{g}_{opt}^{(\boldsymbol{h},n_0)} = \frac{1}{L}\,\mathcal{H}^{H}\left(\mathcal{H}\mathcal{H}^{H}\right)^{-1}\mathbf{e}_0 \tag{4.31b}$$

The time-domain formulation (4.30) makes no assumptions about the structure (i.e., FIR or rational IIR) of the impulse responses corresponding to the synthesis and analysis windows. Consequently, a numerical evaluation of (4.31b) requires, in general, truncation to a finite length.

Relative Sensitivity

In general, the controllable delay n_0 is a design parameter, allowing improved reconstruction sensitivity at the expense of increased delay (see, e.g., Fig. 4.12 below). Our benchmark for comparison between windows with arbitrary reconstruction delay values is the delay-free version of the reconstruction expression (4.27), namely,

$$\widehat{x}[n] = \sum_{k=0}^{L-1} \mathcal{X}_k[n] \equiv \sum_{k=0}^{L-1} X_k[n]\, e^{\,j(2\pi/L)kn} = x[n]$$

which is the one used to recover $x[n]$ from standard dynamic phasors. Since $G_{\text{bench}}(z) = 1$ and thus $\|\,\boldsymbol{g}_{\text{bench}}\,\| = 1$ for this benchmark, it follows that a measure of relative sensitivity for any synthesis window $g[n]$ is provided by the norm $\|\,\boldsymbol{g}\,\|$ of its impulse response. Thus, we define

$$\text{Relative sensitivity} \stackrel{\text{def}}{=} \|\,\boldsymbol{g}\,\| = \Big(\sum_{n} |g[n]|^2\Big)^{1/2} \tag{4.32}$$

In particular, the relative sensitivity of an optimally-robust synthesis window is $\|\,\boldsymbol{g}_{opt}^{(\boldsymbol{h},n_0)}\,\|$, where $\boldsymbol{g}_{opt}^{(\boldsymbol{h},n_0)}$ denotes the solution of (4.30) for a given choice of the analysis window $\boldsymbol{h} = \{h[n]\,;\ 0 \le n < \infty\}$ and reconstruction delay n_0.

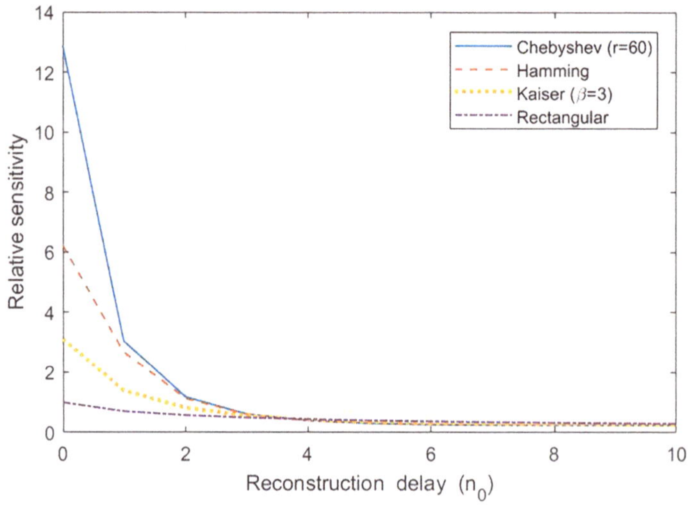

Fig. 4.12 The trade-off between reconstruction sensitivity and delay: comparison of several window functions of length $L = 11$, with a delay-free rectangular window serving as the benchmark.

Windowed-FIR Phasor-banks

The need for truncation of the infinite set of equations (4.30) can be avoided in the important special case called "windowed-FIR" (or W-FIR, for short) namely when the synthesis window $H(z)$ is FIR of length L or less [134, 135]. In this case $h[n] = 0$ for all $n \geq L$, so that the rows of the matrix $\boldsymbol{\mathcal{H}}$ are all mutually-orthogonal. The unique norm-minimizing solution in this special case is, according to (4.31b),

$$\boldsymbol{g}_{opt}^{(\boldsymbol{h},n_0)} = \frac{1}{L}\boldsymbol{\mathcal{H}}^H\left(\boldsymbol{\mathcal{H}}\boldsymbol{\mathcal{H}}^H\right)^{-1}\begin{pmatrix}1\\0\\0\\\vdots\end{pmatrix} = \frac{1}{L\sum_{\ell=0}^{n_0}\left|h[\ell]\right|^2}\begin{pmatrix}h^*[n_0]\\h^*[n_0-1]\\\vdots\\h^*[0]\\0\\\vdots\end{pmatrix} \tag{4.33a}$$

where we used the fact that $\boldsymbol{\mathcal{H}}\boldsymbol{\mathcal{H}}^H$ is diagonal, and its top-left element is $\sum_{\ell=0}^{n_0}\left|h[\ell]\right|^2$.

Thus, the optimally-robust causal synthesis window $G_{opt}^{(\boldsymbol{h},n_0)}(z)$ is also FIR, but of length $n_0 + 1$, viz.,

$$G_{opt}^{(\boldsymbol{h},n_0)}(z) = \frac{1}{L\sum_{\ell=0}^{n_0}\left|h[\ell]\right|^2}\sum_{\ell=0}^{n_0} h^*[n_0-\ell]\,z^{-\ell} \tag{4.33b}$$

When $n_0 = 0$, this optimally-robust choice reduces to the simple synthesis window $G(z) = \frac{1}{Lh[0]}$ that was presented in (4.28).

Notice that the relative sensitivity of a windowed-FIR optimally-robust synthesis window (4.33) is given by the compact expression

$$\text{Relative sensitivity of W-FIR} \stackrel{\text{def}}{=} \| \boldsymbol{g}_{opt}^{(\boldsymbol{h},n_0)} \| = \frac{1}{L\sqrt{\sum_{\ell=0}^{n_0} |h[\ell]|^2}} \tag{4.34}$$

which we used to generate Fig. 4.12. Clearly, relative sensitivity of a windowed-FIR phasor-bank is a monotone decreasing function of the reconstruction delay n_0, as is also evident from the plots in this figure: allowing a modest reconstruction delay, such as $n_0 = 1$, can offer a significant reduction in sensitivity. Notice that within the range $0 \leq n_0 \leq 2$ sensitivity is highly-dependent on the choice of window: the lowest sensitivity is achieved with a rectangular window function.

IIR Phasor-banks [135]

A natural extension to the windowed-FIR case is to consider a rational IIR synthesis window, namely,

$$H(z) = \frac{b(z)}{a(z)} \quad , \qquad a(z) = \sum_{i=0}^{L_a-1} a_i\, z^{-i} \quad , \quad b(z) = \sum_{i=0}^{L_b-1} b_i\, z^{-i} \tag{4.35a}$$

where

$$\max \left\{L_a, L_b\right\} \leq L \tag{4.35b}$$

Notice that this formulation includes windowed-FIR phasor-banks as a special case, obtained by setting $a(z) = 1$. The optimization problem (4.30) admits in this case *constrained-FIR* solutions (C-FIR for short) of the form

$$G_{opt}(z) = \gamma_{opt}(z)\, a(z) \tag{4.35c}$$

where $\gamma_{opt}(z) = \sum_{i=0}^{L_\gamma-1} \gamma_i\, z^{-i}$ is a polynomial in z^{-1}. It will be convenient to introduce the short-hand notations

$$L_g \stackrel{\text{def}}{=} 1 + \deg G(z) \quad , \quad L_\gamma \stackrel{\text{def}}{=} 1 + \deg \gamma(z) \tag{4.35d}$$

We observe that, as a result of these definitions,

$$L_g = L_a + L_\gamma - 1 \geq L_a \tag{4.35e}$$

A unique optimally-robust solution of (4.30) exists for every choice of $L_\gamma \geq 1$.

The C-FIR problem setup (4.35) implies that

$$F(z) \stackrel{\text{def}}{=} G(z)H(z) = \gamma(z)\, b(z) = \text{polynomial of degree } L_b + L_\gamma - 2$$

which allows us to replace (4.30) with the following *finite-dimensional* linearly-constrained quadratic optimization problem

$$\min_{g} \; \| \, g \, \|^2 \tag{4.36a}$$

subject to the constraints

$$g = \mathbf{D}\boldsymbol{\Gamma} \quad \text{and} \quad \mathbf{B}\boldsymbol{\Gamma} = \frac{1}{L}\, \mathbf{e}_0 \tag{4.36b}$$

where we recall that $\mathbf{e}_0 = [1 \;\; 0 \;\; \ldots \;\; 0]^\top$, the column vector $\boldsymbol{\Gamma} = \left[\gamma_i\right]_{i=0}^{L_\gamma - 1}$ consists of coefficients of the polynomial $\gamma(z) = \sum_{i=0}^{L_\gamma-1} \gamma_i\, z^{-i}$, and

$$\mathbf{D} = \begin{pmatrix} 1 & & \mathbf{O} \\ a_1 & \ddots & \\ \vdots & \ddots & 1 \\ \vdots & \ddots & a_1 \\ a_{L_a-1} & \ddots & \vdots \\ & \ddots & \vdots \\ \mathbf{O} & & a_{L_a-1} \end{pmatrix}$$

$$\mathbf{B} = \begin{pmatrix} b_{n_0} & \ldots & b_0 & 0 & \ldots & 0 & 0 & \ldots & 0 & 0 & \ldots \\ 0 & \ldots & 0 & b_{L_b-1} & \ldots & b_0 & 0 & \ldots & 0 & 0 & \ldots \\ 0 & \ldots & 0 & 0 & \ldots & 0 & b_{L_b-1} & \ldots & b_0 & 0 & \ldots \\ & \vdots & & & \vdots & & & & & & \end{pmatrix} \tag{4.36c}$$

Both $\mathbf{D}$ and $\mathbf{B}$ are finite-dimensional matrices, i.e.,

$$\left[\mathbf{D}\right] = L_g \times L_\gamma \quad , \quad \left[\mathbf{B}\right] = K_b \times L_\gamma \tag{4.36d}$$

where

$$K_b \stackrel{\text{def}}{=} 1 + \text{floor}\left(\frac{L_b + L_\gamma - 2 - n_0}{L}\right) \tag{4.36e}$$

Notice that the matrix $\mathbf{B}$ is a subset of truncated rows, selected from the infinite lower-triangular Toeplitz matrix

$$\mathcal{L}(\mathbf{b}) \stackrel{\text{def}}{=} \begin{pmatrix} b_0 & & & & \\ b_1 & b_0 & & \mathbf{O} & \\ \vdots & \ddots & \ddots & & \\ b_{L_b-1} & \dots & b_1 & b_0 & \\ 0 & b_{L_b-1} & \dots & b_1 & b_0 \\ \vdots & & \ddots & & \ddots \quad \ddots \end{pmatrix}$$

The i-th row of $\mathbf{B}$ (where $1 \leq i \leq K_b$) is the $(1 + n_0 + iL)$-th row of $\mathcal{L}(\mathbf{b})$, truncated to length L_γ.

Since $L_b \leq L$, the most likely column-length of the matrix $\mathbf{B}$, for moderate L_γ-values, is either $K_b = 1$ or $K_b = 2$. For instance, when $L_\gamma = n_0 + 1$, then $K_b = 1$ and $\mathbf{B} = \begin{bmatrix} b_{n_0} & \dots & b_0 \end{bmatrix}$. Similarly, if $L_\gamma = L_b + n_0 + 1$, then $K_b = 2$ and

$$\mathbf{B} = \begin{bmatrix} b_{n_0} \dots b_0 & 0 & \dots 0 \\ 0 \ \dots \ 0 & b_{L_b-1} & \dots b_0 \end{bmatrix}$$

Other choices of L_γ-values will result, of course, in different K_b-values, resulting in a variety of truncation patterns for the matrix $\mathbf{B}$.

The unique solution of the finite-dimensional optimization problem (4.36) is

$$g_{opt} \equiv \mathbf{D}\,\boldsymbol{\Gamma}_{opt} = \frac{1}{L}\,\mathbf{D}\left(\mathbf{D}^H\mathbf{D}\right)^{-1}\mathbf{B}^H\left[\mathbf{B}\left(\mathbf{D}^H\mathbf{D}\right)^{-1}\mathbf{B}^H\right]^{-1}\mathbf{e}_0 \tag{4.37a}$$

which also generates a compact expression for relative sensitivity of constrained-FIR phasor-banks, viz.,

$$\begin{aligned} \text{Relative sensitivity of C-FIR} &\stackrel{\text{def}}{=} \| g_{opt} \| \\ &= \frac{1}{L}\sqrt{\mathbf{e}_0^\top\left[\mathbf{B}\left(\mathbf{D}^H\mathbf{D}\right)^{-1}\mathbf{B}^H\right]^{-1}\mathbf{e}_0} \end{aligned} \tag{4.37b}$$

The relative sensitivity of the constrained-FIR scheme is a monotone decreasing function of the parameter L_γ, because the space of valid $G(z)$ choices increases monotonically with L_γ. In the special case when $a(z) = 1$ we obtain $\mathbf{D} = \mathbf{I}$ and $K_b = 1$, which reduces the two expressions (4.37) to their windowed-FIR counterparts (4.33) and (4.34), respectively.

A further extension of these results, which allows $G(z)$ to be rational IIR with a *prescribed set of poles*, is described in [135]. The magnitude and phase of such poles can be used to achieve further reduction in both sensitivity and reconstruction error.

4.5 Guidelines for D-STFT Design

Construction of a discretized STFT scheme involves selection of a window type, as well as setting the values of three design parameters: (a) time-sampling rate F_s, (b) window length T_h, and (c) frequency-discretization parameter L. As already discussed in the introduction to this chapter, we need to consider two different classes of application scenarios: (a) applications that rely solely on time–frequency analysis and (b) applications that involve both analysis and reconstruction, resulting in an invertible transform. The latter is usually implemented as part of a control-loop and thus must be causal and should not introduce excessive delay. Each of these two classes of applications requires a distinct set of design guidelines.

In power system applications, one should design a discrete STFT scheme not for a single waveform, but rather for a whole family of characteristic waveforms, e.g., periodic waveforms with some harmonic distortion and added high-bandwidth transients. The time-sampling rate F_s should be chosen as an integer multiple of the fundamental harmonic frequency f_0 (so that $N_{\text{cycle}} \overset{\text{def}}{=} F_s/f_0$ is an integer) and should be high enough for a faithful representation of the input waveform $x(t)$. This translates to sampling rates in the range of 4,000–20,000 samples/sec [39]. For instance, all real-life examples included in Appendix B of this book use sampling rates in excess of 7,000 samples/sec.

Some design guidelines can be deduced from well-known properties of the *discrete Fourier transform* (DFT). The alternative expression (4.18b) for a discretized STFT, namely, $X_k[n] = \sum_{m=-\infty}^{\infty} \left(h[n-m]\,x[m]\right) e^{-j(2\pi/L)km}$ is, of course, a DFT of the windowed discrete-time waveform $\{h[n-m]\,x[m]\,;\ -\infty < m < \infty\}$ for any given n. Consequently, the properties of $X_k[n_0]$, for a fixed time-instant n_0, are the same as those of a DFT of any finite discrete-time signal segment. In particular [32, 33, 136, 137]:

- The DFT is a frequency-sampled version of the discrete-time Fourier transform (DTFT).
- The DTFT is obtained by replacing the discrete frequency k/L by the continuous-frequency variable f (and setting $L \to \infty$), which results in the expression $\sum_{m=-\infty}^{\infty} \left(h[n-m]\,x[m]\right) e^{-j2\pi fm}$, where $|f| \leq \frac{1}{2}$. This "continuous-frequency spectrum" exhibits in full the effects of windowing, which include *spectral leakage* caused by the sidelobes of $H(j\omega)$, and *limited frequency-resolution*, controlled by the bandwidth of the mainlobe of $H(j\omega)$, the Fourier transform of the continuous-time window function $h(t)$.

- The choice of a window function controls both frequency-resolution and spectral leakage. For instance, using a Kaiser window allows reduction of the sidelobes (by increasing the value of the parameter β), at the expense of increasing the mainlobe bandwidth, which degrades frequency-resolution.
- Time-resolution can be improved, of course, by decreasing the effective length of the window, at the expense of degraded frequency-resolution.
- The "discrete frequency spectrum" described by $X_k[n_0]$, for any given n_0, is a discretized frequency version of the associated DTFT. Thus, the density of the frequency grid, which is determined by the number of subbands L, has only a negligible impact on the time/frequency-resolution of the D-STFT, as long as L is not too small (see discussion in Sect. 4.5.1).

Notice that in order to be able to claim, with any degree of accuracy, that the windowed waveform segment $h(t-\cdot)x(\cdot)$ generated by an *infinite-length window* is effectively time-limited to an interval of length T_h, we need to use a conservative duration metric. Thus, we define the *effective window length* T_h of any continuous-time window function $h(t)$ as

$$T_h \stackrel{\text{def}}{=} \begin{cases} \text{length of } h(t) & \text{finite-length window} \\ \Delta_t^{(PD\star 40)} & \text{infinite-length window} \end{cases} \tag{4.38}$$

where $\Delta_t^{(PD\star 40)}$ denotes a 40 dB prescribed-drop effective window duration, as defined in Sect. B.4.

A digital implementation of the STFT, such as (4.18b), involves a discrete-time window (recall (4.18c)), namely $h[n] = T_s\, h(nT_s)$. For instance, the discretized version of a Hamming window of length T_h is

$$h[n] \stackrel{\text{def}}{=} T_s\, h(nT_s) = \begin{cases} \dfrac{1}{N_h}\left[1 + \dfrac{21}{25}\cos\left(\dfrac{2\pi n}{N_h}\right)\right] & |n| \le \frac{N_h}{2} \\ 0 & \text{else} \end{cases}$$

where we used (B.6), and introduced the normalized window length

$$N_h \stackrel{\text{def}}{=} \frac{T_h}{T_s} \tag{4.39}$$

The same discretization step can be applied to any continuous-time window. We will assume in the sequel that T_h is chosen to be an exact multiple of the time-sampling interval T_s: this will result in an integer value for N_h, which corresponds to the expressions provided in most DSP books.

In summary, sensible D-STFT design always involves a trade-off between several performance objectives, namely,

- Acceptable time–frequency resolution
- Low spectral leakage
- Manageable implementation cost

Time/frequency-resolution of the D-STFT $X_k[n_0]$, for a given sampling rate F_s, depends mainly on the choice of window type and the normalized window length N_h, as we demonstrate in Sect. 4.5.1. Leakage is determined mainly by window type and length, and to a lesser extent by the frequency-discretization parameter L. Implementation cost of the analysis filter-bank (AFB) is proportional to the product $N_h L$. Finally, the set of design objectives used to construct invertible D-STFT schemes should include, in addition to the three objectives listed above, also sensitivity of the reconstruction step to perturbations of subband information.

4.5.1 D-STFT Design for Time–Frequency Analysis

When the sole objective of using a D-STFT is reasonably accurate time–frequency representation, the only performance metrics that need to be considered, in addition to implementation cost, are (a) frequency-resolution, which depends on both the window type and the density of the discretized frequency-grid, (b) spectral leakage, which is caused by sidelobes of the window frequency response $H(j\omega)$, and (c) time-resolution, which is controlled by the length of the window function $h(t)$. The latter also affects agility in responding to fast transients.

Time–Frequency Resolution and Leakage Suppression

A detailed examination of the notions of frequency- and time-resolution of a windowed waveform is provided in Sects. B.5–B.8 of Appendix B, culminating in the following conclusions:

- Time-resolution is proportional to the effective duration of the window function $h(t)$, such as the duration metric $\Delta_t^{(PD\star B)}$.
- Frequency-resolution coincides with an appropriate measure of bandwidth of the window frequency response $H(j\omega) = \mathcal{F}_\omega\{h(\cdot)\}$, such as the A dB bandwidth metric $\Delta_f^{(PD\star A)}$.
- Frequency-resolution is *inversely proportional* to T_h, the effective duration of the window function $h(t)$, while time-resolution is proportional to T_h.
- The product of time-resolution and frequency-resolution, for a given choice of the resolution parameters A and B, is a constant that depends only on the window type. We denote this constant as $\mathcal{K}_{res}(A, B)$.
- A reasonable choice for the resolution parameters is $A = 10\text{db} = B$.

The value of $\mathcal{K}_{res}(A, B)$ is listed in Table 4.1 for several practical choices of the resolution parameters A, B. In particular, we notice that $\mathcal{K}_{res}(10, 10) \approx 1$ for all windows shown in this table. Similarly, $\mathcal{K}_{res}(10, 20) \approx 1.5$ and $\mathcal{K}_{res}(10, 20) \approx 2$

Table 4.1 Product of time- and frequency-resolution

Window type	$\mathcal{K}_{res}(10, 10)$	$\mathcal{K}_{res}(10, 20)$	$\mathcal{K}_{res}(20, 20)$
Rectangular	1.01	1.33	4.83
Kaiser ($\beta = 1$)	1.01	1.35	4.60
Triangular	0.99	1.53	2.02
Kaiser ($\beta = 6$)	0.96	1.53	2.05
Hann	0.98	1.51	1.98
Hamming	0.96	1.52	2.04

for all windows except the rectangular window and the nearly-rectangular Kaiser ($\beta = 1$) window.

Suppression of spectral leakage is achieved by (a) setting $N_h = m\frac{F_s}{f_0}$, which places nulls of $H(j\omega)$ at all harmonic frequencies (except DC), and (b) using windows with reduced sidelobes, such as Hann, Hamming, or Kaiser. Notice, however, that sidelobe reduction always comes at the expense of increased mainlobe width, which results in degraded frequency-resolution.

Density of Frequency Grid

We observe that the D-STFT $X_k[n]$ is the discrete-time/discrete-frequency version of the continuous STFT $X(t\,;\,f_c)$, namely,

$$X_k[n] \;\approx\; X(t\,;\,f_c)\Bigg|_{t=nT_s\,,\;f_c=kF_s/L} \tag{4.40}$$

where we combined (4.16) and (4.18). Thus, increasing the frequency-discretization parameter L only increases the density of the frequency grid on the continuous surface described by $X(t\,;\,f_c)$, without changing the shape of this surface. The impact of frequency-grid density of the D-STFT on frequency-resolution is examined in detail in Sect. B.6 of Appendix B. The main conclusion is that choosing a value of L in the range $2N_h \le L \le 3N_h$ is sufficient to make such impact negligible, as implied by the expression (B.17) for a practical subband density. Any further increase of L results in enhanced spectral leakage artifacts, as we demonstrate below, as well as added implementation cost.

When the waveform of interest is periodic, the D-STFT obtained by using a single-cycle-long rectangular window, and setting $L = N_h$, generates standard (time-invariant) phasors which provide an exact representation of the harmonic content of such a waveform. In this case, all subbands are centered on harmonic frequencies, representing all harmonics from DC to $F_s/2$. Thus, in near-steady-state power system applications one may want to set the window length T_h equal to a single-cycle, so that the normalized window length $N_h \stackrel{\text{def}}{=} T_h/T_s = T_0/T_s \equiv F_s/f_0$

becomes synonymous with $N_{\mathrm{cycle}} = {F_s}/{f_0}$, the number of samples per cycle. Here T_h is the cycle length, and $f_0 = 1/T_0$ is the fundamental harmonic frequency. The number of samples per cycle in power system applications using modern sampling devices, such as used in PMUs, is typically between 100 and 300. If we also set $L = N_h = N_{\mathrm{cycle}}$, the nominal width of each subband remains $F_s/L \equiv f_0$, and the subband-center frequencies are $\{kf_0\,;\ 0 \le k \le L/2\}$.

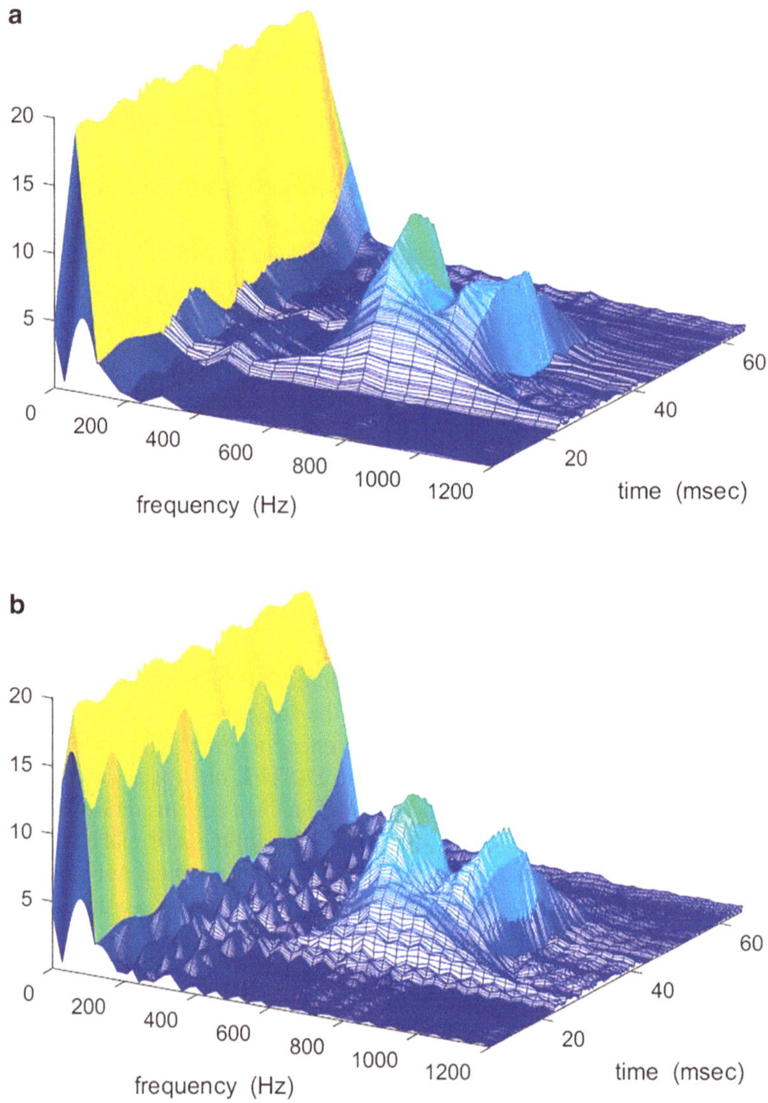

Fig. 4.13 Time–frequency representation of the wind-farm dataset, obtained with a single-cycle-long ($T = 16.7$ msec) Kaiser ($\beta = 6$) window: (a) $L = N_h$ (top) and (b) $L = 2N_h$ (bottom).

These observations are still valid in the presence of in-band transients, such as the paper-mill example. For instance, Fig. 4.2 uses $L = N_h$ discrete frequencies and a *single-cycle-long* rectangular window, so that $T_h = \frac{1}{f_0}$. The discrete frequency grid consists, in this case, entirely of multiples of $\frac{F_s}{L} = \frac{1}{N_h T_s} = \frac{1}{T_h} \equiv f_0$. Using a denser grid in this example would introduce artifacts generated by the sidelobes of the Fourier transform of the window function, but no useful information.

However, a higher density of the frequency grid is required for accurate representation of *inter-harmonic components*. Using a longer window function, i.e., $N_h = m\, N_{\text{cycle}}$ for some positive integer m, results in improved frequency-resolution. For instance, the IEC standard recommends using $L = N_h = 10\, N_{\text{cycle}}$ to achieve frequency-resolution of $f_0/10$ [138–140]. In contrast, accurate representation of transients requires refined time-resolution *and* increased frequency-grid density: these are achieved by setting $N_h = N_{\text{cycle}}$ and $L = m N_h$ for some (moderate) positive integer m, as demonstrated below. In both cases each harmonic of the fundamental frequency f_0 lies exactly on a frequency grid-point.

In principle, increasing the value of L beyond the recommended range results in a denser sampling of the continuous STFT $X(t\,;\,f_c)$ and thus may provide a more detailed time–frequency representation (TFR) of the waveform of interest. However, the added details can be misleading: for instance, increasing the number of frequency subbands from $L = N_h$ to $L = 2N_h$ often introduces a large number of spurious peaks of $\left|X_k[n]\right|$ (Fig. 4.13), caused by spectral leakage. Notice, however, that this increase in L also results in a more accurate depiction of the wide-band transient component in the 500 Hz–1000 Hz range: this improvement involves, of course, doubling the D-STFT implementation cost.

In summary, in the presence of transients one should consider several distinct values of the ratio L/N_h, within the range $1 \leq L/N_h \leq 5$, say, while keeping in mind the trade-off between impact of undesired spectral leakage artifacts and improved representation of wide-band transients. We will refer to the resulting subband pattern as *moderately-dense*.

Parallel TFR Schemes

In general, the choice of window length and frequency-grid density in power system applications is always a compromise between accurate frequency-domain evaluation of steady-state harmonic content and acceptably accurate time–frequency representation of transients. One possible solution to this dilemma is simultaneous operation of several parallel time–frequency representation (TFR) schemes, such as:

- **Slow TFR scheme:** designed for characterization of inter-harmonic components, using a multi-cycle-long window, i.e., $L = N_h \geq 10 N_{\text{cycle}}$.
- **Moderate TFR scheme:** allows depiction of both constant and slowly-varying fundamental and lower-frequency harmonic phasors (a.k.a. "in-band transient"), using a single-cycle-long window and a standard frequency-grid density, i.e., $L = N_h = N_{\text{cycle}}$.

- **Fast TFR scheme:** dedicated to capturing of wide-band transients, using a single-cycle-long window (i.e., setting $N_h = N_{\text{cycle}}$, or even shorter) and an increased frequency-grid density, achieved by setting $L \geq 2N_h$.

An alternative approach, which allows a more flexible trade-off between the last two of these three conflicting objectives is provided by a hybrid subband structure, as described in Sect. 5.4: this subband structure uses constant-bandwidth harmonic-centered subbands to represent the lower-frequency content, and variable-bandwidth subbands to provide a more time-accurate representation of wide-band transients.

Subband Overlap Ratio (SOR)

A different perspective on the role of the frequency-discretization parameter L is provided by considering the overlap between adjacent subbands. Varying L for a fixed choice of window function (which implies a fixed value of the normalized window length N_h) changes the spacing between subband centers and thus controls subband overlap (Fig. 4.14).

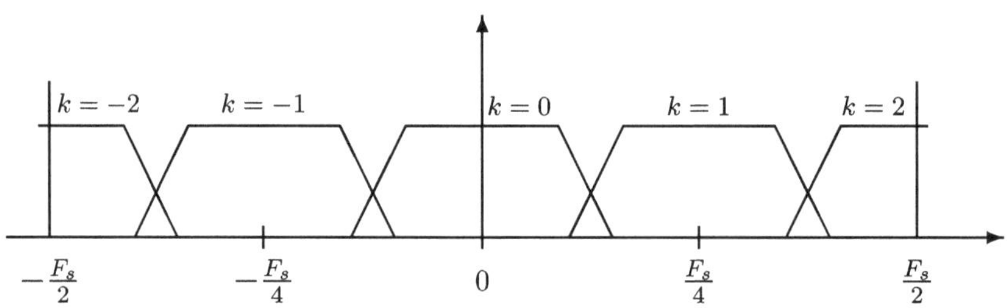

Fig. 4.14 Subband-overlap pattern for $L = 4$: the subband $k = -1$ is the same as $k = 3$, and the subband $k = -2$ is the same as $k = 2$.

In order to quantify the level of subband overlap, we now introduce a *Subband Overlap Ratio* (SOR) metric, viz.,

$$\text{SOR} \stackrel{\text{def}}{=} \frac{\Delta_f^{(PD\star 3)}}{F_s/L} \equiv L\,T_s\,\Delta_f^{(PD\star 3)} = \underbrace{\left[\,T_h\,\Delta_f^{(PD\star 3)}\,\right]}_{\text{window-dependent constant}} \frac{L}{N_h} \tag{4.41}$$

where we used the fact that $T_h = N_h T_s$. Notice that, for a given window function $h(\cdot)$ and time-sampling rate F_s, the SOR is controlled solely by the subband density parameter L. Using the values listed in Table B.2 of Appendix B, we can determine the window-dependent constant $T_h\,\Delta_f^{(PD\star 3)}$ for some of the most

Table 4.2 Value of the window-dependent constant $T_h\ \Delta_f^{(PD\star 3)}$

Window function	$T_h\ \Delta_f^{(PD\star 3)}$
Rectangular	0.891
Kaiser ($\beta = 1$)	0.917
Triangular	1.286
Hamming	1.291
Chebyshev ($r = 60$)	1.324
Kaiser ($\beta = 6$)	1.401

commonly-used window types, as shown in Table 4.2. Notice that a moderately-dense subband pattern can be equivalently characterized as having a moderate SOR value, i.e., in the range $1 \le \mathsf{SOR} \le 7$, or so.

Summary of Design Guidelines

The preceding discussion highlighted several key insights about designing a window-based analysis filter-bank for the sole purpose of time–frequency representation:

- Since the AFB implementation cost is proportional to both N_h and L, the values of these integer parameters should be as small as possible, subject to the constraint $L \ge N_h$, as well as the additional constraints listed below.
- The normalized window length N_h should be chosen as an integer multiple of $\frac{F_s}{f_0} \overset{\text{def}}{=} N_{\mathsf{cycle}}$, the number of samples per cycle of the input waveform $x(\cdot)$: we recall that, by design, N_{cycle} is always an integer. Setting $N_h \ge 2N_{\mathsf{cycle}}$ is needed only for accurate representation of inter-harmonic components and comes at the expense of degraded time-resolution.
- The number of subbands L should also be chosen as an integer multiple of N_{cycle}. Setting $L \ge 2N_h$ is recommended only in the presence of wide-band transients.
- The ratio L/N_h should be kept within a range that ensures moderate SOR.

Since distinct application scenarios may require different choices of the control parameters N_h and L, as discussed earlier in this subsection, one should consider the merits of running several parallel TFR schemes simultaneously. This approach makes it possible to design each TFR scheme for optimal representation of a particular component of interest contained in the input waveform, such as the fundamental (and its harmonics), wide-band transients, or inter-harmonic components.

4.5.2 Offline D-STFT for TFR with Perfect Reconstruction

An appropriately designed subband pattern is the key to construction of filter-banks that exhibit a *perfect reconstruction* (PR) property and thus define an invertible transform, as detailed in Sect. 4.4. However, the PR property by itself imposes almost no constraints on the resulting subband pattern or its SOR. Thus, in power system applications, invertibility of the D-STFT should be introduced as an added constraint, which limits the available choices of analysis window functions and moderately-dense subband patterns.

Perfect reconstruction (PR) of a waveform of interest from its subband components can be achieved by passing the multi-channel output of an analysis filter-bank $\{H_k(z);\ 0 \le k \le L-1\}$ through a suitably-designed synthesis filter-bank $\{G_k(z);\ 0 \le k \le L-1\}$ (recall Fig. 4.11) that satisfies the PR constraint (4.21), viz.,

$$\sum_{k=0}^{L-1} G_k(z)\, H_k(z) \;=\; z^{-n_0} \quad , \qquad\qquad n_0 \ge 0$$

as explained in Sect. 4.4. Notice, however, that the non-decimated D-STFT relies on a special subband-filter relation, namely,

$$H_k(z) \;=\; H\big(e^{-j(2\pi/L)\,k}z\big)$$

and (recall (4.20c), (4.23), and (4.25))

$$G_k(z) \;=\; \begin{cases} G\big(e^{-j(2\pi/L)\,k}z\big) & \text{Uniform DFT} \\ e^{-j(2\pi/L)kn_0}\; G\big(e^{-j(2\pi/L)k}z\big) & \text{Phasor-bank} \end{cases}$$

where $0 \le k \le L-1$, and $H(z), G(z)$ are low-pass filters. This makes it possible to rewrite the PR constraint in terms of the transfer function product $F(z) = H(z)\,G(z)$ and results in an explicit expression for the unique maximally-robust choice of the baseband synthesis filter $G(z)$, as detailed in Sects. 4.4.1–4.4.2.

We conclude that designing an invertible D-STFT of the uniform DFT or phasor-bank type reduces to selection of n_0, L, $H(z)$ and $G(z)$. Usually, the reconstruction delay n_0 is chosen in advance, and the synthesis filter $G(z)$ is determined uniquely by the requirement for maximum reconstruction robustness: this leaves L and $H(z)$ as the only independent design variables that control the subband pattern and its SOR.

The discussion in Sect. 4.5.1 suggests that a reasonable choice of a subband pattern should result in a moderate SOR, as illustrated in Fig. 4.14. Notice that the PR constraint rules out the use of "guard-bands" (i.e., frequency intervals where $H_k((e^{j\omega}) \approx 0$ for all k) between adjacent subbands. This is so because the PR constraint implies a non-vanishing end-to-end frequency response, i.e., $\sum_{k=0}^{L-1} G_k(e^{j\omega})\, H_k((e^{j\omega}) \neq 0$ for every ω. Thus, suppressing subband overlap, i.e., setting $\mathsf{SOR} \to 0$ by introducing "guard-bands" between adjacent subbands, is ruled out by the PR condition.

A similar challenge is associated with very high SOR values, which indicate excessive redundancy in construction of the subband pattern, caused by using an excessively large number of subbands. Efficient numerically-robust inversion of both fixed-bandwidth and variable-bandwidth transforms usually relies on minimizing such redundancy, as explained in Sects. 4.4 and 5.6. A useful high-level perspective on representation redundancy, as well as the non-uniqueness in selecting $G(z)$, is provided by the *theory of frames*.

Frame-theoretic Interpretation of PR Filter-banks

The PR constraint (4.21) for a non-decimated filter-bank, as shown in Fig. 4.11, is equivalent to a pair of time-domain input-output relations, viz.,

$$\mathcal{X}_k[n] \;=\; \sum_{i=-\infty}^{\infty} x[\,i\,]\, h_k[n-i]$$

where $x[\,i\,] \equiv x(i T_s)$ is the sampled input waveform, and

$$x[n-n_0] \;=\; \sum_{k=0}^{L-1} \sum_{m=-\infty}^{\infty} \mathcal{X}_k[m]\, g_k[n-m]$$

Here $h_k[\cdot]$ and $g_k[\cdot]$ are the impulse responses associated with the discrete-time transfer functions $H_k(z)$ and $G_k(z)$, respectively. In the delay-free case (i.e., when $n_0 = 0$), these relations admit a frame-theoretic interpretation, namely,

$$x[\,\cdot\,] \;=\; \sum_{k=0}^{L-1} \sum_{m=-\infty}^{\infty} \mathcal{X}_k[m]\, \phi_{k,m}[\,\cdot\,] \tag{4.42a}$$

and

$$\mathcal{X}_k[m] \;=\; \sum_{i=-\infty}^{\infty} x[\,i\,]\, \psi^*_{k,m}[\,i\,] \;\stackrel{\text{def}}{=}\; \langle\, x[\,\cdot\,],\; \psi_{k,m}[\,\cdot\,]\,\rangle \tag{4.42b}$$

which we recognize as the standard inner-product[6] in the Hilbert space $\ell^2(Z)$ of finite-energy *two-sided* discrete-time signals $\{x[n]\,;\;\; -\infty < n < \infty\}$. Notice

[6] A brief introduction to inner products in Hilbert spaces is provided in Appendix J.

that the decomposition (4.42) involves two distinct sets of discrete-time waveforms, viz.,

$$\phi_{k,m}[n] \stackrel{\text{def}}{=} g_k[n-m] \quad , \qquad \psi_{k,m}[i] \stackrel{\text{def}}{=} h_k^*[m-i] \tag{4.42c}$$

The doubly-indexed waveform sets $\{\phi_{k,m}[\cdot\,];\ 0 \le k \le L-1\,,\ -\infty < m < \infty\}$ and $\{\psi_{k,m}[\cdot\,];\ 0 \le k \le L-1\,,\ -\infty < m < \infty\}$ are known, respectively, as the *synthesis frame* and *analysis frame* implemented by the PR filter-bank. In general, the set of waveforms $\{\phi_{k,m}[\cdot\,]\}$ need not be linearly-independent: when a frame is linearly-independent it is, in fact, a (non-orthogonal) basis. The same statements hold also for the waveform set $\{\psi_{k,m}[\cdot\,]\}$.

Frames capture the inherent redundancy of the perfect reconstruction (PR) constraint (4.21): for any given analysis frame $\{\psi_{k,m}[\cdot\,]\}$, the corresponding synthesis frame $\{\phi_{k,m}[\cdot\,]\}$ is highly non-unique [131, 132]. To qualify as a frame, the analysis waveform set $\{\psi_{k,m}[\cdot\,]\}$ must satisfy the inequalities

$$A \sum_{n=-\infty}^{\infty} \left|x[n]\right|^2 \le \sum_{k=0}^{L-1} \sum_{m=-\infty}^{\infty} \left|\langle\, x[\cdot\,],\ \psi_{k,m}[\cdot\,]\,\rangle\right|^2 \le B \sum_{n=-\infty}^{\infty} \left|x[n]\right|^2$$

for some $0 < A \le B < \infty$, known as *frame bounds*. The frame $\{\psi_{k,m}[\cdot\,]\}$ is called *tight* when $A = B$: in this case $\phi_{k,m}[\cdot\,] = \psi_{k,m}[\cdot\,]$ for all k and m.

The general theory of frames provides several results about robustness of the reconstruction step with respect to (a) perturbations of the subband waveforms $\mathcal{X}_k[\cdot\,]$, and (b) missing subband waveform samples. In particular,

- A stable non-decimated (or partially-decimated) analysis filter-bank (AFB) defines a frame representation of its discrete-time input waveform $x[n]$. There exist multiple choices of the synthesis filter-bank (SFB) that provide perfect reconstruction: each such SFB defines a synthesis frame, known as a *dual frame* [132, 141, 142].
- For any given AFB, the optimally-robust SFB is obtained by choosing the *canonical dual frame,* which is defined uniquely by the AFB frame [132, 141, 142].
- When the AFB is causal, its canonical dual SFB is anti-causal. In particular, the canonical dual of an AFB constructed from a baseband prototype $H(z)$ with impulse response $h[n]$ is an SFB constructed from a baseband prototype $G(z)$ whose impulse response is $g[n] = h^*[-n]$.
- For an AFB with a finite impulse response (FIR), its (non-causal) canonical dual SFB is also FIR *only if* the AFB frame is tight [131, Sect. IV].
- A non-decimated AFB $\{H_k(z)\,;\ 0 \le k \le L-1\}$ defines a tight frame if, and only if, the sum $\sum_{k=0}^{L-1} \left|H_k(e^{j\omega}\right|^2$ is a constant [131, Sect. IV]. Without loss of generality, the AFB can be re-scaled to make this constant equal to unity.
- Tight AFB-SFB frame pairs also result in optimally-robust reconstruction from subband waveforms with missing samples (a.k.a. "erasures") [143].

While these observations motivate using a tight analysis frame and its canonical dual as the synthesis frame in order to optimize reconstruction robustness, they provide no information about the resulting subband pattern and/or its SOR: each one of the infinitely many (tight) frames in $\ell^2(Z)$ results in a different subband pattern, and a different level of sensitivity with respect to perturbations in subband components. This implies that designing an invertible D-STFT is inherently a multi-objective problem, whose optimal solution is determined by the relative weights assigned to the following design criteria: (a) detail level of the resulting time–frequency representation, (b) cost of implementation, and (c) reconstruction robustness.

An example of this multi-criterion design approach is provided by the "lattice-based" construction of a paraunitary QMF two-channel filter-bank [133, Sect. 6.4.3] (see also [144]). In this case, the tight-frame constraints translate into a special filter configuration, known as a *paraunitary QMF lattice*, which guarantees robust reconstruction in finite-precision implementations. However, choosing the optimal values for the QMF lattice coefficients is accomplished by optimizing a separate objective function, which characterizes a desired frequency response for the first (i.e., low-pass) subband filter.

Such classical filter-bank designs for time-frequency analysis always exhibit a moderate level of overlap between adjacent channels. Thus, the most widely-used PR filter-banks tend to have $\mathsf{SOR} \approx 1$, which results in an overlap pattern similar to the one shown in Fig. 4.14, with frequency-grid spacing nearly equal to the 3 dB bandwidth of $H(j\omega)$. For instance, all commonly-used versions of the discrete wavelet transform [133] (see also Sect. 5.6) have $\mathsf{SOR} \approx 1$.

Non-causality of the canonical dual SFB limits the usefulness of frame-theoretic analysis in the Hilbert space $\ell^2(Z)$ of two-sided sequences to applications that rely on *offline processing*. Alternatively, causality can be restored in certain cases (i.e., when the SFB design relies on an FIR baseband synthesis prototype $G(z)$) by delaying the SFB output: this delay is non-adjustable, since it must equal the degree of the FIR filter $G(z)$. These observations motivate our interest in optimally-robust SFB choices subject to a *causality constraint*: such choices are essential for online implementation of PR filter-banks.

4.5.3 Online D-STFT for TFR with Perfect Reconstruction

In contrast to the Z-domain construction of non-causal PR filter-banks that was reviewed in Sect. 4.5.2, practical online implementation of an invertible D-STFT relies on (a) a causal analysis filter-bank (AFB), (b) a causal synthesis filter-bank (SFB) that can provide perfect reconstruction for any desired (non-negative) reconstruction delay n_0, and (c) an explicit characterization of the optimally-robust choice of such an SFB. The standard, window-based, D-STFT implementation as a uniform DFT filter-bank exhibits the first two properties[7] but does not generate an

[7] The only allowed reconstruction delay in this case is $n_0 = 0$ (see Sect. 4.4.1).

optimally-robust SFB. In contrast, phasor-banks (which are described in Sect. 4.4.2) satisfy all three properties: in particular, further reduction of reconstruction sensitivity can be achieved by allowing a non-zero value for the reconstruction delay n_0.

Following the characterization in the preceding subsections, the first step in designing a causal uniform DFT filter-bank, or a causal phasor-bank, amounts to choosing the baseband analysis prototype filter $H(z)$ and the number of subbands L. The corresponding optimally-robust synthesis prototype filter $G(z)$ can then be determined by solving the linearly-constrained quadratic optimization problem (4.30). In particular, the windowed-FIR solution (4.33) and the constrained-FIR solution (4.37a) are special causal SFB choices that provide low-sensitivity perfect reconstruction with a controllable delay $0 \leq n_0 \leq L-1$, while imposing no constraints upon the AFB.

Further improvement in reconstruction robustness can be achieved by carefully readjusting the number of subbands L and the bandwidth of $H(e^{j\omega})$. Notice, however, that such readjustment should not impair the time–frequency selectivity of the resulting D-STFT. This observation suggests imposing a desired range constraint upon the SOR of the analysis subband pattern: attempting to minimize sensitivity without imposing such an SOR constraint can result in highly non-selective AFB designs, such as $H(e^{j\omega}) = 1$ for all ω.

An example of this "careful readjustment" approach is provided by considering a parameterized family of AFB choices, such as used in the design of windowed-FIR phasor-banks: in this case, combining (4.34) with (4.41) results in a constant product property, viz.,

$$\text{Relative sensitivity} \times \text{SOR} = \left(\frac{\Delta_f^{(PD\star 3)}}{F_s}\right) \left[\sum_{\ell=0}^{n_0} \left|h[\ell]\right|^2\right]^{-1/2}$$

Assuming a fixed choice of the reconstruction delay n_0, this product depends only on the synthesis prototype $H(z)$. Also notice that the ratio $\Delta_f^{(PD\star 3)}/F_s$ is, in fact, the 3dB-bandwidth of the discrete-time FIR low-pass filter $H(z) = \sum_{\ell=0}^{N_h} h[\ell]\, z^{-\ell}$.

In other words, the relative sensitivity of a W-FIR scheme is always inversely proportional to the SOR of its discrete-time AFB prototype $H(z)$, as shown in Fig. 4.15. This figure describes the sensitivity-SOR relation resulting from varying both the number of subbands L and the normalized length N_h of a Hamming AFB prototype: to be specific, the ranges used in this figure are $20 \leq L \leq 40$ and $10 \leq N_h \leq 20$. We observe that the mapping from a given integer-valued parameter pair (L, N_h) to a point in the sensitivity-SOR plane (i.e., the image of the (L, N_h)-pair) is *one-to-one*: distinct (L, N_h)-pairs map to distinct image points. Also, notice that decreasing N_h and/or increasing L results in increased SOR (and hence reduced sensitivity), while implementation cost is proportional to the product $N_h L$. We conclude that plotting such trade-off curves makes it possible to determine an acceptable compromise between time–frequency

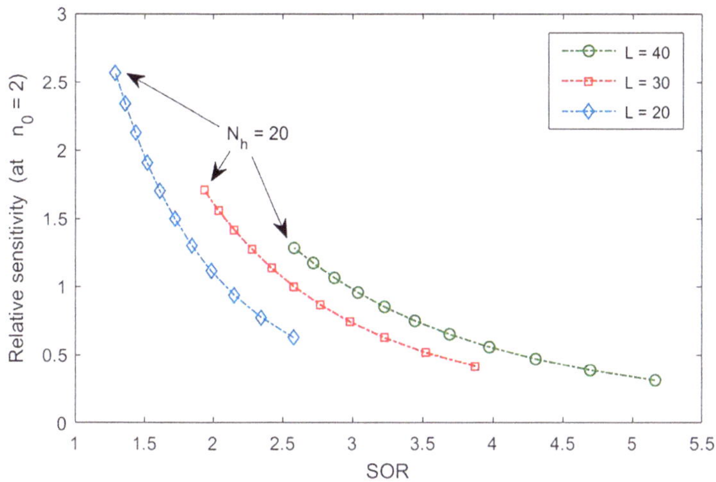

Fig. 4.15 Sensitivity-SOR trade-off for a W-FIR phasor-bank: $F_s/f_0 = 10$, $L = 20{:}10{:}40$, Hamming AFB baseband prototype with $N_h = 20{:}\text{-}1{:}10$.

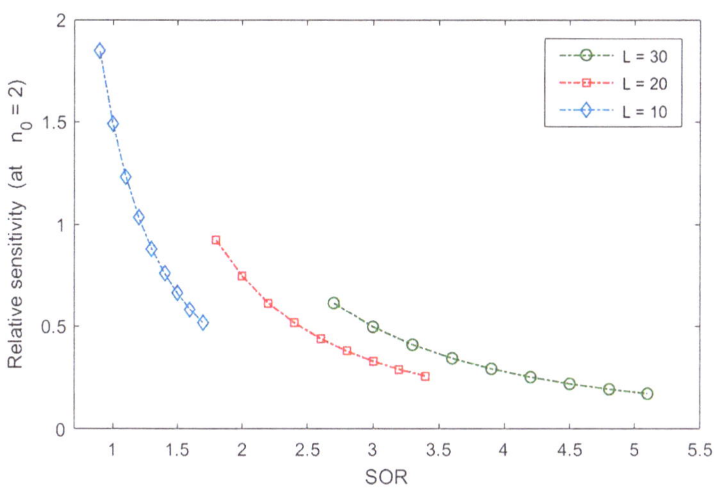

Fig. 4.16 Sensitivity-SOR trade-off for a C-FIR phasor-bank: $F_s/f_0 = 10$, $L_\gamma = 3$, $L = 10{:}10{:}30$, 3-rd order Butterworth AFB baseband prototype with variable 3 dB-cutoff frequency.

selectivity (as captured by the value of SOR), relative reconstruction sensitivity, and overall implementation cost.

Similar results are obtained with the constrained-FIR scheme: varying both the number of subbands and the bandwidth of the IIR baseband AFB prototype $H(z) = \frac{b(z)}{a(z)}$ produces a monotone decreasing sensitivity-SOR relation, as shown in Fig. 4.16. This C-FIR scheme uses a variable-bandwidth 3-rd order low-pass Butterworth filter as the AFB baseband prototype: the corresponding SFB baseband prototype $G(z) = \gamma(z)a(z)$ is constructed via the expression (4.37a), with $L_\gamma = 3$. The ranges of variable design parameters are $\frac{0.9}{20} \leq f_{cb} \leq \frac{1.7}{20}$ for the normalized 3 dB-cutoff frequency of the Butterworth filter, and $L = 10, 20, 30$, for the number of subbands. In this case, too, the sensitivity-SOR relation is monotone decreasing, for either fixed L or fixed f_{cb}. Again, we conclude that plotting such trade-off curves makes it possible to determine an acceptable compromise between time–frequency selectivity, relative reconstruction sensitivity, and overall implementation cost.

Summary of Design Guidelines

The design of a causal, invertible, window-based D-STFT (such as the W-FIR phasor-bank) still involves the guidelines specified at the end of Sect. 4.5.1, as well as the following added considerations:

- The sensitivity-SOR trade-off: minimizing sensitivity should always be subject to a range constraint on the SOR. The lowest sensitivity is achieved at the upper end of the chosen SOR-range.
- The sensitivity-cost trade-off: the lowest sensitivity within a specified SOR-range may be achieved at the expense of increased implementation cost.

Again, we should point out that accurate depiction of each component of a power system waveform – harmonic, inter-harmonic, wide-band transient – requires a dedicated D-STFT design. This observation suggests using multiple TFR schemes operated in parallel, as already proposed in Sect. 4.5.1.

Regarding the design of IIR-based D-STFT schemes, we observe that Figs. 4.15–4.16 suggest a conceptual similarity between W-FIR and C-FIR: the role played by the W-FIR parameter pair (L, N_h) in determining a point in the sensitivity-SOR plane is taken over by the C-FIR parameter pair (L, f_{cb}^{-1}). Nevertheless, the design guidelines specified here, as well as the ones at the end of Sect. 4.5.1, do not apply to fully-optimized C-FIR design because (a) an IIR analysis filter, such as a Butterworth prototype, has no nulls in its stop-band, which means that there is no meaningful IIR counterpart to the special relation $N_h = m\, N_{\text{cycle}}$, and (b) while implementation cost of a W-FIR scheme is proportional to N_h, the cost of a C-FIR scheme is independent of the cut-off frequency f_{cb}. Consequently, fully-optimized design of C-FIR phasor-banks is still an open research problem.

4.6 Discrete Offline Implementation of the STFT

An alternative (batch) method for numerical evaluation of the STFT is obtained by discretizing the *frequency-domain expression* (4.13), namely

$$\mathcal{X}(t\,;\,f_c) \;=\; \int_{-\infty}^{\infty} H\big(j(\omega - 2\pi f_c)\big)\, X(j\omega)\, e^{j\omega t}\, \frac{d\omega}{2\pi} \tag{4.43}$$

where $\mathcal{X}(t\,;\,f_c)$ is the passband-TFR of the waveform $x(\cdot)$. Construction of a discrete-time/discrete-frequency equivalent of (4.43) involves three conceptual steps:

- Discretizing frequency, i.e., setting $f_c = \{k\,\frac{F_s}{L}\,;\; 0 \le k \le K\}$, where $K = \mathsf{floor}\left(\frac{L}{2}\right)$. In other words, we define $K+1$ uniform-width subbands, which are centered at the frequencies $f_{c,k} = k\,\frac{F_s}{L}$.
- Discretizing the integral expression (4.43) by setting a uniform frequency grid in the finite interval $\omega_{min} \le \omega \le \omega_{max}$, with a grid spacing of $\Delta\omega = 2\pi\,\frac{F_s}{M}$, where M is an integral multiple of L.
- Discretizing time, i.e., setting $t = \{nT_s\,;\; 0 \le n \le N_x - 1\}$.

The first step produces the subband output waveforms

$$\mathcal{X}_k(t) \;\stackrel{\text{def}}{=}\; \mathcal{X}(t\,;\,f_{c,k}) \;=\; \int_{-\infty}^{\infty} H\big(j(\omega - 2\pi f_{c,k})\big)\, X(j\omega)\, e^{j\omega t}\, \frac{d\omega}{2\pi}$$

Next, we discretize the integral using the uniform grid $\omega = \{2\pi\, m\,\frac{F_s}{M}\}$. Recall that the time-sampling rate $F_s = \frac{1}{T_s}$ was chosen sufficiently high so as to make $X(j\omega)$ negligible outside the frequency band $|\omega| < 2\pi\,\frac{F_s}{2}$. This suggests setting[8]

$$\omega_{min} \;=\; -\,2\pi\,\frac{F_s}{2} \quad , \qquad \omega_{max} \;=\; 2\pi\,\frac{F_s}{2}$$

which results in the expression

$$\mathcal{X}_k(t) \;\approx\; \frac{F_s}{M} \sum_{m=-m_0}^{M-1-m_0} H\left(j\,2\pi\left[m\,\frac{F_s}{M} \;-\; k\,\frac{F_s}{L}\right]\right)\, X\left(j\,2\pi\, m\,\frac{F_s}{M}\right)\, e^{j\,(2\pi m F_s/M)\,t}$$

where

$$m_0 \;\stackrel{\text{def}}{=}\; \mathsf{floor}\left(\frac{M}{2}\right) \tag{4.44}$$

so that $m = 0$ falls in the middle of the range of summation.

[8] See [152, 160–162] for alternative choices of ω_{min} and ω_{max} .

In the third step we discretize time by setting $t = \{nT_s\}$, which results in the approximation

$$\mathcal{X}_k(nT_s) \approx \frac{F_s}{M} \sum_{m=-m_0}^{M-1-m_0} H\Big(j\,2\pi\Big[m\,\frac{F_s}{M} - k\,\frac{F_s}{L}\Big]\Big)\, X\Big(j\,2\pi\, m\,\frac{F_s}{M}\Big)\, e^{j\,(2\pi/M)\,mn} \tag{4.45}$$

Notice that in this offline scheme the total number (L) of discrete f_c values is upper-bounded by the density of the underlying frequency grid, namely $L \le M$.

The simplest choice is to use the entire discretized frequency grid, so that $L = M$, and thus

$$\mathcal{X}_k(nT_s) \approx \frac{F_s}{M} \sum_{m=-m_0}^{M-1-m_0} H\Big(j\,2\pi\,\frac{m-k}{M}\,F_s\Big)\, X\Big(j\,2\pi\,\frac{m}{M}\,F_s\Big)\, e^{j\,(2\pi/M)\,mn} \tag{4.46}$$

where $-m_0 \le k \le M-1-m_0$. Notice, however, that the guidelines for selection of subband density provided in Sect. 4.5 still apply, which means that a reasonable choice for the parameter L should be a moderate multiple of N_h, the normalized length of the STFT window. Increasing the value of L up to its upper limit (i.e., setting $L = M$) offers no advantage in terms of frequency-resolution: rather, it results in over-abundance of subbands, and thus a very high implementation cost.

Our final step is to replace $X(j\omega)$, the continuous-time Fourier transform of the waveform $\{x(t)\,;\ -\infty < t < \infty\}$, by the "zero-padded" M-point Discrete Fourier Transform (DFT) of the sampled waveform segment $\{x(nT_s)\,;\ 0 \le n \le N_x - 1\}$, namely,

$$X[m] \stackrel{\text{def}}{=} \sum_{n=0}^{N_x-1} x(nT_s)\, e^{-j\,(2\pi/M)\,mn}\ , \qquad -m_0 \le m \le M-1-m_0 \tag{4.47}$$

where the constraint $M \ge N_x$ has to be satisfied in order to avoid aliasing [32, 33]. The DFT expression (4.47) is usually evaluated via the computationally-efficient Fast Fourier Transform (FFT) algorithm. Notice that $X[m]$ is periodic with period M, by its definition: this means that $X[-1] \equiv X[M-1]$, $X[-2] \equiv X[M-2]$, etc., so that the range of m in (4.47) could have been chosen as any M-long interval. However, it will be convenient to continue using the zero-centered range $-m_0 \le m \le M-1-m_0$.

We will refer to (4.47) as the *centered DFT* of the sampled waveform segment $\{x(nT_s)\,;\ 0 \le n \le N_x - 1\}$. Comparing this DFT expression with the Fourier transform (4.9), i.e., $X(j\omega) = \int_{-\infty}^{\infty} x(t)\; e^{-j\omega t}\; dt$, we conclude that $T_s\, X[m] \approx X\big(j\,2\pi\,\frac{m}{M}\,F_s\big)$ for all $-m_0 \le m \le M-1-m_0$. The quality of this approximation depends on the integer-valued parameter M, which controls the density of the frequency grid, as well as on the length of the sampled waveform segment $\{x(nT_s)\,;\ 0 \le n \le N_x - 1\}$ used to determine the DFT $X[m]$.

Increasing N_x improves the frequency-resolution of the DFT by minimizing unwanted artifacts caused by *spectral leakage*, such as false indication of non-negligible magnitudes at frequencies where $|X(j\omega)|$ is negligible.

Thus, the final discrete-time/discrete-frequency version of the integral expression (4.43) is, for $0 \leq n \leq N_x - 1$,

$$\mathcal{X}_k(nT_s) \equiv \mathcal{X}\Big(nT_s\,;k\frac{F_s}{M}\Big) \approx \mathcal{X}_k[n] \;, \qquad -m_0 \leq k \leq M-1-m_0 \tag{4.48a}$$

where we recall that $m_0 = \text{floor}\left(\frac{M}{2}\right)$, and $\mathcal{X}_k[n]$ is given by

$$\mathcal{X}_k[n] \stackrel{\text{def}}{=} \frac{1}{M} \sum_{m=-m_0}^{M-1-m_0} H\Big(j\,2\pi\,\frac{m-k}{M}\,F_s\Big)\;X[m]\,e^{j\,(2\pi/M)\,mn} \tag{4.48b}$$

The discretized baseband-TFR $X_k(nT_s)$ satisfies similar expressions, viz.,

$$X_k(nT_s) \equiv e^{-j\,(2\pi/M)\,kn}\,\mathcal{X}_k(nT_s) \approx X_k[n] \;, \qquad -m_0 \leq k \leq M-1-m_0 \tag{4.49a}$$

where

$$X_k[n] \stackrel{\text{def}}{=} \frac{1}{M} \sum_{m=-m_0}^{M-1-m_0} H\Big(j\,2\pi\,\frac{m-k}{M}\,F_s\Big)\;X[m]\,e^{j\,(2\pi/M)\,(m-k)n} \tag{4.49b}$$

Notice that the discrete STFT expressions (4.48) and (4.49) can only be computed offline: their input is the DFT of the entire sampled waveform record $\{x(nT_s)\,;\;0 \leq n \leq N_x - 1\}$, which must to be evaluated *in advance.*

Matrix Formulation

Equivalently, we can express the offline discrete STFT relation (4.48) in matrix-vector notation, involving a product of three matrices, viz.,

$$\begin{pmatrix} \mathcal{X}_{-m_0}[0] & \mathcal{X}_{-m_0}[1] & \ldots & \mathcal{X}_{-m_0}[N_x-1] \\ \mathcal{X}_{1-m_0}[0] & \mathcal{X}_{1-m_0}[1] & \ldots & \mathcal{X}_{1-m_0}[N_x-1] \\ \mathcal{X}_{2-m_0}[0] & \mathcal{X}_{2-m_0}[1] & \ldots & \mathcal{X}_{2-m_0}[N_x-1] \\ \vdots & \vdots & & \vdots \\ \mathcal{X}_{M-1-m_0}[0] & \mathcal{X}_{M-1-m_0}[1] & \ldots & \mathcal{X}_{M-1-m_0}[N_x-1] \end{pmatrix} = \mathcal{H}\;\boldsymbol{D}_X\;\mathcal{U} \tag{4.50a}$$

where

- $\mathcal{H}$ is an $M \times M$ *Toeplitz matrix*, with elements determined by $H(j\omega)$, the *continuous-time Fourier transform* of the window function $h(t)$, viz.,

$$\mathcal{H} = \begin{pmatrix} H_0 & H_1 & H_2 & \dots & H_{M-1} \\ H_1^* & H_0 & H_1 & & \vdots \\ H_2^* & H_1^* & H_0 & \ddots & \vdots \\ \vdots & & \ddots & \ddots & \vdots \\ H_{M-1}^* & \dots & \dots & \dots & H_0 \end{pmatrix} \tag{4.50b}$$

where $H_\ell \stackrel{\text{def}}{=} H(j2\pi \frac{\ell}{M} F_s)$ for $0 \le \ell \le M-1$, and the superscript * denotes complex conjugation. Also, we rely on the fact that $H(j\omega)$ is conjugate-symmetric, i.e., $H(-j\omega) = H^*(j\omega)$.

- $\boldsymbol{D}_X$ is an $M \times M$ diagonal matrix, with elements determined by the centered DFT $\{X[m]\,;\ -m_0 \le m \le M-1-m_0\}$, viz.,

$$\boldsymbol{D}_X = \begin{pmatrix} X[-m_0] & 0 & \dots & 0 & & 0 \\ 0 & X[1-m_0] & \ddots & & & \vdots \\ \vdots & \ddots & \ddots & \ddots & & \vdots \\ 0 & & \ddots & X[0] & \ddots & \vdots \\ \vdots & & & \ddots & \ddots & 0 \\ 0 & \dots & \dots & \dots & 0 & X[M-1-m_0] \end{pmatrix} \tag{4.50c}$$

The centered DFT can be obtained, for instance, by applying the MATLAB function `fft(x,M)` to the sampled waveform record $\{x(nT_s)\,;\ 0 \le n \le N_x - 1\}$, followed by the `fftshift` function.

- $\mathcal{U}$ is an $M \times N_x$ matrix, similar to the well-known inverse DFT-matrix [32,33], viz.,

$$\mathcal{U} = \frac{1}{M} \begin{pmatrix} 1 & w_M^{(-m_0)} & w_M^{2(-m_0)} & \dots & w_M^{(N_x-1)(-m_0)} \\ 1 & w_M^{(1-m_0)} & w_M^{2(1-m_0)} & \dots & w_M^{(N_x-1)(1-m_0)} \\ 1 & w_M^{(2-m_0)} & w_M^{2(2-m_0)} & \dots & w_M^{(N_x-1)(2-m_0)} \\ \vdots & \vdots & \vdots & \ddots & \vdots \\ 1 & w_M^{(M-1-m_0)} & w_M^{2(M-1-m_0)} & \dots & w_M^{(N_x-1)(M-1-m_0)} \end{pmatrix} \tag{4.50d}$$

$$\equiv \mathcal{U}_{\mathsf{IDFT}} \begin{pmatrix} 1 & 0 & \cdots & \cdots & 0 \\ 0 & w_M^{-m_0} & \ddots & & \vdots \\ \vdots & \ddots & w_M^{-2m_0} & \ddots & \vdots \\ \vdots & & \ddots & \ddots & 0 \\ 0 & \cdots & \cdots & 0 & w_M^{-(N_x-1)m_0} \end{pmatrix}$$

where we define $w_M \stackrel{\text{def}}{=} e^{j(2\pi/M)}$, and

$$\mathcal{U}_{\mathsf{IDFT}} = \frac{1}{M} \begin{pmatrix} 1 & 1 & 1 & \cdots & 1 \\ 1 & w_M & w_M^2 & \cdots & w_M^{N_x-1} \\ 1 & w_M^2 & w_M^4 & \cdots & w_M^{2(N_x-1)} \\ \vdots & \vdots & \vdots & \ddots & \vdots \\ 1 & w_M^{(M-1)} & w_M^{2(M-1)} & \cdots & w_M^{(N_x-1)(M-1)} \end{pmatrix} \tag{4.50e}$$

Since $w_M^{-i} = w_M^{M-i}$, the matrix $\mathcal{U}$ can also be viewed as a row-permuted version of the inverse DFT-matrix $\mathcal{U}_{\mathsf{IDFT}}$.

When N_x is large (say, more than 1000 samples), the discretized frequency grid is sufficiently dense, so that choosing $M > N_x$ offers no advantages. In particular, increasing M has little impact on the frequency-resolution of the TFR $\mathcal{X}_k[n]$.

Frequency Marginal Property

The choice $M = N_x$ also results in a useful relation between the time-summation of the baseband-TFR $X_k[n]$, and the DFT of the sampled waveform segment $\{x(nT_s)\,;\ 0 \le n \le M-1\}$, viz.,

$$\begin{aligned} \sum_{n=0}^{M-1} X_k[n] &= \sum_{m=-m_0}^{M-1-m_0} H\Big(j\,2\pi\,\frac{m-k}{M}\,F_s\Big)\; X[m] \left[\frac{1}{M}\sum_{n=0}^{M-1} e^{j\,(2\pi/M)\,(m-k)n}\right] \\ &= \sum_{m=-m_0}^{M-1-m_0} H\Big(j\,2\pi\,\frac{m-k}{M} F_s\Big)\; X[m]\,\delta[m-k] \\ &= X[k] \end{aligned} \tag{4.51}$$

where $\delta[m]$ denotes the *Kronecker delta function*, and we used the unity DC-gain property $H(0) = 1$. This relation is a discretized *offline equivalent* of (4.10): it involves the finite summation $\sum_{n=0}^{M-1} X_k[n]$ and the DFT $\{X[k]\,;\ -m_0 \leq k \leq M-1-m_0\}$ of the finite sampled waveform record $\{x(nT_s)\,;\ 0 \leq n \leq M-1\}$. In contrast, the online version (4.19) involves the infinite summation $\sum_{n=-\infty}^{\infty} X_k[n]$ and the DTFT of the corresponding infinite waveform record.

Computational Cost

The online scheme (4.20) consists of L filters, each one of length N_h, carrying out the convolutions $h_k[n] \circledast x(nT_s)$ for $0 \leq k \leq L-1$. Since the waveform $x(\cdot)$ is real-valued, while $h_k[n]$ is complex-valued, the number of real multiplications used to determine the output $\mathcal{X}_k[n]$ (for a given k and n) is $2N_h$. Thus, the cost of the entire scheme is $2LN_h$ real multiplications per input sample. Actually, since the STFT is conjugate-symmetric (in f_c), we only need to evaluate $\mathcal{X}_k[n]$ for $0 \leq k \leq M-1-m_0$, which cuts the computational cost in half, to LN_h real multiplications per input sample. Also, recall that the number of subbands (L) is a design parameter, constrained only by the requirement $L \geq N_h$.

In contrast, the computational cost of the offline scheme (4.50) is significantly higher. One reason for this difference is that the offline scheme typically uses many more subbands: the expression (4.48) involves M subbands, where $M \geq N_x$. The number of subbands can be reduced (from M to some $L < M$) by thinning out the set of k-values (i.e., the set of discrete f_c values) used in (4.50). We now observe that evaluating $\mathcal{X}_k[n]$ via (4.48) for a given k and n requires $8M$ real multiplications (the equivalent of 2 complex multiplications). Since the added cost of evaluating the FFT $X[m]$ is negligible, the overall computational cost of the offline scheme is $\mathcal{O}(4LM)$ real multiplications per input sample, where we assume that only $\frac{L}{2}$ discrete frequency values are used, due to conjugate symmetry. This is still quite higher than the cost of the online scheme, because $M \gg N_h$.

Inversion of the Transform

The matrix expression (4.50) defines a mapping from the FFT $\{X[m]\,;\ -m_0 \leq m \leq M-1-m_0\}$ to the STFT $\{\mathcal{X}_k[n]\,;\ -m_0 \leq k \leq M-1-m_0\,,\ 0 \leq n \leq N_x-1\}$. This mapping admits a (non-unique) inverse under mild regularity conditions. In particular, when $M = N_x$: (i) the matrix $\boldsymbol{\mathcal{U}}$ is square and always non-singular, and (ii) the matrix $\boldsymbol{\mathcal{H}}$ is also square, and almost always non-singular. Consequently, the diagonal matrix $\boldsymbol{D}_X$ can be recovered from the time–frequency representation $\{\mathcal{X}_k[n]\,;\ -m_0 \leq k \leq M-1-m_0\,,\ 0 \leq n \leq N_x-1\}$. A more comprehensive discussion of STFT invertibility is provided in Sect. 4.4.

Chapter 5 Beyond Phasors: Variable-Bandwidth Transforms

Time-frequency representations (TFR) that employ variable-bandwidth subbands have found numerous applications in the analysis of acoustic, seismic, and biomedical waveforms containing intermittent aperiodic transients. A variable-bandwidth TFR can provide superior time-resolution for high frequency components (such as wide-band fault transients) at the expense of reduced frequency-resolution. The extensive technical literature on this subject lists a plethora of variable-bandwidth transforms. Some of the better known ones are:

- Generalized (variable-bandwidth) STFT (see, e.g., [145–147])
- Continuous Wavelet Transform (CWT) [148, Chap. 2] (see also [149])
- Discrete Wavelet Transform (DWT) [148, Chap. 3] (see also [150])
- Discrete Orthonormal S-Transform (DOST) [151]
- S-transform [152]

The first four of these distinct approaches share a common foundation: a *constant-Q subband pattern*. In broad terms, this means that their frequency-discretized implementations consist of a single-input/multiple-output bank of *band-pass filters* $\{H_k(s)\,;\;1 \le k \le K\}$, such that both the center frequency and the bandwidth of the magnitude response $|H_k(j\omega)|$ are proportional to Γ^k, where $\Gamma > 1$ is a design parameter known as a *scaling factor*, which is used to control subband density. The primary difference between such contending time-frequency analysis schemes – in addition to the value selected for the scaling factor Γ – is their choice of the subband transfer functions $\{H_k(s)\,;\;1 \le k \le K\}$, as detailed in Sect. 5.1.

In this chapter we advocate a design methodology that generates a constant-Q subband pattern first, without relying on any information about the intended choice of subband filters. The only information used to construct a subband pattern consists of the scaling factor Γ and the total number of subbands K: see Sect. 5.2 for a detailed construction procedure. This approach enables customized design of a subband pattern to address the needs of any specific application. Once the subband

H. Lev-Ari, A. M. Stanković, *Dynamic Phasors in Energy Processing Systems*, Power Electronics and Power Systems, https://doi.org/10.1007/978-3-031-50020-6_6

pattern has been determined, a set of subband filters $\{H_k(s)\,;\ 1 \le k \le K\}$ can be designed to match the given subband pattern, as demonstrated in Sect. 5.3. Moreover, we point out again, as we did in Chap. 4, that critical down-sampling of subband outputs is hardly the only possibility: non-decimated implementation – which is essential in many power system applications – is always a viable, and often desirable, option [131].

We should reiterate here a distinction made in Chap. 4 between two very different categories of applications, namely:

- **Category AO (analysis only):** applications that involve only generation of a time-frequency representation (TFR) from a given input waveform.
- **Category AR (analysis and reconstruction):** applications that rely on perfect reconstruction of the original input waveform from its TFR, which requires invertibility of the waveform-to-TFR map.

Of the five transform approaches listed earlier, only the generalized-STFT and the CWT allow flexible control of frequency-resolution, time-resolution, inter-subband leakage, and output rate of the resulting TFR. In contrast, use of the DWT or DOST is justified primarily for category AR applications, since each one of these transforms has an easy to compute unique inverse. However, the DWT and DOST lack the flexibility offered by the CWT and the generalized-STFT and thus are less suited for category AO applications: both have a pre-determined subband structure, corresponding to the unalterable choice $\Gamma = 2$ that results in a very low subband density (i.e., a single subband per octave), and both rely on critical down-sampling of their subbands. Consequently, any well-designed constant-Q transform is at least as effective as the DWT when used in applications of the AO category [153].

Using a constant-Q transform in *power system applications* presents a unique design challenge, originating from two conflicting sets of specifications, namely:

- Accurate representation of dominant harmonics requires constant-bandwidth subbands, centered on the harmonics of interest – say, odd harmonics up to the 11th[1] – whose width is comparable with the fundamental frequency f_0.
- Agile detection of the onset of a wide-band transient requires improved time-resolution, which mandates using considerably wider high-frequency subbands.

Clearly, a conventional constant-Q transform, such as DWT or CWT, cannot satisfy both sets of requirements, as often observed in the power system literature (see, e.g., [112, Sects. 4.3.7–4.3.8]). We propose in Sect. 5.4 a hybrid approach that addresses this design challenge by using constant-bandwidth lower-frequency subbands alongside constant-Q subbands for the higher frequency range.

[1] These harmonics are identified by IEEE Standard 519-2014 as dominant in power system current waveforms [154, Tables 2 and 3].

The S-transform – a popular choice with power system practitioners – is a special variable-bandwidth time-frequency representation that employs *uniformly-spaced subband center frequencies*, and uses a scalable Gaussian window to generate the entire set of subband filters $\{H_k(s)\}$. Since the bandwidth of each subband is made proportional to its center frequency, the resulting $H_k(s)$ is a constant-Q filter. This also means that the resulting subband pattern is highly inefficient, due to excessive overlap between high-index subbands, as detailed in Sect. 5.5.

The wide variety of subband filter choices that can be used to generate a constant-Q time-frequency representation (TFR) for a given input waveform $x(\cdot)$ is severely reduced in the context of category AR applications, which rely on efficient *error-free reconstruction* of $x(\cdot)$ from its TFR. One convenient approach to construction of an invertible transform is via waveform representation in a suitably chosen orthogonal basis. However, in order to be applicable to time-frequency analysis, all members of such a basis must be *localized in both frequency and time*. This added constraint further limits the available choices of invertible constant-Q transforms for category AR applications. We describe in Sect. 5.6 two of the best known transforms of this kind: the Discrete Wavelet Transform (DWT), and the Discrete Orthogonal S-Transform (DOST).

We should point out that associating a TFR with an orthogonal basis requires removing all redundancy, which is achieved by sampling each subband output at its critical rate. Since this step reduces subband processing rates and increases overall delay, accepting such a compromise makes sense only if transform invertibility is an absolutely necessary feature. Otherwise, one should use a more flexible choice, such as a generalized-STFT or a continuous wavelet transform: this often overlooked key observation is elucidated also in [149, 150].

Finally, we observe that the squared magnitude of a TFR, otherwise known as a *spectrogram*, can also be interpreted in terms of a *Wigner–Ville distribution* (VWD), as explained in Sect. 5.7. To be specific, the VWD of the k-th subband output is a smoothed version of the WVD of the input waveform $x(\cdot)$, where the smoothing kernel for each subband is determined by the impulse response of the corresponding subband filter $H_k(s)$. The shape of this smoothing kernel provides a means for quantitative comparison of distinct subband filter families.

5.1 Generalized Transforms with Frequency-Dependent Bandwidth

Linear transforms with frequency-dependent bandwidth are a natural extension of the fixed-bandwidth STFT, which was presented in Chap. 4. Starting from the expressions (4.5) and (4.6), i.e.,

$$\mathcal{X}(t\,;\,f_c) \stackrel{\text{def}}{=} X(t\,;\,f_c)\,e^{j2\pi f_c t} = \int_{-\infty}^{\infty}\left[h(\tau)\,e^{j2\pi f_c \tau}\right]x(t-\tau)\,d\tau$$

which is the *passband version* of the STFT (see also Fig. 4.7), we now consider the *generalized-STFT* (in passband form), viz.,

$$\mathcal{X}(t\,;\,f_c) \;=\; \int_{-\infty}^{\infty} h(\tau\,;\,f_c)\,x(t-\tau)\,d\tau \;\equiv\; h(t\,;\,f_c) \circledast x(t) \tag{5.1a}$$

where $\circledast$ denotes convolution, and $h(t\,;\,f_c)$ is a *band-pass filter* whose bandwidth is a function of its center frequency f_c. The equivalent *baseband version* is, of course,

$$X(t\,;\,f_c) \;\stackrel{\text{def}}{=}\; e^{-j2\pi f_c t}\,\mathcal{X}(t\,;\,f_c) \tag{5.1b}$$

so that $|X(t\,;\,f_c)| = |\mathcal{X}(t\,;\,f_c)|$. It will be helpful to interpret f_c as a *parameter*, rather than as a variable, i.e., we consider $\mathcal{X}(t\,;\,f_c)$ (and $X(t\,;\,f_c)$) as a parameterized univariate function of the continuous-time variable "t." As in Chap. 4, we consider here only the case $f_c > 0$.

Notice that, when both $x(t)$ and $h(t\,;\,f_c)$ are *real-valued*, the generalized-STFT output $\mathcal{X}(t\,;\,f_c)$ will be severely impacted by leakage from negative frequency components of the input waveform $x(t)$. For instance, consider the sinusoidal waveform $x(t) = 2\cos\omega_1 t$: in this case $\mathcal{X}(t\,;\,f_c) = H(j\omega_1\,;\,f_c)\,e^{j\omega_1 t} + H(-j\omega_1\,;\,f_c)\,e^{-j\omega_1 t}$, where $H(-j\omega\,;\,f_c) = H^*(j\omega\,;\,f_c)$, and

$$H(s\,;\,f_c) \;\stackrel{\text{def}}{=}\; \mathcal{L}\{h(t\,;\,f_c)\} \tag{5.1c}$$

Consequently, $|\mathcal{X}(t\,;\,f_c)| = |H(j\omega_1\,;\,f_c) + H^*(j\omega_1\,;\,f_c)\,e^{-2j\omega_1 t}|$ which oscillates, as a function of time, between zero and $2|H(j\omega_1\,;\,f_c)|$. This undesirable situation can be avoided if either $x(t)$ or $h(t\,;\,f_c)$ has negligible negative frequency content. Consequently, one can use two different approaches to avoid excessive leakage from negative frequency components:

- Replace the input waveform $x(t)$ by its analytic version $\widehat{x}(t) = x(t) + j\mathcal{H}\{x(t)\}$ (recall (3.39a)). By definition, the negative frequency content of an analytic waveform is zero.
- Choose an analytic, or nearly-analytic, filter $h(t\,;\,f_c)$. We say that a filter is *nearly-analytic* if it has negligible magnitude response for $\omega < 0$, namely, $|H(j\omega\,;\,f_c)|\Big|_{\omega<0} \ll \max_\omega |H(j\omega\,;\,f_c)|$.

In this chapter – and, in fact, in the entire book – we focus mainly on the case of real-valued input waveforms. Thus, we can only rely on the second approach, i.e., use an $h(t\,;\,f_c)$ filter that is analytic, or nearly-analytic.

One useful category of transforms with variable bandwidth is obtained by setting

$$h(t\,;\,f_c) \;=\; \frac{1}{\sigma(f_c)}\,g\left(\frac{t}{\sigma(f_c)}\right)e^{j2\pi f_c t} \;\Longleftrightarrow\; H(s\,;\,f_c) = G\big(\sigma(f_c)(s-j2\pi f_c)\big) \tag{5.2}$$

where $H(s\,;\,f_c) \stackrel{\text{def}}{=} \mathcal{L}\{h(t\,;\,f_c)\}$, $g(t)$ is a *low-pass filter* whose transfer function $G(s) \stackrel{\text{def}}{=} \mathcal{L}\{g(t)\}$ does not depend on the parameter f_c and $\sigma(f_c)$ is an f_c-dependent scaling factor. We will refer to transforms that rely on (5.2) as belonging to the *low-pass-prototype category*. Notice that the filter $h(t\,;\,f_c)$ given by (5.2) is nearly-analytic by construction, provided that the low-pass prototype $G(j\omega)$ has reasonably low sidelobes. This category of variable-bandwidth transforms was first suggested in [126] and was analyzed in detail in [127], with a special focus on the case $\sigma(f_c) \sim \frac{1}{f_c}$.

Another useful category of variable-bandwidth transforms is obtained by setting

$$h(t\,;\,f_c) = \frac{1}{\sigma(f_c)}\, g\Big(\frac{t}{\sigma(f_c)}\Big) \quad \Longleftrightarrow \quad H(s\,;\,f_c) = G\big(\sigma(f_c)s\big) \tag{5.3}$$

where $g(t)$ is a *band-pass filter* whose transfer function $G(s) \stackrel{\text{def}}{=} \mathcal{L}\{g(t)\}$ does not depend on the parameter f_c, and, as before, $\sigma(f_c)$ is an f_c-dependent scaling factor. We will refer to transforms that rely on (5.3) as belonging to the *band-pass-prototype category*. Notice that the scaling factor $\sigma(f_c)$ controls both the bandwidth and center frequency of the filter $H(s\,;\,f_c)$. The well-known Continuous Wavelet Transform[2] [132, 148], viz.,

$$T(a,t) = \int_{-\infty}^{\infty} x(\tau)\, \psi^*\Big(\frac{\tau - t}{a}\Big)\, d\tau$$

comprises a special sub-family within this category. It is obtained from (5.3) by making the choices $a = \sigma(f_c)$ and $g(t) = \psi^*(-t)$, where the "mother wavelet" function $\psi(t)$ satisfies several additional regularity constraints. The explicit relation between the scaling factor "a" and the center frequency f_c, namely, $a \sim \frac{1}{f_c}$, follows implicitly from the definition of the continuous wavelet transform (see also the discussion in Sects. 5.2 and 5.3). Notice that, in order to avoid excessive leakage, transforms in this category must employ analytic or nearly-analytic choices of the band-pass prototype $g(t)$, such as Morlet or Morse wavelets: see Sect. 5.3.2 for further detail.

Filter-bank Implementation

When used in practical applications, the continuous-time/continuous-frequency transform (5.1a) can be evaluated only for a finite number of distinct frequencies, say, $\{f_{c,k}\,;\; 1 \le k \le K\}$. The *frequency-discretized version* of (5.1a) is

$$\mathcal{X}_k(t) = h_k(t) \circledast x(t) \quad , \qquad 1 \le k \le K \tag{5.4a}$$

[2] The MATLAB function "cwt" is a discretized version of the continuous wavelet transform with several built-in choices of the mother wavelet $\psi(t)$ (see, e.g., [149]).

where

$$\mathcal{X}_k(t) \stackrel{\text{def}}{=} \mathcal{X}(t\,;\,f_{c,k}) \quad , \quad h_k(t) \stackrel{\text{def}}{=} h(t\,;\,f_{c,k}) \tag{5.4b}$$

This frequency-discretized version of the variable bandwidth transform (5.1a) can be interpreted and implemented as a filter-bank, consisting of K subbands (Fig. 5.1). Again, recall that we interpret $\mathcal{X}_k(t)$ as a parameterized univariate function (of the time variable "t") that depends on the parameter $f_{c,k}$.

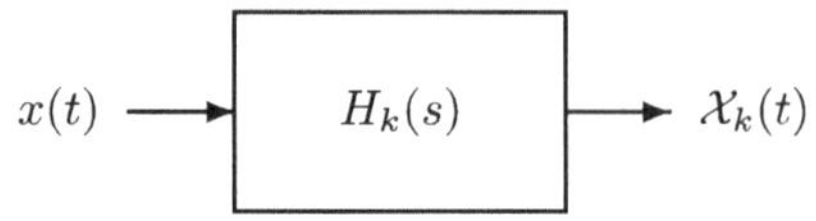

Fig. 5.1 Filtering interpretation of the discretized variable-bandwidth transform.

Each subband component $\mathcal{X}_k(t)$ is generated by a dedicated band-pass filter $H_k(s)$. The magnitude response $|H_k(j\omega)|$ of this subband filter is centered at the frequency $f_{c,k}$, and its bandwidth varies with the subband index "k." Here, too, we can define an equivalent *baseband version*, viz.,

$$X_k(t) \stackrel{\text{def}}{=} e^{-j2\pi f_{c,k}t}\,\mathcal{X}_k(t) \tag{5.4c}$$

so that $|X_k(t)| = |\mathcal{X}_k(t)|$ and $\arg\{X_k(t)\} = \arg\{\mathcal{X}_k(t)\} - 2\pi f_{c,k}t$.

The frequency-discretized version of the band-pass-prototype category characterization (5.3) is, of course,

$$h_k(t) = \frac{1}{\sigma_k}\,g\left(\frac{t}{\sigma_k}\right) \quad \Longleftrightarrow \quad H_k(s) = G\big(\sigma_k\,s\big) \tag{5.5a}$$

where $\sigma_k = \sigma(f_{c,k})$. Similarly, the frequency-discretized version of the low-pass-prototype category characterization (5.2) is

$$h_k(t) = \frac{1}{\sigma_k}\,g\left(\frac{t}{\sigma_k}\right)\,e^{j2\pi f_{c,k}t} \quad \Longleftrightarrow \quad H_k(s) = G\big(\sigma_k(s - j2\pi f_{c,k})\big) \tag{5.5b}$$

The time-scaling parameters $\{\sigma_k\,;\;\; 1 \le k \le K\}$ control the subband pattern generated by selecting $g(t)$ and using the fundamental relation (5.5), as explained in Sect. 5.3.

The frequency-discretized transform (5.4a) can be implemented either in the time domain or in the frequency domain. A time-domain implementation relies on a discrete-time version of the continuous-time impulse response $h_k(t)$ of the k-th subband filter, which is obtained via the mapping (4.17b), viz.,

$$h_k[n] = T_s\,h_k(nT_s) \tag{5.6a}$$

where $T_s = {}^1\!/_{F_s}$, and F_s is the waveform sampling rate. The resulting *non-decimated* discrete version of the transform is

$$\mathcal{X}_k(t)\Big|_{t=nT_s} \approx h_k[n] \circledast x(nT_s) \tag{5.6b}$$

The details of frequency-domain implementation, including a non-decimated discrete version, are examined next.

Frequency-domain Version

The time-domain definition (5.1a) of the variable-bandwidth transform also admits an equivalent *frequency-domain* version, viz.,

$$\mathcal{F}_\omega\Big\{\mathcal{X}(t\,;\,f_c)\Big\} = H(j\omega\,;\,f_c)\,\mathcal{F}_\omega\Big\{x(t)\Big\} = H(j\omega\,;\,f_c)\,X(j\omega)$$

where $\mathcal{F}_\omega\{\,\cdot\,\}$ denotes the continuous-time Fourier transform, i.e.,

$$\mathcal{F}_\omega\Big\{x(t)\Big\} \stackrel{\text{def}}{=} \int_{-\infty}^{\infty} x(t)\,e^{-j\omega t}\,dt \tag{5.7}$$

Notice that $\mathcal{F}_\omega$ applies to the time variable t, mapping it into the radial-frequency variable ω. Next, applying an inverse Fourier transform to this identity generates the explicit frequency-domain representation

$$\mathcal{X}(t\,;\,f_c) = \int_{-\infty}^{\infty} H(j\omega\,;\,f_c)\,X\big(j\omega\big)\,e^{j\omega t}\,\frac{d\omega}{2\pi} \tag{5.8a}$$

The frequency-discretized version of this relation is obtained by using the specific center frequencies $\{f_{c,k}\,;\ 1 \le k \le K\}$, resulting in the representation

$$\mathcal{X}_k(t) = \int_{-\infty}^{\infty} H_k(j\omega)\,X\big(j\omega\big)\,e^{j\omega t}\,\frac{d\omega}{2\pi} \tag{5.8b}$$

which offers a convenient starting point for constructing an offline numerical implementation, using the Fourier transform of the input waveform as a starting point. In particular, setting $h_k(t) = \frac{1}{a_k}\,\psi^*\big(-\frac{t}{a_k}\big)$, where $\psi(t)$ is the "mother wavelet," recovers the "Fourier transform version" of the wavelet transform [148, Sect. 2.10], viz.,

$$T(a_k, t) = \int_{-\infty}^{\infty} \Psi^*(ja_k\,\omega)\,X\big(j\omega\big)\,e^{j\omega t}\,\frac{d\omega}{2\pi}$$

where $\Psi(j\omega) \stackrel{\text{def}}{=} \mathcal{F}_\omega\{\psi(t)\}$ denotes the Fourier transform of the wavelet $\psi(t)$.

The *baseband equivalent* of the frequency-domain expression (5.8a) is

$$X(t\,;\,f_c) \;=\; \int_{-\infty}^{\infty} H(j\omega\,;\,f_c)\,X\big(j\omega\big)\,e^{j\,(\omega-2\pi f_c)\,t}\,\frac{d\omega}{2\pi} \tag{5.9a}$$

and the corresponding frequency-discretized version is

$$X_k(t) \;=\; \int_{-\infty}^{\infty} H_k(j\omega)\,X\big(j\omega\big)\,e^{j\,(\omega-2\pi f_{c,k})\,t}\,\frac{d\omega}{2\pi} \tag{5.9b}$$

Both the baseband and passband versions of the transform can be implemented efficiently by relying on standard techniques for digital evaluation of integrals.

Numerical evaluation of (5.8) and (5.9) involves replacement of the continuous-time Fourier transform $X(j\omega)$ by the "zero-padded" M-point Discrete Fourier Transform (DFT) of the sampled waveform segment $\{x(nT_s)\,;\; 0 \le n \le N_x - 1\}$, namely (recall (4.47)),

$$X[m] \;\stackrel{\text{def}}{=}\; \sum_{n=0}^{N_x-1} x(nT_s)\,e^{-j\,(2\pi/M)\,mn} \;, \qquad -\,m_0 \le m \le M-1-m_0 \tag{5.10a}$$

where

$$m_0 \;\stackrel{\text{def}}{=}\; \text{floor}\left(\frac{M}{2}\right) \tag{5.10b}$$

and the constraint $M \ge N_x$ has to be satisfied in order to avoid aliasing [32, 33]. We will refer to (5.10) as a *centered DFT* of the sampled waveform segment: it is usually evaluated via the computationally-efficient Fast Fourier Transform (FFT) algorithm.[3] Notice that $X[m]$ is periodic in m with period M: this means that $X[-1] \equiv X[M-1]$, $X[-2] \equiv X[M-2]$, etc., so that the range of m in (5.10) could have been chosen as any M-long interval. However, it will be convenient to continue using the zero-centered range $-m_0 \le m \le M-1-m_0$.

The centered DFT is designed to approximate $X(j2\pi f)$ within the finite frequency interval $|f| < \frac{F_s}{2}$, based on the assumption that the sampling rate F_s was chosen so as to make $X(j2\pi f)$ negligible for $|f| > \frac{F_s}{2}$. Comparing the DFT expression (5.10) with (5.7), we conclude that

$$X[m] \;\approx\; F_s\,X\big(j2\pi\,m\,\tfrac{F_s}{M}\big) \qquad \text{for} \qquad -\,m_0 \le m \le M-1-m_0 \tag{5.11a}$$

This approximation uses the discretized frequency grid

$$\text{discretized } \omega \;=\; \{2\pi\,m\,\tfrac{F_s}{M}\,;\; -m_0 \le m \le M-1-m_0\} \tag{5.11b}$$

[3] A centered DFT is obtained in MATLAB by using the "fft" function, followed by "fftshift."

which is, indeed, contained within the interval $|f| \leq \frac{F_s}{2}$. The quality of approximation depends on the parameter M, which controls the frequency-grid density, as well as on the length of the sampled waveform segment $\{x(nT_s)\,;\ 0 \leq n \leq N_x - 1\}$ used to determine the centered DFT $\{X[m]\}$. Discretizing (5.8b) and (5.9b) on this frequency grid results in numerical approximation expressions, viz.,

$$\begin{aligned} \mathcal{X}_k(t)\Big|_{t=nT_s} &\approx \frac{1}{M} \sum_{m=-m_0}^{M-1-m_0} H_k\left(j2\pi\, m\, \tfrac{F_s}{M}\right) X[m]\, e^{j2\pi \frac{m}{M} n} \\ X_k(t)\Big|_{t=nT_s} &\approx \frac{1}{M} \sum_{m=-m_0}^{M-1-m_0} H_k\left(j2\pi\, m\, \tfrac{F_s}{M}\right) X[m]\, e^{j2\pi \left(\frac{m}{M} - \frac{f_{c,k}}{F_s}\right) n} \end{aligned} \tag{5.12}$$

which can be used with any choice of the center frequencies $\{f_{c,k}\,;\ 1 \leq k \leq K\}$.

Subband Pattern Properties

In this chapter we focus mainly on frequency-discretized transforms that exhibit one (or both) of the following *constant-ratio* properties:

- **Constant-Q:** each subband filter $H_k(s)$ has the same Q-factor, namely, the ratio

$$\text{Q-factor} \stackrel{\text{def}}{=} \frac{\text{center frequency of } |H_k(j\omega)|}{\text{3 db bandwidth of } |H_k(j\omega)|} \tag{5.13a}$$

  is independent of the subband index "k," which implies that the 3 dB bandwidth increases with subband index.
- **Geometrical spacing:** the sequence $\{f_{c,k}\,;\ 1 \leq k \leq K\}$ of subband centers frequencies is geometric, namely, the ratio

$$\Gamma \stackrel{\text{def}}{=} \frac{f_{c,k}}{f_{c,k-1}} \tag{5.13b}$$

  is independent of the subband index "k," which implies that the subband spacing $f_{c,k} - f_{c,k-1}$ increases with the subband index.

When both properties are satisfied, the sequence of subband widths is also geometric, with the same constant-ratio Γ as in (5.13b). In contrast, both the bandwidth and spacing of STFT subbands are *uniform*, as described in Chap. 4.

We present in Sects. 5.2–5.3 a conceptual framework and design guidelines for transforms that satisfy the two constant-ratio properties (5.13a,b). We examine in Sect. 5.4 the challenges encountered when applying such transforms to power system waveforms. Then, we introduce in Sect. 5.5 the S-transform, which satisfies the constant-Q property (5.13a), but has uniform subband spacing (instead of geometrical spacing), i.e., $f_{c,k} - f_{c,k-1}$ is independent of the subband index "k."

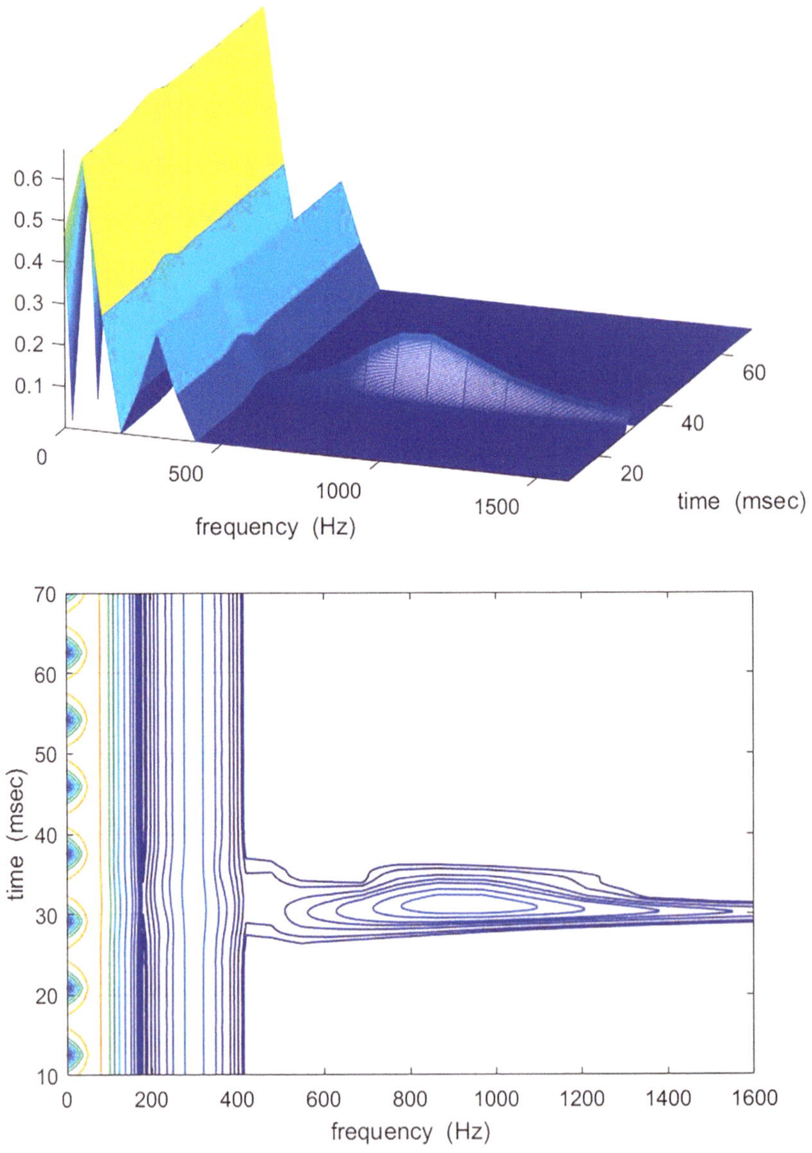

Fig. 5.2 Generalized spectrogram (synthetic-short-transient example), using 20 subbands with non-uniform width: time-frequency mesh plot (top), and contour plot (bottom). See Sect. L.2.1 for details about this example.

Application to Time-frequency Analysis

The most common application of both the STFT and its frequency-dependent variations is as a time-varying power spectrum based on the squared magnitude $|\mathcal{X}_k(t)|^2$, namely,

$$\text{generalized spectrogram of } x(\cdot) \stackrel{\text{def}}{=} \left|\mathcal{X}(t\,;\,f_c)\right|^2 \equiv \left|X(t\,;\,f_c)\right|^2 \tag{5.14}$$

For instance, Fig. 5.2 shows a generalized spectrogram for the synthetic-short-transient example: notice the short transient ($\approx$10 msec) centered at 900 Hz (= 15th harmonic), as well as the time-invariant fundamental and 5th harmonic components. Notice that one can use either the baseband-TFR $X(t\,;\,f_c)$ or the passband-TFR $\mathcal{X}(t\,;\,f_c)$ to determine the generalized spectrogram: the former was used exclusively in the introduction of the short-time Fourier transform in Chap. 4 (see, e.g., (4.2)). In contrast, wavelet-based transforms use a modified passband-based generalized spectrogram, known as a "scalogram" [148]. The increased bandwidth and improved time-resolution of the higher subbands in a constant-Q transform can provide accurate indication of the onset of fast wide-band transients: see Sect. 5.4 for illustrative examples.

In some applications, the phase $\arg\{\mathcal{X}_k(t)\}$ can provide useful information about the occurrence of sharp transitions and/or discontinuities in the input waveform [148]. Since $\arg\{\mathcal{X}_k(t)\}$ depends on the phase response of $H_k(s)$, the most accurate phase information is obtained by selecting subband filters with a zero phase response. This is achieved, for both subband filter categories of (5.5), when the prototype frequency response $G(j\omega)$ is real-valued and non-negative (see, e.g., the Morlet and Morse wavelet examples in Sect. 5.3.2).

The selection of the subband pattern parameters Γ and K must also take into account the desired time-resolution in the lowest and highest subband, as well as the impact of boundary effects: these are described in detail in the following section. This is particularly important in power system applications, which experience both low-frequency transients, such as voltage sag/swell or transformer inrush, as well as wide-band high-frequency transients, such as lightning strokes or capacitor switching.

5.2 Constant-Q Subband Pattern

A constant-Q subband pattern defines a partition of a given frequency interval $f_{\text{min}} \le f \le f_{\text{max}}$ into subbands that have the same width on a *logarithmic frequency scale.* In other words, the k-th subband is the frequency sub-interval

$$\Gamma^{k-1} f_{\text{min}} \le f \le \Gamma^k f_{\text{min}} \quad , \qquad 1 \le k \le K \tag{5.15a}$$

for some $\Gamma > 1$. Notice that $f_{\text{max}} \equiv \Gamma^K f_{\text{min}}$, so that

$$\Gamma = \left(\frac{f_{\text{max}}}{f_{\text{min}}}\right)^{1/K} \tag{5.15b}$$

which implies that we must have $f_{\text{min}} > 0$. The subband index "k" is sometimes replaced by the so-called *scale level* index $r \stackrel{\text{def}}{=} K - k$. In this case, the highest-index (i.e., K-th) subband is called "scale 0," the one below it "scale 1," and so on [148–150]. Notice that applying a constant-Q transform to a *sampled waveform* limits the available frequency range of the transform to $0 \leq f \leq \frac{F_s}{2}$, where F_s is the sampling rate. This observation suggests setting $f_{\text{max}} = \frac{F_s}{2}$ in digital implementations of constant-Q transforms.

The coefficient Γ, which is known as the *scaling factor* of the constant-Q subband pattern, determines the density of subbands, i.e., the number of subbands per frequency octave (a.k.a. "voices per octave" [148]), viz.,

$$\text{subband density} \stackrel{\text{def}}{=} \frac{1}{\log_2 \Gamma} \equiv \frac{\log 2}{\log \Gamma} \quad \text{subbands/octave} \tag{5.15c}$$

Notice that

$$\frac{\text{width of } k\text{-th subband}}{\text{width of } (k-1)\text{-st subband}} = \frac{\left(\Gamma^k - \Gamma^{k-1}\right) f_{\text{min}}}{\left(\Gamma^{k-1} - \Gamma^{k-2}\right) f_{\text{min}}} = \Gamma$$

as well as

$$\frac{\text{center frequency of } k\text{-th subband}}{\text{center frequency of } (k-1)\text{-st subband}} = \frac{\left(\Gamma^k + \Gamma^{k-1}\right) f_{\text{min}}/2}{\left(\Gamma^{k-1} + \Gamma^{k-2}\right) f_{\text{min}}/2} = \Gamma$$

We observe that the percentage of the frequency range $0 \leq f \leq f_{\text{max}}$ occupied by the highest-index subband is

$$\frac{\left(\Gamma^K - \Gamma^{K-1}\right) f_{\text{min}}}{f_{\text{max}}} = 1 - \frac{1}{\Gamma}$$

which increases with Γ (Fig. 5.3).

The choice of Γ controls the density and width of subbands and thus has a direct impact on the achievable time- and frequency-resolution. For instance, the scaling factor used to construct the discrete wavelet transform (DWT) is $\Gamma = 2$, which means that the highest-index (= "scale zero") subband of the DWT occupies 50% of the available frequency range, and the subband density is one subband per octave. In contrast, the discretized continuous wavelet transform (= the "cwt" function in MATLAB) provides for any desirable (integer-valued) subband density, resulting in $1 < \Gamma \leq 2$. In general, lowering the value of Γ increases the Q-factor, which improves frequency-resolution at the expense of impairing time-resolution, as we presently demonstrate.

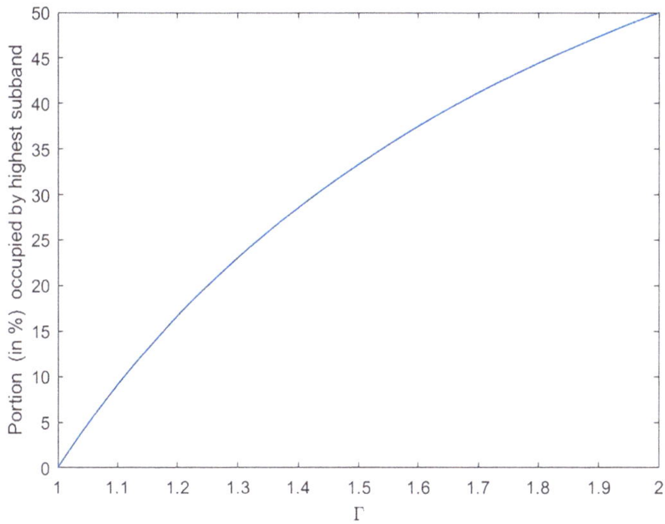

Fig. 5.3 Portion of frequency range assigned to the highest-index subband, as a function of the scaling factor Γ.

Finally, since $f_{\text{min}} > 0$, a constant-Q subband pattern cannot cover the entire frequency range $0 \leq f \leq f_{\text{max}}$. Consequently, one may wish to augment this subband pattern with a special "DC-subband," i.e., a low-pass filter with a normalized cutoff frequency that matches the lower edge of the first subband, i.e., $f_{\text{dc-cutoff}} = f_{\text{min}}$. In this case, the total number of subbands, including the DC-subband, will be $K + 1$.

Subband Overlap, the Q-factor and Frequency-resolution

We will assume from now on that the set of subband filters $\{H_k(s)\,;\ 1 \leq k \leq K\}$ has been designed to *match a given subband pattern*, in the sense that it meets the following mild restrictions:

(R1) The ratio between the 3 dB bandwidth of $H_k(s)$ and the width of the k-th subband is a k-independent constant

(R2) Each $H_k(s)$ is a band-pass filter with unity peak magnitude,[4] i.e., $\max_\omega |H_k(j\omega)| = 1$.

(R3) The magnitude response $|H_k(j\omega)|$ intersects both edges of the k-th subband ($f_{e,k-1} \equiv \Gamma^{k-1} f_{\text{min}}$ and $f_{e,k} \equiv \Gamma^k f_{\text{min}}$), at a prescribed k-independent level of $-A$ dB.

[4] Wavelet characterizations tend to use a different normalization, motivated by considerations of signal-to-noise ratio, namely $\int |H_k(j\omega)|^2 \frac{d\omega}{2\pi} = 1$.

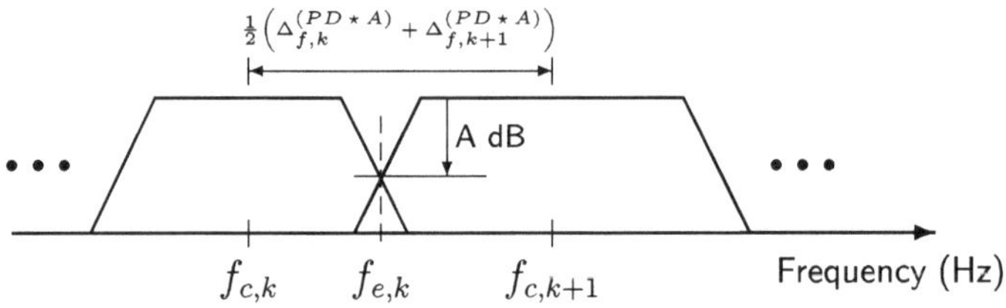

Fig. 5.4 Subband structure for a constant-Q pattern.

These *compliance constraints*, which are automatically satisfied by the filter design methods we describe in Sect. 5.3, make it possible to relate the width of the k-th subband to the $A\,\mathsf{dB}$ bandwidth of the subband filter $H_k(s)$. The key concept here is the point of intersection (located at the frequency $f_{e,k}$) between the subband magnitude responses $|H_k(j\omega)|$ and $|H_{k+1}(j\omega)|$ (Fig. 5.4).

As a direct consequence of the constraints on $H_k(s)$, we find that the level of overlap between adjacent subbands is independent of k, i.e., the subband-overlap ratio (SOR) has the same value for every pair of adjacent subbands. First, we need to modify the SOR definition (4.41), which was based on the assumption of uniformly-spaced subbands of equal width, so that it can be applied to any (non-uniform) subband pattern. Since the magnitude response curves of each pair of adjacent subbands intersect $A\,\mathsf{dB}$ below the peak, the spacing between the center frequencies satisfies the relation (see Fig. 5.4)

$$f_{c,k+1} - f_{c,k} = \frac{1}{2}\left(\Delta_{f,k}^{(PD\star A)} + \Delta_{f,k+1}^{(PD\star A)}\right)$$

where

$$f_{c,k} \stackrel{\text{def}}{=} \text{center frequency of } k\text{-th subband} = \tfrac{1}{2}\left(\Gamma^k + \Gamma^{k-1}\right) f_{\min} \tag{5.16a}$$

and $\Delta_{f,k}^{(PD\star A)}$ is the $A\,\mathsf{dB}$ bandwidth of $|H_k(j\omega)|$. Choosing $A = 3\,\mathsf{dB}$ as a benchmark motivates the definition

$$\mathsf{SOR}_k \stackrel{\text{def}}{=} \frac{1}{2}\,\frac{\Delta_{f,k}^{(PD\star 3)} + \Delta_{f,k+1}^{(PD\star 3)}}{f_{c,k+1} - f_{c,k}} \tag{5.16b}$$

so that $\mathsf{SOR}_k = 1$ when the magnitude response curves of adjacent subbands intersect exactly at the $3\,\mathsf{dB}$ level. Notice that this expression reduces to (4.41) for a uniform frequency grid with frequency-independent subband widths.

Next, we observe that the width of each subband is proportional to its center frequency, viz.,

$$\frac{\text{center frequency of } k\text{-th subband}}{\text{width of } k\text{-th subband}} = \frac{\left(\Gamma^k + \Gamma^{k-1}\right) f_{\min}/2}{\left(\Gamma^k - \Gamma^{k-1}\right) f_{\min}} = \frac{1}{2}\cdot\frac{\Gamma+1}{\Gamma-1}$$

Since, by the constraints placed on $H_k(s)$, the 3 dB bandwidth of this filter is proportional to the width of the k-th subband, it now follows that the ratio $\frac{f_{c,k}}{\Delta_{f,k}^{(PD\star 3)}}$ (= the Q-factor of the k-th subband) is also a k-independent constant of the subband pattern. The Q-factor of any band-pass filter is defined as (recall (5.13a))

$$\text{Q-factor} \stackrel{\text{def}}{=} \frac{\text{center frequency}}{\text{3 db bandwidth}}$$

Applying this definition to the k-th subband, we find that

$$\frac{f_{c,k}}{\Delta_{f,k}^{(PD\star 3)}} = \frac{\Delta_{f,k}^{(PD\star A)}}{\Delta_{f,k}^{(PD\star 3)}} \frac{\left(\Gamma^k + \Gamma^{k-1}\right) f_{\text{min}}/2}{\left(\Gamma^k - \Gamma^{k-1}\right) f_{\text{min}}} = \frac{\rho_f^{(PD\star A)}}{2} \frac{\Gamma + 1}{\Gamma - 1}$$

where

$$\rho_f^{(PD\star A)} \stackrel{\text{def}}{=} \frac{\Delta_{f,K}^{(PD\star A)}}{\Delta_{f,K}^{(PD\star 3)}} \equiv \frac{\Delta_{f,k}^{(PD\star A)}}{\Delta_{f,k}^{(PD\star 3)}} \quad \text{for all } k \tag{5.17}$$

is a k-invariant coefficient that depends only on our choice of the prototype filter $h(t)$, and we used the fact that, according to Fig. 5.4,

$$\Delta_{f,k}^{(PD\star A)} = f_{e,k} - f_{e,k-1} = \left(\Gamma^k - \Gamma^{k-1}\right) f_{\text{min}}$$

In particular, when the subband filter $h_k(t)$ has finite length $T_{h,k}$ then, according to Sect. B.4, $\Delta_{f,k}^{(PD\star A)} = C_f^{(PD\star A)} T_{h,k}^{-1}$, which implies that $\rho_f^{(PD\star A)} = C_f^{(PD\star A)}/C_f^{(PD\star 3)}$, where the coefficient $C_f^{(PD\star A)}$ is a monotone increasing function of A (see Tables B.2, B.3, and B.4).

In summary, the k-invariant Q-factor of a constant-Q filter-bank satisfies the key relation

$$Q = \frac{1}{2} \rho_f^{(PD\star A)} \frac{\Gamma + 1}{\Gamma - 1} \tag{5.18}$$

where "A" is the level of intersection (in dB below the peak) between the magnitude responses of each pair of adjacent subbands, as illustrated in Fig. 5.4. In turn, this relation generates a simple expression for the subband-overlap ratio: combining the SOR definition (5.16b) with the relation $\Delta_{f,k}^{(PD\star 3)} = \frac{f_{c,k}}{Q}$, we conclude that

$$\text{SOR}_k = \frac{1}{2Q} \frac{f_{c,k+1} + f_{c,k}}{f_{c,k+1} - f_{c,k}} = \frac{1}{2Q} \frac{\Gamma + 1}{\Gamma - 1}$$

so that

$$\mathsf{SOR}_k = \left[\rho_f^{(PD\star A)}\right]^{-1} \qquad \text{for all } k \tag{5.19a}$$

We denote this k-invariant quantity as SOR: it satisfies the relation

$$(2Q)\ \mathsf{SOR} = \frac{\Gamma+1}{\Gamma-1} \qquad \Longleftrightarrow \qquad \Gamma = \frac{(2Q)\ \mathsf{SOR}+1}{(2Q)\ \mathsf{SOR}-1} \tag{5.19b}$$

which connects the three constants Γ, Q, and SOR.

It is also well-known that the Q-factor is directly related to frequency-resolution (see, e.g., [146]). To be specific, this relation exhibits a "constant product" behavior for each subband, viz.,

$$\frac{\text{frequency-resolution}}{\text{center frequency}} \cdot Q = \text{filter-dependent constant} \tag{5.20}$$

This result follows from combining: (a) the relation (B.15b), which connects resolution with prescribed-drop bandwidth, i.e., $F_{res}^{(CD\star 4)} = \Delta_f^{(PD\star 10)}$, and (b) the definition (5.13a) of the Q-factor. We conclude that the constant value of the product in (5.20) is, in fact, the ratio $\frac{\Delta_f^{(PD\star 10)}}{\Delta_f^{(PD\star 3)}} \equiv \rho_f^{(PD\star 10)}$. Examination of Tables B.2 and B.3 reveals that the value of this constant, for subband filters that are derived from classical (FIR) windows, is typically in the range 1.67–1.82.

Time-resolution

The time-resolution of each subband is inversely proportional to its width: the best (i.e., shortest) resolution is achieved in the highest-index (=K-th) subband. Since the subband width increases geometrically with the subband index "k," the resolution decreases (i.e., gets better) in the same manner, so that

$$\frac{\text{time-resolution of the } k\text{-th subband}}{\Gamma^{K-k}} = \text{time-resolution of the } K\text{-th subband}$$

The analysis of time-resolution in Sect. B.7 shows that the product of time-resolution and filter bandwidth is a filter-dependent constant. To be specific, we have defined in Sect. B.8 the filter-dependent constant

$$\mathcal{K}_{res}(A, B) \stackrel{\text{def}}{=} \Delta_f^{(PD\star A)}\ T_{res}^{(CD\star B)}$$

and examined its value for several combinations of A and B values.

Table 5.1 Comparison of $\mathcal{K}_{res}(3, 10)$ for several filter prototypes

Filter prototype	$\mathcal{K}_{res}(3, 10)$
Gaussian	0.531
Triangular	0.566
Rectangular	0.606
Kaiser ($\beta = 1$)	0.605
Kaiser ($\beta = 6$)	0.546
Hann	0.561
Hamming	0.543

The key step in all filter design methods discussed in Sect. 5.3 is setting the A dB bandwidth of $|H_K(j\omega)|$ equal to the width of the K-th subband, which is $f_{\mathsf{max}}\left(1 - \frac{1}{\Gamma}\right)$. For the sake of concreteness, we will use here the special choices $B = 10$, motivated by Fig. B.5, as well as $A = 3$, which is motivated, in part, by the constraint (R3). These choices lead to the result

$$\text{time-resolution of the } K\text{-th subband} = \frac{\mathcal{K}_{res}(3, 10)}{f_{\mathsf{max}}} \cdot \frac{\Gamma}{\Gamma - 1} \tag{5.21}$$

The value of $\mathcal{K}_{res}(3, 10)$ can be determined from the data in Tables B.2 and B.7, and is listed in Table 5.1: notice that $\mathcal{K}_{res}(3, 10) \approx 0.5$ for all filter prototypes. For instance, when $\Gamma = 1.3$ and $f_{\mathsf{max}} = \frac{F_s}{2}$, the time-resolution of the top subband is $\approx 4.3\, T_s$, where $T_s = {}^1/_{F_s}$ is the sampling interval used in digital implementation of the constant-Q transform.

Since time-resolution gets worse with decreasing subband index, specifying a desired resolution for the 1-st subband induces an upper bound on the total number of subbands K. Another bound on K is associated with the effect of waveform boundaries, as explained below.

Lower Bound on Subband Density

Subband density controls the level of detail provided by a time-frequency representation (TFR). Sufficient subband density is needed to avoid "scalloping" – a TFR distortion that occurs when a peak of the continuous-frequency/continuous-time magnitude TFR $|\mathcal{X}(t\,;\,f_c)| \equiv |X(t\,;\,f_c)|$ (recall (5.1a)) falls between two subband center frequencies. However, given the limitation on TFR detail associated with frequency-resolution of subband filters, excessive increase of subband density should be avoided, as it increases cost without improving TFR quality.

An estimate of the *lowest acceptable subband density* can be obtained by considering the frequency-resolution analysis carried out in Sects. B.5–B.6. In terms of Fig. B.1, two nearby peaks of a continuous TFR can still be observed in the frequency-discretized version of this TFR if a subband center frequency, say $f_{c,k}$, falls in the "valley" between those two peaks, and the adjacent subband center frequencies – i.e., $f_{c,k+1}$ and $f_{c,k-1}$ – coincide (approximately) with the peaks.

Using $\Delta_{f,k}^{(PD\star 10)}$ as a measure of frequency-resolution in the k-th subband – as suggested in Sect. B.5 – and $\Delta_{f,k}^{(PD\star A)}$ as the nominal width of the same subband (recall Fig. 5.4), we obtain an approximate (implicit) characterization of the lowest acceptable subband density, namely $2\,\Delta_{f,k}^{(PD\star A)} \leq \Delta_{f,k}^{(PD\star 10)}$. Dividing by $\Delta_{f,k}^{(PD\star 3)}$ results in an *equivalent k-independent constraint*, namely $2\,\rho_f^{(PD\star A)} \leq \rho_f^{(PD\star 10)}$ (recall (5.17)).

Next, we observe that this inequality defines a Q-dependent upper bound for the scaling factor Γ, i.e., $\Gamma \leq \Gamma_{cr}$, where

$$\Gamma_{cr} \stackrel{\text{def}}{=} \frac{4\,Q + \rho_f^{(PD\star 10)}}{4\,Q - \rho_f^{(PD\star 10)}} \tag{5.22a}$$

and we used the relation (5.18) to express $\rho_f^{(PD\star A)}$ in terms of Q and Γ. In turn, applying (5.15c) generates a lower bound on subband density, viz.,

$$\text{acceptable subband density} \geq \left[\,\log_2 \Gamma_{cr}\,\right]^{-1} \text{ subbands/octave} \tag{5.22b}$$

which is a monotone increasing function of Q. For instance, since $\rho_f^{(PD\star 10)} = 2.273/1.291 = 1.761$ for a Hamming prototype, we find that $\left[\,\log_2 \Gamma_{cr}\,\right]^{-1}$ increases from 3.5 to 5.5 subbands/octave when the Q-factor increases from 4.5 to 7. This Q-dependent bound should be contrasted with the built-in rigid lower bound of 4 subbands/octave used in the MATLAB function "cwt".

Upper Bound on the Number of Subbands

When a constant-Q transform is applied to a finite waveform record, say $\{x(t)\,;\ 0 \leq t \leq T_x\}$, the resulting subband components $\{X_k(t)\,;\ 1 \leq k \leq K\}$ exhibit distortion near the two boundaries of the finite time interval. The length of the impacted portion of the time interval $[\,0\ \ T_x\,]$ depends on the effective duration of the subband impulse response $h_k(t)$, which decreases with increasing "k." Thus, the lowest-index subband (i.e., $k = 1$) is the one most affected by this unwanted side-effect.

To be specific, the subband component $X_1(t)$ exhibits a boundary effect in the two boundary-adjacent intervals $0 \leq t \leq \frac{T_{h,1}}{2}$ and $T_x - \frac{T_{h,1}}{2} \leq t \leq T_x$. The relative width of the combined impacted interval, viz.,

$$\varepsilon \equiv \text{boundary-effect factor} \stackrel{\text{def}}{=} \frac{T_{h,1}}{T_x}$$

can be used as a metric for the severity of boundary-effect impact on the subband component $X_1(t)$. A straightforward calculation shows that this metric increases with the total number of subbands.

Indeed, if we identify $T_{h,1}$ with the 3 dB effective duration of the impulse response $h_1(t)$, as defined in Sect. B.4, then

$$T_{h,1} = \frac{\mathcal{K}_h^{(PD\star 3)}}{\Delta_f^{(PD\star 3)}} = \frac{2\,\mathcal{K}_h^{(PD\star 3)}}{F_s\left(1-\frac{1}{\Gamma}\right)}\,\Gamma^{K-1}$$

and, consequently,

$$\varepsilon \equiv \frac{T_{h,1}}{T_x} = \frac{2\,\mathcal{K}_h^{(PD\star 3)}}{1-\frac{1}{\Gamma}}\left(\frac{T_s}{T_x}\right)\Gamma^{K-1}$$

where we rely on the simplifying (albeit not strictly necessary) assumptions that the 3 dB bandwidth of the lowest-index subband is $f_{\mathsf{max}}(1-\frac{1}{\Gamma})/\Gamma^{K-1}$ and, as suggested earlier, $f_{\mathsf{max}} = \frac{F_s}{2}$. Thus, in order not to exceed a specified level of the boundary-effect factor ε, we must limit the total number of subbands, viz.,

$$K \le 1 + \frac{\log\left(\varepsilon\left(1-\frac{1}{\Gamma}\right)\left(\frac{T_x}{T_s}\right)\right) - \log\left(2\,\mathcal{K}_h^{(PD\star 3)}\right)}{\log \Gamma}$$

As an example this upper bound is shown in Fig. 5.5 for a waveform record of $\frac{T_x}{T_s} = 5000$ samples, and using a typical value of $\mathcal{K}_h^{(PD\star 3)} = 0.5$ (compare

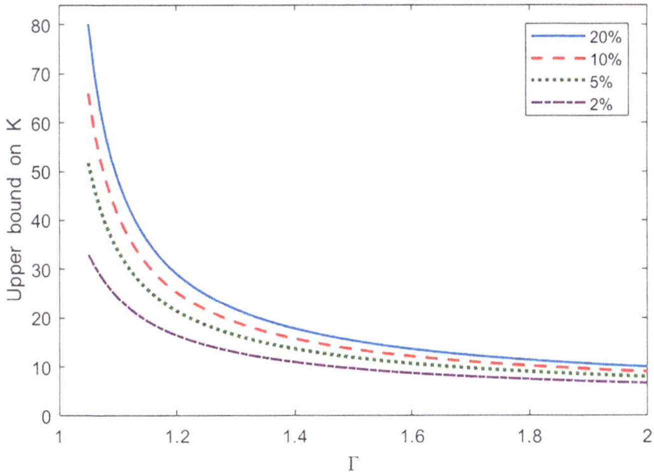

Fig. 5.5 Upper bound on the number of subbands for $\varepsilon = 2\%, 5\%, 10\%, 20\%$ ($\frac{T_x}{T_s} = 5000$ and $\mathcal{K}_h^{(PD\star 3)} = 0.5$).

with the values presented in Table B.2). Notice that, for $\Gamma > 1.35$ and $\varepsilon \leq 20\%$, the allowed number of subbands is under 20.

Examples of Constant-Q Subband Patterns

- **Constant-Q spectral transform for musical analysis** [146]: this subband pattern[5] is obtained by setting $\Gamma = 2^{1/24} = 1.029$, $f_{\mathsf{min}} = 165\,\mathsf{Hz}$, and $Q = 34$ (which results, via (5.19b), in $\mathsf{SOR} = 1.018$). Notice that the choice of Γ corresponds to a musical quarter-tone interval, resulting in a density of 24 subbands-per-octave. There is no "DC-subband," and the total number of subbands is 150.
- **Constant-Q spectral transform for EEG analysis** [147]: this subband pattern is obtained by setting $\Gamma = 2^{1/2} = 1.414$ and $f_{\mathsf{min}} = 0.46\,\mathsf{Hz}$. The Q-factor is not specified: however, for the given value of Γ, and assuming $\mathsf{SOR} \approx 1$, we expect $Q \approx 3$. Notice that the choice of Γ results in a density of 2 subbands per octave. There is no "DC-subband," and the total number of subbands is 14.
- **Discrete wavelet transform (DWT)** [133, 148]: this subband pattern is obtained by setting $\Gamma = 2$, $\mathsf{SOR} = 1$ (which results in $Q = 1.5$), and $f_{\mathsf{min}} = 2^{-L} F_s$, where L is the total number of subbands (including the "DC-subband"), and F_s is the waveform sampling rate. This subband pattern has a density of one subband per octave, and its highest-index subband occupies half of the entire frequency range, from $\frac{F_s}{4}$ to $\frac{F_s}{2}$. The cutoff frequency of the "DC-subband" is, of course, $f_{\text{dc-cutoff}} = f_{\mathsf{min}} = 2^{-L} F_s$. The total number of subbands is $K + 1 \equiv L$.
- **Discretized continuous wavelet transform (CWT)** [148, 149]: this subband pattern is obtained by choosing a desired subband density "d," and setting $\Gamma = 2^{1/d}$, where d is a positive integer. There is no DC-subband, and the Q-factor depends on the filter prototype used and the SOR (see (5.18), (5.19b)).
- **Tunable-Q wavelet transform (TQWT) applied to speech analysis** [169]: the subband pattern is obtained by setting $\Gamma = 1.2$, $f_{\mathsf{min}} = 97.9\,\mathsf{Hz}$, and $Q = 3$ (which results, via (5.19b), in $\mathsf{SOR} = 11/3 = 1.833$). Notice that the choice of Γ results in 3.8 subbands-per-octave. The total number of subbands is 23.

We observe that subband density in these examples ranges from $d = 24$ in the musical analysis example, all the way down to $d = 2$ in the EEG analysis example, and $d = 1$ for the discrete wavelet transform. The corresponding Q-factor ranges from $Q = 34$ to $Q = 1.5$. Thus, the standard DWT cannot provide the level of detail (= frequency-resolution) that is needed for musical analysis (as already observed in [146]), or EEG analysis. On the other hand, low-density schemes, such as the DWT, can be quite useful in applications that require superior time-resolution of high frequency transients.

---

[5] Actually, this is not a pure constant-Q pattern: the lower subbands in this application have $Q = 34$, while the upper subbands are designed with $Q = 68$.

Discrete Implementation with Subband-dependent Down-sampling

Digital implementations of the variable-bandwidth transform (5.1a), such as the time-domain scheme (5.6) or the frequency-domain scheme (5.12), involve discretization of the time variable "t." This means that the transform is evaluated on a grid $\{(nT_s\,,\, f_{c,k})\,;\ 0 \leq n \leq N-1\,,\, 0 \leq k \leq K\}$ in the time-frequency plane, where we consider only non-negative frequencies, and we also include the special "DC-subband," which corresponds to the subband index $k = 0$. The resulting discretized time-frequency representation $\{\mathcal{X}_k(nT_s)\}$ is redundant, in the sense that its time-index "n" can be further decimated (i.e., down-sampled) without loss of information.

The fundamental rule of down-sampling filter-bank subbands is that the resulting combined output rate cannot be lower than the input rate (i.e., F_s samples/sec) of the filter-bank. Violation of this rule results in loss of information, causing a distortion of the TFR $\{\mathcal{X}_k(nT_s)\}$. *Critical sampling* describes the case when the combined output rate is exactly equal to F_s samples/sec. This rule is used, for instance, in [132, 133] to deduce the maximal decimation factor that still allows for design of a uniform DFT filter-bank with the perfect reconstruction (PR) property.

Critical sampling of a constant-Q filter-bank is achieved when the k-th subband is down-sampled by a subband-dependent factor of D_k, where

$$D_k = \left[\text{relative width of } k\text{-th subband}\right]^{-1}$$

According to (5.15), the width of the k-th subband is

$$\left(\Gamma^k - \Gamma^{k-1}\right) f_{\text{min}} = \Gamma^k\left(1 - \tfrac{1}{\Gamma}\right) \frac{f_{\text{max}}}{\Gamma^K} = \frac{1 - \frac{1}{\Gamma}}{\Gamma^{K-k}} \cdot \frac{F_s}{2}$$

where we assume, as usual, that $f_{\text{max}} = \frac{F_s}{2}$. Thus the relative width of this subband, compared to the available frequency range ($0 \leq f \leq \frac{F_s}{2}$), is

$$\left(\frac{F_s}{2}\right)^{-1} \left(\frac{1 - \frac{1}{\Gamma}}{\Gamma^{K-k}} \cdot \frac{F_s}{2}\right) = \frac{1 - \frac{1}{\Gamma}}{\Gamma^{K-k}}$$

which implies that

$$D_k = \frac{\Gamma^{K-k}}{1 - \frac{1}{\Gamma}} \quad , \qquad 1 \leq k \leq K \tag{5.23}$$

In particular, when $\Gamma = 2$, this subband-dependent decimation factor is $D_k = 2^{K-k+1}$, which is precisely the sampling scheme used in construction of the Discrete Wavelet Transform (DWT) [132, 133, 148].

The combined output rate of a critically-sampled constant-Q scheme is

$$\sum_{k=1}^{K} \frac{F_s}{D_k} = F_s \sum_{k=1}^{K} \frac{\Gamma^{K-k}}{1 - \frac{1}{\Gamma}} = F_s - \frac{F_s}{\Gamma^K}$$

where we used the down-sampling factor specified in (5.23). This combined output rate is missing the "DC-subband," which has a cutoff frequency of $f_{\min}$ and, therefore, a critically-sampled rate of

$$2\, f_{\min} = 2\,\frac{f_{\max}}{\Gamma^K} = \frac{F_s}{\Gamma^K}$$

Including this subband with the K constant-Q subbands results in a combined output rate of F_s samples/sec, which confirms the "critically-sampled" status of the decimation scheme (5.23).

Critical sampling establishes a one-to-one relation between the sampled input waveform $\{x(nT_s)\,;\ 0 \le n \le N-1\}$ and the corresponding fully-decimated subband outputs. It is a key ingredient in the construction of invertible constant-Q transforms, such as the DWT and the Discrete Orthonormal S-Transform (DOST) (see Sect. 5.6), which have numerous coding and compression applications. In contrast, fast-response power system applications such as computer-based protection and fault detection, as well as closed-loop applications such as system modeling and control, are adversely impacted by the added delay caused by decimation.

Therefore, following the non-decimation policy that has been described in detail in Sects. 4.2, 4.3, and 4.4, we continue to emphasize in this chapter implementation choices that provide the highest available refresh rate in all subbands, in order to achieve the best time-resolution and processing speed, as well as a fast response to onset of transients. Ensuring a refresh rate that equals the waveform sampling rate requires using a non-decimated (= sample-by-sample) processing scheme, in which subband outputs are re-evaluated as soon as a new input waveform sample becomes available.

5.3 Design of Constant-Q Subband Filters

Following the selection of a constant-Q subband pattern, the next task is to design the individual subband transfer functions $\{H_k(s)\,;\ 1 \le k \le K\}$. In principle, each transfer function can be designed independently of the other ones, as long as it satisfies the three constraints (R1)–(R3) of Sect. 5.2. Consequently, there are many options for designing the subband filters: some of these designs are by-products of pursuing specific objectives, such as the construction of an orthonormal basis for the space of finite energy waveforms. However, when our only objective is construction of a time-frequency analysis tool, we should opt for a less constrained design approach. Using a single prototype to determine $H_k(s)$ for every k offers sufficient flexibility, while maintaining moderate conceptual and computational cost.

It will be convenient to use the highest-index (i.e., K-th) subband as the starting point for the design process, which means that we will use the *scale level* index $r \overset{\text{def}}{=} K - k$ as a subband identifier (instead of the subband index "k"). The key relation shared by all design methods described in this section is the *time-scaling property*

$$h_{K-r}(t) = \frac{1}{\Gamma^r} h_K\left(\frac{t}{\Gamma^r}\right) \quad , \qquad 0 \le r \le K-1 \tag{5.24a}$$

which induces an equivalent frequency-scaling property, viz.,

$$H_{K-r}(s) = H_K\left(\Gamma^r s\right) \quad , \qquad 0 \le r \le K-1 \tag{5.24b}$$

This property guarantees that the constraint (R1) is satisfied regardless of our choice of $H_K(s)$.

The scaling-properties (5.24) imply that the constraints (R1)–(R3) are satisfied by the entire set of subband filters $\{H_k(s)\,;\ 1 \le k \le K\}$ if, and only if, the highest-index subband filter $H_K(s)$ satisfies constraints (R2) and (R3). Indeed, we observe that

$$\max_\omega \left|H_{K-r}(j\omega)\right| = \max_\omega \left|H_K(j\,\Gamma^r\omega)\right| = \max_\omega \left|H_K(j\omega)\right|$$

so that imposing the amplitude-scaling constraint $\max_\omega \left|H_K(j\omega)\right| = 1$ guarantees that (R2) is satisfied by $H_k(s)$ for all k.

In summary, the only remaining design task is to construct a *compliant* impulse response $h_K(t)$ for the highest-index subband, i.e., one that satisfies the two constraints (R2) and (R3). This is achieved by:

- adjusting the amplitude of $h_K(t)$ so that $\max_\omega |H_K(j\omega)| = 1$, and
- adjusting the time-scaling of $h_K(t)$ to ensure that the constraint (R3) is satisfied in the K-th subband.

The remaining subbands will automatically satisfy (R2) and (R3) by virtue of the scaling relations (5.24).

5.3.1 Design from a Low-Pass Prototype

A compliant highest-index (a.k.a. "top") subband filter can be constructed for any given constant-Q subband pattern by using a single low-pass prototype and one scaling parameter. To be specific, the impulse response of such a top-subband filter is given by the expression

$$h_K(t) \;=\; \frac{1}{\sigma_K}\, g\left(\frac{t}{\sigma_K}\right)\, e^{\,j2\pi f_{c,K}\, t} \tag{5.25a}$$

where $f_{c,K} = \frac{1}{2}\, f_{\max}\,(1+\frac{1}{\Gamma})$ is the center frequency of the top subband, and $g(t)$ is the real-valued impulse response of any desired low-pass filter. We can assume without loss of generality that $\int g(t)\, dt = 1 = G(0)$, where $G(s) = \mathcal{L}\{g(t)\}$ is the transfer function associated with this impulse response: this ensures that $h_K(t)$ satisfies the constraint (R2).

Since the scaling factor Γ and the center frequency $f_{c,K}$ are determined by the given constant-Q subband pattern, the only remaining design task is to adjust the control parameter σ_K to satisfy the constraint (R3), namely,

$$\text{A dB bandwidth of } H_K(j\omega) \;\equiv\; \frac{\Delta_f^{(PD\star A)}}{\sigma_K} \;=\; f_{\max}\left(1-\frac{1}{\Gamma}\right)$$

where $\Delta_f^{(PD\star A)}$ is the A dB bandwidth of the prototype filter $g(t)$. We conclude that (R3) is satisfied when

$$\sigma_K \;=\; \frac{\Delta_f^{(PD\star A)}}{f_{\max}\left(1-\frac{1}{\Gamma}\right)} \tag{5.25b}$$

which completes the design process of a compliant top-subband filter $H_K(s)$.

According to (5.24), the complete set of subband filters is given by the expression $h_{K-r}(t) \;=\; \frac{1}{\Gamma^r}\, h_K\left(\frac{t}{\Gamma^r}\right)$, namely,

$$h_{K-r}(t) \;=\; \frac{1}{\Gamma^r\sigma_K}\, g\left(\frac{t}{\Gamma^r\sigma_K}\right)\, e^{\,j2\pi f_{c,K-r}\, t}\;, \qquad 0 \le r \le K-1 \tag{5.26a}$$

and the corresponding transfer functions are

$$H_{K-r}(s) \;\stackrel{\text{def}}{=}\; \mathcal{L}\{h_{K-r}(t)\} \;=\; G\Big(\Gamma^r\sigma_K\big(s - j2\pi f_{c,K-r}\big)\Big) \tag{5.26b}$$

where

$$f_{c,K-r} \;=\; \frac{1}{\Gamma^r}\, f_{c,K} \;=\; \frac{\Gamma+1}{2\,\Gamma^{r+1}}\, f_{\max} \tag{5.26c}$$

This confirms that $|H_{K-r}(j\omega)|$ is centered at $\omega = 2\pi f_{c,K-r}$, and its bandwidth is inversely proportional to $\Gamma^r\sigma_K$.

For digital implementation purposes, the discrete-time version of the subband filter $h_{K-r}(t)$ is (recall (5.6))

$$h_{K-r}[n] \stackrel{\text{def}}{=} T_s\, h_{K-r}(nT_s) = \frac{T_s}{\Gamma^r \sigma_{\mathsf{K}}}\, g\left(\frac{nT_s}{\Gamma^r \sigma_{\mathsf{K}}}\right) e^{j2\pi\eta_{K-r}n} \qquad (5.27)$$

where F_s is the sampling rate, $T_s \stackrel{\text{def}}{=} \frac{1}{F_s}$ is the sampling interval, and $\eta_{K-r} \stackrel{\text{def}}{=} \frac{f_{c,K-r}}{F_s}$ is the normalized center frequency of the $(K-r)$-th subband.

Example 5.1 (Design from a Gaussian prototype): Consider a given constant-Q subband pattern, with $\Gamma = 1.3$ and $f_{\mathsf{max}} = F_s/2$, where F_s is the intended sampling rate of the input waveform. The chosen prototype is a Gaussian window, viz.,

$$g(t) = \frac{1}{\sqrt{2\pi}}\, e^{-\frac{t^2}{2}} \quad \Longleftrightarrow \quad G(j\omega) = e^{-\frac{\omega^2}{2}}$$

and its A dB bandwidth is $\frac{1}{\pi}\sqrt{\frac{A}{10}\ \ln 10}$ (see Sect. B.4). If we choose $A = 3$ then the corresponding bandwidth is $\Delta_f^{(PD\star 3)} = 0.265$ (see Table B.2), and thus

$$\sigma_{\mathsf{K}} = \frac{\Delta_f^{(PD\star 3)}}{f_{\mathsf{max}}\left(1 - \frac{1}{\Gamma}\right)} = 2.297\ T_s$$

where $T_s = \frac{1}{F_s}$ is the sampling interval. The resulting subband filter impulse responses are

$$h_{K-r}(t) = \frac{0.174}{T_s\,\Gamma^r}\, e^{-\left[\left(0.308/\Gamma^r\right)\left(t/T_s\right)\right]^2} e^{j2\pi f_{c,K-r}\, t}, \qquad 0 \le r \le K-1$$

where

$$f_{c,K-r} = \frac{\left(1 + \frac{1}{\Gamma}\right) f_{\mathsf{max}}}{2\,\Gamma^r} = \frac{0.442}{\Gamma^r}\, F_s$$

The equivalent discrete-time version is, according to (5.27),

$$h_{K-r}[n] = \frac{0.174}{\Gamma^r}\, e^{-\left(0.308\,n/\Gamma^r\right)^2} e^{j2\pi\eta_{K-r}n}, \qquad 0 \le r \le K-1$$

where $\eta_{K-r} \stackrel{\text{def}}{=} \frac{f_{c,K-r}}{F_s} = \frac{0.442}{\Gamma^r}$ are the normalized subband center frequencies.

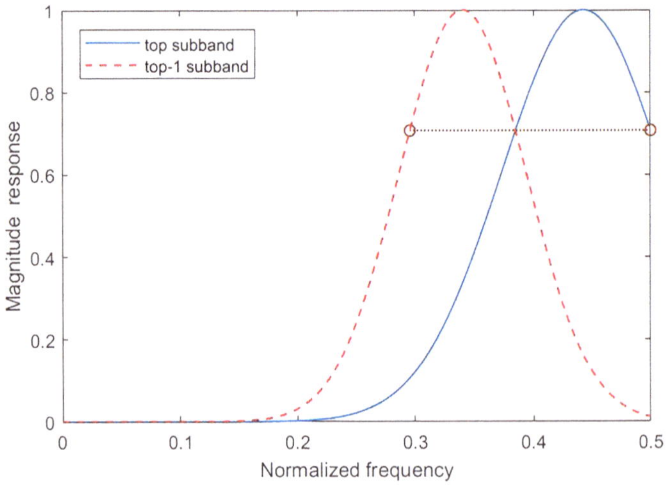

Fig. 5.6 Subband pattern, showing the K-th and $(K-1)$-th subbands, for a low-pass Gaussian prototype: $-3\,\mathrm{dB}$ level indicated by dotted line.

The magnitude responses of the top two subbands are shown in Fig. 5.6: notice that adjacent subbands intersect at the $-3\,\mathrm{dB}$ level. Since $h_k(t)$ is conjugate symmetric, the frequency response $H_k(j\omega)$ is real-valued, and so the phase response of each subband is zero. ■

Example 5.2 (Design from a Hamming prototype): Consider again the constant-Q subband pattern of Example 5.1, namely $\Gamma = 1.3$ and $f_{\mathsf{max}} = F_s/2$, where F_s is the intended sampling rate of the input waveform. Notice that, as in the Gaussian prototype example,

$$f_{c,K-r} = \frac{\left(1+\frac{1}{\Gamma}\right) f_{\mathsf{max}}}{2\,\Gamma^r} = \frac{0.442}{\Gamma^r}\,F_s$$

Now, the chosen prototype is a Hamming window, viz.,

$$g(t) = \begin{cases} \dfrac{1}{T_g}\left[1 + \dfrac{21}{25}\cos\left(\dfrac{2\pi t}{T_g}\right)\right] & |t| \le \frac{T_g}{2} \\ 0 & \text{else} \end{cases}$$

as defined in (B.6). According to Table B.2, the $3\,\mathrm{dB}$ bandwidth of this low-pass filter is $\Delta_f^{(PD\star3)} = 1.291\;T_g^{-1}$, which means that the width-matching constraint (R3) reduces in this case to

$$1.291\ T_g^{-1} = f_{\text{max}}(1 - \tfrac{1}{\Gamma}) = 0.115\ F_s \quad \Longrightarrow \quad T_g = 11.189\ T_s$$

Notice that, in this example, the window-length parameter T_g has the same role, and the same value, as the scaling parameter σ_K.

The resulting subband filter impulse responses (for $0 \le r \le K-1$) are

$$h_{K-r}(t) = \begin{cases} \dfrac{1}{\Gamma^r T_g}\left[1 + \dfrac{21}{25}\cos\left(\dfrac{2\pi t}{\Gamma^r T_g}\right)\right] e^{j2\pi f_{c,K-r}\,t} & |t| \le \dfrac{\Gamma^r T_g}{2} \\ 0 & \text{else} \end{cases}$$

and the equivalent discrete-time version is

$$h_{K-r}[n] = \begin{cases} \dfrac{1}{\Gamma^r}\left(\dfrac{T_s}{T_g}\right)\left[1 + \dfrac{21}{25}\cos\left(\dfrac{2\pi n}{\Gamma^r}\cdot\dfrac{T_s}{T_g}\right)\right] e^{j2\pi\eta_{K-r}\,n} & |n| \le \text{floor}\left(\dfrac{\Gamma^r T_g}{2T_s}\right) \\ 0 & \text{else} \end{cases}$$

where, of course, $\frac{T_s}{T_g} = 11.189^{-1} = 0.0894$, $\Gamma = 1.3$, and $\eta_{K-r} \stackrel{\text{def}}{=} \frac{f_{c,K-r}}{F_s} = \frac{0.442}{\Gamma^r}$. The magnitude responses of the top two subbands are shown in Fig. 5.7: notice that adjacent subbands intersect at the $-3\,\text{dB}$ level. Since $h_k(t)$ is conjugate symmetric, the frequency response $H_k(j\omega)$ is real-valued, which means that the phase response of each subband can only alternate between zero and $180°$, due to sign changes of $H_k(j\omega)$. ■

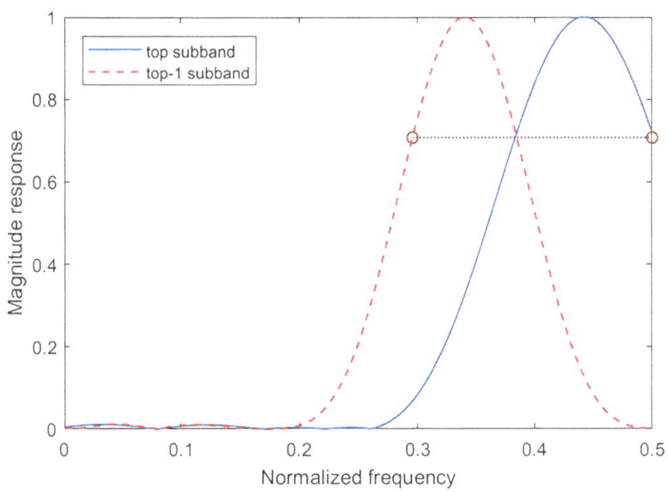

Fig. 5.7 Subband pattern, showing the K-th and $(K-1)$-th subbands, for a low-pass Hamming prototype: $-3\,\text{dB}$ level indicated by dotted line.

In summary, any low-pass prototype can be used in the construction, via (5.25), of subband filters that are compliant with a given constant-Q subband pattern. As pointed out earlier, such subband filters are nearly-analytic by construction when the chosen low-pass prototype $g(\cdot)$ has sufficiently low sidelobes. The key feature of the design approach presented in this section is a complete separation between adjustment of the top-subband filter center frequency – which depends only on the modulating frequency $f_{c,K}$ – and adjustment of this filter's bandwidth, which depends only on the scaling factor σ_K.

5.3.2 Design from an Adjustable Band-pass Prototype

The starting point for the design of a compliant highest-index (a.k.a. "top") subband filter, based on an *adjustable band-pass prototype*, is the scaling relation

$$h_K(t) \;=\; \frac{1}{\sigma_K}\, g\left(\frac{t}{\sigma_K}\,;\,\Theta\right) \quad\Longleftrightarrow\quad H_K(s) \;=\; G(\sigma_K\, s\,;\,\Theta) \tag{5.28a}$$

where $g(t\,;\,\Theta)$ is the impulse response of a given band-pass filter with adjustable parameter(s) Θ, and $G(s\,;\,\Theta) \stackrel{\text{def}}{=} \mathcal{L}\{g(t\,;\,\Theta)\}$ is the Laplace transform of this impulse response. We recall that the band-pass prototype $g(t\,;\,\Theta)$ must be analytic, or nearly-analytic, in order to minimize leakage from negative frequency components of the input waveform. We will assume that the magnitude response of this band-pass filter has unity peak value, namely,

$$\max_{\omega}\, \big|\, G(j\omega\,;\,\Theta)\,\big| = 1 \tag{5.28b}$$

for all values of Θ. This ensures that $h_K(t)$ satisfies the constraint (R2).

Since $h_K(t)$ has (at least) two adjustable parameters, it should be possible, in general, to make it into a compliant subband filter: this means we need to adjust these parameters so that the A dB frequencies of $|H_K(j\omega)|$ coincide with the two edges of the top-subband in the given subband pattern. The examples we provide below (using the choice $A = 3$) serve to illustrate the design process.

Example 5.3 (Design from a Morlet-wavelet prototype): Consider the same constant-Q subband pattern we used in Example 5.1, namely $\Gamma = 1.3$ and $f_{\text{max}} = F_s/2$, so that

$$f_{c,K} \;=\; \frac{1}{2}\, f_{\text{max}} \left(1 + \frac{1}{\Gamma}\right) \;=\; 0.442\, F_s$$

The chosen prototype is a Morlet wavelet[6] [148, 150], viz.,

$$g(t\,;\,f_m) \;=\; \frac{1}{\sqrt{2\pi}}\, e^{-\frac{t^2}{2}}\, e^{j2\pi f_m t} \quad \Longleftrightarrow \quad G(j2\pi f\,;\,f_m) \;=\; e^{-2\pi^2 (f-f_m)^2}$$

where the modulation frequency f_m can be adjusted at will. This choice of a band-pass prototype is nearly-analytic when $|G(0;\,f_m)| \ll 1$ or, equivalently, when $f_m \geq 3/2\pi \approx 0.5\,\mathsf{Hz}$. Notice that, for $f_m = 0$, this prototype coincides with the Gaussian window of Example 5.1.

Now, we observe that, according to (5.28a),

$$H_K\,(j2\pi f) \;=\; G\left(j2\pi\,\sigma_{\mathsf{K}} f\,;\,f_m\right) \;=\; e^{-2\pi^2\left(\sigma_{\mathsf{K}} f - f_m\right)^2} \;=\; e^{-2\sigma_{\mathsf{K}}{}^2\pi^2\left(f-\frac{f_m}{\sigma_{\mathsf{K}}}\right)^2}$$

is centered at the frequency $f = \frac{f_m}{\sigma_{\mathsf{K}}}$: this suggests the parameter choice $f_m = \sigma_{\mathsf{K}}\, f_{c,K}$ where, as before, $f_{c,K} = \frac{1}{2}\, f_{\mathsf{max}} \left(1+\frac{1}{\Gamma}\right)$. The resulting impulse response of the top subband filter is

$$h_K(t) \;=\; \frac{1}{\sigma_{\mathsf{K}}\sqrt{2\pi}}\; e^{-\frac{t^2}{2\sigma_{\mathsf{K}}{}^2}}\; e^{j2\pi f_{c,K} t}$$

which is identical to the low-pass-based expression (5.25a) for a Gaussian prototype.

Thus, the only remaining design task is to adjust the control parameter σ_{K} to satisfy the constraint (R3), namely,

$$\text{3 dB bandwidth of } |H_K(j\omega)| \;\equiv\; \frac{\Delta_f^{(PD\star 3)}}{\sigma_{\mathsf{K}}} \;=\; f_{\mathsf{max}}\left(1-\frac{1}{\Gamma}\right)$$

where $\Delta_f^{(PD\star 3)}$ is the 3 dB bandwidth of the prototype filter $g(t\,;\,f_m)$. Since the 3 dB bandwidth of this band-pass filter is independent of its center frequency, it is given by $\Delta_f^{(PD\star 3)} = 0.265$ (see Table B.2). We conclude that, as in Example 5.1, the constraint (R3) is satisfied when

$$\sigma_{\mathsf{K}} \;=\; \frac{\Delta_f^{(PD\star 3)}}{f_{\mathsf{max}}\left(1-\frac{1}{\Gamma}\right)} \;=\; 2.297\; T_s \tag{5.29}$$

which is the same value we found in Example 5.1. This completes the design process of a compliant top-subband filter $H_K(s)$.

---

[6] The usual definition of a Morlet wavelet includes a correction term, which serves to ensure a zero-mean property. However, this property is not required in the setting of our design problem.

The resulting subband filter impulse responses are (for $0 \leq r \leq K-1$),

$$h_{K-r}(t) = \frac{1}{\Gamma^r \sigma_K} g\left(\frac{t}{\Gamma^r \sigma_K}\,;\,\sigma_K f_{c,K}\right) = \frac{0.174}{T_s \Gamma^r}\, e^{-\left[\left(0.308/\Gamma^r\right)\left(t/T_s\right)\right]^2} e^{j2\pi f_{c,K}\left(t/\Gamma^r\right)}$$

Since $\frac{f_{c,K}}{\Gamma^r} = f_{c,K-r}$, this is identical to the subband filter expression obtained in Example 5.1. For digital implementation purposes, we need to use the equivalent discrete-time version, viz.,

$$h_{K-r}[n] = \frac{0.174}{\Gamma^r}\, e^{-\left(0.308\, n/\Gamma^r\right)^2} e^{j2\pi \eta_{K-r} n}\,, \qquad 0 \leq r \leq K-1$$

where $\eta_{K-r} \stackrel{\text{def}}{=} \frac{f_{c,K-r}}{F_s} = \frac{0.442}{\Gamma^r}$ are the normalized subband center frequencies. ■

Example 5.4 (Design from a Morse-wavelet prototype): Consider the same constant-Q subband pattern we used in Example 5.1, namely $\Gamma = 1.3$ and $f_{\text{max}} = F_s/2$. The chosen prototype is a Morse wavelet [150, 155], usually defined in terms of its frequency response, viz.,

$$G(j\omega\,;\beta,\gamma) = \begin{cases} \left(\frac{\beta}{e\gamma}\right)^{-\frac{\beta}{\gamma}} \omega^\beta\, e^{-\omega^\gamma} & \omega \geq 0 \\ 0 & \omega < 0 \end{cases}$$

where β and γ are adjustable design parameters. This real-valued function of ω has a single peak (Fig. 5.8), located at

$$\omega = \left(\frac{\beta}{\gamma}\right)^{1/\gamma} \stackrel{\text{def}}{=} \omega_{\text{peak}}$$

We notice that $G(j\omega\,;\beta,\gamma)$ satisfies the constraint (R2), namely,

$$\left|G(j\,\omega_{\text{peak}}\,;\beta,\gamma)\right| \equiv G(j\,\omega_{\text{peak}}\,;\beta,\gamma) = 1$$

Also notice that the Morse wavelet is analytic by definition, and thus perfectly suited for generating low-leakage TFRs of real-valued input waveforms.

Let $f_\ell(\beta,\gamma)$ and $f_h(\beta,\gamma)$ denote, respectively, the lower and upper 3 dB frequencies of the frequency-domain Morse wavelet $G(j\omega\,;\beta,\gamma)$, for given values of β and γ. The fundamental relation (5.28), i.e., $H_K(j\omega) = G(j\sigma_K\omega\,;\beta,\gamma)$, suggests that we use the three parameters σ_K, β, and γ to match the 3 dB passband of $|H_K(j\omega)|$, which is

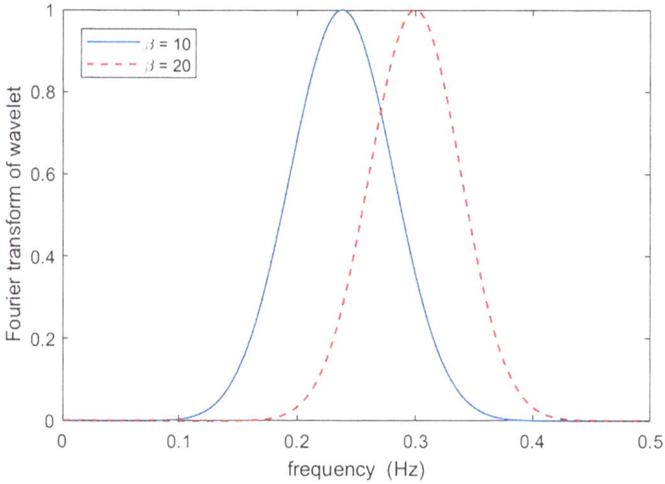

Fig. 5.8 Frequency-domain Morse wavelet, for $\gamma = 3$.

$$\frac{f_\ell(\beta,\gamma)}{\sigma_K} \le f \le \frac{f_h(\beta,\gamma)}{\sigma_K}$$

with the frequency interval of the top-subband. Thus, the constraint (R3) translates into two equations in three unknown parameters, viz.,

$$f_\ell(\beta,\gamma) = \sigma_K \frac{f_{\max}}{\Gamma} \quad \text{and} \quad f_h(\beta,\gamma) = \sigma_K f_{\max}$$

where Γ and $f_{\max}$ are defined by the given subband pattern. Dividing these two equation by each other, we obtain a σ_K-independent equation, viz.,

$$\rho(\beta;\gamma) \stackrel{\text{def}}{=} \frac{f_h(\beta,\gamma)}{f_\ell(\beta,\gamma)} = \Gamma$$

which relates the unknown parameters β and γ to the desired subband scaling factor Γ.

The *band-ratio* $\rho(\beta;\gamma)$ can be evaluated numerically as a function of β, for any desired value of γ : here we will consider only the two frequently-used choices $\gamma = 2$ and $\gamma = 3$ [150, 155]:

- When $\gamma = 2$, the solution of the equation $\rho(\beta;2) = \Gamma \equiv 1.3$ is $\beta = 20.15$ (see top plot in Fig. 5.9). The design process is completed by evaluating the upper 3 dB frequency function $f_h(\beta, 2)$ (see bottom plot in Fig. 5.9), and setting

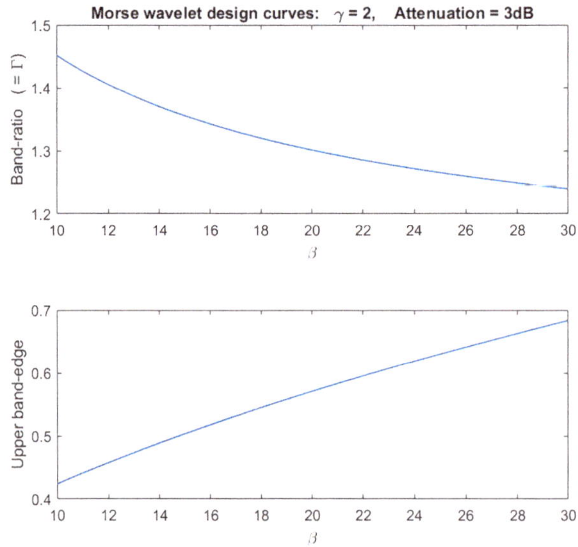

Fig. 5.9 Morse-wavelet design curves for $\gamma = 2$.

$$\sigma_K = \frac{f_h(20.15, 2)}{f_{max}} = \frac{0.573}{F_s/2} = 1.146\,T_s$$

The end result is

$$H_K(j\omega) = \left(\frac{\beta}{e\gamma}\right)^{-\frac{\beta}{\gamma}}(\sigma_K\,\omega)^\beta\, e^{-(\sigma_K\omega)^\gamma} = 1.853\cdot 10^{-6}\left[1.146\frac{\omega}{F_s}\right]^{20.15} e^{-(1.146\,\omega/F_s)^2}$$

- Similarly, when $\gamma = 3$, the solution of the equation $\rho(\beta; 3) = 1.3$ is $\beta = 13.50$ (see top plot in Fig. 5.9). This allows us to evaluate $f_h(\beta, 3)$ (see bottom plot in Fig. 5.10), and set

$$\sigma_K = \frac{f_h(13.50, 3)}{f_{max}} = \frac{0.297}{F_s/2} = 0.594\,T_s$$

The end result is

$$H_K(j\omega) = \left(\frac{\beta}{e\gamma}\right)^{-\frac{\beta}{\gamma}} \left(\sigma_K\,\omega\right)^\beta e^{-\left(\sigma_K\omega\right)^\gamma} = 0.1035\left[0.594\frac{\omega}{F_s}\right]^{13.50} e^{-(0.594\,\omega/F_s)^3}$$

The remaining subband filters are obtained via the fundamental relation (5.24), viz.,

$$H_{K-r}(j\omega) = H_K\,(j\,\Gamma^r\,\omega)$$

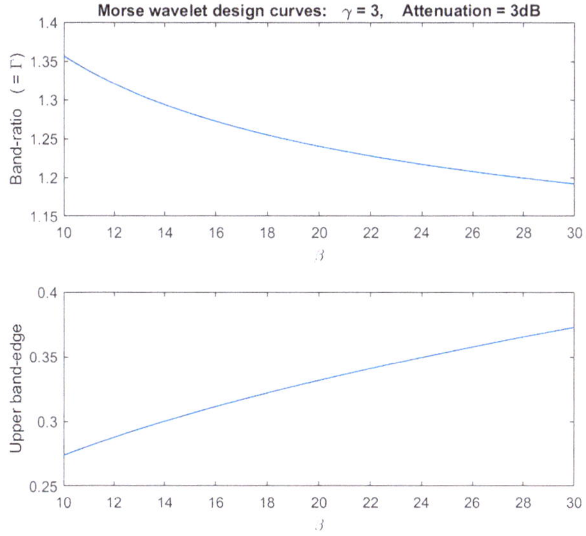

Fig. 5.10 Morse-wavelet design curves for $\gamma = 3$.

The most direct way to obtain the associated impulse responses $h_{K-r}(t)$ is to evaluate numerically the inverse Fourier transform of these frequency-domain expressions. ■

The same design approach can be used with any adjustable band-pass prototype, such as a Butterworth or Chebyshev band-pass filter. For instance, several such prototypes have been proposed, and implemented in commercial software tools for analysis of geophysical data [156]. However, such conjugate-symmetric frequency responses violate the "nearly-analytic" requirement and thus generate excessive leakage from negative frequencies, as compared with a Morse wavelet. The same observation applies to all real-valued wavelet functions, including Coiflets and Daubechies wavelets.

5.3.3 A Unified Design Method

The preceding discussion in this section has established a general method for design of compliant subband filters from either a low-pass or a band-pass prototype. The proposed design approach relies on the time-scaling property (5.24), which ensures that the compliance constraint (R1) is satisfied for all subbands. This also implies that the constraints (R2) and (R3) are satisfied by all subbands if, and only if, they are satisfied by the top subband. Thus, the only design task left is to construct a compliant top subband. Finally, since the compliance constraint (R2) is easily satisfied by a suitable normalization of $h_K(t)$, the task of designing a compliant top subband filter reduces to making $h_K(t)$ compliant with (R3).

The techniques described in Sects. 5.3.1–5.3.2 can be combined into a single design approach, based on a unified characterization of the top subband filter, viz.,

$$h_K(t) \;=\; \frac{1}{\sigma_K}\, g\Big(\frac{t}{\sigma_K}\,;\,\Theta\Big)\, e^{\,j2\pi f_m t} \quad\Longleftrightarrow\quad H_K(s) \;=\; G\big(\sigma_K\,(s - j2\pi f_m)\,;\,\Theta\big) \tag{5.30a}$$

where $g(t\,;\Theta)$ is the impulse response of a given prototype filter, either low-pass or band-pass, with adjustable parameter(s) Θ, and $G(s\,;\Theta) \overset{\text{def}}{=} \mathcal{L}\{g(t\,;\Theta)\}$ is the transfer function associated with this impulse response. We also assume that the magnitude response of this prototype filter has been normalized to a unity peak value, namely,

$$\max_{\omega}\,\big|\,G(j\omega\,;\Theta)\,\big| = 1 \tag{5.30b}$$

This ensures that $h_k(t)$ satisfies the constraint (R2) for all $1 \le k \le K$.

The compliance constraint (R3) translates in this case into a set of two (nonlinear) equations in the adjustable parameters Θ, σ_K, and f_m, viz.,

$$\begin{aligned} f_h(A\,;\Theta)\,\sigma_K^{-1} \;+\; f_m \;&=\; f_{\text{max}} \\ f_\ell(A\,;\Theta)\,\sigma_K^{-1} \;+\; f_m \;&=\; \frac{f_{\text{max}}}{\Gamma} \end{aligned} \tag{5.31a}$$

where

- $f_\ell(A\,;\Theta)$ and $f_h(A\,;\Theta)$ denote, respectively, the lower and upper A dB frequencies of the prototype transfer function $G(s\,;\Theta)$ of (5.28a), for given values of A and Θ.
- Θ is the set of adjustable parameters of the prototype transfer function $G(s\,;\Theta)$. Notice that Θ does not include time-scaling, which is controlled separately by the parameter σ_K.
- The parameters Γ and f_{max} are defined by the given subband pattern.

Notice that the set of equations (5.31a) is *linear* in σ_K^{-1} and f_m, which means it admits a unique closed-form solution for every value of Θ (and A), viz.,

$$\begin{aligned} \sigma_K \;&=\; \frac{f_h(A\,;\Theta) \;-\; f_\ell(A\,;\Theta)}{f_{\text{max}}\,(1 - \frac{1}{\Gamma})} \\ f_m \;&=\; \frac{\Gamma^{-1}\rho(A\,;\Theta) \;-\; 1}{\rho(A\,;\Theta) \;-\; 1}\; f_{\text{max}} \end{aligned} \tag{5.31b}$$

where

$$\rho\,(A\,;\Theta) \;\stackrel{\text{def}}{=}\; \frac{f_h\,(A\,;\Theta)}{f_\ell\,(A\,;\Theta)} \tag{5.31c}$$

denotes the *band-ratio* of the prototype filter. This solution fully coincides with the one obtained in Example 5.3 (Morlet), where the parameter Θ did not exist, and we chose $A = 3$. On the other hand, in Example 5.4 (Morse) we were able to adjust the parameter β so that the band-ratio $\rho\,(3\,;\Theta)$ was exactly equal to the desired Γ. In this case the unified solution (5.31b) reduces to $f_m = 0$ (because $\rho\,(3;\Theta) = \Gamma$) and $\sigma_{\mathsf{K}} = \frac{f_h\,(3\,;\Theta)}{f_{\max}}$: this matches exactly the results obtained in Example 5.4.

Since the unique solution (5.30) does not rely on adjustability of the prototype filter, it can be used to design subband filters from non-adjustable prototypes such as the Ricker wavelet (a.k.a. "Mexican hat"), as well as dyadic multi-resolution wavelets (e.g., Daubechies), for any desired value of the passband edge attenuation parameter A. Moreover, we can even enforce the choice $f_m = 0$ by either: (a) appropriate selection of A, which results in non-unity subband-overlap ratio (SOR), or (b) maintaining the choice $A = 3$ and unity SOR, but accepting a value of Γ that deviates from our desired choice. The following example illustrates these different options.

Example 5.5 (Design from a Ricker-wavelet prototype): Consider the same constant-Q subband pattern we used in Example 5.1, namely $\Gamma = 1.3$ and $f_{\max} = F_s/2$. The chosen prototype is a Ricker wavelet, also known as a Mexican-hat function [148, 156], viz.,

$$g(t) \;=\; \frac{C}{\sqrt{2\pi}}\,(1 - t^2)\,e^{-\frac{t^2}{2}} \quad\Longleftrightarrow\quad G(j\omega) \;=\; C\,\omega^2\,e^{-\frac{\omega^2}{2}}$$

which has a single peak (at $\omega = \sqrt{2}$), and no adjustable parameters (i.e., Θ does not exist). Unity peak gain is achieved with $C = e/2$. We can now choose between three distinct implementation approaches:

- Set $A = 3$ and use the solution (5.31b), viz.,

$$\sigma_{\mathsf{K}} \;=\; \frac{f_h\,(3) \;-\; f_\ell\,(3)}{f_{\max}(1 - \frac{1}{\Gamma})} \;=\; 1.606\,T_s$$

$$f_m \;=\; \frac{\Gamma^{-1} f_h(3) \;-\; f_\ell(3)}{f_h(3) \;-\; f_\ell\,(3)}\, f_{\max} \;=\; 0.298\,F_s$$

- Adjust the value of A so that $f_m = 0$ for the desired value of Γ. When $\Gamma = 1.3$, this requires setting $A = 0.3$ (see Table 5.2),[7] and results in $\sigma_{\mathrm{K}} = \frac{f_h(0.3)}{f_{\max}} = 0.511\, T_s$.
- Maintain the choice $A = 3$ and $f_m = 0$. This results in $\Gamma = \frac{f_h(3)}{f_\ell(3)} = 2.333$ (see Table 5.2), which means that the top subband will occupy nearly 60% of the available frequency range. The resulting scaling factor is $\sigma_{\mathrm{K}} = \frac{f_h(3)}{f_{\max}} = 0.649\, T_s$. ■

Table 5.2 Ricker wavelet: passband characteristics for $f_m = 0$

A	0.3	0.6	1	2	3	4
$f_h(A)$	0.255	0.268	0.281	0.305	0.324	0.341
$f_\ell(A)$	0.196	0.185	0.173	0.154	0.139	0.127
$\rho(A)$	1.301	1.452	1.621	1.988	2.333	2.676

5.4 Time-Frequency Analysis of Power System Waveforms: A Hybrid-TFR Approach

As pointed out in the introduction to this chapter, time-frequency analysis of power system waveforms poses a special challenge because it involves several very different objectives:

- Accurate representation of dominant harmonics requires uniformly-spaced subbands, centered on the harmonics of interest, with sufficient frequency-resolution, and decent suppression of leakage from sidelobes.
- Capturing in-band transients (in the dominant harmonics) requires good time-resolution. A sensible compromise between these two requirements is having subbands whose width is f_0, which results in time-resolution $\Delta_t \approx T \equiv {}^1\!/_{f_0}$, as detailed in Appendix B.
- Capturing wide-band transients requires a much higher time-resolution. This can be achieved by using wider subbands at the higher frequency range, say, above $11 f_0$ or so. Using constant-Q subbands is one possibility for achieving this objective.

The first two of these objectives can be satisfactorily achieved by a constant-bandwidth TFR scheme. As an illustration, consider an STFT of the paper-mill example (Fig. 5.11): it shows significant fundamental and 5th harmonics, both of which provide a clear indication of an in-band transient. While the continuous

[7] In this case, $f_h(0.3) = 0.255 \approx 1.3\, f_\ell(0.3)$, since $\Gamma = 1.3$.

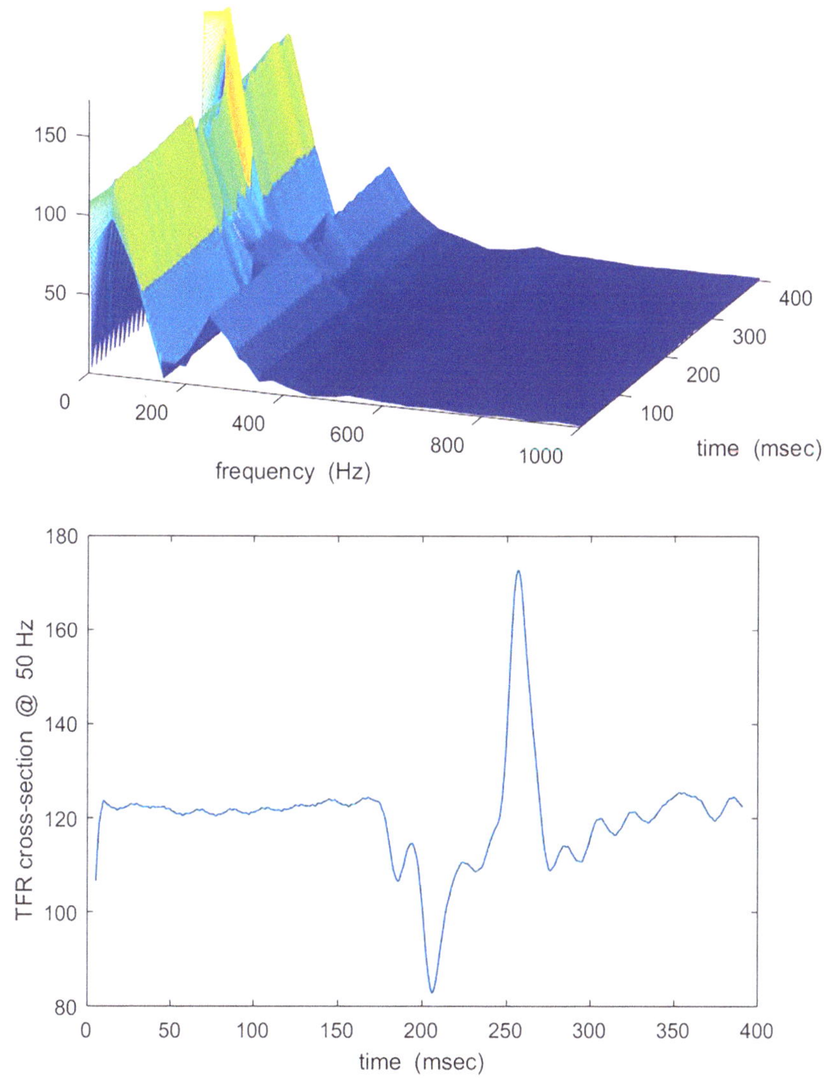

Fig. 5.11 STFT of "phase a" current in the paper-mill example ($f_0 = 50$ Hz, Hamming window): complete TFR (top) and a TFR cross-section at the fundamental frequency (bottom).

wavelet transform (CWT) provides a largely similar representation in the higher frequency range (Fig. 5.12, top), its depiction of time variation in the fundamental harmonic exhibits excessive smoothing (Fig. 5.12, bottom). This is caused by the very narrow width of the subband that contains the fundamental: the width of this subband in the paper-mill example is approximately 10 Hz, which results in poor time-resolution (≈100 msec). Since 4 subbands-per-octave is the lowest density

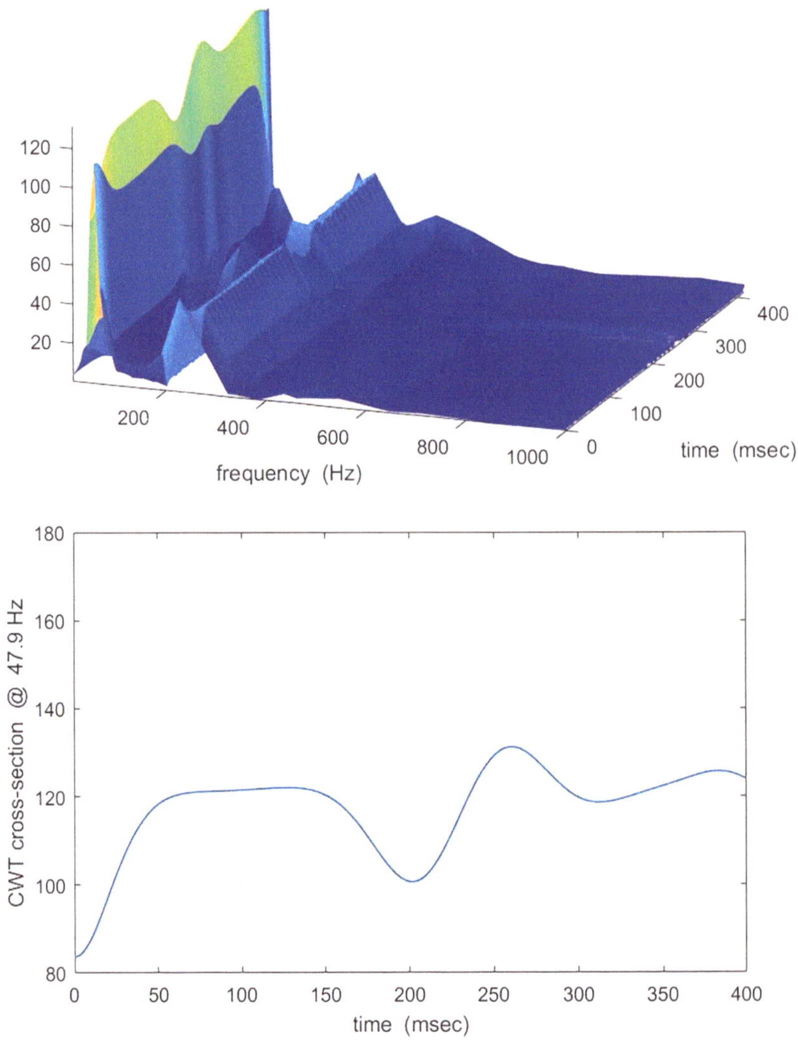

Fig. 5.12 Continuous wavelet transform (CWT) of "phase a" current in the paper-mill example ($f_0 = 50\,\text{Hz}$), using 4 subbands-per-octave: complete TFR (top) and a TFR cross-section near the fundamental frequency (bottom).

allowed for the "cwt" function in MATLAB, time-resolution at the fundamental frequency cannot be significantly improved. A similar conclusion is suggested by other real-life examples, such as the wind-farm dataset (Fig. 5.13). In summary, a constant-bandwidth (CB) transform, like the STFT, is the preferred choice for reliable representation of both steady-state harmonic content and in-band transients.

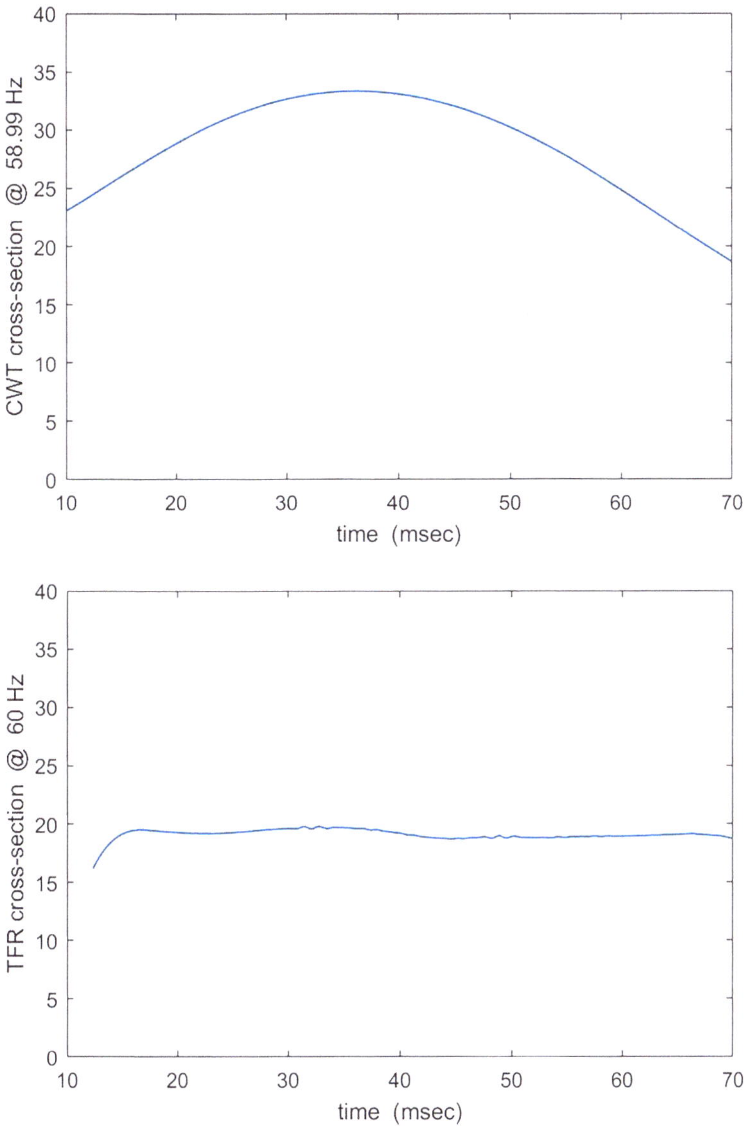

Fig. 5.13 TFR cross-section at the fundamental harmonic (wind-farm example, "phase a" current): CWT (top) and STFT (bottom).

However, the situation is reversed in the presence of a high-frequency wide-band transient. It turns out that a constant-Q (CQ) transform, like the CWT, provides a more insightful depiction of such transients, due to its improved time-resolution at higher frequencies. The wind-farm example provides an illustration of this capability: both Fig. 5.14 (STFT) and Fig. 5.15 (CWT) display a transient peak

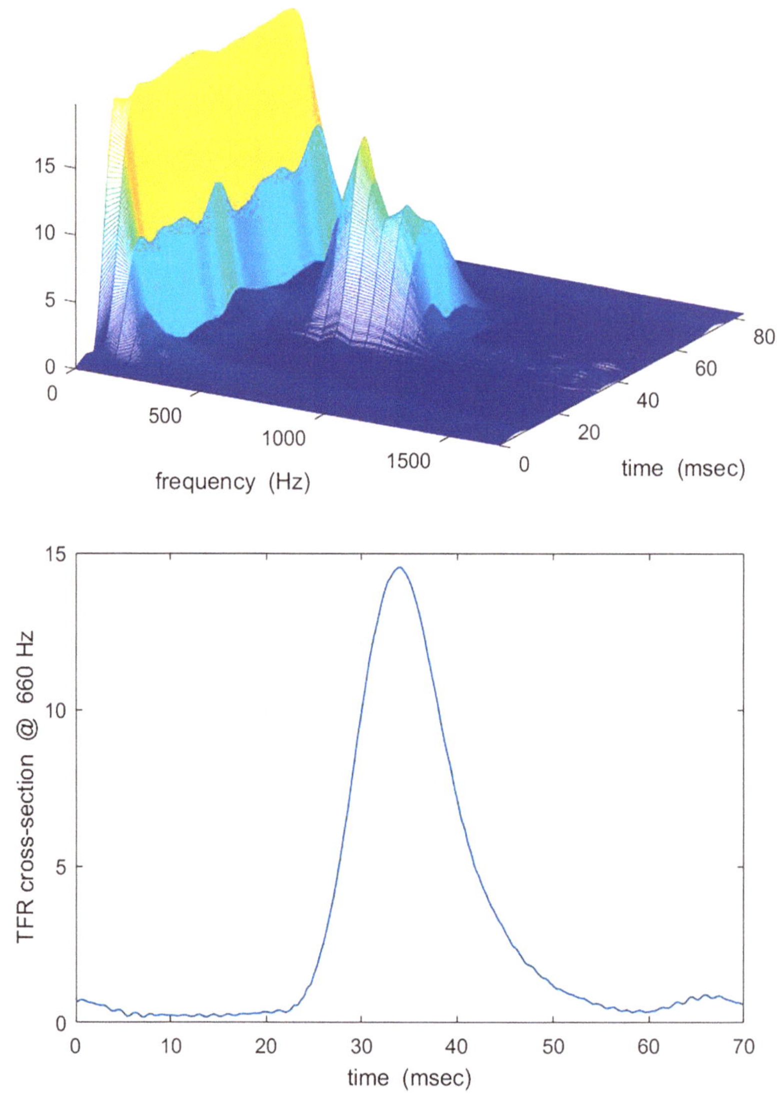

Fig. 5.14 STFT of "phase a" current in the wind-farm example (Hamming window, $f_0 = 60\,\mathrm{Hz}$): complete TFR (top) and a TFR cross-section at the transient peak frequency (bottom).

concentrated in the time interval 23–50 msec and the frequency range 500–1000 Hz, which is generated by switching on a bank of capacitors. However, the TFR cross-section through this peak is somewhat wider in the STFT-generated TFR (Fig. 5.14, bottom) than in its CQ counterpart (Fig. 5.15, bottom). We conclude that CQ transforms are a better choice for representing wide-band high-frequency transients.

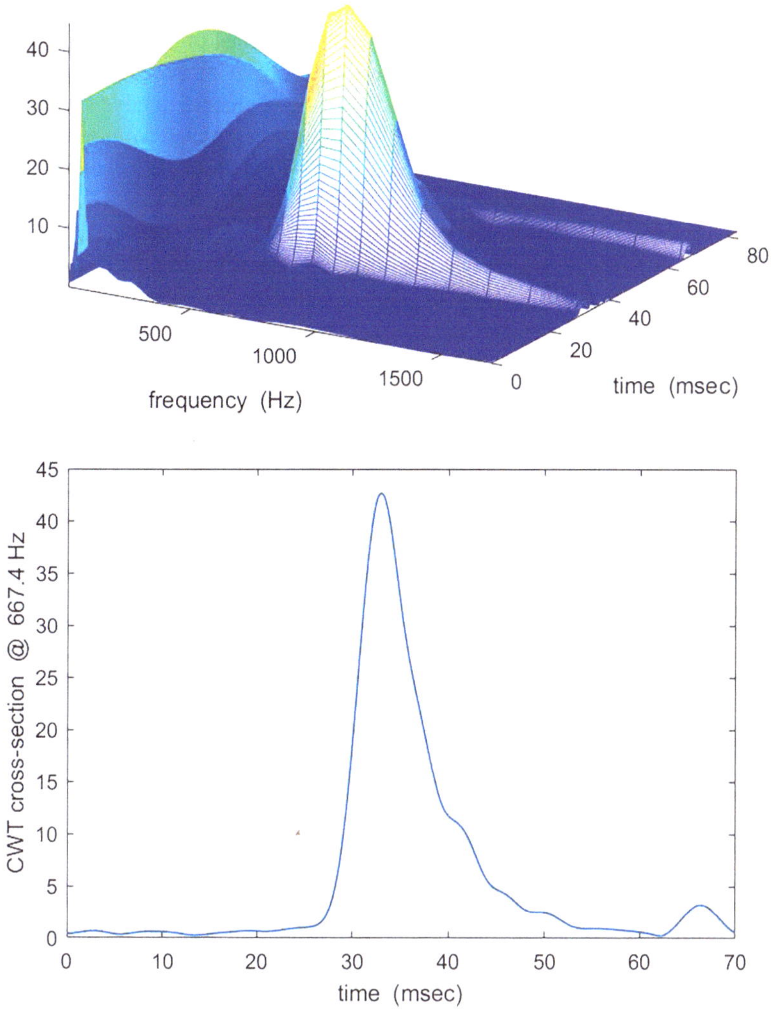

Fig. 5.15 CWT of "phase a" current in the wind-farm example (Morse wavelet, 6 subbands-per-octave, $f_0 = 60\,\text{Hz}$): complete TFR (top) and a TFR cross-section at the transient peak frequency (bottom).

We propose to address this dichotomy by adopting a *hybrid-TFR approach*, which splits in two the frequency range of interest: a constant-bandwidth transform (e.g., the STFT) is used in the lower-frequency range, while a constant-Q transform (generalized-STFT or CWT) is used in the upper range. We will use the terms *CB-module* and *CQ-module* to describe these two distinct components of the hybrid-TFR. Guidelines for application-specific design of each module are provided in Sects. 5.4.1, 5.4.2, 5.4.3, 5.4.4, and 5.4.5 below.

Of the five transform approaches listed in the introduction to this chapter, only the generalized-STFT and the CWT allow flexible control of frequency-resolution, time-resolution, inter-subband leakage, and output rate of the resulting TFR. In particular, when the primary goal is detailed time-frequency analysis or accurate localization of signal transients, the quality of features extracted from power system waveforms via either CWT or the generalized-STFT is substantially superior to that provided by the DWT or DOST.[8]

The main difference between CWT/generalized-STFT techniques and the other transform approaches – which directly impacts the resulting TFR quality – is the underlying design philosophy of each transform. For instance, the primary design objective of DWT schemes is construction of an orthonormal basis for the space of finite energy waveforms, as detailed in Sect. 5.6. Feasible DWT prototype subband-filter choices are constrained by the requirements of satisfying a dyadic dilation equation and providing a controllable degree of smoothness, neither of which correlates well with good frequency-resolution and/or suppression of sidelobe leakage. Consequently, the resulting subband pattern has a low, non-adjustable, subband density[9] (i.e., a single subband per octave), which compromises frequency-resolution and also makes it hard to match dominant harmonics with DWT subbands. In addition, DWT relies on critical down-sampling of its subbands, which degrades its time-resolution.

In contrast, CWT/generalized-STFT schemes provide an adjustable, and much higher, subband density: typical values are in the range of 4–10 subbands-per-octave (see [148, 149], as well as Sect. 5.4.3). Furthermore, STFT windows, such as Hamming, Blackman or Chebyshev, are designed specifically to achieve the objective of superior leakage suppression, and the same is true for the Morse prototype used by the CWT.

In addition, typical filter prototypes used in DWT implementations, such as Daubechies wavelets, lack the "nearly-analytic" property. This necessitates converting each acquired real-valued input waveform $x(\cdot)$ into its analytic version $\widehat{x}(t) = x(t) + j\mathcal{H}\{x(t)\}$, applying the DWT separately to the real and imaginary parts of $\widehat{x}(t)$, and combining the resulting two DWT outputs into a single TFR.

Figure 5.16 provides an illustration of some DWT-based time-frequency features for the paper-mill example ($f_0 = 50\,\mathsf{Hz}$): the DWT subband that contains this fundamental frequency is $27.3\,\mathsf{Hz} \leq f \leq 54.7\,\mathsf{Hz}$ (Fig. 5.16, top-left). We observe that the STFT counterpart, shown in the top-right panel of Fig. 5.16, provides a significantly more accurate depiction of the onset of an in-band transient, which starts at $t \approx 180\,\mathsf{msec}$, as well as the steady-state regime within the interval that precedes this transient. Similarly, the 5th harmonic, which falls within the relatively wide DWT subband $219\,\mathsf{Hz} \leq f \leq 438\,\mathsf{Hz}$ (Fig. 5.16, bottom-left), is much easier

[8] An informative discussion of the key differences between CWT and DWT can be found in [157].

[9] See Sect. 5.4.3 for a characterization of subband-density values used in generating high-fidelity time-frequency representations.

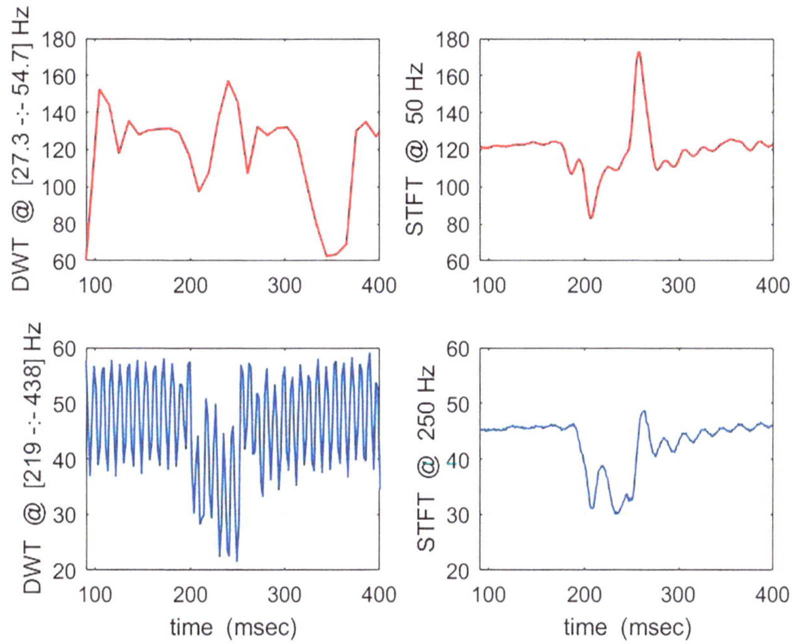

Fig. 5.16 DWT-STFT comparison for the paper-mill example ($f_0 = 50$ Hz): fundamental harmonic (top) and 5th harmonic (bottom). DWT output levels were rescaled to match those of the STFT outputs.

to discern in the appropriate STFT subband, centered at $f = 5f_0 = 250$ Hz. Similar observations apply to scenarios that exhibit wide-band transients in addition to dominant harmonics, such as in the wind-farm example.

We conclude that a spectrogram generated by a well-designed generalized-STFT or CWT is preferable, in most power system applications, to feature extraction via the DWT. Consequently, we limit our discussion in the remainder of this section to hybrid-TFR schemes whose CQ-modules employ adequate subband density (at least 4 subbands-per-octave), analytic/near-analytic prototype filters, and non-decimated subband processing.

5.4.1 Design of the CB-module

Section 4.5 presented several options for customizing the design of a constant-bandwidth STFT. Since the objective of the CB-module is accurate depiction of dominant harmonics, both in steady-state and in the presence of slowly-varying transients, we propose using the "moderate TFR scheme" from Sect. 4.5 as our design guideline. To be specific, this means employing harmonic-centered subbands, viz.,

$$\text{k-th subband :} \quad \left|f - kf_0\right| \leq \frac{f_0}{2} \quad , \qquad 0 \leq k \leq K_{\mathsf{CB}}$$

and using a single-cycle-long window as the prototype low-pass filter. A reasonable choice for the number of CB subbands, based in part on the recommendations of IEEE Standard 519 [154], is $7 \leq K_{\mathsf{CB}} \leq 11$.

The subband filters for this module are given, of course, by (4.16b), namely,

$$h_k(t) = h(t)\, e^{jk2\pi f_0 t} \quad , \qquad 0 \leq k \leq K_{\mathsf{CB}}$$

where $h(t)$ is a single-cycle-long window, i.e., $T_h = {}^1\!/_{f_0}$. The 3 dB bandwidth of each subband filter $h_k(t)$, as given by Table B.2, is

$$\Delta_{f,k}^{(PD\star3)} = \frac{C_f^{(PD\star3)}}{T_{h,k}} \equiv C_f^{(PD\star3)} f_0 \qquad \text{for all } k$$

where the coefficient $C_f^{(PD\star3)}$ relates the 3 dB bandwidth of a given window function $h(t)$ to the window duration T_h, and we rely on the fact that $T_{h,k} = T_h = {}^1\!/_{f_0}$. Notice that when $C_f^{(PD\star3)}$ is larger than unity, as is the case for windows such as Hamming, Hann or Kaiser, the resulting window bandwidth exceeds f_0, which means that magnitude responses of adjacent subbands in the CB-module intersect at a level $A_{\mathsf{CB}} < 3$. For instance, Hamming-based subband filters intersect at a level of $A_{\mathsf{CB}} = 1.75$ dB (see Fig. 5.17 below).

5.4.2 CB/CQ-Junction Constraints

The *CB/CQ-junction* is the point of transition between the frequency ranges of the CB-module and the CQ-module, i.e., the intersection point between the magnitude responses of the highest subband in the CB-module and the lowest subband in the CQ-module. The frequency of this transition point can be controlled by adjusting the parameter f_{min}, which defines the lower edge of the nominal frequency range covered by the CQ-module.

A convenient way to describe this degree of freedom is in terms of the *junction edge ratio* (JER), viz.,

$$\mathsf{JER} \overset{\text{def}}{=} \frac{f_{\mathsf{min}}}{(K_{CB} + {}^1\!/_2)\, f_0} \equiv \frac{\text{lowest CQ subband-edge frequency}}{\text{highest CB subband-edge frequency}} \tag{5.32a}$$

We now recall that the subband filters of a CQ-module are designed to intersect at a prescribed level A_{CQ}, as shown in Fig. 5.4 and specified by the compliance constraint (R3). This means that one has several options for choosing a JER-value (and, thus, an f_{min}-value), namely,

- Set $\mathsf{JER} = 1$, which results in a level of intersection at the CB/CQ-junction that is different from either A_{CB} or A_{CQ}.

Table 5.3 JER as a function of K_{CB} for a Hamming CB-window function

K_{CB}	7	8	9	10	11
JER ($A_{\text{CQ}} = 0.3$)	0.9609	0.9655	0.9691	0.9721	0.9745
JER ($A_{\text{CQ}} = 3$)	1.0194	1.0171	1.0153	1.0139	1.0127

- Set $\text{JER} = 1$ and design the CQ-module to have the same intersection level as the CB-module, i.e., set A_{CQ} to be equal to A_{CB}. This approach results in uniform subband overlap for the entire CB+CQ filter-bank.
- Determine JER so that the junction intersection level matches A_{CB}: this results in $\text{JER} < 1$ when $A_{\text{CQ}} > A_{\text{CB}}$, but $\text{JER} > 1$ when $A_{\text{CQ}} < A_{\text{CB}}$.
- Determine JER so that the junction intersection level matches A_{CQ}: this results in $\text{JER} > 1$ when $A_{\text{CQ}} > A_{\text{CB}}$, but $\text{JER} < 1$ when $A_{\text{CQ}} < A_{\text{CB}}$.

Once a JER-value has been determined, the corresponding f_{min}-value, viz.,

$$f_{\text{min}} = \text{JER}\,(K_{\text{CB}} + 1/2)\,f_0 \tag{5.32b}$$

provides a needed input for the CQ-module design step, which is described in the following subsection.

It should be pointed out that even the last two of these four design options generate $\text{JER} \approx 1$, so that all four design options generate very similar subband patterns. For instance, the 4-th option induces the constraint $f_{\text{min}} = K_{\text{CB}}\,f_0 + \frac{1}{2}\,\Delta_{f,\text{CB}}^{(PD\star A_{\text{CQ}})}$, where $\Delta_{f,\text{CB}}^{(PD\star A)} = C_{f,\text{CB}}^{(PD\star A)}\,T_h^{\,-1}$ is the A dB bandwidth of the CB-window function $h(t)$, with the value of $C_{f,\text{CB}}^{(PD\star A)}$ specified in Tables B.2–B.4. This constraint results in the simple relation

$$\text{JER} = 1 + \frac{C_{f,\text{CB}}^{(PD\star A_{\text{CQ}})} - 1}{2\,K_{\text{CB}} + 1} \tag{5.33}$$

Evaluating this expression for $7 \le K_{\text{CB}} \le 11$ and $0.3 \le A_{\text{CQ}} \le 3$ reveals that $0.0127 \le |\text{JER} - 1| \le 0.0391$: see Table 5.3, which is based on $C_{f,\text{CB}}^{(PD\star 0.3)} = 0.4138$ and $C_{f,\text{CB}}^{(PD\star 3)} = 1.291$ (the values for a Hamming CB-window). This supports our observation that $\text{JER} \approx 1$ is a reasonable approximation in most practical cases.

5.4.3 Design of the CQ-module

Design of any constant-Q filter-bank starts with construction of its subband pattern, as detailed in Sect. 5.2. The input variables for this step are (a) the frequency range $[\,f_{\text{min}}\;\; f_{\text{max}}\,]$ covered by the filter-bank, (b) K_{CQ} – the desired number of subbands, and (c) A_{CQ} – the desired level of intersection between magnitude responses of adjacent subbands. While the value of f_{min} depends primarily on K_{CB} (the number of subbands in the CB-module – recall (5.32) and (5.33)), the top frequency f_{max} can be selected based on prior knowledge about the nature of transients contained in the waveform of interest. When such information is not readily available, one can resort to the default choice $f_{\text{max}} = F_s/2$. The number of subbands in the CQ-module serves as a design variable that should be chosen subject to an application-specific upper bound $K_{\text{CQ}} \le K_{\text{CQ,max}}$. The upper bound $K_{\text{CQ,max}}$ depends on A_{CQ}, as well as the frequency variables f_0, f_{min} and f_{max}, as specified in (5.36) below.

Construction of a subband pattern for the CQ-module starts with the key relation (5.15), namely,

$$\Gamma = \left(\frac{f_{\text{max}}}{f_{\text{min}}}\right)^{1/K_{\text{CQ}}}$$

Next, the scaling factor Γ is used to determine:

- **Subband density**, which is equal to $\left[\log_2 \Gamma\right]^{-1}$ subbands-per-octave.
- **Subband center frequencies:** according to (5.26c),

$$f_{c,K_{\text{CQ}}-r} = \frac{\Gamma+1}{2\,\Gamma^{r+1}}\, f_{\text{max}} \quad , \qquad 0 \le r \le K_{\text{CQ}}-1 \tag{5.34}$$

- **Subband frequency ranges:** in particular, the range of the lowest CQ subband is $f_{\text{min}} \le f \le \Gamma f_{\text{min}}$.

Information about subband frequency ranges can be used to formulate a *junction bandwidth constraint*, which is described below.

CQ Subband-density Constraint

The CQ-module is designed to provide a gradually increasing subband width, which controls the time/frequency-resolution trade-off in each subband. Thus, it is reasonable to require that nominal width of the lowest CQ subband – i.e., $(\Gamma-1)\,f_{\text{min}}$ – is at least as large as the bandwidth of the filter used in the highest CB-subband, as illustrated in Fig. 5.17. Since the former is equal to the A_{CQ}-bandwidth of the corresponding subband filter, we will use the same prescribed-drop level to define filter bandwidth for the highest CB-subband, viz.,

$$\Delta_{f,\text{CB}}^{(PD\star A_{\text{CQ}})} = C_{f,\text{CB}}^{(PD\star A_{\text{CQ}})}\, T_h^{-1} \equiv C_{f,\text{CB}}^{(PD\star A_{\text{CQ}})}\, f_0$$

where we assume that the prototype CB-module filter is a window of length $T_h = 1/f_0$: a detailed analysis of finite-length windows is provided in Appendix B.

The resulting constraint is $(\Gamma-1)\,f_{\text{min}} \ge C_{f,\text{CB}}^{(PD\star A_{\text{CQ}})}\, f_0$, which generates an upper bound on CQ subband-density, because

$$(\Gamma-1)\, f_{\min} \geq C_{f,\mathrm{CB}}^{(PD\star A_{\mathrm{CQ}})} f_0 \Longrightarrow \Gamma \equiv 2^{1/\text{subband-density}} \geq 1+C_{f,\mathrm{CB}}^{(PD\star A_{\mathrm{CQ}})} \frac{f_0}{f_{\min}}$$

and so,

$$\text{CQ subband-density} \leq \left[\log_2\left(1 + C_{f,\mathrm{CB}}^{(PD\star A_{\mathrm{CQ}})} \frac{f_0}{f_{\min}}\right)\right]^{-1} \text{subbands/octave} \tag{5.35}$$

When $0.3 \leq A_{\mathrm{CQ}} \leq 3$ – which, for a Hamming window prototype, results in $0.4138 \leq C_{f,\mathrm{CB}}^{(PD\star A_{\mathrm{CQ}})} \leq 1.291$ – we find that this upper bound on CQ subband-density is typically within the range $[4 \quad 20]$, assuming we abide by the constraint $7 \leq K_{\mathrm{CB}} \leq 11$ that was proposed earlier for the number of CB subbands.

This subband-density constraint generates an equivalent bound on the choice of K_{CQ}, the number of CQ subbands. Since, by definition,

$$\begin{aligned}\text{CQ subband-density} &\stackrel{\text{def}}{=} \frac{\text{number of CQ-subbands}}{\text{number of CQ frequency octaves}} \\ &\equiv \frac{K_{\mathrm{CQ}}}{\log_2\left(f_{\max}/f_{\min}\right)}\end{aligned}$$

it follows that

$$K_{\mathrm{CQ}} \leq K_{\mathrm{CQ.max}} \stackrel{\text{def}}{=} \text{floor}\left(\frac{\log_2\left(\frac{f_{\max}}{f_{\min}}\right)}{\log_2\left(1 + C_{f,\mathrm{CB}}^{(PD\star A_{\mathrm{CQ}})} \frac{f_0}{f_{\min}}\right)}\right) \tag{5.36}$$

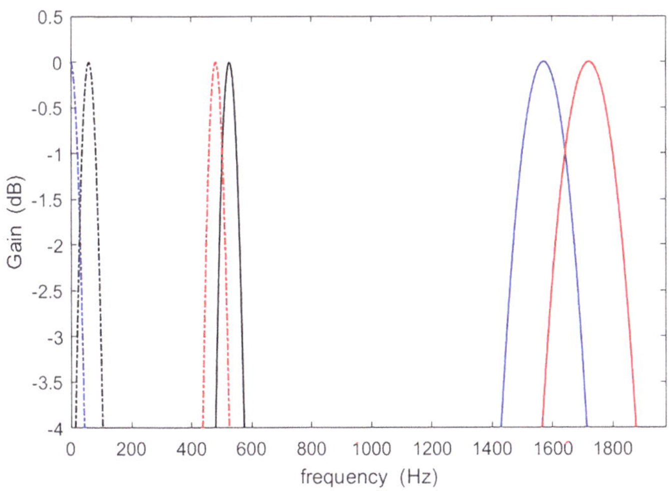

Fig. 5.17 Subband pattern of a 9-Hamming/14-Hamming hybrid-TFR scheme, subject to both CB/CQ-junction constraints (with $A_{\mathrm{CQ}} = 1\mathrm{dB}$), displaying: (a) last and first two of the nine CB subbands (dash-dot line), and (b) first and last two of the fourteen CQ subbands (solid line). Frequency parameters: $f_0 = 60\,\mathrm{Hz}$ and $f_{\max} = 1800\,\mathrm{Hz}$, resulting in $f_{\min} = 502.6\,\mathrm{Hz}$.

Figure 5.17 demonstrates the consequences of enforcing both CB/CQ-junction constraints, namely, $K_{\mathsf{CQ}} \leq K_{\mathsf{CQ,max}}$ and the junction edge constraint (5.33). Setting $K_{\mathsf{CB}} = 8$, $f_{\mathsf{max}} = 1800\,\mathsf{Hz}$, and $A_{\mathsf{CQ}} = 1\,\mathsf{dB}$ in this (wind-farm, $f_0 =$ 60Hz) example results in $f_{\mathsf{min}} = 502.6\,\mathsf{Hz}$ and $K_{\mathsf{CQ,max}} = 14$. The corresponding subband pattern, generated with $K_{\mathsf{CQ}} = 14$, has monotone non-decreasing subband widths, and a $1\,\mathsf{dB}$ subband intersection level at the CB/CQ-junction. The short-hand label "9-Hamming/14-Hamming" indicates the number of subbands and the filter prototype used in the CB-module/CQ-module, respectively.

Notice that the bandwidth of the first CQ subband is very nearly the same as the bandwidth of the last CB-subband: this is a by-product of setting K_{CQ} equal to its upper bound. Moreover, this property holds regardless of the A_{CQ}-value used to determine $K_{\mathsf{CQ,max}}$ via (5.36), which also means that the resulting distinct subband patterns of the CQ-module – one pattern for each (A_{CQ}, $K_{\mathsf{CQ,max}}$)-pair, where A_{CQ} can assume any value – all share the same Q-factor. This observation forms the basis for an effective CQ-module design method: see Sect. 5.4.4 for a detailed discussion.

Design of Subband Filters

Subband filters for the CQ-module can be designed from a low-pass prototype according to the procedure described in Sect. 5.3.1. The key relation is (5.26a), i.e.,

$$h_{K-r}(t) = \frac{1}{\Gamma^r \sigma_{\mathsf{K}}}\, g\left(\frac{t}{\Gamma^r \sigma_{\mathsf{K}}}\right) e^{j2\pi f_{c,K-r}\, t}\,, \qquad 0 \leq r \leq K-1$$

where we set $K = K_{\mathsf{CQ}}$ and the subband center frequencies $f_{c,K-r}$ are as given in (5.34). The bandwidth-scaling coefficient σ_K is given by the expression

$$\sigma_K = \frac{\Delta_f^{(PD\star A)}}{f_{\mathsf{max}}\left(1 - \frac{1}{\Gamma}\right)} \tag{5.37}$$

The function $g(\cdot)$ is the impulse response of a prototype low-pass filter of choice, scaled to have unity DC gain, and $\Delta_f^{(PD\star A)}$ is the $A\,\mathsf{dB}$ bandwidth of this prototype filter. For instance, if we use the Hamming window as a prototype, and choose the CQ subband intersection level as $A_{\mathsf{CQ}} = 3\,\mathsf{dB}$, then (recall (B.6))

$$g(x) = \begin{cases} 1 + \frac{21}{25}\cos\left(2\pi x\right) & |x| \leq \frac{1}{2} \\ 0 & \text{else} \end{cases}$$

so that $T_g = 1$ and $\Delta_f^{(PD\star 3)} = 1.291\, T_g^{-1} = 1.291$. In this case, the $(K{-}r)$-th subband impulse response $h_{K-r}(t)$ has length $\Gamma^r \sigma_K$.

Alternatively, one can use a band-pass prototype (e.g., a Morlet or Morse wavelet) as a starting point for designing subband filters, as detailed in Sect. 5.3.2. The key relation for this approach is (5.28a), i.e.,

$$h_K(t) \;=\; \frac{1}{\sigma_K}\, g\left(\frac{t}{\sigma_K}\,;\,\Theta\right)$$

where we also set $K = K_{CQ}$. In this case, the bandwidth-scaling coefficient σ_K is still given by the expression (5.37), the function $g(\cdot)$ is the impulse response of a prototype band-pass filter of choice, scaled to have unity DC gain, and $\Delta_f^{(PD\star A)}$ is the A dB bandwidth of this band-pass prototype filter. Finally, we recall that both design approaches can be merged into a single unified procedure, as detailed in Sect. 5.3.3.

5.4.4 Iso-Q Curves: A Versatile Tool for CQ-module Design

The final step in designing the subband pattern of a CQ-module amounts to choosing the parameter A_{CQ}, which controls the upper bound $K_{CQ,max}$, as well as selecting the number of CQ subbands $K_{CQ} \le K_{CQ,max}$. In other words, once the CB-module design step has been completed, and f_{max} has been selected, we are left with the task of choosing a point in the (A_{CQ}, K_{CQ})-plane.

Each such choice completely determines a CQ-module subband pattern. In particular, choosing an (A_{CQ}, K_{CQ})-pair determines the Q-factor of the resulting CQ-module and, thus, controls frequency-resolution (a continuous-frequency property) in each subband of this module. It turns out that distinct (A_{CQ}, K_{CQ})-pairs can give rise to *the same Q-factor*. The set of all such pairs is a hyperbola-like curve (Fig. 5.18), which we call an *iso-Q curve*. On the other hand, the location of an (A_{CQ}, K_{CQ})-point on a given iso-Q curve controls the density of our discretized frequency grid and thus impacts the level of detail displayed by $X_k(\cdot)$ – the discrete frequency TFR – as explained in Sect. 5.2. In summary, a CQ-module subband pattern is completely determined by (a) choosing an iso-Q curve of interest and (b) selecting an (A_{CQ}, K_{CQ})-point on this curve. We shall use this viewpoint throughout this section as a foundation for orderly CQ-module design.

An explicit expression for (A_{CQ}, K_{CQ})-points on an iso-Q curve of interest can be obtained by expressing the scaling factor Γ in terms of the Q-factor and the bandwidth parameter ρ_f via (5.18), namely,

$$\Gamma(A_{CQ}; Q) \;=\; \frac{2Q + \rho_f(A_{CQ})}{2Q - \rho_f(A_{CQ})}$$

Next we express K_{CQ} in terms A_{CQ} and Γ via the expression (5.15b), i.e.,

$$K_{CQ} \;=\; \frac{\log_2\left(\frac{f_{max}}{f_{min}(A_{CQ})}\right)}{\log_2 \Gamma(A_{CQ}; Q)}$$

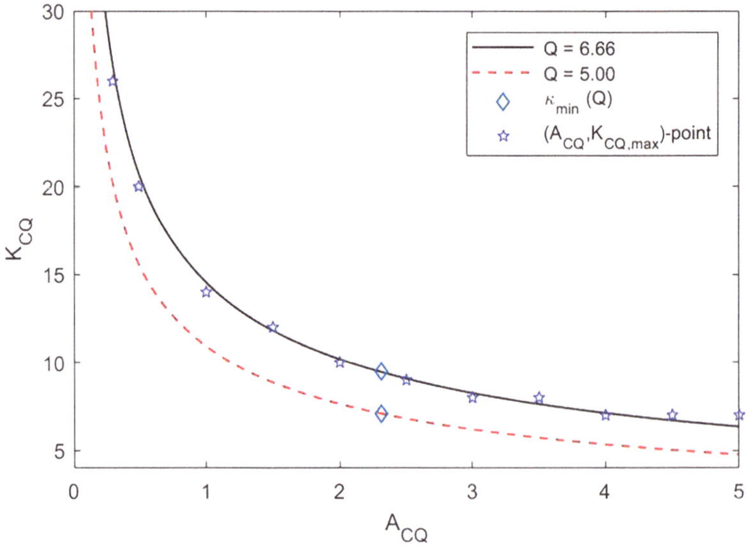

Fig. 5.18 Two iso-Q curves and eleven selected $(A_{\mathrm{CQ}}, K_{\mathrm{CQ,max}})$-points. Frequency parameters (for both the wind-farm and the synthetic-short-transient examples): $f_0 = 60\,\mathrm{Hz}$ and $f_{\max} = 1800\,\mathrm{Hz}$. See (5.39) for a definition of $\kappa_{\min}(Q)$.

However, since the right-hand side of this expression need not be an integer, the final form of this relation becomes $K_{\mathrm{CQ}} = \mathrm{round}\left(\kappa\left(A_{\mathrm{CQ}}; Q\right)\right)$, where

$$\kappa\left(A_{\mathrm{CQ}}; Q\right) \stackrel{\mathrm{def}}{=} \frac{\log_2\left(\frac{f_{\max}}{f_{\min}(A_{\mathrm{CQ}})}\right)}{\log_2\left(\frac{2Q+\rho_f(A_{\mathrm{CQ}})}{2Q-\rho_f(A_{\mathrm{CQ}})}\right)} \tag{5.38}$$

is a functional representation of the iso-Q curve, with $f_{\min}(A_{\mathrm{CQ}})$ determined via the JER expressions (5.32) and (5.33). Figure 5.18 shows two such iso-Q curves, for two distinct values of the Q-factor: $Q = 6.66$ and $Q = 5.00$. Notice that the upper bound on K_{CQ}, indicated in this figure by several star-shaped markers located at selected $(A_{\mathrm{CQ}}, K_{\mathrm{CQ,max}})$-points, practically coincides with the iso-Q curve for $Q = 6.66$.

As we pointed out earlier, the Q-factor of a CQ-module controls subband filter bandwidth and thus determines the trade-off between time- and frequency-resolution for each subband. Consequently, iso-Q subband patterns generate very similar TFRs, which differ only in small details, due to a different level of overlap between subbands. Figure 5.19 illustrates this property: although the two designs differ greatly in their $(A_{\mathrm{CQ}}, K_{\mathrm{CQ}})$-pair values, the corresponding TFRs are almost identical.

CQ-modules that correspond to $(A_{\mathrm{CQ}}, K_{\mathrm{CQ}})$-points located on an iso-Q curve with a lower Q-value, such as $Q = 5$ in Fig. 5.18, should be considered when time-resolution of high frequency subbands is of prime concern. However, an excessive reduction of the Q-factor may result in severe distortion of the resulting time-

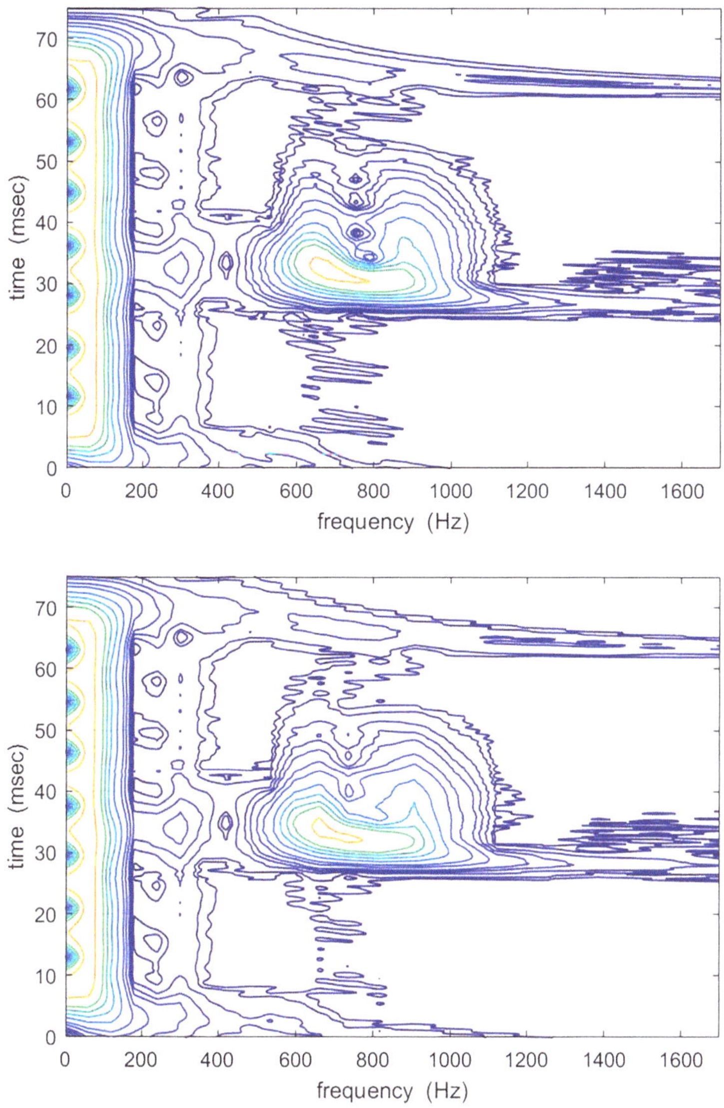

Fig. 5.19 Contour plots of hybrid-TFR (wind-farm example, "phase a" current) generated by two distinct iso-Q subband patterns (both with $Q = 6.43$), using a Hamming prototype: $(A_{CQ}, K_{CQ}) = (0.30, 26)$ (top), and $(A_{CQ}, K_{CQ}) = (1.36, 12)$ (bottom). Frequency parameters: $f_0 = 60\,\text{Hz}$ and $f_{max} = 1800\,\text{Hz}$.

frequency representation: this effect is due to degraded frequency-resolution, as well as increased leakage from adjacent subbands, both of which tend to "smear out" frequency detail. For instance, using the parameter values $A_{CQ} = 0.5$ and $K_{CQ} = 13$ to reduce the Q-factor in the wind-farm example from $Q = 6.43$ to $Q = 4.17$ results in a significant degradation in frequency-resolution: compare the contour plot in Fig. 5.20 with the ones in Fig. 5.19. We propose imposing an

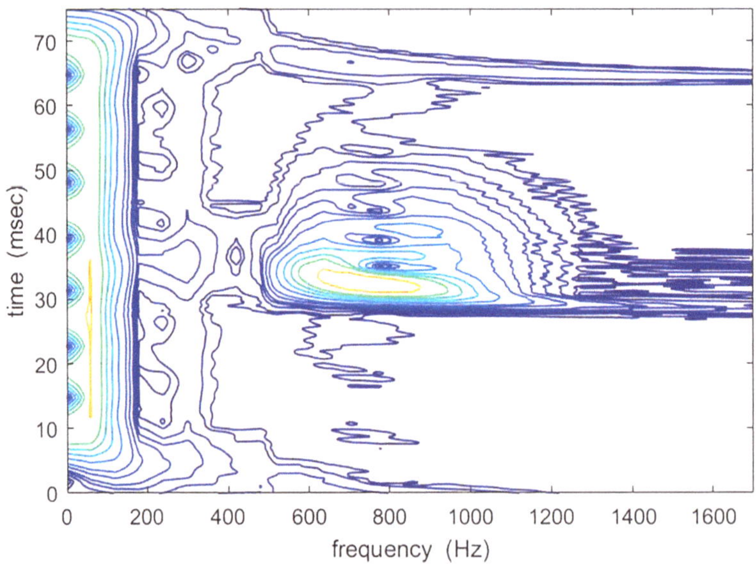

Fig. 5.20 Contour plot a of hybrid-TFR (wind-farm example, "phase a" current), generated by a "reduced-Q" subband pattern (with $Q = 4.17$), using a Hamming prototype: $(A_{\mathsf{CQ}}, K_{\mathsf{CQ}}) = (0.5\,,13)$. Frequency parameters: $f_0 = 60\,\mathsf{Hz}$ and $f_{\mathsf{max}} = 1800\,\mathsf{Hz}$.

empirical lower bound on the value of Q, in order to prevent excessive frequency smearing: a suitable bound in our examples appears to be $Q \approx 4.5$.

Once a desired Q-value has been selected, which determines a particular iso-Q curve and, thus, a particular trade-off between time- and frequency-resolution, the next step in the CQ-module design process is selection of a particular $(A_{\mathsf{CQ}}, K_{\mathsf{CQ}})$-point on this curve. We observe that points in the upper-left segment of an iso-Q curve result in more CQ subbands and thus increased operation cost. On the other hand, choosing an $(A_{\mathsf{CQ}}, K_{\mathsf{CQ}})$-point in the lower-right segment of such a curve reduces the number of subbands – which determines the density of our discretized frequency grid – thus causing loss of fine detail,[10] such as shape of the "ravine" seen in Fig. 5.19, located at $f_c \approx 750\,\mathsf{Hz}$ within the time interval $40\,\mathsf{msec} \leq t \leq 50\,\mathsf{msec}$. Thus, in order to prevent excessive loss of detail, one should consider imposing a lower bound on the number of subbands.

One way of generating such a bound is by exploiting the lower bound (5.22) on subband density, which was derived in Sect. 5.2: it induces a *Q-dependent* approximate lower bound for the number of subbands, i.e., $K_{\mathsf{CQ}} \geq \mathsf{ceil}\big(\kappa_{\mathsf{min}}(Q)\big)$, where

[10] Notice the difference in detail between the top and bottom plots of Fig. 5.19, caused by substantially different densities of the discretized frequency grid.

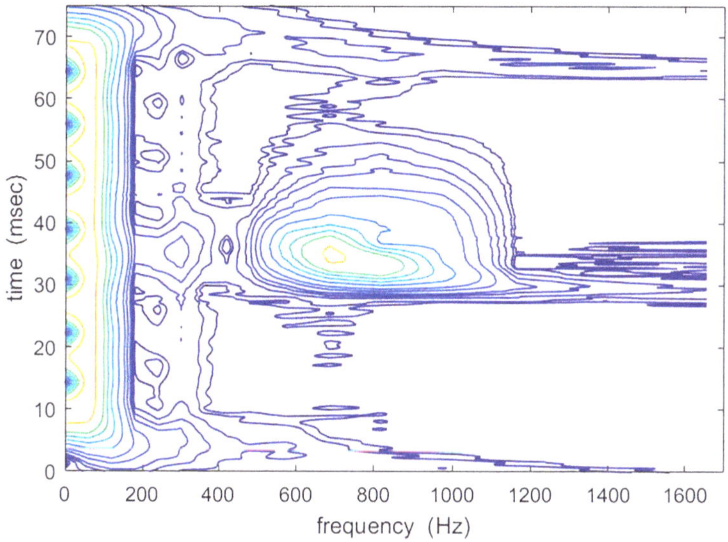

Fig. 5.21 Contour plot a of hybrid-TFR (wind-farm example, "phase a" current), generated by an overly-sparse subband pattern (with $Q = 6.54$), using a Hamming prototype: $(A_{\mathrm{CQ}}, K_{\mathrm{CQ}}) = (4, 7)$. Frequency parameters: $f_0 = 60\,\mathrm{Hz}$ and $f_{\max} = 1800\,\mathrm{Hz}$, resulting in $\kappa_{\min}(Q) = 9.1$.

$$\kappa_{\min}(Q) \stackrel{\text{def}}{=} \frac{\log_2\left(\frac{f_{\max}}{f_{\min}(A_{\mathrm{CQ}})}\right)}{\log_2\left(\Gamma_{cr}(Q)\right)} \tag{5.39a}$$

and $\Gamma_{cr}(Q)$ is as defined in (5.22a), namely,

$$\Gamma_{cr}(Q) \stackrel{\text{def}}{=} \frac{4\,Q + \rho_f^{(PD\star 10)}}{4\,Q - \rho_f^{(PD\star 10)}} \tag{5.39b}$$

In other words, each iso-Q curve has its own lower bound on K_{CQ}, as indicated by the "diamond" markers in Fig. 5.18. Violating this lower bound – i.e., choosing $K_{\mathrm{CQ}} < \kappa_{\min}(Q)$ – results in loss of fine detail, as can be seen in Fig. 5.21: in particular, the "ravine" seen in Fig. 5.19 (located at $f_c \approx 750\,\mathrm{Hz}$) is completely absent from the contour plot in Fig. 5.21, which uses $K_{\mathrm{CQ}} = 7$ even though $\kappa_{\min}(Q) = 9.1$. We conclude that acceptable choices of an $(A_{\mathrm{CQ}}, K_{\mathrm{CQ}})$-point should be restricted to the upper-left segment of the iso-Q curve, namely, left of the $\kappa_{\min}(Q)$ marker.

The other criterion that should be taken into account when making a specific $(A_{\mathrm{CQ}}, K_{\mathrm{CQ}})$ choice is cost (per input sample) of operating the CQ-module, which increases linearly with K_{CQ} (or very nearly so) along each iso-Q curve (Fig. 5.22). This property follows from the observation that $T_{h,K}$ – the length of the finite-length window $h_K(t)$, associated with the K-th subband – is, according to Sect. B.4, given by the expression

$$T_{h,K} = \frac{\mathcal{C}_{f,\mathsf{CQ}}^{(PD\star A_{\mathsf{CQ}})}}{\Delta_{f,\mathsf{CQ}}^{(PD\star A_{\mathsf{CQ}})}} \equiv \frac{\mathcal{C}_{f,\mathsf{CQ}}^{(PD\star A_{\mathsf{CQ}})}}{f_{\max}(1 - 1/\Gamma)}$$

where we rely on the relation between the nominal width of a subband and the A dB bandwidth of the corresponding subband filter, as illustrated in Fig. 5.4. Next, notice that the cost of operating the k-th subband filter equals the discrete-time length of this FIR filter, which is

$$\mathsf{ceil}\left(\frac{T_{h,k}}{T_s}\right) = \mathsf{ceil}\left(F_s\,\frac{\mathcal{C}_{f,\mathsf{CQ}}^{(PD\star A_{\mathsf{CQ}})}}{f_{\max}(1 - 1/\Gamma)}\,\Gamma^{K_{\mathsf{CQ}}-k}\right)\,, \qquad 1 \le k \le K_{\mathsf{CQ}}$$

The total cost (per input sample) of operating the CQ-module is obtained by adding up the operating costs of the individual subband filters, viz.,

$$\text{Operation cost (per sample) of a CQ-}\textbf{module} = \sum_{r=0}^{K_{\mathsf{CQ}}-1} F_s\,\frac{\mathcal{C}_{f,\mathsf{CQ}}^{(PD\star A_{\mathsf{CQ}})}}{f_{\max}(1 - 1/\Gamma)}\,\Gamma^{r}$$

$$= \mathcal{C}_{f,\mathsf{CQ}}^{(PD\star A_{\mathsf{CQ}})}\,\frac{F_s}{f_{\max}}\left(\Gamma^{K_{\mathsf{CQ}}} - 1\right)\frac{\Gamma}{(\Gamma - 1)^2} = \mathcal{C}_{f,\mathsf{CQ}}^{(PD\star A_{\mathsf{CQ}})}\left(\frac{F_s}{f_{\min}} - \frac{F_s}{f_{\max}}\right)\frac{\Gamma}{(\Gamma - 1)^2}$$

where we used the fact that $\Gamma^{K_{\mathsf{CQ}}} = f_{\max}/f_{\min}$ and, for simplicity, ignored the rounding operation $\mathsf{ceil}(\cdot)$. Plotting this cost as a function of K_{CQ} – for points along an iso-Q curve – reveals a nearly linear behavior (see Fig. 5.22).

Examination of Figs. 5.18 and 5.22 suggests that a reasonable choice for the number of subbands in this example is $10 \le K_{\mathsf{CQ}} \le 15$ which, equivalently, corresponds to the parameter choice $1 \le A_{\mathsf{CQ}} \le 2$. This range of K_{CQ} values somewhat exceeds the lower bound (5.39), at the expense of a modest increase in operation cost. In addition, we observe that operation cost of the CQ-module is directly proportional to the waveform sampling rate F_s, which should discourage excessive over-sampling. For instance, a satisfactory representation of the wind-farm transient is obtained with $f_{\max} = 1800\,\mathsf{Hz}$ (see Figs. 5.23 and 5.24 below), which implies that a sampling rate of 3600 samples/sec would have been sufficient to generate this TFR.[11]

[11] The actual sampling rate of the wind-farm dataset is 7680 samples/sec: see Sect. L.2.2.

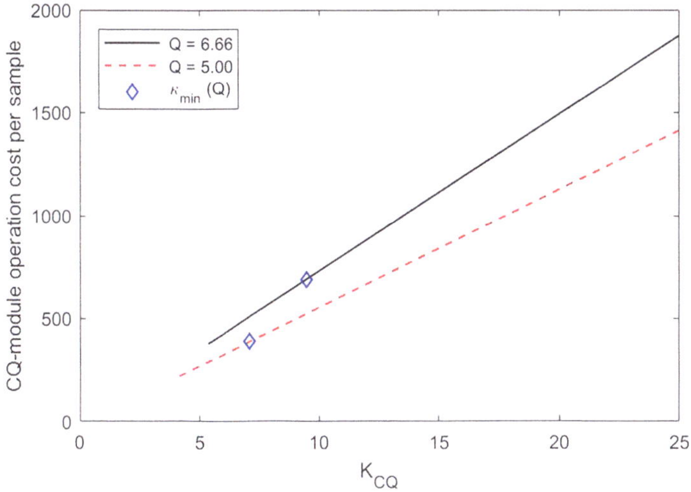

Fig. 5.22 CQ-module operation cost (per input sample) along an iso-Q curve. Frequency parameters (for both wind-farm and synthetic-short-transient): $f_0 = 60\,\text{Hz}$ and $f_{\text{max}} = 1800\,\text{Hz}$.

5.4.5 The Combined CB/CQ Transform

Putting together the two modules generates a TFR that covers the entire frequency range of interest, from DC to f_{max}. In the wind-farm example one version of this hybrid-TFR consists of nine constant-width subbands (because $K_{\text{CB}} = 8$) and fourteen constant-Q subbands (with $A_{\text{CQ}} = 1\,\text{dB}$ and $K_{\text{CQ}} = K_{\text{CQ,max}} = 14$), all generated from a Hamming window prototype (Fig. 5.23). The nine harmonics contained in the CB portion of this TFR are, of course, identical to the ones obtained from a standard Hamming-based STFT: compare the TFR in the lower-frequency range $0 \le f \le f_{\text{min}}$ of Fig. 5.23 with the same range in Fig. 5.14. In particular, notice how well this TFR scheme portrays the fundamental harmonic, both in steady-state and in the presence of an in-band transient.

Similar results are obtained when a Morse prototype – with $\gamma = 3$, $\beta = 27$, and $A_{\text{CQ}} = 1\,\text{dB}$ – is used to implement the CQ-module (Fig. 5.24). Both the TFR and its cross-section are practically identical to those shown in Fig. 5.23, which confirms our observation that choice of a CQ prototype has a very minor impact on TFR quality, once the CQ subband pattern has been finalized. In fact, the only noticeable difference between the Morse-based and Hamming-based TFRs in this example is the presence of a small ripple in the TFR cross-section plot of Fig. 5.23, which is caused by the sidelobes of the Hamming window. Finally, a comparison with Fig. 5.14 (uniform STFT) underscores the advantages of using a constant-Q subband pattern for portrayal of wide-band transients: our two hybrid-TFR schemes (i.e., Figs. 5.23 and 5.24) display a taller transient peak, while also achieving a better time-resolution in the TFR cross-section for this peak.

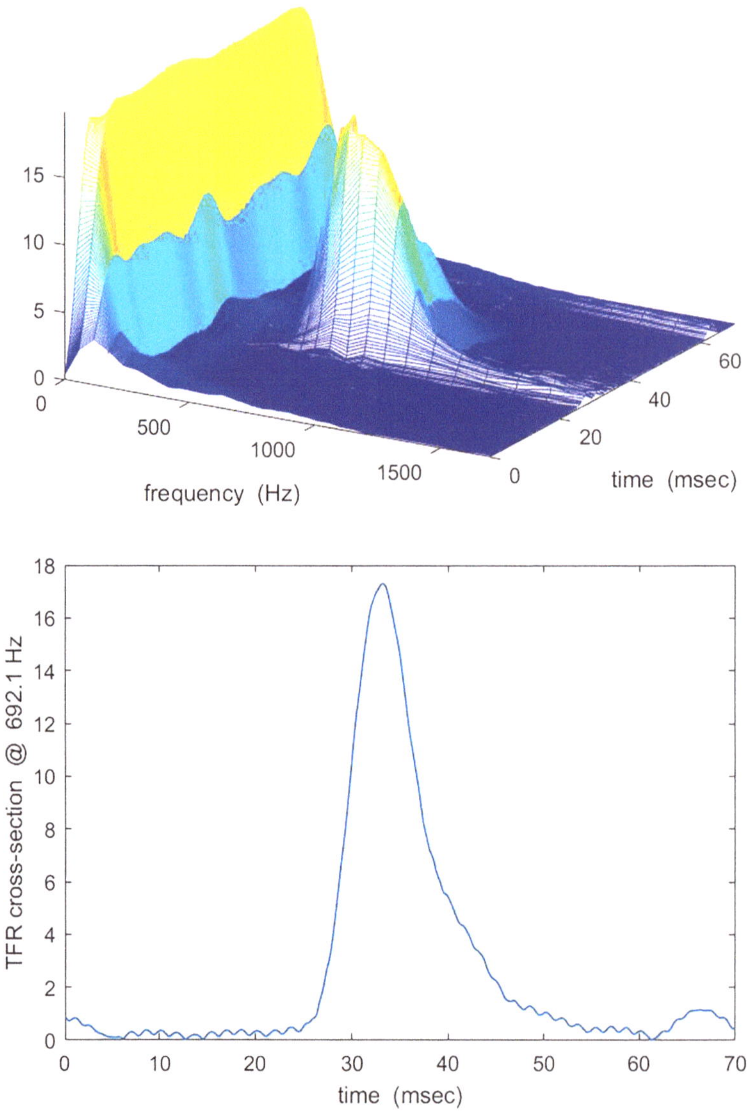

Fig. 5.23 Hybrid-TFR, using a 9-Hamming/14-Hamming subband pattern (with $A_{CQ} = 1\,\text{dB}$), applied to "phase a" current of the wind-farm example: complete TFR (top) and a TFR cross-section at the transient peak frequency (bottom). Frequency parameters: $f_0 = 60\,\text{Hz}$ and $f_{max} = 1800\,\text{Hz}$, resulting in $Q = 6.40$ and $\kappa_{min}(Q) = 9.2$.

Reducing the Q-factor in this (wind-farm) example from $Q = 6.4$ to $Q = 5.0$, by changing from $A_{CQ} = 1\,\text{dB}$ to $A_{CQ} = 0.6\,\text{dB}$ results in a barely noticeable improvement in time-resolution: compare Fig. 5.25, bottom with Fig. 5.24, bottom. On the other hand, this "reduced-Q" scheme suffers from a significantly degraded

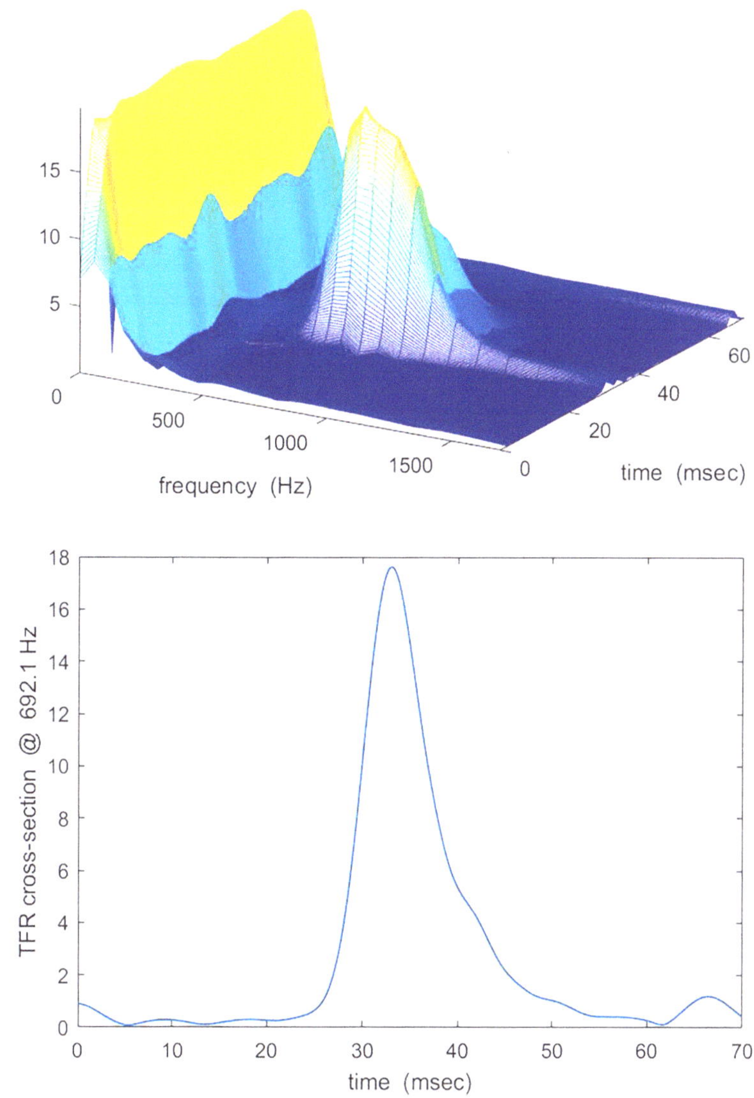

Fig. 5.24 Hybrid-TFR scheme, using a 9-Hamming/14-Morse subband pattern (with $A_{CQ} = 1\,\text{dB}$, $\gamma = 3$, $\beta = 27$), applied to "phase a" current of the wind-farm example: complete TFR (top) and a TFR cross-section at the transient peak frequency (bottom). Frequency parameters: $f_0 = 60\,\text{Hz}$ and $f_{\text{max}} = 1800\,\text{Hz}$, resulting in $Q = 6.48$ and $\kappa_{\text{min}}(Q) = 8.9$.

frequency-resolution: compare the contour plot in Fig. 5.25 (top) with the ones in Fig. 5.19. These observations suggest that a "reduced-Q" approach should be used with caution: the improvement in time-resolution that can be achieved by a reduction of the Q-factor may be in some cases outweighed by a serious degradation in frequency-resolution.

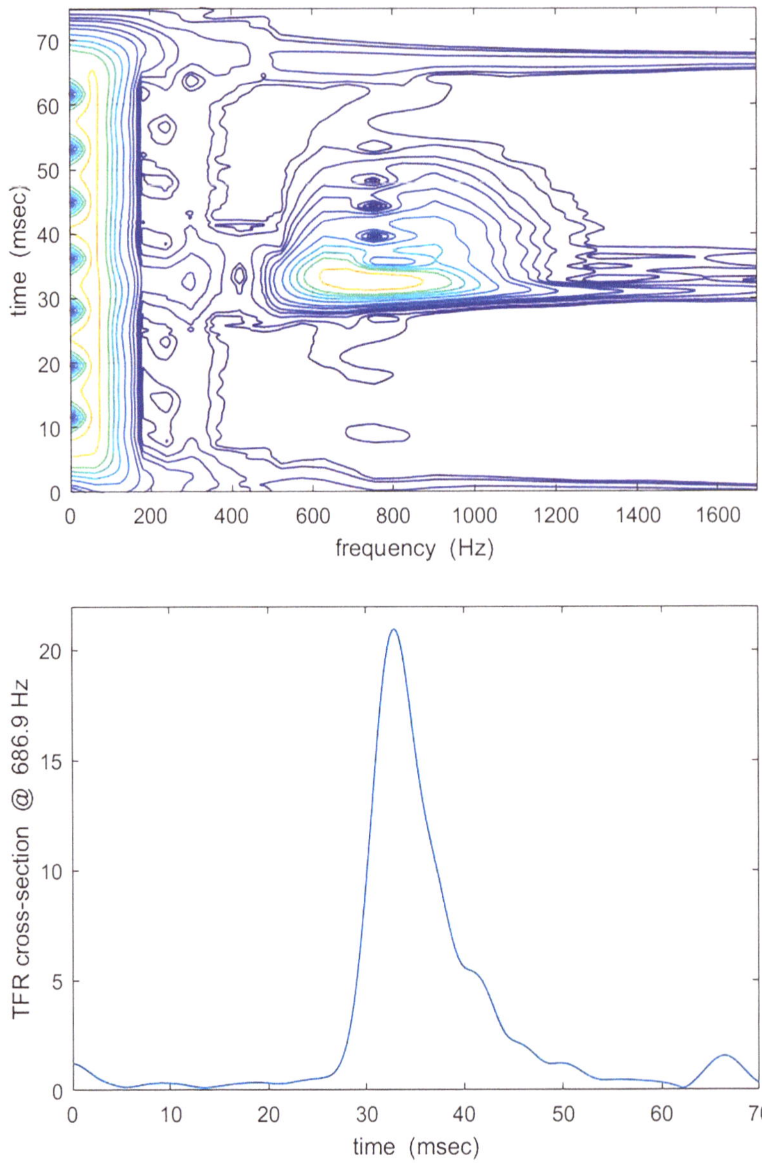

Fig. 5.25 "Reduced-Q" hybrid-TFR, using a 9-Hamming/14-Morse subband pattern (with $A_{\text{CQ}} = 0.6\,\text{dB}$, $\gamma = 3$, $\beta = 27$), applied to "phase a" current of the wind-farm example: contour plot (top) and a TFR cross-section at the transient peak frequency (bottom). Frequency parameters: $f_0 = 60\,\text{Hz}$ and $f_{\text{max}} = 1800\,\text{Hz}$, resulting in $Q = 5.00$ and $\kappa_{\text{min}}(Q) = 6.9$.

In general, a combined CB/CQ transform is completely determined by the following design variables:

- The fundamental frequency f_0 and the number of CB subbands, i.e., $K_{CB} + 1$
- The top frequency f_{max}, the level of intersection A_{CQ} between adjacent CQ subbands, and the number of CQ subbands, which is K_{CQ}
- The window type chosen for each one of the two modules, namely, one CB-prototype and one CQ prototype, which need not be the same

This means we can fully characterize any CB/CQ transform by the 7-element set $\big(f_0, K_{CB} + 1, \text{CB-window type}\big)/\big(f_{max}, A_{CQ}, K_{CQ}, \text{CQ-window type}\big)$. The CB-window is always a low-pass filter, but the CQ-window prototype can be either low-pass or band-pass, as discussed in Sect. 5.4.3. Notice that design of the CB-module must be completed before one can determine f_{min}, which serves as input to the CQ-module design step.

In summary, feature extraction from a hybrid-TFR transform is more accurate and reliable as compared with that achieved via traditional constant-bandwidth (CB) or constant-Q (CQ) transforms. The quality of TFR information depends primarily on the combined CB/CQ subband pattern, and, to a lesser degree, on the choice of prototype filters used in construction of the two modules of the transform. A well-designed hybrid-TFR scheme can provide accurate information about (a) dominant harmonic components, (b) in-band transients, and (c) wide-band transients, as demonstrated by the single-phase examples in this section.

5.4.6 Hybrid-TFR for Polyphase Waveforms

We should point out that time-frequency analysis in power system applications involves polyphase current and voltage waveforms, where each single-phase waveform gives rise to a distinct TFR. One can also view this representation as a pair of *polyphase TFRs*, one for current, and the other one for voltage.

In a three-phase system each polyphase TFR consists of a set of three complex waveforms (one for each phase) associated with every subband, viz.,

$$X_k(t) = \left[\, X_k^{(a)}(t) \quad X_k^{(b)}(t) \quad X_k^{(c)}(t) \,\right]$$

A scalar spectrogram that combines information from all three-phase components can be generated from such a polyphase TFR by using any vector-norm of interest. For instance, using the p-norm, viz.,

$$\left\| X_k(t) \right\|_p = \left[\, |X_k^{(a)}(t)|^p + |X_k^{(b)}(t)|^p + |X_k^{(c)}(t)|^p \,\right]^{1/p} \qquad 1 \le p \le \infty$$

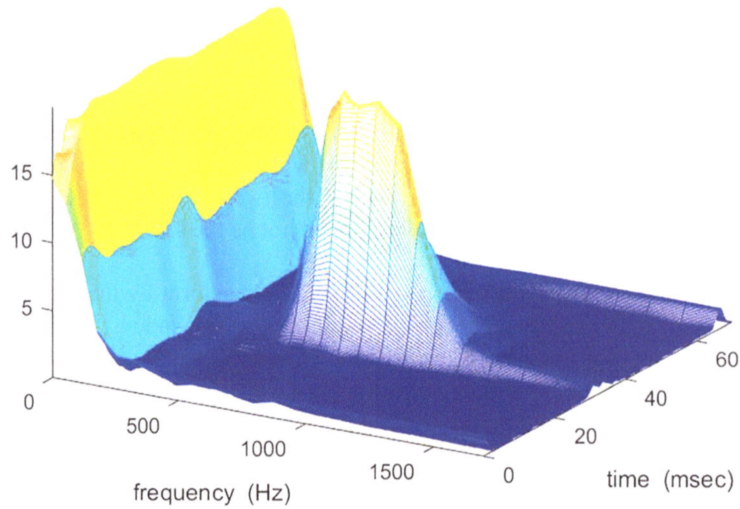

Fig. 5.26 A "p-norm" spectrogram (with $p = \infty$) representing the 3-phase current of our wind-farm example, using a 9-Hamming/14-Morse subband pattern (with $A_{\mathsf{CQ}} = 1\,\mathsf{dB}$, $\gamma = 3$, $\beta = 27$). Frequency parameters: $f_0 = 60\,\mathsf{Hz}$ and $f_{\mathsf{max}} = 1800\,\mathsf{Hz}$, resulting in $Q = 6.48$ and $\kappa_{\mathsf{min}}(Q) = 8.9$.

results in a scalar spectrogram, such as the one shown in Fig. 5.26. Notice how this approach preserves information about both the lower harmonic content and the higher frequency wide-band transient: compare this spectrogram with the one shown in Fig. 5.24.

Alternatively, one could also evaluate a separate spectrogram for each one of the *symmetrical sequence components* of $X_k(t)$, which means that we apply the mapping (2.18a) to each one of the $K_{\mathsf{CB}} + K_{\mathsf{CQ}}$ subband filter outputs. The resulting spectrograms for the positive and negative sequence components of a 3-phase current are shown in Fig. 5.27. Notice that the positive sequence component (top spectrogram) captures well the lower harmonic content but shows a highly-attenuated transient peak, while the reverse is true for the negative sequence component. Similar information can also be obtained, for any three-phase waveform, from the (α, β)-components generated by the Clarke transform.

Clearly, combining voltage and current information from the corresponding polyphase TFRs, as well as using such joint information to generate features of interest for monitoring and event detection applications, is still an open research problem. The same statement applies, in general, to time-frequency analysis of any multivariate (i.e., vector-valued) waveform, as evident from the scant publication record on this challenging subject: see, e.g., the discussion in [158, 159].

A different approach to address the "polyphase TFR challenge" is to construct a TFR of a *scalar waveform* that combines, in some physically meaningful way, polyphase current and voltage information, such as the *instantaneous power* $p(t) =$

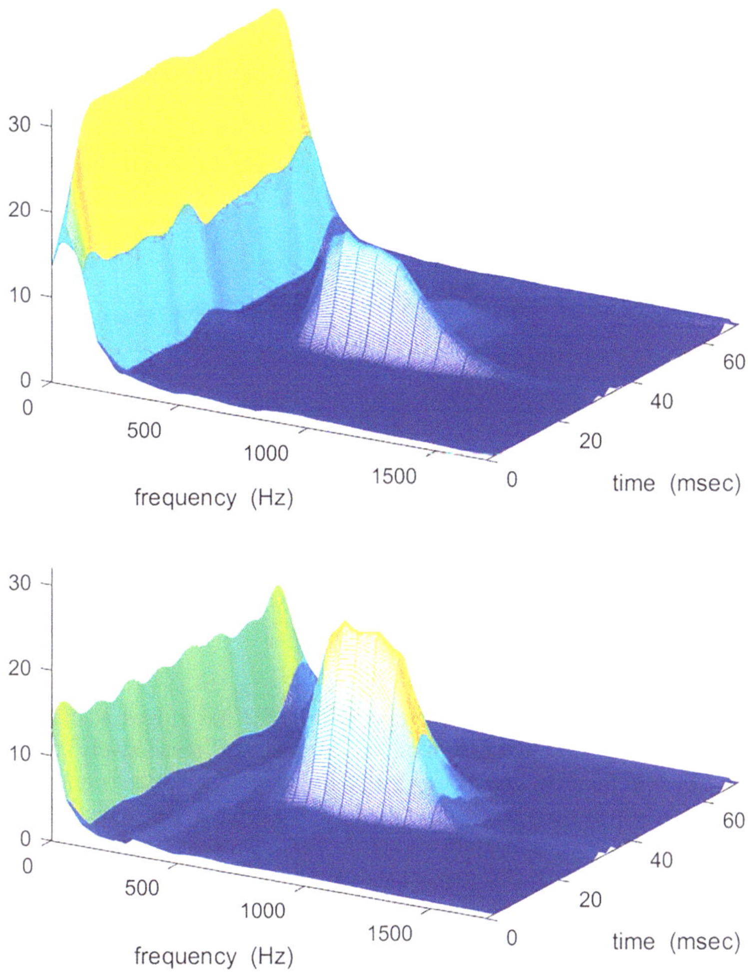

Fig. 5.27 Symmetrical sequence component spectrograms representing the 3-phase current of our wind-farm example, using a 9-Hamming/14-Morse subband pattern (with $A_{\mathsf{CQ}} = 1\,\mathsf{dB}$, $\gamma = 3$, $\beta = 27$): positive sequence (top) and negative sequence (bottom). Frequency parameters: $f_0 = 60\,\mathsf{Hz}$ and $f_{\mathsf{max}} = 1800\,\mathsf{Hz}$, resulting in $Q = 6.48$ and $\kappa_{\mathsf{min}}(Q) = 8.9$.

$v(t)\, i^{\top}(t)$. The resulting spectrogram, evaluated for our synthetic-short-transient example,[12] exhibits characteristics similar to the one generated from a current waveform: compare Fig. 5.28 with the spectrogram shown in Fig. 5.2. However, notice the differences in frequency content between these two figures:

[12] See Sect. L.2.1 for a detailed description of this example.

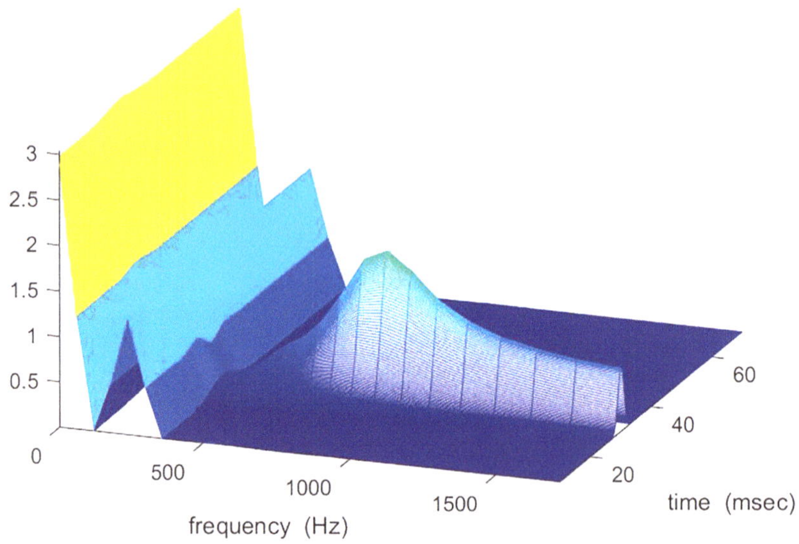

Fig. 5.28 Instantaneous power spectrogram (synthetic-short-transient example), using a 9-Hamming/14-Hamming subband pattern (with $A_{\mathrm{CQ}} = 1\,\mathrm{dB}$). Frequency parameters: $f_0 = 60\,\mathrm{Hz}$ and $f_{\max} = 1800\,\mathrm{Hz}$, resulting in $Q = 6.48$.

- The current-based spectrogram shown in Fig. 5.2 has ridges at $60\,\mathrm{Hz}$ and $300\,\mathrm{Hz}$, i.e., the 1st and 5th harmonic frequencies, respectively. In contrast, the $p(t)$-based spectrogram of Fig. 5.28 has ridges at $0\,\mathrm{Hz}$ (i.e., at DC) and $240\,\mathrm{Hz}$, which correspond to differences of the original harmonic frequencies.
- The current-based spectrogram shown in Fig. 5.2 exhibits a transient peak at $692\,\mathrm{Hz}$. In contrast, the transient peak of the $p(t)$-based spectrogram in Fig. 5.28 is much taller and somewhat wider, with its center at $909\,\mathrm{Hz}$.

We conclude that while $p(t)$-based spectrograms may be helpful in capturing wide-band transients, they require careful interpretation when used to determine harmonic content of voltage and current waveforms.

It is worthwhile noticing that the output of the lowest subband in a $p(t)$-based spectrogram (i.e., the "DC subband") is, of course, a low-pass filtered version of the instantaneous power waveform: it provides a reasonably close approximation of the dynamic active power $P(t)$, which was defined in (2.5) in terms of a rectangular window. We introduce in Chap. 9 a family of dynamic power quality metrics (DPQM) which, like $P(t)$, are constant in steady-state operation, and still have narrow (low-pass) bandwidth in the presence of transients. It turns out that such metrics – which rely on voltage and current information from a wide range of frequencies – can provide accurate time-domain information about onset and duration of both in-band and wide-band transients.

Finally, we observe that modern digital signal processing devices can readily handle the computational load and processing rate required in real-time power system applications.

5.5 The S-transform

The *S-transform* is a special case of the variable-bandwidth transform (5.1a) that we introduced in Sect. 5.1. Its original definition [152], expressed in the notation of Sect. 5.1, is

$$S(t\,;\,f_c) \overset{\text{def}}{=} |f_c| \int_{-\infty}^{\infty} g\big(f_c(t-\tau)\big)\, x(\tau)\, e^{-j2\pi f_c \tau}\, d\tau \tag{5.40a}$$

where $g(t)$ is *Gaussian*, viz.,

$$g(t) \;=\; \frac{1}{\sqrt{2\pi}}\, e^{-t^2/2} \qquad \Longleftrightarrow \qquad G(s) \;=\; e^{s^2/2} \tag{5.40b}$$

Following the practice in previous sections, we consider here only positive frequencies, so that $f_c > 0$. Changing the integration variable to $\xi = t - \tau$ results in the alternative expression

$$S(t\,;\,f_c) \;=\; e^{-j2\pi f_c t}\; f_c \int_{-\infty}^{\infty} g\,(f_c\,\xi)\; x(t-\xi)\; e^{j2\pi f_c \xi}\; d\xi$$

This suggests introducing a *passband version* of this transform, namely,

$$\mathcal{S}(t\,;\,f_c) \overset{\text{def}}{=} e^{j2\pi f_c t}\; S(t\,;\,f_c) \;=\; h(t\,;\,f_c) \circledast x(t) \tag{5.41a}$$

where

$$h(t\,;\,f_c) \;=\; f_c\; g\big(f_c\, t\big)\, e^{j2\pi f_c t} \qquad \Longleftrightarrow \qquad H(s\,;\,f_c) \;=\; G\left(\frac{s - j2\pi f_c}{f_c}\right) \tag{5.41b}$$

is a constant-Q band-pass filter, centered at the frequency $f = f_c$. The authors of [152] refer to the passband-to-baseband mapping $S(t\,;\,f_c) = e^{-j2\pi f_c t}\, \mathcal{S}(t\,;\,f_c)$ as a "phase correction," since it removes the linear phase component of $\arg\{\mathcal{S}(t\,;\,f_c)\}$.

Since the passband version (5.41) matches exactly the definition (5.2), we conclude that the S-transform belongs in the low-pass-prototype category of variable-bandwidth transforms. In addition, we observe that the passband filter $H(s\,;\,f_c)$ in (5.41b) satisfies the constant-Q property (5.13a), namely

$$\text{Q-factor} \overset{\text{def}}{=} \frac{\text{center frequency of } |H(j\omega\,;\,f_c)|}{\text{3 db bandwidthof } |H(j\omega\,;\,f_c)|} \;=\; \frac{f_c}{0.265\, f_c} \;=\; 3.77$$

where we used the expression for 3 dB bandwidth of a Gaussian window provided in Table B.2. Notice that the bandwidth of $|H(j\omega; f_c)|$ shrinks to zero as $f_c \to 0$. This singularity can be avoided by imposing a lower limit on the parameter f_c, i.e., $f_c \geq f_{\text{min}} > 0$.

Filter-bank Interpretation of the S-transform

The *frequency-discretized version* of the S-transform, as provided in [152, Eq. (11)], relies on uniformly-spaced subband center frequencies, viz.,

$$f_{c,k} = k\,\frac{F_s}{M} \quad , \qquad 1 \leq k \leq K \tag{5.42a}$$

where F_s is the intended sampling rate of the waveform of interest, and M is the number of samples obtained from the acquired finite segment of this waveform. Applying an S-transform to a sampled waveform limits the available range of f_c (and, consequently, of the subband center frequencies $f_{c,k}$) to $f_{\text{min}} < f_c < \frac{F_s}{2}$. In turn, this constraint imposes an upper limit on the subband index, i.e.,

$$K \leq \text{floor}\left(\frac{M}{2}\right) \tag{5.42b}$$

Notice that this choice of subband center frequencies does not satisfy the geometric spacing property (5.13b): this sets the original S-transform apart from the rest of the constant-Q transforms that we examined in the preceding sections. Finally, since $f_{c,1} \geq f_{\text{min}} > 0$, the subbands of the S-transform cannot cover the entire frequency range $[0 \quad \frac{F_s}{2}]$. Similar to the approach suggested in Sect. 5.2, one can augment this subband pattern with a special "dc-subband," i.e., a low-pass filter with a normalized cutoff frequency that matches the lower edge of the first subband, say, $f_{\text{dc-cutoff}} \approx f_{\text{min}}$.

The resulting frequency-discretized version of (5.41) is

$$\mathcal{S}_k(t) = h_k(t) \circledast x(t) \quad , \qquad 1 \leq k \leq K \tag{5.43a}$$

where

$$\mathcal{S}_k(t) = \mathcal{S}(t\,; f_{c,k}) \quad , \qquad h_k(t) = h(t\,; f_{c,k}) \tag{5.43b}$$

This can be interpreted as a filter-bank, consisting of K subbands. Each subband component $\mathcal{S}_k(t)$ is generated by a band-pass filter $H_k(s)$ (Fig. 5.29), where

$$h_k(t) = \frac{f_{c,k}}{\sqrt{2\pi}}\, e^{-(f_{c,k}\,t)^2/2}\, e^{j2\pi f_{c,k}\,t} \quad \Longleftrightarrow \quad H_k(j2\pi f) = e^{-2\pi^2\left(\frac{f}{f_{c,k}}-1\right)^2} \tag{5.43c}$$

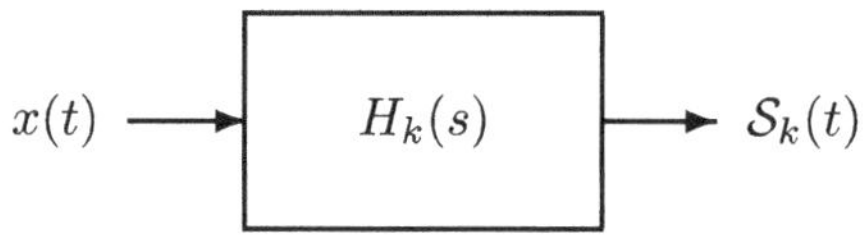

Fig. 5.29 Filtering interpretation of the frequency-discretized S-transform.

The magnitude response $|H_k(j2\pi f)|$ of this subband filter is centered at the frequency $f = f_{c,k}$, and its bandwidth increases with the subband index "k."

Since the subbands of the S-transform are uniformly-spaced, while their width increases linearly with the subband index, the overlap between subbands also increases with "k." Indeed, the subband-overlap ratio (SOR) of the S-transform is

$$\mathsf{SOR}_k = \frac{1}{2}\frac{0.265\,(f_{c,k+1}+f_{c,k})}{f_{c,k+1}-f_{c,k}} = 0.265\left(k+\tfrac{1}{2}\right)$$

where we used (5.42a) and the SOR definition (5.16b). Clearly, the SOR increases linearly as a function of the subband index k. Since a typical waveform record may consist of several thousands of samples (or more), the range of SOR values across the set of subbands is quite large. For instance, when $M = 1,024$, the subband index ranges from $k = 1$ to $k = 512$, which implies that

$$\mathsf{SOR}_1 = 0.4 \qquad \text{and} \qquad \mathsf{SOR}_{512} = 136$$

This means that high-index subbands provide an extremely redundant time-frequency representation, while low-index subbands are separated by "frequency gaps" whose content is suppressed by the S-transform.

The filter-bank structure described in Fig. 5.29 can be used to provide a digital implementation of the S-transform, either online or offline, via the discrete-time relation $h_k[n] = T_s\, h_k(nT_s)$ (recall (5.6)), viz.,

$$h_k[n] = \frac{\eta_k}{\sqrt{2\pi}}\, e^{-(\eta_k n)^2/2}\, e^{j2\pi\eta_k n}\,, \quad 1 \le k \le K \tag{5.44a}$$

where η_k is the *normalized center frequency* of the k-th subband, viz.,

$$\eta_k \stackrel{\text{def}}{=} \frac{f_{c,k}}{F_s} \tag{5.44b}$$

Consequently, the implementable (discrete-time/discrete-frequency) time-domain version of the S-transform is

$$\mathcal{S}_k(nT_s) \equiv S_k(nT_s)\, e^{j2\pi\eta_k n} = h_k[n] \circledast x(nT_s) \tag{5.44c}$$

where $T_s = {}^1/_{F_s}$, and F_s is the waveform sampling frequency.

Frequency-domain (offline) Implementation of the S-transform

The S-transform is, clearly, a special case of the low-pass-prototype category, as is evident from comparing (5.41b) with (5.2). This means we can insert the S-transform subband-filter expression (5.43c), i.e., $H_k(j2\pi f) = e^{-2\pi^2(\frac{f}{f_{c,k}}-1)^2}$, into the frequency-domain relations (5.8b) and (5.9b). The resulting passband and baseband expressions for the S-transform are, respectively,

$$\mathcal{S}_k(t) = \int_{-\infty}^{\infty} e^{-2\pi^2(\frac{f}{f_{c,k}}-1)^2} X(j2\pi f)\, e^{j2\pi f t}\, df \tag{5.45a}$$

and

$$\begin{aligned} S_k(t) &= \int_{-\infty}^{\infty} e^{-2\pi^2(\frac{f}{f_{c,k}}-1)^2} X(j2\pi f)\, e^{j2\pi(f-f_{c,k})t}\, df \\ &= \int_{-\infty}^{\infty} e^{-2\pi^2(\frac{\zeta}{f_{c,k}})^2} X\big(j2\pi(\zeta + f_{c,k})\big)\, e^{j2\pi\zeta t}\, d\zeta \end{aligned} \tag{5.45b}$$

where $f_{c,k} = k\,\frac{F_s}{M}$, as defined in (5.42a). Using the discretized frequency grid (5.11), we obtain numerical approximation expressions, as in (5.12), viz.,

$$\mathcal{S}_k(t)\Big|_{t=nT_s} \approx \frac{1}{M} \sum_{m=-m_0}^{M-1-m_0} \underbrace{e^{-2\pi^2(\frac{m}{k}-1)^2}}_{H_{k,m}} X[m]\, e^{j2\pi\,\frac{m}{M}\,n} \tag{5.46a}$$

for the passband version, as well as

$$\begin{aligned} S_k(t)\Big|_{t=nT_s} &\approx \frac{1}{M} \sum_{m=-m_0}^{M-1-m_0} e^{-2\pi^2(\frac{m}{k}-1)^2} X[m]\, e^{j2\pi\,\frac{m-k}{M}\,n} \\ &= \frac{1}{M} \sum_{\ell=-m_0-k}^{M-1-m_0-k} e^{-2\pi^2(\frac{\ell}{k})^2} X[\ell+k]\, e^{j2\pi\,\frac{\ell}{M}\,n} \end{aligned} \tag{5.46b}$$

for the baseband version. Notice that, as in Sect. 4.6, these discretized expressions can only be computed offline: they rely on the DFT of the entire sampled waveform record $\{x[n];\ 0 \le n \le N_x - 1\}$, which must be evaluated in advance. This observation also implies that the constraint $M \ge N_x$ must be satisfied in order to avoid aliasing [32, 33]. Finally, we observe that (5.46) is only one of several possible discretizations of the frequency-domain expressions (5.45). Alternative discretization schemes, which use a modified window function and/or a different (non-centered) range of summation, have been presented in [152, 160–162].

Since we discuss variable-bandwidth transforms only for the case $f_c > 0$, we need to evaluate $\mathcal{S}_k(nT_s)$ (and/or $S_k(nT_s)$) only for $0 \le k \le K$. Notice, however, that (5.45) and (5.46) become singular for $k = 0$, because $f_{c,0} = 0$. One possible remedy is to introduce a special "DC-subband" to augment the subband structure of the S-transform, such as the averaging operation

$$\mathcal{S}_0(nT_s) \stackrel{\text{def}}{=} \frac{1}{N_x} \sum_{i=0}^{N_x-1} x(iT_s) \equiv \mathcal{S}_0(nT_s)$$

which was proposed in [152, 162]. However, since our interest is primarily in dynamic phenomena, we shall focus in the sequel on the "DC-free" range $1 \leq k \leq K$.

Equivalently, we can express the offline discrete S-transform relation (5.46a) in matrix notation, involving a product of three matrices, viz.,

$$\begin{pmatrix} \mathcal{S}_1(0) & \mathcal{S}_1(T_s) & \dots & \mathcal{S}_1\big((N_x-1)T_s\big) \\ \mathcal{S}_2(0) & \mathcal{S}_2(T_s) & \dots & \mathcal{S}_2\big((N_x-1)T_s\big) \\ \vdots & \vdots & & \vdots \\ \mathcal{S}_K(0) & \mathcal{S}_K(T_s) & \dots & \mathcal{S}_K\big((N_x-1)T_s\big) \end{pmatrix} \approx \boldsymbol{\mathcal{H}}\, \boldsymbol{D}_X\, \boldsymbol{\mathcal{U}} \tag{5.47a}$$

where

- $\boldsymbol{\mathcal{H}}$ is a $K \times M$ matrix, viz.,

$$\boldsymbol{\mathcal{H}} = \begin{pmatrix} H_{1,-m_0} & H_{1,1-m_0} & \dots & H_{1,M-1-m_0} \\ H_{2,-m_0} & H_{2,1-m_0} & \dots & H_{2,M-1-m_0} \\ \vdots & \vdots & \vdots & \vdots \\ H_{K,-m_0} & H_{K,1-m_0} & \dots & H_{K,M-1-m_0} \end{pmatrix} \tag{5.47b}$$

  with elements $H_{k,m} = e^{-2\pi^2(\frac{m}{k}-1)^2}$.
- $\boldsymbol{D}_X$ is an $M \times M$ diagonal matrix, with elements determined by the centered DFT $\{X[m];\ -m_0 \leq m \leq M-1-m_0\}$, viz.,

$$\boldsymbol{D}_X = \begin{pmatrix} X[-m_0] & 0 & \dots & 0 & & 0 \\ 0 & X[1-m_0] & \ddots & & & \vdots \\ \vdots & \ddots & \ddots & \ddots & & \vdots \\ 0 & & \ddots & X[0] & \ddots & \vdots \\ \vdots & & & \ddots & \ddots & 0 \\ 0 & \dots & \dots & \dots & 0 & X[M-1-m_0] \end{pmatrix} \tag{5.47c}$$

The centered DFT can be obtained, for instance, by applying the MATLAB function fft(x,M) to the sampled waveform record $\{x(nT_s)\,;\; 0 \le n \le N_x - 1\}$, followed by the fftshift operation.

- $\mathcal{U}$ is an $M \times N_x$ matrix, the same as defined in (4.50d)–(4.50e), viz.,

$$\mathcal{U} = \frac{1}{M} \begin{pmatrix} 1 & w_M^{(-m_0)} & w_M^{2(-m_0)} & \cdots & w_M^{(N_x-1)(-m_0)} \\ 1 & w_M^{(1-m_0)} & w_M^{2(1-m_0)} & \cdots & w_M^{(N_x-1)(1-m_0)} \\ 1 & w_M^{(2-m_0)} & w_M^{2(2-m_0)} & \cdots & w_M^{(N_x-1)(2-m_0)} \\ \vdots & \vdots & \vdots & \ddots & \vdots \\ 1 & w_M^{(M-1-m_0)} & w_M^{2(M-1-m_0)} & \cdots & w_M^{(N_x-1)(M-1-m_0)} \end{pmatrix}$$

$$\equiv \mathcal{U}_{\mathsf{IDFT}} \begin{pmatrix} 1 & 0 & \cdots & \cdots & 0 \\ 0 & w_M^{-m_0} & \ddots & & \vdots \\ \vdots & \ddots & w_M^{-2m_0} & \ddots & \vdots \\ \vdots & & \ddots & \ddots & 0 \\ 0 & \cdots & \cdots & 0 & w_M^{-(N_x-1)m_0} \end{pmatrix} \tag{5.47d}$$

where $w_M \stackrel{\text{def}}{=} e^{\,j(2\pi/M)}$, and

$$\mathcal{U}_{\mathsf{IDFT}} = \frac{1}{M} \begin{pmatrix} 1 & 1 & 1 & \cdots & 1 \\ 1 & w_M & w_M^2 & \cdots & w_M^{N_x-1} \\ 1 & w_M^2 & w_M^4 & \cdots & w_M^{2(N_x-1)} \\ \vdots & \vdots & \vdots & \ddots & \vdots \\ 1 & w_M^{(M-1)} & w_M^{2(M-1)} & \cdots & w_M^{(N_x-1)(M-1)} \end{pmatrix} \tag{5.47e}$$

Since $w_M^{-i} = w_M^{M-i}$, the matrix $\mathcal{U}$ is, in fact, a row-permuted version of the inverse DFT-matrix $\mathcal{U}_{\mathsf{IDFT}}$.

The cost of this frequency-domain scheme is essentially the same as that of (4.48b), except that there is no frequency thinning (i.e., $L = M$): when $M = N_x$, the overall cost is $\mathcal{O}(4M^2)$.

Also, as in Sect. 4.6, setting $M = N_x$ results in a useful relation between $\sum_{n=0}^{M-1} S_k(nT_s)$, the time-summation of the baseband S-transform, and the DFT of the sampled waveform segment $\{x(nT_s)\,;\; 0 \le n \le M-1\}$, viz.,

$$\begin{aligned}\sum_{n=0}^{M-1} S_k(nT_s) &= \sum_{m=-m_0}^{M-1-m_0} H_{k,m}\, X[m] \left[\frac{1}{M}\sum_{n=0}^{M-1} e^{j\,(2\pi/M)\,(m-k)n}\right] \\ &= \sum_{m=-m_0}^{M-1-m_0} H_{k,m}\, X[m]\,\delta[m-k] \\ &= X[k] \end{aligned} \tag{5.48}$$

where $\delta[m]$ denotes the *Kronecker delta function*, and we used the unity DC-gain property $H_{k,k} = 1$.

5.6 Time-Frequency Representation Via Orthogonal Decomposition

The preceding sections of this chapter presented a flexible conceptual framework for the construction of variable bandwidth transforms. We have demonstrated that a wide variety of subband filter choices can be used to generate a constant-Q time-frequency representation (TFR) $X(t;\, f_c)$ for a given waveform $x(\cdot)$. However, this abundance of choices is severely reduced by adding the requirement for an efficient error-free reconstruction of $x(\cdot)$ from its TFR $X(t;\, f_c)$. One convenient scheme for construction of an invertible transform is in terms of an *inner-product* between the given input waveform, and a suitably chosen orthogonal basis. However, in order to be applicable to time-frequency analysis, all members of this basis must be localized in both frequency and time. This added constraint on the orthogonal basis further limits the available choices of invertible constant-Q transforms. In fact, the majority of constant-Q transform applications tend to rely on a single choice: the celebrated Discrete Wavelet Transform (DWT) [132, 133, 164].

The notion of an inner-product of waveforms relies on the introduction of an appropriate *waveform space*.[13] Since variable-bandwidth transforms cannot be constructed on a finite-length (sliding) window, we must instead consider waveforms defined on the infinite interval $-\infty < t < \infty$. This observation leads to the introduction of a space of *finite-energy waveforms*, i.e., waveforms that satisfy the constraint $\int_{-\infty}^{\infty} |x(t)|^2\, dt < \infty$. This is, of course, the defining property of $\mathcal{L}^2(-\infty, \infty)$ (also known as $\mathcal{L}^2(R)$), the space of complex-valued square-integrable waveforms on the infinite interval $R \stackrel{\text{def}}{=} -\infty < t < \infty$. The inner-product in this *separable Hilbert space*[14] is

$$\langle\, x(\cdot)\,,\, y(\cdot)\,\rangle = \int_{-\infty}^{\infty} x(t)\, y^*(t)\, dt \equiv \int_{-\infty}^{\infty} X(j\omega)\, Y^*(j\omega)\, \frac{d\omega}{2\pi} \tag{5.49a}$$

[13] See Appendix J for a brief introduction to waveform spaces.

[14] See Appendix J for an additional information about Hilbert spaces.

where the superscript * denotes complex conjugation, and $X(j\omega) \stackrel{\text{def}}{=} \mathcal{F}_\omega\{x(t)\}$, $Y(j\omega) \stackrel{\text{def}}{=} \mathcal{F}_\omega\{y(t)\}$ are the Fourier transforms of $x(t)$ and $y(t)$.

For digital processing purposes, the inner-product is approximated by a finite sum, viz.,

$$\frac{1}{T_s}\langle\, x(\cdot)\,,\, y(\cdot)\,\rangle \;\approx\; \sum_{n=0}^{N-1} x(nT_s)\, y^*(nT_s) \;\equiv\; \frac{1}{N}\sum_{r=0}^{N\;\;1} X[r]\, Y^*[r] \tag{5.49b}$$

where $X[r] = \sum_{n=0}^{N-1} x(nT_s)e^{-j\frac{2\pi}{N}nr}$ is the DFT (recall (4.47)) of the finite record of waveform samples $\{x(nT_s)\,;\; 0 \le n \le N\}$, and, similarly, $Y[r]$ is the DFT of $\{y(nT_s)\,;\; 0 \le n \le N\}$. Notice that this implementation of the inner-product is necessarily an *offline operation.*

Time-frequency analysis relies on choosing an orthogonal basis that provides *localization in both frequency and time.* This requires basis waveforms $\{\psi_{k,m}(\cdot)\}$ that are doubly-indexed, so that each individual basis waveform corresponds to a particular location (a.k.a. "tile" or "atom") in the time-frequency plane. We describe in this section two distinct invertible variable-bandwidth transform schemes that rely on particular choices of an orthonormal basis for $\mathcal{L}^2(R)$ and possess the desirable property of time-frequency localization: (a) multi-resolution wavelet basis and (b) discrete orthogonal S-transform (DOST).

Given an orthonormal basis $\{\psi_{k,m}(\cdot)\}$ of $\mathcal{L}^2(R)$ that has the time-frequency localization property, we can decompose any waveform in this space in terms of its projections on the basis waveforms, viz.,

$$x(t) \;=\; \sum_{k,m} X_{k,m}\;\psi_{k,m}(t) \tag{5.50a}$$

where

$$X_{k,m} \;\stackrel{\text{def}}{=}\; \langle\, x(\cdot)\,,\, \psi_{k,m}(\cdot)\,\rangle \;=\; \int_{-\infty}^{\infty} x(t)\;\psi^*_{k,m}(t)\;dt \tag{5.50b}$$

The decomposition (5.50) establishes an invertible one-to-one correspondence between any finite-energy waveform $x(\cdot)$ and the set $\{X_{k,m}\}$ of *transform coefficients* that represent this waveform in the basis $\{\psi_{k,m}(\cdot)\}$. The map $x(\cdot) \longrightarrow \{X_{k,m}\}$ is known as the *forward transform*, while the map $\{X_{k,m}\} \longrightarrow x(\cdot)$ is the *inverse transform*. In other words, one can use the set of transform coefficients $\{X_{k,m}\}$ as an equivalent (error-free) representation of the waveform $x(\cdot)$: the coefficient $X_{k,m}$ describes a component of the waveform $x(\cdot)$ associated with the time-frequency atom defined by the k-th frequency subband and the m-th time interval.

In addition, the resulting set of transform coefficients should be sparse: this facilitates computational efficiency, allowing computing resources to be allocated to evaluation of a few dominant coefficients, while avoiding computation of (infinitely many) negligible transform coefficients. However, there is no single scheme that can efficiently handle all transients, without regard to the specific domain of the application. This means that selecting a variable-bandwidth transform for analysis and representation of transient phenomena must always be guided by specific domain knowledge.

Decomposition in an orthonormal basis always exhibits a *Parseval property*, namely,

$$\langle\, x(\cdot)\,,\; y(\cdot)\,\rangle \;\equiv\; \int_{-\infty}^{\infty} x(\tau)\, y^*(\tau)\, d\tau \;=\; \sum_k \sum_m X_{k,m}\, Y_{k,m}^* \tag{5.51}$$

where $X_{k,m} = \langle\, x(\cdot)\,,\, \psi_{k,m}(\cdot)\,\rangle$ and $Y_{k,m} = \langle\, y(\cdot)\,,\, \psi_{k,m}(\cdot)\,\rangle$ are the decomposition coefficients of the finite-energy waveforms $x(\cdot)$ and $y(\cdot)$, respectively. Notice, however, that this global relation does not provide any information about the *local behavior* of the product $x(t)\, y^*(t)$. This should be contrasted with dynamic phasors, which satisfy the *local Parseval relation* (2.3).

5.6.1 Multi-Resolution Wavelet Bases

One useful scheme of orthogonal decomposition in $\mathcal{L}^2(R)$ is provided by multi-resolution wavelet bases. It is directly applicable to time-frequency analysis, because wavelet basis functions typically have limited bandwidth and time-duration.

In general, a multi-resolution wavelet basis of $\mathcal{L}^2(R)$ is a doubly-indexed set of orthonormal waveforms $\{\psi_{k,m}(t)\,;\; -\infty < k < \infty\,,\; -\infty < m < \infty\}$ that satisfy a *dilation-and-shift* relation, namely,

$$\psi_{k,m}(t) \;=\; \Gamma^{k/2}\, \psi\left(\Gamma^k t - m T_0\right) \tag{5.52}$$

where $\Gamma > 1,\, T_0 > 0$ are design parameters. The entire basis is constructed from a single specially-designed waveform $\psi\,(t) \equiv \psi_{0,0}(t)$, known as the *mother wavelet*. We also observe that (5.52) is equivalent to the two relations

$$\psi_{k,0}(t) \;=\; \Gamma^{k/2}\, \psi\left(\Gamma^k t\right) \tag{5.53a}$$

and

$$\psi_{k,m}(t) \;=\; \psi_{k,0}\left(t - m\,\tfrac{T_0}{\Gamma^k}\right) \tag{5.53b}$$

The conjugate-reversed waveform $\psi_{k,0}^*(-t)$ can be interpreted as the impulse response of a k-th subband filter, with higher k-values corresponding to higher frequency subbands, which is consistent with the subband indexing convention used in the preceding sections of this chapter.[15]

Indeed, we observe that the forward transform map, from $x(\cdot)$ to $X_{k,m}$, i.e.,

$$X_{k,m} = \langle\, x(\cdot)\,,\, \psi_{k,m}(\cdot)\,\rangle = \int_{-\infty}^{\infty} x(\tau)\, \psi_{k,0}^{*}\left(\tau - m\,\tfrac{T_0}{\Gamma^k}\right)\, d\tau$$

is closely related to a sampled version of the frequency-discretized variable-bandwidth transform (5.4). To be specific, we can express the forward map $x(\cdot) \longrightarrow X_{k,m}$ as a subband filtering operation, viz.,

$$X_{k,m} = \psi_{k,0}^{*}(-t) \circledast x(t)\Big|_{t\,=\,m\,\frac{T_0}{\Gamma^k}}$$

and we observe that the subband filters $\{\psi_{k,0}^{*}(-t)\,;\; -\infty < k < \infty\}$ satisfy the compliance constraints (R1) and (R3) of Sect. 5.2.

Notice, however, that there is a minor discrepancy between the standard normalization of wavelets (to unity $\mathcal{L}^2$-norm) and the unity peak-gain property included in the constraint (R2): the peak gain of the unity-norm subband filter $\psi_{k,0}^{*}(-t)$ is, according to (5.53a),

$$\max_{\omega}\, \big|\Psi_{k,0}(j\omega)\big| = \Gamma^{-k/2}\, \max_{\omega}\, \big|\Psi(j\omega)\big| \neq 1$$

where $\Psi_{k,0}(j\omega) = \mathcal{F}_{\omega}\{\psi_{k,0}(t)\}$ and $\Psi(j\omega) = \mathcal{F}_{\omega}\{\psi(t)\}$ denote, respectively, the Fourier transforms of $\psi_{k,0}(t)$ and $\psi(t)$. This means we can incorporate the multi-resolution wavelet basis into our subband filter framework, which requires satisfying the compliance constraints (R1)–(R3), by defining subband filters $\{h_k(t)\,;\; -\infty < k < \infty\}$ that are suitably re-normalized versions of $\psi_{k,0}^{*}(-t)$, viz.,

$$h_k(t) \stackrel{\text{def}}{=} \frac{\Gamma^{k/2}}{G_{\psi}}\, \psi_{k,0}^{*}(-t) \equiv \frac{1}{G_{\psi}}\, \Gamma^k\, \psi^{*}\left(-\Gamma^k\, t\right) \tag{5.54a}$$

where $G_{\psi} \stackrel{\text{def}}{=} \max_{\omega} \big|\Psi(j\omega)\big|$. This results in the desired unity peak-gain property $\max_{\omega} \big|H_k(j\omega)\big| = 1$, as required by the compliance constraint (R2), as well as the scaled-correspondence relation

$$\mathcal{X}_k\left(m\,\tfrac{T_0}{\Gamma^k}\right) = \frac{\Gamma^{k/2}}{G_{\psi}}\, X_{k,m} \tag{5.54b}$$

[15] In contrast, standard presentations of multi-resolution wavelets, such as in [132, 133], use a reverse indexing scheme (i.e., with "$-k$" replacing our "k"), so that higher k-values become associated with *lower-frequency subbands*.

Furthermore, the impulse responses defined in (5.54a) satisfy the key scaling relation (5.24) for all $-\infty < k < \infty$, i.e.,

$$h_k(t) \;=\; \frac{1}{\Gamma^{K-k}}\, h_K\!\left(\frac{t}{\Gamma^{K-k}}\right) \;\equiv\; \Gamma^k\, h_0(\Gamma^k t)$$

The converse is also true: any choice of subband filters $h_k(t)$ that satisfy this key scaling relation for every $-\infty < k < \infty$ must exhibit the relation (5.53a), albeit with a modified normalization.

The dilation-and-shift relation (5.52) is only one of a larger set of necessary and sufficient conditions that are needed to ensure completeness and orthogonality of the wavelet basis $\{\psi_{k,m}(t)\}$. These involve a number of constraints that must be satisfied by the mother wavelet $\psi(t)$. These constraints are usually specified in the wavelets literature (see, e.g., [132, 133, 148]) for the *dyadic multi-resolution* case (i.e., constructed with $\Gamma = 2$), as follows:

- **Dilation equation:** the mother wavelet $\psi(t)$ and an associated *scaling function* $\phi(t)$ must satisfy the coupled *dilation equations*

$$\phi(t) \;=\; \sum_m c_m\, \phi(2t - mT_0)$$

$$\psi(t) \;=\; \sum_m d_m\, \phi(2t - mT_0)$$

  for some shift $T_0 > 0$, where $\{c_m\}$ and $\{d_m\}$ are suitably chosen sets of coefficients. In particular, the 2×2 *alias-component matrix*

$$M(z) \;=\; \begin{pmatrix} C(z) & D(z) \\ C(-z) & D(-z) \end{pmatrix}$$

  where

$$C(z) \;\stackrel{\text{def}}{=}\; \frac{1}{2}\sum_m c_m\, z^{-m} \qquad \text{and} \qquad D(z) \;\stackrel{\text{def}}{=}\; \frac{1}{2}\sum_m d_m\, z^{-m}$$

  must be non-singular, in the sense that $\det M(e^{j\theta}) \neq 0$, for all $|\,\theta\,| \leq \pi$.
- **Orthogonality:** a necessary condition for mutual orthogonality of the wavelet waveforms $\{\psi_{k,m}(\cdot)\}$ is that the alias-component matrix $M(z)$ is *paraunitary* (a.k.a. "para-Hermitian"), namely, $M^{-1}(z) = \left[M(\frac{1}{z^*})\right]^H$.
- **Completeness:** a number of additional constraints have been used to formulate a necessary and sufficient condition for orthogonality and completeness of the wavelet waveform set $\{\psi_{k,m}(\cdot)\}$: see, e.g., [132, 133].

- **Regularity:** The smoothness of the mother wavelet $\psi(\cdot)$ is controlled by the level of differentiability of $C(e^{j\theta})$ for $\theta \to \pi$ [132, 163]. Regularity translates into a "vanishing moments" property of the mother wavelet waveform, i.e., $\int_{-\infty}^{\infty} t^r \, \psi(t) \, dt = 0$ for all $0 \leq r \leq r_{max}$, where r_{max} serves as a measure of smoothness for $\psi(\cdot)$.

In particular, *Daubechies wavelets* are obtained by choosing $C(z)$ as a suitably constrained polynomial. These wavelets are maximally-smooth among all valid choices associated with a polynomial $C(z)$ of a given degree [132, 163].

The wavelet coefficients $X_{k,m}$ of any given finite-energy waveform $x(\cdot)$ satisfy a recursion in the resolution index k that can be interpreted in terms of discrete-time paraunitary quadrature-mirror-filter (QMF) banks with a binary-tree structure [133]. The fundamental building block of this architecture is a discrete-time two-channel paraunitary QMF filter-bank, with $C(z)$ and $D(z)$ serving as its low-pass and high-pass transfer functions, respectively. These observations lead to a computationally-efficient (albeit slightly inaccurate[16]) discrete-time implementation of the multi-resolution wavelet transform, known as the *Discrete Wavelet Transform* (DWT), which is usually applied to a sampled version of the input waveform $x(\cdot)$. The DWT inaccuracy can be reduced by suitable prefiltering of the discrete-time input $x(nT_s)$ (see, e.g., [165–167]).

As mentioned earlier in this section, the Parseval property (5.51) is shared by all orthogonal decompositions, including those obtained via a multi-resolution wavelet basis, namely,

$$\int_{-\infty}^{\infty} x(\tau) \, y^*(\tau) \, d\tau \; = \; \sum_k \sum_m X_{k,m} \, Y_{k,m}^*$$

where $X_{k,m} = \langle \, x(\cdot) \, , \, \psi_{k,m}(\cdot) \, \rangle$ and $Y_{k,m} = \langle \, y(\cdot) \, , \, \psi_{k,m}(\cdot) \, \rangle$ are the wavelet coefficients of the waveforms $x(\cdot)$ and $y(\cdot)$, respectively. However, as pointed out earlier, this global Parseval relation does not provide any information about the *local behavior* of the product $x(t) \, y^*(t)$.

This shortcoming of the multi-resolution wavelet decomposition is of particular concern in power system applications. We show in Sect. 9.1 that the local Parseval identity (2.3), which involves dynamic phasors of the voltage and current waveforms $v(\cdot)$ and $i(\cdot)$, provides a useful characterization of time-localized electrical power flow. In contrast, the equivalent wavelet-based expression, i.e., $\sum_k \sum_m \, V_{k,m} \, I_{k,m}^*$ where $V_{k,m}$ and $I_{k,m}$ are the wavelet coefficients of $v(\cdot)$ and $i(\cdot)$, cannot provide any information about time-localized power flow, although the m-dependent quantity $\sum_k \, V_{k,m} \, I_{k,m}^*$ can be used as an approximation.

[16] The cause of this inaccuracy is a common oversight, as explained in detail in [164] where it is designated as a "wavelet crime."

5.6.2 Discrete Orthonormal S-transform (DOST) Basis

DOST is an orthogonal basis with the desired time-frequency localization property. Like wavelets, it provides (a) constant-Q subbands, (b) geometrically-spaced subband center frequencies, and (c) uniformly-spaced time shifts for each subband, with the shift interval inversely proportional to the subband width.

Equation (5) of [151] defines DOST as a discrete-time transform that maps a finite sequence of input waveform samples, say $\{x(nT_s)\,;\ 0 \le n \le N-1\}$, into a set of transform coefficients, viz.,

$$\text{DOST output} = \sum_{n=0}^{N-1} x(nT_s)\ S_{\nu,\beta,\tau}[nT_s]$$

where $S_{\nu,\beta,\tau}[nT_s]$ is an adjustable discrete-time waveform: the association of the DOST output with a particular location in the time-frequency plane is controlled by the integer-valued parameters ν (center frequency), β (bandwidth) and τ (time shift). The DOST can also be interpreted as a *discrete-time inner-product* (recall (5.49b)) between the sampled input waveform and $\left[\, S_{\nu,\beta,\tau}[nT_s]\,\right]^*$. The DOST discrete-time/discrete-frequency orthonormal basis $\{\psi_{k,m}[n]\}$ can then be obtained by a suitable restriction of the allowed ν and β values [151, 168]. Notice that we are following here standard DSP practice by using "n" to denote a discrete-time instant (rather than "k," as in [151]). Similarly, we use "m" (instead of Stockwell's notation "τ") as an indicator of the discrete-time interval associated with the basis waveform $\psi_{k,m}[n]$. We also assume that $N = 2^K$ for some positive integer K.

Thus, inserting our notation changes into [151, Eq. (6)], we now define the DOST *adjustable basis waveform*

$$\begin{aligned} B_{\nu,\beta,m}[n] &\stackrel{\text{def}}{=} \left[\, S_{\nu,\beta,m}[nT_s]\,\right]^* \\ &= \frac{j\,e^{j\pi m}}{\sqrt{\beta}}\ \frac{e^{j(2\nu+\beta-1)\left(\frac{n}{N}-\frac{m}{\beta}\right)\pi} - e^{j(2\nu-\beta-1)\left(\frac{n}{N}-\frac{m}{\beta}\right)\pi}}{2\sin\left[\left(\frac{n}{N}-\frac{m}{\beta}\right)\pi\right]} \end{aligned}$$

which can be rewritten in the compact form

$$B_{\nu,\beta,m}[n] = \frac{(-1)^{m+1}}{\sqrt{\beta}}\ \frac{\sin\left[\beta\left(n - m\frac{N}{\beta}\right)\frac{\pi}{N}\right]}{\sin\left[\left(n - m\frac{N}{\beta}\right)\frac{\pi}{N}\right]}\ e^{j(2\nu-1)\left(n-m\frac{N}{\beta}\right)\frac{\pi}{N}} \tag{5.55}$$

where we used the fact that $e^{j\pi m} = (-1)^m$ and m is an integer. Here N is the length of the sampled input waveform, and the integer $0 \le m \le \beta - 1$ controls the time shift of this basis waveform and also adds an m-dependent sign flip.

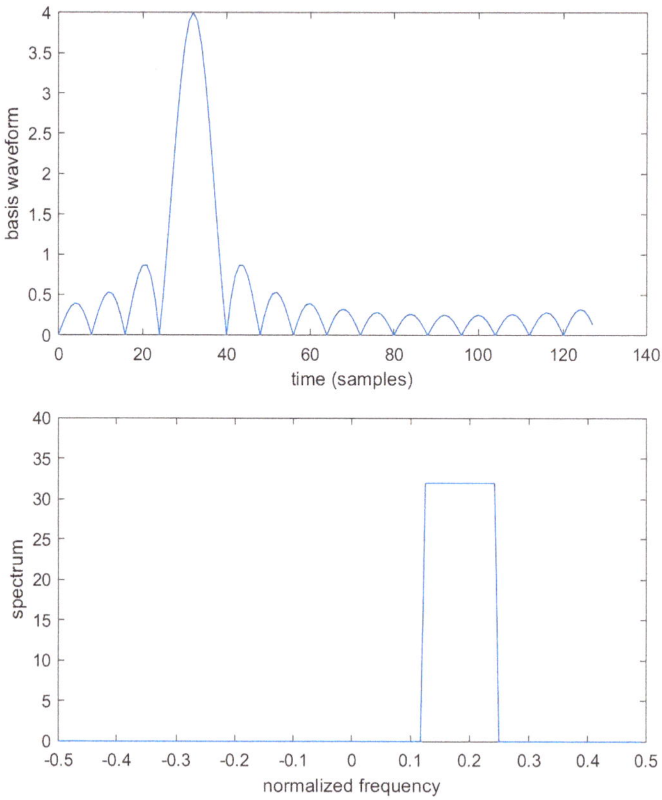

Fig. 5.30 Basis waveform (absolute value, top) and its spectrum (bottom) of the DOST adjustable basis waveform $B_{\nu,\beta,m}$: $N = 128$, $\beta = 16$, $\nu = \frac{3}{2}\beta = 24$, and $m = 4$.

The center frequency and bandwidth of the discrete Fourier transform (DFT) of $B_{\nu,\beta,m}$ are controlled by the integer indices ν and β, respectively. We observe that the DFT of $B_{\nu,\beta,m}[n]$, evaluated at the normalized frequency $\frac{r}{N}$, is

$$\sum_{n=0}^{N-1} B_{\nu,\beta,m}[n]\, e^{-j\frac{2\pi}{N}nr} = \begin{cases} \frac{(-1)^m N}{\sqrt{\beta}}\, e^{-j\frac{2\pi}{\beta}mr} & \nu - \frac{\beta}{2} \leq r \leq \nu + \frac{\beta}{2} - 1 \\ 0 & \text{else} \end{cases} \tag{5.56}$$

which is bandlimited to the DFT-index range $\nu - \frac{\beta}{2} \leq r \leq \nu + \frac{\beta}{2} - 1$ and has a constant magnitude within this range, as clearly evident from Fig. 5.30. Notice that the basis waveform $B_{\nu,\beta,m}[n]$ is very similar to a discrete-time version of the Nyquist wavelet (a.k.a. "sinc wavelet" [132]): however, they differ in their phase response, as demonstrated in [151].

The DOST orthonormal basis is obtained by suitably restricting the allowed values of β and ν. This is achieved by making both parameters dependent on the subband index k, as follows [151, 168]:

- For $|k| \leq 1$ the only allowed shift is $m = 0$, and the corresponding basis waveforms are

$$\psi_{k,0}[n] = \begin{cases} 1 & k = 0 \\ e^{j\frac{2\pi}{N}n} & k = 1 \\ e^{-j\frac{2\pi}{N}n} & k = -1 \end{cases} \tag{5.57a}$$

- For $2 \leq k \leq K - 1$ (where we recall that $K = \log_2 N$), the allowed shifts are $0 \leq m \leq \beta - 1$, with $\beta = 2^{k-1}$. The corresponding basis waveforms are

$$\psi_{k,m}[n] = B_{\nu,\beta,m}[n]\Big|_{\beta=2^{k-1} \text{ and } \nu=\frac{3\beta}{2}} \tag{5.57b}$$

- For $-(K-1) \leq k \leq -2$ the allowed shifts are $0 \leq m \leq \beta - 1$, with $\beta = 2^{|k|-1}$. The corresponding basis waveforms are

$$\psi_{k,m}[n] = B_{\nu,\beta,m}[n]\Big|_{\beta=2^{|k|-1} \text{ and } \nu=1-\frac{3\beta}{2}} \tag{5.57c}$$

- For $k = -K$ the only allowed shift is $m = 0$, and the corresponding basis waveform is

$$\psi_{-K,0} = e^{j\pi n} \equiv (-1)^n \tag{5.57d}$$

Notice that subbands are defined on a discretized frequency grid, which divides the normalized frequency range $[-0.5 \quad 0.5)$ into N equal intervals.

The expression given for $\psi_{k,m}[n]$ when $2 \leq |k| \leq K - 1$ becomes singular when $n = m\frac{N}{\beta} \stackrel{\text{def}}{=} n_s$, resulting in a $\frac{0}{0}$ operation. Notice that n_s is always an integer, because $N = 2^K$, while $\beta = 2^{|k|-1}$ and $|k| \leq K - 1$. When $0 \leq n_s \leq N - 1$, the singularity can be addressed by using the limiting value [151, 168]

$$\lim_{n \to n_s} \psi_{k,m}[n] = \lim_{n \to n_s} B_{\nu,\beta,m}[n]\Big|_{\beta=2^{|k|-1}} = (-1)^{m+1}\sqrt{\beta}$$

instead of the original expression (5.55).

The basis waveforms $\{\psi_{k,m}[n]\}$ are conjugate-symmetric in the subband index k, namely, $\psi^*_{k,m}[n] = \psi_{-k,m}[n]$ [168]. This is so because the parameter $\beta = 2^{|k|-1}$ is independent of the sign of k, and for $k \geq 2$,

$$\nu_+ = \frac{3\beta}{2} \implies 2\nu_+ - 1 = 3\beta - 1$$

while for $k \leq -2$

$$\nu_- = 1 - \frac{3\beta}{2} \quad \Longrightarrow \quad 2\nu_- - 1 = -(3\beta - 1)$$

Thus, according to (5.55),

$$B^*_{\nu_+,\beta,m}[n] = B_{\nu_-,\beta,m}[n]$$

which completes the verification of the time-domain conjugate symmetry property. Applying a DFT, we obtain a frequency-domain version of the same property, viz.,

$$\psi_{-k,m}[n] = \psi^*_{k,m}[n] \quad \Longleftrightarrow \quad \Psi_{-k,m}[r] = \Psi^*_{k,m}[-r] \tag{5.58}$$

which holds for all $1 \leq |k| \leq K - 1$, all n and all r. Consequently, when the input waveform $x(\cdot)$ is real-valued, its DOST coefficients

$$X_{k,m} = \sum_{n=0}^{N-1} x(nT_s)\, \psi^*_{k,m}[n] \tag{5.59}$$

are also conjugate-symmetric, i.e., $X^*_{k,m} = X_{-k,m}$.

Since the range of the subband index is $-K \leq k \leq K - 1$, the total number of subbands is $2K = 2\log_2 N$. Consequently, the total number of basis waveforms is, according to the subband-specific definitions (5.57),

$$4 + 2\sum_{k=2}^{K-1} 2^{k-1} = 4 + 2\sum_{\ell=1}^{K-2} 2^{\ell} = 4 + 2\left(2^{K-1} - 2\right) = 2^K \equiv N$$

This means that the DOST maps a given sampled input waveform $x(\cdot)$, consisting of N samples, into a set $\{X_{k,m}\}$ of N transform coefficients.

The cost of evaluating a complete set $\{X_{k,m}\}$ of DOST coefficients is $\mathcal{O}(N^2)$ operations [168], as compared with the $\mathcal{O}(N^3)$ of the original S-transform scheme. A fast DOST version, which requires only $\mathcal{O}(N \log_2 N)$ operations for either the forward or the inverse transform, was introduced in [168]. It relies on converting the inner-product expression (5.59) into its frequency-domain equivalent via the identity (5.49b), i.e.,

$$X_{k,m} = \sum_{n=0}^{N-1} x(nT_s)\, \psi^*_{k,m}[n] \equiv \frac{1}{N} \sum_{r=-\frac{N}{2}}^{\frac{N}{2}-1} X[r]\, \Psi^*_{k,m}[r] \tag{5.60}$$

where $X[r]$ and $\Psi_{k,m}[r]$ denote the DFTs of $x(nT_s)$ and $\psi_{k,m}[n]$, respectively. The former can be efficiently evaluated via the FFT algorithm, while the latter is very sparse, because of the strictly-bandlimited nature of $\Psi_{k,m}[r]$ (recall Fig. 5.30). Indeed, for $2 \leq k \leq K-1$,

$$\frac{1}{N}\,\Psi_{k,m}[r] = \begin{cases} \frac{(-1)^m}{\sqrt{\beta}}\, e^{-j\frac{2\pi}{\beta}mr} & \beta \leq r \leq 2\beta - 1 \\ 0 & \text{else} \end{cases} \tag{5.61a}$$

where $\beta = 2^{k-1}$ and $0 \leq m \leq \beta - 1$. Also,

$$\frac{1}{N}\,\Psi_{k,0}[r] = \begin{cases} \delta[r] & k = 0 \\ \delta[r-1] & k = 1 \\ \delta[r+\frac{N}{2}] & k = -K \end{cases} \tag{5.61b}$$

where $\delta[\cdot]$ is the Kronecker delta function. Finally, the conjugate symmetry property (5.58), i.e., $\Psi_{-k,m}[r] = \Psi^{*}_{k,m}[-r]$, provides a complete characterization of $\Psi_{k,m}[r]$ for all negative subband indices, except $k = -K \equiv -\frac{N}{2}$.

In summary, we observe that DOST shares several key features of the DWT: (a) both are purely discrete-time/discrete-frequency schemes, defined over a finite-length time interval $0 \leq n \leq N-1$, and using a discretized normalized frequency grid, with a $\frac{1}{N}$ grid interval; (b) both use a subband pattern with a density of one subband per octave or, equivalently, a scaling factor $\Gamma = 2$; and (c) in both cases the highest-index subband occupies half of the available frequency range.

5.7 Wigner–Ville Interpretation of Spectrograms and Scalograms

The variable-bandwidth transform (5.1a) defines a time-frequency representation (TFR) $\mathcal{X}(t\,;f_c)$ of a given input waveform $x(t)$. The squared TFR magnitude can be used to characterize *time-frequency energy distribution* (a.k.a. "instantaneous power spectrum") of the waveform $x(\cdot)$ (recall (5.14)), namely,

$$\text{generalized spectrogram of } x(\cdot) \stackrel{\text{def}}{=} \left|\mathcal{X}(t\,;f_c)\right|^2 \equiv \left|X(t\,;f_c)\right|^2 \tag{5.62}$$

The spectrogram defined by a fixed-bandwidth transform is usually plotted with linear time and frequency scales, while plots of generalized spectrograms obtained from a constant-Q transform often employ a logarithmic frequency scale.

The *Wigner–Ville distribution* (WVD), which is defined for a given waveform $x(\cdot)$ as

$$W_x(t,f) \stackrel{\text{def}}{=} \int x\left(t+\tfrac{\tau}{2}\right) x^{*}\left(t-\tfrac{\tau}{2}\right) e^{-j2\pi f\tau}\, d\tau \tag{5.63}$$

is another well-known time-frequency energy distribution metric. Unfortunately, the WVD exhibits a number of undesirable characteristics:

- Multi-component signals, such as multi-harmonic waveforms, generate sizable cross-terms. For instance, a waveform consisting of two frequencies, say f_1 and f_2, will generate a large spurious WVD peak at $0.5(f_1 + f_2)$, in addition to the peaks generated at f_1 and f_2. Furthermore, if the f_1-component vanishes (i.e., is turned off) for $t > -t_0$ and the f_2-component vanishes for $t < t_0$, the spurious peak will show up in the intermediate interval ($|t| < t_0$) in which both components are completely turned off.
- Interpretation of the WVD as an energy distribution is compromised by possible occurrence of localized negative values of $W_x(t, f)$.
- The range of relative signal levels that can be reliably observed in a WVD (a.k.a. "dynamic range") is smaller than that achieved with a generalized spectrogram [170].

Several WVD smoothing schemes have been proposed to address these issues [106, 128, 171].

Generalized Spectrograms Interpreted as Smoothed WVDs

It turns out that the generalized spectrogram (5.62) is, in fact, a smoothed version of the WVD, i.e., a special case of the so-called *smoothed Wigner distribution* [106, 128]. This observation follows from combining two standard properties of the WVD with the characterization of subband filters that was provided in Sect. 5.3. For the sake of conceptual and notational simplicity of our presentation, we will focus here on (5.4), the frequency-discretized version of the transform, namely,

$$\mathcal{X}_k(t) = h_k(t) \circledast x(t) \quad , \quad X_k(t) = \mathcal{X}_k(t)\, e^{-j2\pi f_{c,k} t}$$

According to [106, Sect. 12.2, Property W7], the corresponding Wigner–Ville distributions satisfy a similar relation (i.e., convolution in the time variable), viz.,

$$W_{\mathcal{X}_k}(t, f) = W_{h_k}(t, f) \underset{t}{\circledast} W_x(t, f) \overset{\text{def}}{=} \int_{-\infty}^{\infty} W_{h_k}(t-\theta, f)\, W_x(\theta, f)\, d\theta$$

Integration of $W_{\mathcal{X}_k}(t, f)$ with respect to the frequency variable "f" results, according to [106, Sect. 12.2, Property W3], in $|\mathcal{X}_k(t)|^2 = \int_{-\infty}^{\infty} W_{\mathcal{X}_k}(t, f)\, df$, and thus provides an explicit expression for the generalized spectrogram in terms of the WVDs of $h_k(\cdot)$ and $x(\cdot)$, i.e.,

$$\left|\mathcal{X}_k(t_c)\right|^2 \equiv \left|X_k(t_c)\right|^2 = \int_{-\infty}^{\infty}\int_{-\infty}^{\infty} W_{h_k}(t_c - t, f)\, W_x(t, f)\, dt\, df \qquad (5.64)$$

The same conclusions apply also to the continuous-frequency/continuous-time generalized spectrogram $|\mathcal{X}(t_c\,;\,f_c)|^2$.

Notice that we have adopted here the convention of using $(t_c\,,\,f_c)$ as the location variables of the generalized spectrogram, in order to distinguish them from $(t,\,f)$, the variables of the WVD. We will continue to use this notational convention in the remainder of this section. Since the smoothing kernel $W_{h_k}(t,\,f)$ typically has a single peak at or near $(0,\,f_{c,k})$, the expression (5.64) relies primarily on the local content of $W_x(t,\,f)$ in the vicinity of $(t_c\,,\;f_{c,k})$ to determine the generalized spectrogram $|\mathcal{X}_k(t_c)|^2$. This observation justifies the interpretation of $|\mathcal{X}_k(t_c)|^2$ as a measure of energy associated with the time-frequency "bin" centered at $(t_c\,,\;f_{c,k})$.

In most cases, the subband filter $h_k(t)$ admits a "modulated low-pass" representation, viz.,

$$h_k(t) \;=\; g_k(t)\, e^{j2\pi f_{c,k}\,t} \qquad \Longrightarrow \qquad H_k(j\omega) \;=\; G_k\big(j(\omega - 2\pi f_{c,k})\big) \tag{5.65}$$

where $f_{c,k}$ is the center frequency of the k-th subband, and $g_k(t)$ is the (possibly complex-valued) impulse response of a low-pass filter, known as the *baseband equivalent* of $h_k(t)$. In particular, this is true in the following cases:

- **Fixed-bandwidth transform (STFT):** according to (4.16),

$$h_k(t) \;=\; \underbrace{h(t)}_{g_k(t)}\; e^{j2\pi f_{c,k}\,t}$$

  where $h(t)$ is a finite length window, symmetric to the origin. In this case the time-duration of $g_k(t) \equiv h(t)$ and the bandwidth of its Fourier transform depend only on the choice of the fixed window function $h(t)$.
- **Variable-bandwidth transform designed from a low-pass prototype:** according to (5.5b),

$$h_k(t) \;=\; \underbrace{\frac{1}{\sigma_k}\, g\left(\frac{t}{\sigma_k}\right)}_{g_k(t)}\; e^{j2\pi f_{c,k}\,t} \qquad \Longrightarrow \qquad G_k(j\omega) \;=\; G(j\sigma_k\,\omega)$$

  Notice that the time-duration of $g_k(t)$ is proportional to the control parameter σ_k, while the bandwidth of $G_k(j\omega)$ is inversely-proportional to the same parameter.
- **Variable-bandwidth transform designed from an analytic band-pass prototype:** a filter is said to be *analytic* when its frequency response vanishes for all negative frequencies (see discussion in Sect. 3.4). Every analytic band-pass filter $h_k(t)$ admits the modulated low-pass representation (5.65): the resulting $g_k(t)$ is known as the *baseband equivalent* (also "complex envelope") of $h_k(t)$ [83, 84]. In addition, when $H_k(j\omega)$ is real-valued its baseband equivalent

$G_k(j\omega) = H_k\big(j(\omega + 2\pi f_{c,k})\big)$ is also real-valued, so that $g_k(t)$ is conjugate-symmetric with respect to the origin.

- **S-transform:** according to (5.43b),

$$h_k(t) \;=\; \underbrace{\frac{f_{c,k}}{\sqrt{2\pi}}\, e^{-(f_{c,k}\,t)^2/2}}_{g_k(t)}\; e^{j2\pi f_{c,k}\,t} \quad\Longrightarrow\quad G_k(j\omega) \;=\; e^{-(\frac{\omega}{f_{c,k}})^2/2}$$

Here the bandwidth of $G_k(j\omega)$ is proportional to the control parameter $f_{c,k}$, while the time-duration of $g_k(t)$ is inversely-proportional to the same parameter.

- **Discrete orthonormal S-transform (DOST)**: the DOST basis waveforms are analytic by construction (see, e.g., bottom plot of Fig. 5.30) and thus admit a modulated low-pass representation of the form (5.65).

According to [106, Sect. 12.2, Property W2], the modulated low-pass relation (5.65) implies that $W_{h_k}(t, f) \;=\; W_{g_k}(t, f - f_{c,k})$. Incorporating this fact into (5.64) results in a bivariate convolution expression for the generalized spectrogram, viz.,

$$\begin{aligned}\big|\mathcal{X}_k(t_c)\big|^2 \;\equiv\; \big|X_k(t_c)\big|^2 \;&=\; \int_{-\infty}^{\infty}\int_{-\infty}^{\infty} W_{g_k}(t_c - t,\, f - f_{c,k})\, W_x(t, f)\, dt\, df \\ &\equiv\; \int_{-\infty}^{\infty}\int_{-\infty}^{\infty} W_{g_k}(t, -f)\, W_x(t_c - t,\, f_{c,k} - f)\, dt\, df\end{aligned} \tag{5.66}$$

In all practical cases the baseband-equivalent impulse response $g_k(t)$ is centered at the time-origin, and the corresponding frequency response $G_k(j\omega)$ is centered at the frequency-origin. Consequently, the peak of the smoothing kernel $W_{g_k}(t, -f)$ is centered at the time-frequency origin (see, e.g., Fig. 5.31). We conclude that the generalized spectrogram $\big|\mathcal{X}_k(t_c)\big|^2$ is determined mainly by the content of $W_x(t, f)$ in the vicinity of $(t_c\,,\, f_{c,k})$, weighted by the baseband-equivalent kernel $W_{g_k}(\cdot, \cdot)$.

The generalized spectrogram expression (5.66) is a special case of the so-called smoothed Wigner distribution [106], with $W_{g_k}(\cdot, \cdot)$ serving as the smoothing kernel. We observe that, as a consequence of this smoothing operation, the generalized spectrogram (5.66) is free of the three main shortcomings of the WVD, i.e., presence of cross-terms, reduced dynamic range, and possible occurrence of negative values. This improvement comes at the cost of moderately lower resolution. However, the change in resolution is typically less than 50% [170], which means that, in most engineering applications, the advantages of eliminating cross-terms, reduced dynamic range and non-positivity definitely outweigh the resolution compromise.

The shape of the bivariate function $W_{g_k}(t, f)$ and its effective extent in the time and frequency dimensions determine the degree of smoothing applied to $W_x(t, f)$. The contour plot of $W_{g_k}(t, f)$ can be used to compare alternative choices of

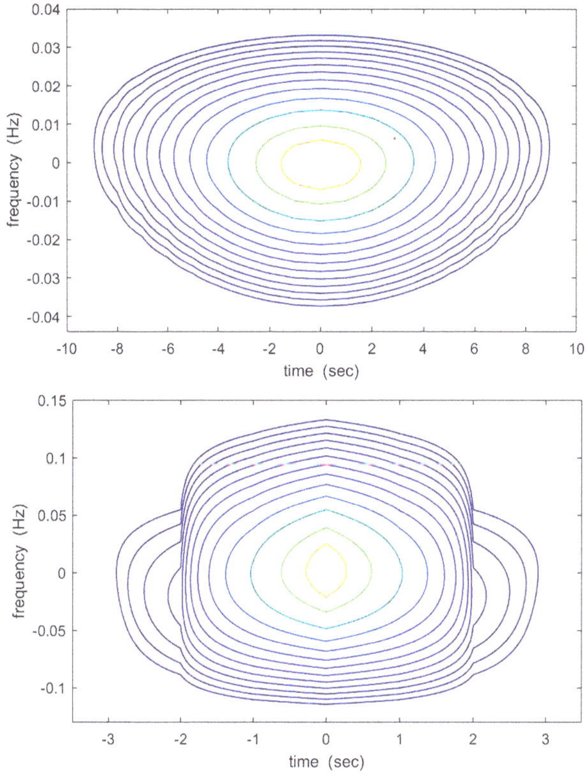

Fig. 5.31 Contour plot of the baseband-equivalent smoothing kernel defined by a Morse wavelet: (**a**) $\gamma = 5$ and $\beta = 50$ (top), (**b**) $\gamma = 2.5$ and $\beta = 6$ (bottom).

$g_k(t)$ (see, e.g., [150, 155]). In most cases, the smoothing kernel $W_{g_k}(t, f)$ has a single control parameter, which means that its effective time-duration and frequency-bandwidth cannot be adjusted individually. For instance, the smoothing kernel associated with the S-transform is

$$W_{g_k}(t, f) = \frac{f_{c,k}}{\sqrt{\pi}} e^{-(f_{c,k}t)^2} e^{-\left(2\pi \frac{f}{f_{c,k}}\right)^2}$$

so that its frequency bandwidth is proportional to $f_{c,k}$, while its effective time-duration is inversely-proportional to the same control parameter. The Morse wavelet is a rare exception: it has two control parameters (γ and β), which provide some added flexibility in controlling the shape of the smoothing kernel (Fig. 5.31). However, since these parameters simultaneously adjust the shape of both $g_k(t)$ *and* $G_k(j\omega)$, the fundamental "uncertainty principle" [33, 86] still applies, setting a lower bound on the product of effective time-duration and frequency bandwidth for a Morse wavelet.

The energy distribution of a wavelet transform is usually described in terms of a *scalogram*, which is often promoted as a novel alternative to the spectrogram. However, a scalogram is merely a different way of drawing a generalized spectrogram: the logarithmic frequency axis of a generalized (constant-Q) spectrogram is replaced by a logarithmic *scale axis*, drawn in reverse, from high to low scale values [148]. In other words, scalograms use a (shifted) logarithmic frequency axis, i.e.,

$$-\log(\mathsf{scale}) = \log f - \log f_{\psi,c}$$

where $f_{\psi,c}$ is the center frequency of $\Psi(j\omega)$, the Fourier transform of the mother wavelet $\psi(t)$: this follows from the one-to-one *scale-frequency* correspondence $\mathsf{scale} = \frac{f_{\psi,c}}{f}$ [148, Eq. (2.23)].

The preceding observation suggests that scalograms can also be interpreted as smoothed versions of Wigner–Ville distributions. Indeed, starting from (5.54a), the subband filter characterization of a multi-resolution wavelet basis, viz.,

$$h_k(t) = \frac{1}{G_\psi} \Gamma^k \psi^*\left(-\Gamma^k t\right)$$

we obtain the passband smoothing kernel expression

$$W_{h_k}(t, f) = \frac{\Gamma^k}{G_\psi^2} W_\psi\left(-\Gamma^k t, \frac{f}{\Gamma^k}\right)$$

The resulting generalized spectrogram, according to (5.64), is

$$\begin{aligned} \left|\mathcal{X}_k(t_c)\right|^2 &= \int_{-\infty}^{\infty}\int_{-\infty}^{\infty} W_{h_k}(t_c - t, f)\, W_x(t, f)\, dt\, df \\ &= \frac{\Gamma^k}{G_\psi^2} \int_{-\infty}^{\infty}\int_{-\infty}^{\infty} W_\psi\left(\Gamma^k(t - t_c), \frac{f}{\Gamma^k}\right) W_x(t, f)\, dt\, df \end{aligned}$$

This is identical (except for the multiplicative constant $\frac{\Gamma_k}{G_\psi^2}$) to the WVD "affine smoothing" expression for the scalogram [172, Eq. 24)], with the specific scale factor choice $a = \Gamma^{-k}$.

Time-frequency Sampling, Tiling Diagrams, and Resolution

Digital implementation of a variable-bandwidth transform involves discretization of the time variable "t." This means that the transform is evaluated on a grid $\{(nT_s, f_{c,k});\ 0 \le n \le N-1,\ 0 \le k \le K\}$ in the time-frequency plane, where we consider only non-negative frequencies, and we also include the special "DC-subband," which corresponds to the subband index $k = 0$. The resulting discretized time-frequency representation $\{\mathcal{X}_k(nT_s)\}$ is redundant, in the sense that its time-index "n" can be further decimated (i.e., down-sampled) without loss of information.

The most compact discretized transforms, known as "critically-sampled," are obtained when each subband is down-sampled by a factor that is inversely proportional to the relative width of this subband, as detailed in Sect. 5.2. Critical sampling is also a necessary condition for constructing invertible transforms that are based on orthogonal decomposition, such as DWT and DOST.

In the special case of a discretized *fixed-bandwidth transform*, the width of each subband is $\frac{F_s}{L}$, where L is the total number of subbands in the frequency range $-\frac{F_s}{2} \leq f \leq \frac{F_s}{2}$ (recall (4.15)). This allows each subband to be decimated by a factor of L, so that the resulting critically-sampled grid is $\{(m(LT_s),\, k(\frac{F_s}{L}))\,;\ 0 \leq n \leq \frac{N}{L} - 1\ ,\, 0 \leq k \leq K\}$. Such a uniform time-frequency sampling pattern can also be described by a so-called *tiling diagram* [173], where each tile (a.k.a. "Heisenberg box" [148]) is centered on a grid point, and has dimensions $LT_s \times \frac{F_s}{L}$ (Fig. 5.32). Notice that the area of each such tile is $F_s\, T_s = 1$.

On the other hand, critically-sampled discretization of a constant-Q transform results in a non-uniform grid. In this case, higher subbands are represented by a higher output rate, so that the temporal separation of grid point decreases geometrically as a function of the subband index "k," while the subband width increases in the same proportion, as detailed in Sect. 5.2. This means that tile dimensions also change geometrically as a function of "k," while retaining the unit-area property (Fig. 5.33).

It is a common mistake to assume that the TFR $\mathcal{X}_k(t_c)$ depends entirely on the $W_x(t, f)$ information contained within the tile centered at $(t_c, f_{c,k})$. In reality, the tiles in Figs. 5.32 and 5.33 represent only the sampling interval and frequency

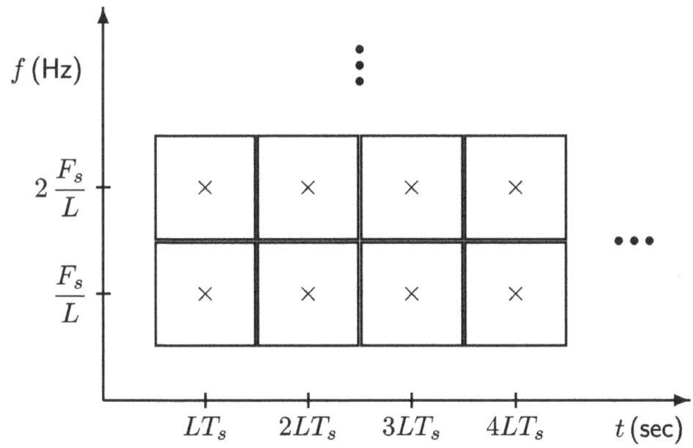

Fig. 5.32 Sampling grid and tiling diagram for a fixed-bandwidth subband pattern: $\frac{L}{2}$ uniform subbands in $0 \leq f \leq \frac{F_s}{2}$ (grid points indicated by $\times$).

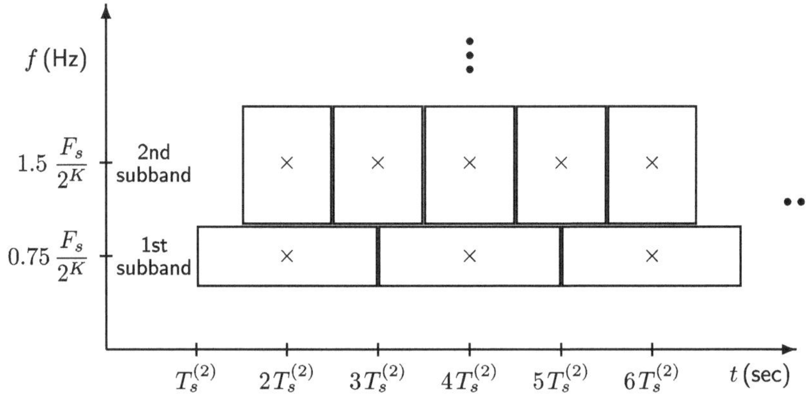

Fig. 5.33 Sampling grid and tiling diagram for a constant-Q subband pattern: K subbands generated with $\Gamma = 2$ in $0 < f \le \frac{F_s}{2}$ (grid points indicated by $\times$, and $T_s^{(2)} \stackrel{\text{def}}{=} 2^{K-1}\, T_s$ is the sampling interval of the 2nd subband).

width for each subband. The tiles *are not meant to represent the actual partitioning of energy*, which depends on both the sampling grid and the choice of a subband filter. In other words, the actual time- and frequency-resolution of the generalized spectrogram, as defined in Sects. B.5 and B.7 of Appendix B, do not match the tile dimensions, and the product time-resolution $\times$ frequency-resolution is not exactly equal to the (unity) area of a tile, as is evident from Table B.8. Nevertheless, it has been a common practice to use tile dimensions as an approximation for these resolution metrics (see, e.g., [170]). In contrast to the purely qualitative nature of the tiling diagram, the smoothing relations (5.64) and (5.66) provide an accurate quantitative characterization of the dependence of the generalized spectrogram $|\mathcal{X}_k(t_c)|^2$ on the content of the WVD $W_x(t, f)$ in the vicinity of the grid point $(t_c\,, f_{c,k})$.

Part II

Modeling and Control Applications of Dynamic Phasors

"It is a poor project to propose plans which are in no danger of being tried."

Thomas Beddoes, 1807

Chapter 6 Power Electronic Converters

This chapter describes some approaches to applying frequency-selective averaging to modeling the dynamic behavior of pulse-width modulated (PWM) and line-commutated converters. We start with the case that is considered practical when the switching frequency is considerably (say two orders of magnitude) higher than the highest frequency of the waveform being synthesized at the converter's output (say low frequency AC or even DC). As expected, the modeling procedure is reasonably straightforward, and only issues such as unbalance in polyphase AC systems could make it challenging. In the second section, we look at the case when the switching frequency is closer to the highest frequency being synthesized or observed, colloquially labeled as low-frequency switching (the labels clearly refer to relative separation among frequencies and not to their actual values). A different type of challenge is encountered in the third section when modeling systems with semi-controllable switches (such as thyristors) or converters in which uncontrollable switches (such as diodes) play a key role. In the final section, we briefly review the application of four different nonlinear control design procedures to the nonlinear dynamic phasor model derived in the preceding section and observe the versatility and utility of dynamic phasors models.

In this book we have been using the one-sided form of the Fourier series, as defined by (1.4) – this is simple and in agreement with the standard convention in the electric energy industry. It turns out, however, that the two-sided version (1.2) is the most convenient *accounting* form for the multiplicative nonlinearity which is perhaps the most common one in energy conversion systems. We thus use the two-sided version of the Fourier series exclusively both in this chapter and in the next one.

Consider a simple example of two waveforms $x(\cdot)$ and $y(\cdot)$, each with the dominant fundamental and a small fifth harmonic. From simple trigonometry, it is clear that their product will have large DC and second harmonic, "medium-sized" fourth and sixth, and a small tenth harmonic. It may then make sense to make the *modeling* decision to keep the DC and the second harmonic in the simplest

H. Lev-Ari, A. M. Stanković, *Dynamic Phasors in Energy Processing Systems*,
Power Electronics and Power Systems, https://doi.org/10.1007/978-3-031-50020-6_7

dynamical model and to have a refinement ready (if necessary) that additionally contains the fourth and sixth harmonics.

In the general case, the dynamic k-th Fourier coefficient of the product of any two waveforms $x(\cdot)$ and $y(\cdot)$ is given by a "two-sided" *index-domain convolution* of the respective dynamic Fourier coefficients, namely,

$$\langle x\, y \rangle_k (t) = \sum_{\ell=-\infty}^{\infty} \langle x \rangle_{k-\ell} (t)\ \langle y \rangle_\ell (t) \tag{6.1}$$

Notice that the dynamic coefficients $\langle x \rangle_k (t)$ and $\langle y \rangle_k (t)$ are defined via (2.2c) for both positive and negative index values. Indeed, the full range of harmonic index values is required for evaluation of the index-domain convolution (6.1). The one-sided counterpart of the expression (6.1) is much more complicated – this observation serves as the primary motivation for our decision to use two-sided Fourier coefficients throughout Chaps. 5 and 6.

6.1 Applications to PWM-Modulated Converters

This section illustrates standard modulation of power converter switching using examples from [175]. Voltage-sourced power electronic converters (VSCs) dominate a number of application domains, including emerging power systems that integrate renewable and conventional (electromechanical) sources. These components offer a great promise in control of power systems, including speed and accuracy of the controlled response. An important aspect when modeling components like VSC is that their model has to be embedded in a larger system model that may contain more conventional components like electromechanical sources, transformers, and transmission lines. The model aims at accurately capturing large-signal behavior, while being amenable to analysis, useful in fast simulations, and serving as a basis of control design. The fundamental difficulties stem from the hybrid nature of the converter dynamics that amalgamates continuous-time dynamics (associated with the voltages and currents on capacitors and inductors) with inherently discrete events (associated with the switching actions).

Models based on dynamic phasors fill an important gap between quasi-static models (which lack accuracy when voltage stability, transient stability, or other dynamic phenomena are of interest [5, 174]) and detailed time-domain simulation models ("EMTP"-like) (whose analysis is challenging because of their time-varying nature). Dynamic phasor models are time-invariant (autonomous) and typically nonlinear. Their accuracy is customizable and can approach the detailed (switched) time-domain models by adjusting the number of harmonics and other discrete harmonics included in the model. In steady-state they provide a new class of quasi-static models that may prove useful in analysis of slower phenomena such as optimized power flow.

One key aspect of deriving dynamic phasor models is the selection of discrete frequency components (often harmonics) that are present in the system. This selection is typically based on experience and on the availability of models for other components in the system. It can be modified and corrected based on the frequency analysis of detailed time-domain simulations. An increase of the number of discrete frequency components typically makes the model more accurate at the price of increasing system size and internal intricacy. This is essentially an engineering decision where multiple factors may be at play, and rich and growing literature on dynamic phasor models can serve as guidance.

As an illustration, this section derives a continuous-time large-signal model of the Unified Power Flow Controller (UPFC) under unbalanced operation. Our models describe dynamics of various frequency components of the waveforms of interest by decomposing them into dynamical positive, negative, and zero-sequence components (which are mapped into DC and second harmonic after the d q transformation that is a concatenation of Clarke and Park's transformations).

In this example we follow [175] and assume that the generator speed has only the DC component, the AC currents have only the fundamental ($\ell = 1$) *in positive and negative sequence*, and the DC link voltage and the electrical torque have the DC and second harmonic (the same holds for real and reactive power). Options to increase the model accuracy would include for example 6-th and 12-th harmonics in the $d - q$ reference frame and would be justified if one wanted to study the effects of UPFC harmonics. Similarly, more harmonics in the DC link voltage could be useful in the design of UPFC itself. In this example we will consider switching to be ideal; the capabilities of dynamic phasors to model converters that substantially deviate from this assumption are explored in [46] and discussed later in this chapter. A justification for use of models with ideal switches in control-oriented studies is elucidated in [176]. Our modeling emphasis is on capturing system-level effects of UPFC in relatively short transients, so we add further simplifying assumptions such as constant mechanical torque and field voltage. While each assumption can be relaxed in a straightforward way, perhaps a more important aspect is system-wide coherence of the assumptions for a particular study.

The foundational reference on UPFC is [177], expanded in more detail in [178, 179]. Additional background information can be found in many references, including [175]. We use a single-machine infinite-bus benchmark power system from [49] for model validation, similar to the one used in [177]. With our emphasis on UPFC modeling, we prefer simplicity over performance in other components such as controllers.

6.1.1 Time-Domain Model of the UPFC

Consider a UPFC which is a part of the system shown in Fig. 6.1 (adapted from [49, Example 13.4]). The model of the UPFC (shunt branch with subscript s and series with subscript r) in the $d - q$ reference frame [180] is

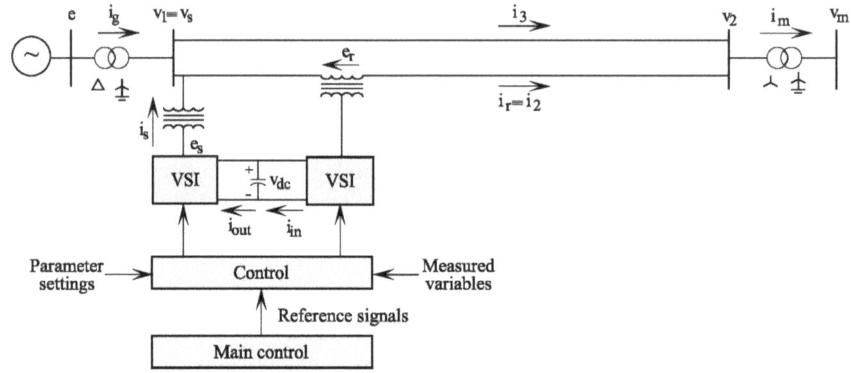

Fig. 6.1 Basic system configuration.

$$
\begin{aligned}
\frac{di_{sd}}{dt} &= \frac{\omega_0}{L_s}(e_{sd} - v_{sd} - R_s i_{sd}) + \omega_r i_{sq} \\
\frac{di_{sq}}{dt} &= \frac{\omega_0}{L_s}(e_{sq} - v_{sq} - R_s i_{sq}) - \omega_r i_{sd} \\
\frac{di_{rd}}{dt} &= \frac{\omega_0}{L_{2e}}(v_{sd} - e_{rd} - v_{2d} - R_{2e} i_{rd}) + \omega_r i_{rq} \\
\frac{di_{rq}}{dt} &= \frac{\omega_0}{L_{2e}}(v_{sq} - e_{rq} - v_{2q} - R_{2e} i_{rq}) - \omega_r i_{rd}
\end{aligned}
\tag{6.2a}
$$

where R_s and R_{2e} are resistances in the shunt and series branch, respectively, and where in each of the axes (d and q) m denotes the modulation index:

$$
e_d = m_d v_{dc} \quad , \quad e_q = m_q v_{dc} \tag{6.2b}
$$

and the real and reactive power are given by

$$
P = e_d i_d + e_q i_q \quad , \quad Q = e_d i_q - e_q i_d \tag{6.2c}
$$

The dynamics of the DC link is given by

$$
\begin{aligned}
\frac{dv_{dc}}{dt} &= \frac{1}{C} i_{dc} = \frac{1}{C}(i_{in} - i_{out}) - \frac{v_{dc}}{C R_{loss}} = \frac{1}{C}\frac{P_r - P_s}{v_{dc}} - \frac{v_{dc}}{C R_{loss}} \\
&= \frac{1}{C}(m_{rd} i_{rd} + m_{rq} i_{rq} - m_{sd} i_{sd} - m_{sq} i_{sq}) - \frac{v_{dc}}{C R_{loss}}
\end{aligned}
\tag{6.3}
$$

Please note that the DC link equation could be written in terms of v_{dc}^2 which would make the homogeneous part linear; we do not pursue that option further here.

Please note that in this particular example connections of transformers in addition prevent the zero-sequence currents from flowing through the line with the UPFC. In a more general case, these equations could be added (one for series and one for the shunt branch). With standard components, they would be decoupled from the quantities in $d - q$ axes and would not affect the DC link dynamics in the case of tree-legged inverters; more general cases would require additional modeling considerations.

6.1.2 Dynamical Phasor Model of the UPFC

The dynamical phasor model is derived from the time-domain model (6.2) via the averaging expression (2.12). We assume in this section that the time variation of ω_r across a single period is negligible, so that the UPFC phasor model becomes

$$\begin{aligned}
\frac{d}{dt}\langle i_{sd}\rangle_k &= -jk\omega_0\langle i_{sd}\rangle_k + \frac{\omega_0}{L_s}\left(\langle e_{sd}\rangle_k - \langle v_{sd}\rangle_k - R_s\langle i_{sd}\rangle_k\right) + \omega_r\langle i_{sq}\rangle_k \\
\frac{d}{dt}\langle i_{sq}\rangle_k &= -jk\omega_0\langle i_{sq}\rangle_k + \frac{\omega_0}{L_s}\left(\langle e_{sq}\rangle_k - \langle v_{sq}\rangle_k - R_s\langle i_{sq}\rangle_k\right) - \omega_r\langle i_{sd}\rangle_k \\
\frac{d}{dt}\langle i_{rd}\rangle_k &= -jk\omega_0\langle i_{rd}\rangle_k + \omega_r\langle i_{rq}\rangle_k + \frac{\omega_0}{L_{2e}}\left(\langle v_{sd}\rangle_k - \langle e_{rd}\rangle_k - \langle v_{2d}\rangle_k - R_{2e}\langle i_{rd}\rangle_k\right) \\
\frac{d}{dt}\langle i_{rq}\rangle_k &= -jk\omega_0\langle i_{rq}\rangle_k - \omega_r\langle i_{rd}\rangle_k + \frac{\omega_0}{L_{2e}}\left(\langle v_{sq}\rangle_k - \langle e_{rq}\rangle_k - \langle v_{2q}\rangle_k - R_{2e}\langle i_{rq}\rangle_k\right)
\end{aligned} \tag{6.4a}$$

where

$$\begin{aligned}
\langle e_d\rangle_k &= \sum_{\ell=-2,0,2}\langle m_d\rangle_\ell\langle v_{dc}\rangle_{k-\ell} \\
\langle e_q\rangle_k &= \sum_{\ell=-2,0,2}\langle m_q\rangle_\ell\langle v_{dc}\rangle_{k-\ell}
\end{aligned} \tag{6.4b}$$

and we used the fact that the only non-vanishing $\langle v_{dc}\rangle_k$, $\langle m_d\rangle_k$, $\langle m_q\rangle_k$, $\langle e_d\rangle_k$, and $\langle e_q\rangle_k$ are for $k = -2, 0, 2$. Also, from equation (6.2c),

$$\begin{aligned}
\langle P\rangle_k &= \sum_{\ell=-2,0,2}\langle e_d\rangle_\ell\langle i_d\rangle_{k-\ell} + \sum_{\ell=-\infty}^{\infty}\langle e_q\rangle_\ell\langle i_q\rangle_{k-\ell} \\
\langle Q\rangle_k &= \sum_{\ell=-2,0,2}\langle e_d\rangle_\ell\langle i_q\rangle_{k-\ell} - \sum_{\ell=-\infty}^{\infty}\langle e_q\rangle_\ell\langle i_d\rangle_{k-\ell}
\end{aligned} \tag{6.4c}$$

The k-th harmonic of the DC link voltage for $k = 0$ and $k = 2$ satisfies the dynamic equation

$$
\begin{aligned}
\frac{d}{dt}\langle v_{dc}\rangle_0 &= \frac{1}{C}\left(\langle m_{rd}\rangle_0 \langle i_{rd}\rangle_0 + \langle m_{rq}\rangle_0 \langle i_{rq}\rangle_0\right) - \frac{1}{C}\left(\langle m_{sd}\rangle_0 \langle i_{sd}\rangle_0 + \langle m_{sq}\rangle_0 \langle i_{sq}\rangle_0\right) + \\
&+ \frac{2}{C}\left(\Re\left(\langle m_{rd}\rangle_2 \langle i_{rd}\rangle_2^* + \langle m_{rq}\rangle_2 \langle i_{rq}\rangle_2^*\right)\right) - \\
&- \frac{2}{C}\left(\Re\left(\langle m_{sd}\rangle_2 \langle i_{sd}\rangle_2^* + \langle m_{sq}\rangle_2 \langle i_{sq}\rangle_2^*\right)\right) - \frac{1}{CR_{loss}}\langle v_{dc}\rangle_0 \\
\frac{d}{dt}\langle v_{dc}\rangle_2 &= -\left(\frac{1}{CR_{loss}} + j2\omega_0\right)\langle v_{dc}\rangle_2 + \\
&+ \frac{1}{C}\left(\langle m_{rd}\rangle_0 \langle i_{rd}\rangle_2 + \langle m_{rq}\rangle_0 \langle i_{rq}\rangle_2\right) + \frac{1}{C}\left(\langle m_{rd}\rangle_2 \langle i_{rd}\rangle_0 + \langle m_{rq}\rangle_2 \langle i_{rq}\rangle_0\right) - \\
&- \frac{1}{C}\left(\langle m_{sd}\rangle_0 \langle i_{sd}\rangle_2 + \langle m_{sq}\rangle_0 \langle i_{sq}\rangle_2\right) - \frac{1}{C}\left(+\langle m_{sd}\rangle_2 \langle i_{sd}\rangle_0 + \langle m_{sq}\rangle_2 \langle i_{sq}\rangle_0\right)
\end{aligned}
\tag{6.5}
$$

where $\Re$ denotes the real part of a complex number, and the superscript * denotes complex conjugation. The equations are clearly nonlinear in general, involving multiplications of states and control inputs.

A quick count reveals that the phasor dynamic model has more states than the corresponding time-domain model, since equations are complex for non-DC components, and thus each corresponds to two real equations. In the end, there are 12 states for the currents (4 real, 4 complex that each have real and imaginary parts, or magnitude and phase) and three states for the DC link voltage (one real, one complex). Thus, the overall model for the UPFC is of the 15-th order. While this may seem large compared with the fifth-order time-domain model, we want to emphasize that the phasor model is time-invariant (autonomous) and thus much more amenable to analysis. In the following section we show that the phasor model is fast and accurate to simulate. In a steady-state, the dynamic phasor model (6.4a)–(6.5) can yield a complete equivalent circuit, simply by setting all time derivatives to zero.

6.1.3 Validation of the Dynamic Phasor Model for UPFC

Details about the benchmark example shown in Fig. 6.1 and used for validation of the dynamic phasor model are in [175]; two unbalanced modes of operation are considered in large-signal transient simulations: (1) voltage asymmetry with 5% negative sequence voltage and (2) one phase to ground fault at the beginning of the line parallel to the one with the UPFC, where after the fault the affected line is disconnected.

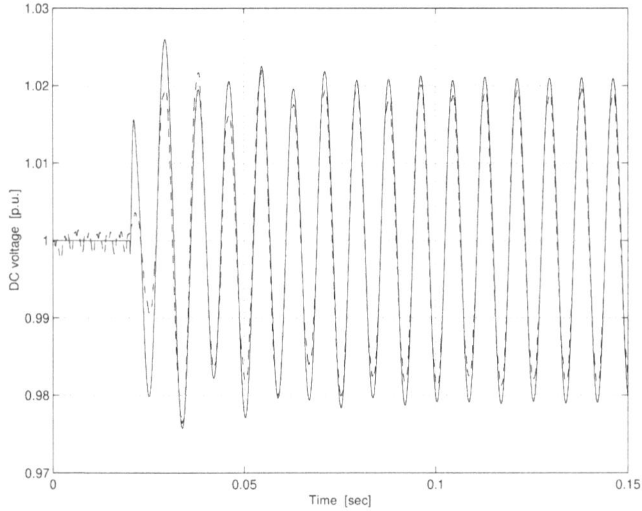

Fig. 6.2 DC link capacitor voltage following the onset of unbalance at $t = 0.02$ sec: time-domain switched simulation (dashed line) and phasor model (solid line).

We first present the simulation results in the case of voltage asymmetry that starts at time $t = 0.02$ sec. The solid line in all the figures is the prediction of the dynamic phasor model, while the dashed line corresponds to the detailed switched ("48-pulse") time-domain model (see, for example, [176, 181]). In Fig. 6.2 we show the DC link capacitor voltage: the outstanding agreement is evident.

It turns out that in this example the simulation with the phasor model is typically at least an order of magnitude faster than the simulation of the switched model (no effort was made to speed up either of these Simulink implementations). An explanation for this behavior is provided by the relatively slow variation of key harmonics. In Fig. 6.3 we display the components of the DC link voltage – the DC component in the top panel and the real (solid line) and imaginary (dashed line) components of the second harmonics (bottom panel).

Next, we consider the case of the one phase-to-ground fault described earlier – in Fig. 6.4 we display the DC link voltage. We again observe a very good overall agreement, even in this very abrupt transient.

This section derived a dynamical phasor model for unbalanced UPFC. The same methodology is of course applicable to related (and simpler) voltage-sourced converters like STATCOM and SSSC. The main theoretical benefit is the time-invariant nature of this approximate large-signal model. The section presented only one in a family of possible models, as a different choice of retained harmonics would result in a different model. This nested nature of the dynamic phasor approach is particularly valuable when models of power converters are needed with a variable level of detail.

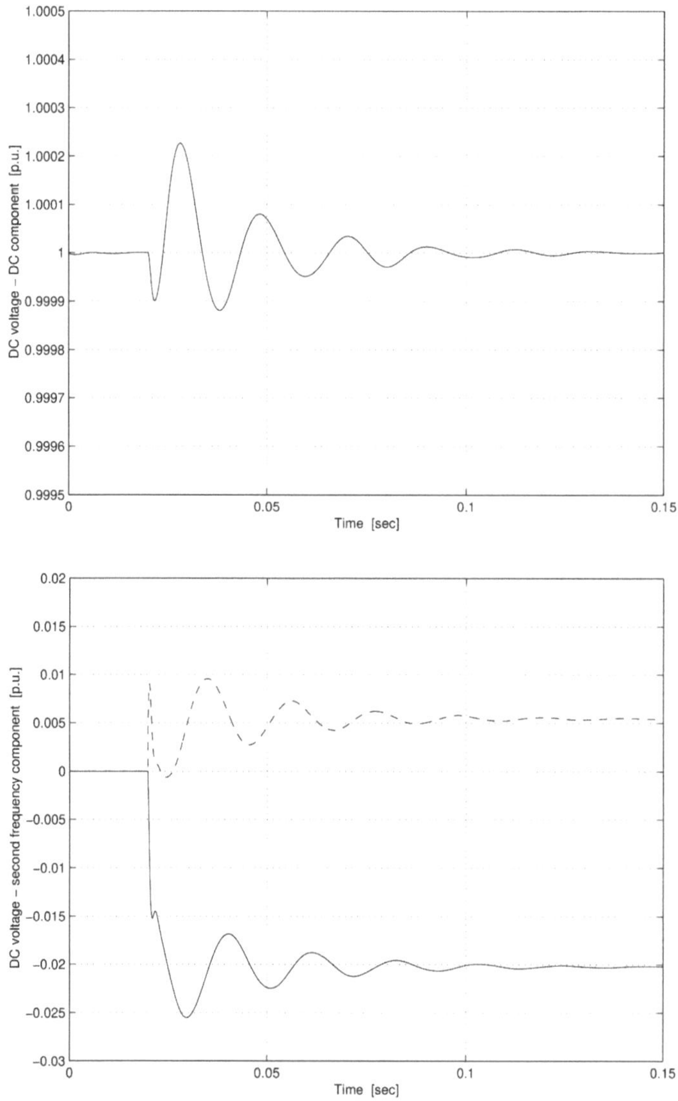

Fig. 6.3 DC link voltage components following the onset of unbalance at $t = 0.02$ sec: DC component (top panel) and the real part (solid line) and the imaginary part (dashed line) of the second harmonic (bottom panel).

6.2 Multi-frequency Dynamic Phasor Models

In this section we are looking into modeling of power electronic converters when the switching frequency is not significantly higher than the natural response frequencies of (passive) components (also known as the small ripple condition in the literature [182]). As a consequence, many variables of interest, including the switching

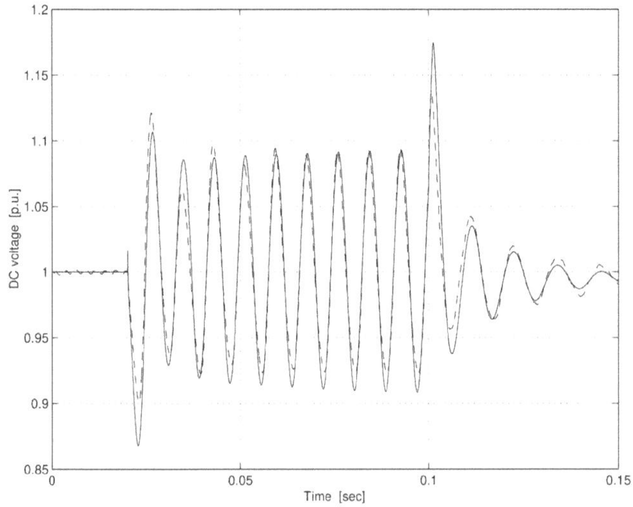

Fig. 6.4 DC voltage during a one phase to ground fault: time-domain switched simulation (dashed line) and phasor model (solid line).

function, will have ripple in addition to the desired (typically DC) component in appropriate coordinates. This colloquially labeled "slow" switching may occur not only due to physical limitations of switches but also because of system-level goals such as reduced heat dissipation and shaping of the electromagnetic signature (EMI). In this section we follow derivations presented in [46, 183] and illustrated on a boost converter example. We also address the performance of both open- and closed-loop models and explore the option of reducing models (say by retaining only "standard" DC components) in a way that still maintains information about ripple quantities.

Consider the case when the switching function $q(t)$ and the variable $x(t)$ can be well approximated by the sum of their index-0 (DC component) and index-1 (fundamental phasor) averages, with all other index-k components negligible:

$$q(t) \approx Q_0 + Q_{-1}e^{-j\omega_0 t} + Q_1 e^{j\omega_0 t} \tag{6.6a}$$

$$x(t) \approx X_0 + X_{-1}e^{-j\omega_0 t} + X_1 e^{j\omega_0 t} \tag{6.6b}$$

Applying the well-known discrete convolution result for phasor components of a product of two variables, we get

$$\langle qx \rangle_0 = Q_0 X_0 + Q_{-1} X_1 + Q_1 X_{-1} \tag{6.7a}$$

$$\langle qx \rangle_1 = Q_0 X_1 + Q_1 X_0 \tag{6.7b}$$

$$\langle qx \rangle_{-1} = Q_0 X_{-1} + Q_{-1} X_0 \tag{6.7c}$$

We assume that all other components of the product (such as the second harmonic) are negligible. Note that quantities like X_1 and Q_{-1} are in general complex, and positive and negative indices are complex conjugates of each other. Thus

$$Q_1 = Q_1^R + jQ_1^I = Q_{-1}^* = (Q_{-1}^R + jQ_{-1}^I)^* \tag{6.8a}$$

$$X_1 = X_1^R + jX_1^I = X_{-1}^* = (X_{-1}^R + jX_{-1}^I)^* \tag{6.8b}$$

where the superscripts R and I denote the *real* and the *imaginary* parts of the defined quantities and the asterisk (*) denotes complex conjugation. It is then straightforward to show that

$$\langle qx \rangle_0 = Q_0 X_0 + 2(Q_1^R X_1^R + Q_1^I X_1^I) \tag{6.9a}$$

$$\langle qx \rangle_1^R = Q_0 X_1^R + X_0 Q_1^R \tag{6.9b}$$

$$\langle qx \rangle_1^I = Q_0 X_1^I + X_0 Q_1^I \tag{6.9c}$$

In the cases of interest to us, $q(t)$ is a binary switching function and $x(t)$ is a state variable such as a capacitor voltage or an inductor current. The first of the above expressions is useful in the index-0 model, whereas the last two are necessary for constructing the index-1 model.

In many applications of dynamic phasors, it is of interest to quantify the error introduced by truncating the higher order indices. For this purpose, we need a simply computable but fairly tight bounds on the error that would be useful in modeling, analysis, and design. Some results and discussions in this direction are presented, for example, in [9, 184], but the practically useful bounds still seem elusive. We are, however, encouraged by numerous references that demonstrate the utility of dynamic phasors on reasonably low-order models even in the absence of theoretical guarantees.

6.2.1 A Multi-Frequency Averaged Model of the Boost Converter

The schematic of the boost converter is given in Fig. 6.5. A natural state-space model for this converter, assuming continuous conduction mode, can be written as

$$\frac{di(t)}{dt} = \frac{1}{L}\Big(v_{in}(t) - \big[1 - q(t)\big]v(t)\Big) \tag{6.10a}$$

$$\frac{dv(t)}{dt} = \frac{1}{C}\left(\big[1 - q(t)\big]i(t) - \frac{v(t)}{R}\right) \tag{6.10b}$$

where $q(t)$=1 when the switch is on (closed) and $q(t)$=0 when it is off (open).

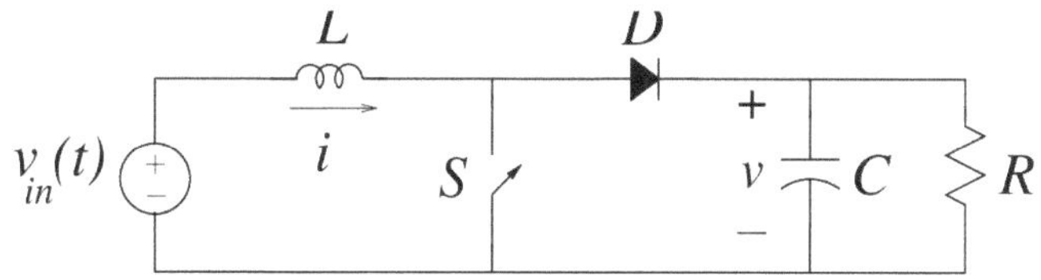

Fig. 6.5 Boost converter schematic.

After assuming that $v_{in}(t)$ is constant and equal to V_{in} and taking the index-0 average of both sides, we get

$$\frac{dI_0}{dt} = \frac{1}{L}\big(V_{in} - V_0 + \langle qv\rangle_0\big) \tag{6.11a}$$

$$\frac{dV_0}{dt} = \frac{1}{C}\left(I_0 - \langle qi\rangle_0 - \frac{V_0}{R}\right) \tag{6.11b}$$

where we suppress the time dependence to reduce notational clutter. Substituting (6.9a) into (6.11a) and (6.11b) for $\langle qv\rangle_0$ and $\langle qi\rangle_0$ and simplifying, we obtain

$$\frac{dI_0}{dt} = \frac{1}{L}\left(V_{in} - Q_0' V_0 + 2Q_1^R V_1^R + 2Q_1^I V_1^I\right) \tag{6.12a}$$

$$\frac{dV_0}{dt} = \frac{1}{C}\left(Q_0' I_0 - 2Q_1^R I_1^R - 2Q_1^I I_1^I - \frac{V_0}{R}\right) \tag{6.12b}$$

where $Q_0' \overset{\text{def}}{=} 1 - Q_0$. These equations describe the dynamics of the index-0 averages of the boost converter. If we were to exclude the index-1 (ripple) terms in (6.6a) and (6.6b), then the above equations would simply reduce to the conventional state-space averaged (SSA) equations. That would be justifiable in most practical converters, and there would be little difference between the evolution of (V_0, I_0) and actual state variables of the switching converter. Here, however, we proceed to derive a model that includes the fundamental ripple component in the states and the switching function.

Applying the same rules, we obtain the following *complex* differential equations:

$$\frac{dI_1}{dt} = -j\omega_0 I_1 + \frac{1}{L}\big(- V_1 + \langle qv\rangle_1\big) \tag{6.13a}$$

$$\frac{dV_1}{dt} = -j\omega_0 V_1 + \frac{1}{C}\left(I_1 - \langle qi\rangle_1 - \frac{V_1}{R}\right) \tag{6.13b}$$

Using (6.9b) and (6.9c) and simplifying, we obtain a total of four equations for real and imaginary parts of the fundamental phasor of the state variables:

$$\frac{dI_1^R}{dt} = \omega_0 I_1^I + \frac{1}{L}\Big(- Q_0' V_1^R + V_0 Q_1^R \Big) \tag{6.14a}$$

$$\frac{dI_1^I}{dt} = -\omega_0 I_1^R + \frac{1}{L}\Big(- Q_0' V_1^I + V_0 Q_1^I \Big) \tag{6.14b}$$

$$\frac{dV_1^R}{dt} = \omega_0 V_1^I + \frac{1}{C}\Big(Q_0' I_1^R - I_0 Q_1^R - \frac{V_1^R}{R} \Big) \tag{6.14c}$$

$$\frac{dV_1^I}{dt} = -\omega_0 V_1^R + \frac{1}{C}\Big(Q_0' I_1^I - I_0 Q_1^I - \frac{V_1^I}{R} \Big) \tag{6.14d}$$

Please note the explicit frequency dependence of the averaged model described by these equations.

The structure of coupling among differential equations is best seen in a state-space form:

$$\frac{d}{dt}\left[\begin{array}{c} \langle \bullet \rangle_0 \\ \hline \langle \star \rangle_1 \end{array}\right] = \underbrace{\left[\begin{array}{c|c} \mathbf{A}_1 & \mathbf{A}_2 \\ \hline \mathbf{A}_3 & \mathbf{A}_4 \end{array}\right]}_{\mathbf{A}} \left[\begin{array}{c} \langle \bullet \rangle_0 \\ \hline \langle \star \rangle_1 \end{array}\right] + \left[\begin{array}{c} \mathbf{b} \\ \hline \mathbf{0} \end{array}\right] V_{in} \tag{6.15a}$$

The six state variables are here divided into two sets, with $\langle \bullet \rangle_0$ and $\langle \star \rangle_1$ representing the index-0 and index-1 state variables, respectively:

$$\langle \bullet \rangle_0 \overset{\text{def}}{=} \begin{bmatrix} I_0 \\ V_0 \end{bmatrix} \tag{6.15b}$$

$$\langle \star \rangle_1 \overset{\text{def}}{=} \begin{bmatrix} I_1^R \\ I_1^I \\ V_1^R \\ V_1^I \end{bmatrix} \tag{6.15c}$$

By comparing (6.15a) with the six coupled differential equations in (6.12a), (6.12b), and (6.14a)–(6.14d), we identify the matrices $\mathbf{A}_1$ through $\mathbf{A}_4$ and the vector $\mathbf{b}$ as

$$\mathbf{A}_1 = \begin{bmatrix} 0 & -\frac{Q_0'}{L} \\ \frac{Q_0'}{C} & -\frac{1}{RC} \end{bmatrix} \tag{6.15d}$$

$$\mathbf{A}_2 = \begin{bmatrix} 0 & 0 & \frac{2Q_1^R}{L} & \frac{2Q_1^I}{L} \\ -\frac{2Q_1^R}{C} & -\frac{2Q_1^I}{C} & 0 & 0 \end{bmatrix} \tag{6.15e}$$

$$\mathbf{A}_3 = \begin{bmatrix} 0 & \frac{Q_1^R}{L} \\ 0 & \frac{Q_1^I}{L} \\ -\frac{Q_1^R}{C} & 0 \\ -\frac{Q_1^I}{C} & 0 \end{bmatrix} \tag{6.15f}$$

$$\mathbf{A}_4 = \begin{bmatrix} 0 & \omega_0 & -\frac{Q_0'}{L} & 0 \\ -\omega_0 & 0 & 0 & -\frac{Q_0'}{L} \\ \frac{Q_0'}{C} & 0 & -\frac{1}{RC} & \omega_0 \\ 0 & \frac{Q_0'}{C} & -\omega_0 & -\frac{1}{RC} \end{bmatrix} \tag{6.15g}$$

$$\mathbf{b} = \begin{bmatrix} \frac{1}{L} \\ 0 \end{bmatrix} \tag{6.15h}$$

Equations (6.15a)–(6.15h) comprise our *multi-frequency averaged* (MFA) model. This model possesses a considerable structure which can be gleaned from visual inspection of the matrices in (6.15d)–(6.15h). It turns out that this structure carries over into a modal separation between the index-0 dynamics and the index-1 dynamics, which will guide us in simplifying the full model.

For the boost converter used for model verification (with parameters given in the next subsection), the eigenvalues of the 6×6 system matrix $\mathbf{A}$ in (6.15a) are

$$\lambda_{1,2} = (-1.4205 \pm j2.2229) \times 10^4 \text{ rad/sec} \tag{6.16a}$$

$$\lambda_{3,4} = (-1.4205 \pm j4.4651) \times 10^4 \text{ rad/sec} \tag{6.16b}$$

$$\lambda_{5,6} = (-1.4205 \pm j8.6187) \times 10^4 \text{ rad/sec} \tag{6.16c}$$

Note that all eigenvalues have the same real part, so model reduction using singular perturbation ideas is not straightforward. It turns out that only the index-0 averages contribute significantly to creating the first set of eigenvalues $\lambda_{1,2}$, while only the index-1 averages contribute significantly to the next two pairs of eigenvalues, $\lambda_{3,4}$ and $\lambda_{5,6}$. The contribution of state variables to system modes is assessed using *participation factors* [185]. In this particular case, the evidence for modal decoupling can be seen from the eigenvalues of the full 6×6 system. $\mathbf{A}$ can be closely estimated by those of the submatrices $\mathbf{A}_1$ and $\mathbf{A}_4$ separately [46].

Next, consider the system realization in state-space form given in (6.15a) with zero input (i.e., $V_{in} = 0$). The resulting system may be represented by the block diagram in Fig. 6.6. At this point we are interested in reduced but *enhanced* index-0 model that approximately takes into account the limited interaction between the index-0 and index-1 subsystems. The Selective Modal Analysis (SMA) reduction

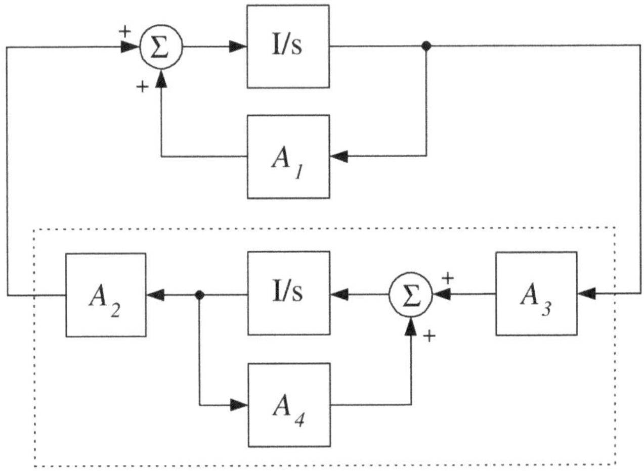

Fig. 6.6 Interaction of index-0 and index-1 subsystems.

procedure described, for example, in [185] focuses on approximating the part of the subsystem inside the dotted region in Fig. 6.6, by approximating its transfer function

$$\mathbf{H}(s) = \mathbf{A}_2\left(s\mathbf{I} - \mathbf{A}_4\right)^{-1}\mathbf{A}_3 \tag{6.17}$$

This transfer function represents the effect of the less important variables (the index-1 variables in this case) on the important variables (the index-0 variables). In singular perturbation approaches [186], one represents the system inside the dotted region by its DC gain

$$\mathbf{H}(0) = -\mathbf{A}_2\mathbf{A}_4^{-1}\mathbf{A}_3 \tag{6.18}$$

The justification for this step is that the modes of interest are much slower than the bandwidth of $\mathbf{H}(s)$, so that we are justified in replacing the dynamic block by its static gain. This reasoning is not valid in our case since the real parts of all of the eigenvalues are equal.

The idea behind SMA is to replace dynamics that give rise to $\mathbf{H}(s)$ by a constant matrix $\mathbf{M}$ whose effect on the modes of interest is almost identical to that of the replaced dynamics. This requirement translates into the following constraint:

$$\mathbf{M}\left[\widehat{\mathbf{v}}_1 | \widehat{\mathbf{v}}_2\right] = \left[\mathbf{H}(\widehat{\lambda}_1)\widehat{\mathbf{v}}_1 | \mathbf{H}(\widehat{\lambda}_2)\widehat{\mathbf{v}}_2\right] \tag{6.19}$$

where $\widehat{\lambda}_1$ and $\widehat{\lambda}_2$ are the eigenvalues computed $\mathbf{A}_1$ (predominately influenced by the important index-0 variables) and $\widehat{\mathbf{v}}_1$ and $\widehat{\mathbf{v}}_2$ are the corresponding eigenvectors of $\mathbf{A}_1$. In this particular example the resulting $\mathbf{M}$ is diagonal and is given by [46]

$$\mathbf{M} = \begin{bmatrix} -\beta & 0 \\ 0 & \beta \end{bmatrix} \tag{6.20}$$

where β is given by the expression

$$\beta = \frac{2\zeta}{\pi^2} \cdot \frac{\omega_{\mathrm{LC}}^2\Big[1 - \cos(2\pi Q_0)\Big]}{\omega_{\mathrm{LC}}^2 - 4\Big[\big(\omega_{\mathrm{LC}}\langle q\rangle'_0\big)^2 - \zeta^2\Big]} \tag{6.21}$$

where $\omega_{\mathrm{LC}} \overset{\mathrm{def}}{=} 1/\sqrt{LC}$ and $\zeta \overset{\mathrm{def}}{=} 1/(2RC)$. The resulting reduced-order model can now be written in state-space form as

$$\frac{d}{dt}\begin{bmatrix} I_0 \\ V_0 \end{bmatrix} = \Big[\mathbf{A}_1 + \mathbf{M}\Big]\begin{bmatrix} I_0 \\ V_0 \end{bmatrix} + \mathbf{b}V_{in} \tag{6.22}$$

We shall refer to this model as the *condensed* MFA (CMFA) model.

6.2.2 Validation of the Multi-frequency Averaged Model for Boost Converter

Let us consider the boost converter of Fig. 6.5 with the following component values: $V_{in} = 5\,\mathrm{V}$, $L = 100\,\mu\mathrm{H}$, $C = 4.4\,\mu\mathrm{F}$ $R = 8\,\Omega$, $f_s = 1/T = 10\,\mathrm{kHz}$, and $D = 0.5$.

To perform open-loop simulation (from control standpoint) of PWM converters based on averaged models derived in the previous subsection, we can use a standard circuit simulator such as PSpice. The duty ratio enters these models through the averages of $q(t)$. If the duty ratio takes the constant value D, the index-0 and index-1 averages of $q(t)$ are constants given by $Q_0 = D$, $Q_1^R = \frac{1}{2\pi}\sin(2\pi D)$, $Q_1^I = \frac{1}{2\pi}(\cos(2\pi D) - 1)$. For slowly varying duty ratio (whether varied in open loop or in closed loop), the same expressions could be used with D replaced by the time varying duty ratio, as will be seen later.

We start by showing averaged and switched simulations for the boost converter are shown in Fig. 6.7, where the capacitor (output) voltage and the inductor current are displayed. We observe a steady-state error in the average values of the state variables obtained by the standard state-space averaging (SSA) method. At the same time, we note that the index-0 averages computed by our sixth-order MFA model approach the correct average values. This is due to the fact that the index-0 averages are influenced by the ripple in the state variables, as seen in (6.12a) and (6.12b).

We also show results for open-loop operation of the reduced model (CMFA) in the case of capacitor voltage $v(t)$ in Fig. 6.8. It is clear that the index-0 average from the second-order CMFA model tracks very well the original index-0 average computed from the sixth-order MFA model. The advantage of the CMFA model is that we do not need to calculate the index-1 averages, so it is very similar to the standard SSA model in terms of computational costs [46].

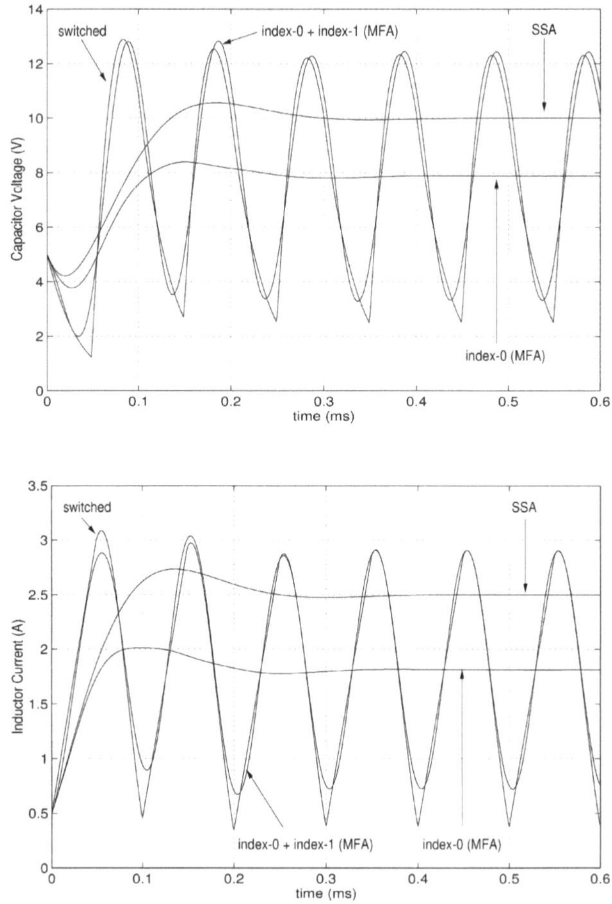

Fig. 6.7 Comparison of open-loop simulations for a boost converter with $V_{in} = 5\,\text{V}$, $L = 100\,\mu\text{H}$, $C = 4.4\,\mu\text{F}$, $R = 8\,\Omega$, $f_s = 10\,\text{kHz}$, and $D = 0.5$: (**a**) capacitor voltage and (**b**) inductor current.

Next, we turn our attention to closed-loop operation. The basic idea is to use state error signals (depending on the deployed sensors) to produce the needed modulating signal $m(t)$. This signal is in turn compared with a sawtooth signal $s(t)$ to produce the switching signal $q(t)$, which is used as the gating signal for the controlled switch in the converter. For the multi-frequency averaged models, we need to be able to generate the index-0 and index-1 averages of $q(t)$. There is a plethora of controllers proposed for power electronic converters; we use a simple proportional controller with algorithmic details described in [46].

In closed-loop regulation of PWM DC–DC converters, one usually generates error signals and processes them with an error amplifier that produces a modulating

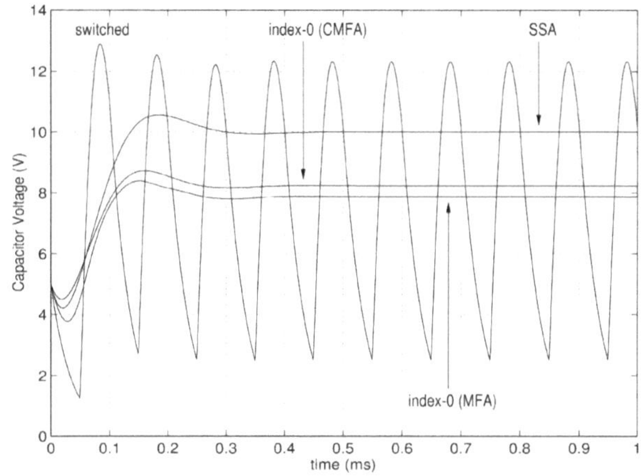

Fig. 6.8 Comparison of conventional and frequency-selective averages for a boost converter with $V_{in} = 5\,\text{V}$, $L = 100\,\mu\text{H}$, $C = 4.4\,\mu\text{F}$, $R = 8\,\Omega$, $f_s = 10\,\text{kHz}$, and $D = 0.5$.

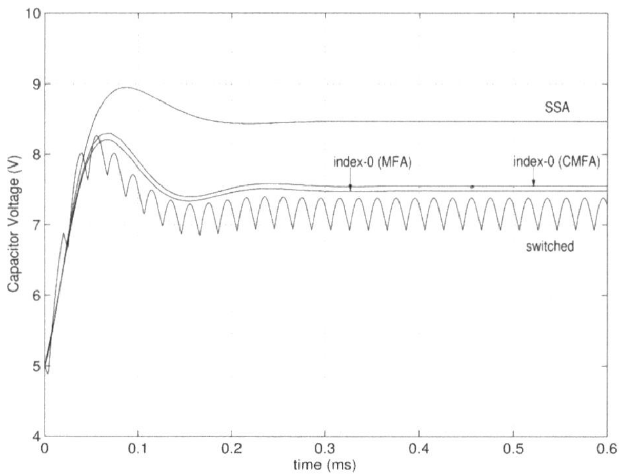

Fig. 6.9 Comparison of closed-loop simulations for a boost converter with $V_{in} = 5\,\text{V}$, $L = 50\,\mu\text{H}$, $C = 4.4\,\mu\text{F}$, $R = 28\,\Omega$, and $f_s = 50\,\text{kHz}$.

signal $m(t)$. The modulating signal is compared with a sawtooth $s(t)$ to produce the switching signal $q(t)$, which is used as the gating signal for the controlled switch in the converter. The operation of the comparator may be described by representing the switching function as $q(t) = h(m(t) - s(t))$, where $h(\bullet)$ represents the Heaviside step function. To use the averaged models that we have developed, we have to be able to generate the index-0 and index-1 averages of $q(t)$.

The results of the simulation for the boost converter are shown in Fig. 6.9. In this figure, we show the output voltage from the switched model, as well as from the

SSA, MFA, and CMFA models. As noted in [187], state-space SSA is not able to approach the correct steady-state value. However, there is a significant improvement with our MFA/CMFA models and the underlying switched waveforms. The residual discrepancy between the index-0 and the switched waveform is caused by the higher harmonics in the modulating signal $m(t)$ that are not included in our model (which only considers the DC and fundamental components).

6.3 Modeling of Converters with Semi-controllable Switches

This section develops simple but accurate continuous-time, large-signal dynamic phasor models for a Thyristor-Controlled Series Capacitor (TCSC). The TCSC represents the class of thyristor-based power converters, whose main characteristics are that the switching occurs only once per period of the fundamental AC waveform and that the turn-off transition is not explicitly controlled.

The dynamic phasor models are an attractive alternative to sampled-data models because of simpler representation, modularity, and direct compatibility with the models that are routinely used to represent the dynamics of generators and other power system components. In this section we rely on results originally presented in [188, 189].

Series capacitive compensation in AC transmission systems has the potential to increase power transfer capability and enhance transient stability. However, the fixed series compensation can also lead to subsynchronous resonance (SSR), which can cause generator damage from the torsional stress on the shaft [190, 191]. With the increased penetration of renewable sources interfaced through inverters, new resonance phenomena are emerging in similar frequency ranges.

Thyristor-Controlled Series Capacitors (TCSCs) have found applications as adjustable series capacitive compensators [192–194]. In addition to regulating the power flow, TCSCs aim to damp power oscillations and mitigate SSR [195–198]. Analytical techniques for studying SSR with TCSCs are challenging due to the presence of hybrid dynamics involving both continuous and discrete variables. Such analytical difficulties are common in all flexible AC transmission systems (FACTS) whose operation is based on thyristor switching. Alternative approaches to understanding SSR rely mostly on detailed time-domain simulations, using standard software tools exemplified by the Electromagnetic Transients Program (EMTP) [195–198]. Such simulations have the potential to be accurate when based on large and detailed models but tend to offer little insight, especially in close-to-unstable conditions.

Figure 6.10 shows the circuit diagram of a single-phase TCSC. The system consists of a fixed capacitor in parallel with a Thyristor Controlled Reactor (TCR). The control means is the phase delay γ of the thyristor firing pulses relative to the zero crossings of some reference waveform. In some applications the variations of the control input can be modeled in a straightforward way as variations in the effective capacitive/inductive reactance at the fundamental frequency. However, studies aimed at SSR and similar phenomena require a more comprehensive model. Details on TCSC characteristics can be found in [199–201].

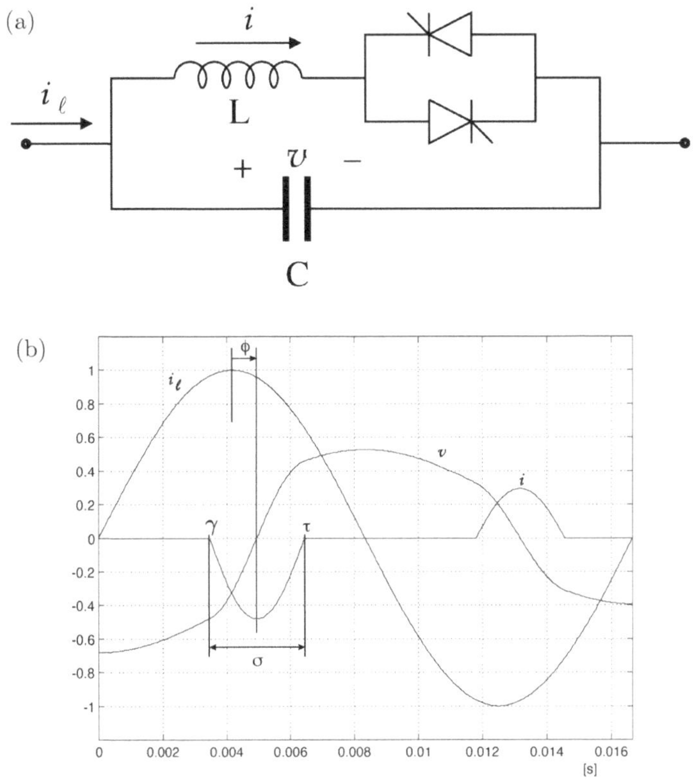

Fig. 6.10 (**a**) TCSC circuit diagram and (**b**) transient waveforms.

The capacitor voltage v and reactor current i are natural state variables for the TCSC circuit shown in Fig. 6.10a. The corresponding state-space model is then

$$C\,\frac{dv}{dt} = i_\ell - i$$
$$L\,\frac{di}{dt} = q\,v \tag{6.23}$$

where q is a switching function: $q = 1$ when one thyristor is on and $q = 0$ when both thyristors are off.

6.3.1 Dynamic Phasor Models for TCSC

The fundamental (index-1) dynamic phasor model of TCSC is

$$
\begin{aligned}
C\,\frac{dV_1}{dt} &= I_\ell \;-\; I_1 \;-\; j\,\omega_0\,C\,V_1 \\
L\,\frac{dI_1}{dt} &= <q\,v>_1 \;-\; j\,\omega_0\,L\,I_1
\end{aligned}
\tag{6.24}
$$

where I_ℓ is the 1-phasor for the line current i_ℓ. We refer to (6.24) as the full-order model. The key difficulty in making this a state-space model is in approximately averaging the time-varying nonlinear term qv. The reason why our strong assumption about the frequency content of the line current may still result in an effective model at SSR frequencies is that an approximate expression for $< qv >_1$ only assumes the line current can be well-represented as a sinusoid over the interval during which qv is nonzero, which is just a portion of a half-cycle. The sinusoid is, however, allowed to vary from one such interval to the next.

The first step in reduction of (6.24) comes from time scale reasoning, since for typical parameter values [188] I_1 has the fast dynamics relative to that of V_1. We can then assume [189] that I_1 can be written as

$$
I_1 \approx \frac{V_1}{j\omega_0 L_{\mathit{eff}}(\sigma)} \tag{6.25}
$$

where $L_{\mathit{eff}}(\sigma)$ is the effective inductance of the TCR branch in the quasi-steady-state that corresponds to the prevailing conduction angle σ. We will make this reasoning more precise shortly. Substituting (6.25) in the top equation of (6.24), we obtain

$$
C\,\frac{dV_1}{dt} = I_\ell - \Big(j\omega_0 C + \frac{1}{j\omega_0 L_{\mathit{eff}}(\sigma)}\Big)V_1 = I_\ell - j\omega_0 C_{\mathit{eff}}(\sigma)V_1 \tag{6.26}
$$

where $j\omega_0 C_{\mathit{eff}}(\sigma)$ is the effective quasi-steady-state admittance of the TCSC defined by the term in parentheses in the first equation of (6.26). Assuming a sinusoidal line current conduction angle σ in the steady-state obtained with $dV_1/dt = 0$ in (6.26), we can evaluate the ratio $I_\ell / V_1 = j\omega_0 C_{\mathit{eff}}(\sigma)$. The steady-state analysis of the TCSC in Fig. 6.10, with a sinusoidal line current i_ℓ, yields the following formidable-looking expression for the equivalent capacitance $C_{\mathit{eff}}(\sigma)$ [189]:

$$
C_{\mathit{eff}}(\beta) = C\left[1 - \frac{2}{\pi}\left(\mathcal{S}\Big[\beta + \frac{\sin(2\beta)}{2}\Big] + 2\Big(\frac{\mathcal{S}}{\eta}\Big)^2\Big[\tan(\beta) - \eta\tan(\eta\beta)\Big]\cos^2\beta\right)\right]^{-1} \tag{6.27a}
$$

where η and $\mathcal{S}$ are given by

$$\eta = \frac{1}{\omega_0 \sqrt{LC}}, \quad \mathcal{S} = \frac{1}{1-\eta^{-2}} \equiv \frac{\eta^2}{\eta^2 - 1} \tag{6.27b}$$

and ω_0 is the line frequency. Note that this is just a memoryless nonlinearity. The prevailing conduction angle σ will be approximated as in [188]:

$$\sigma = \sigma^{\text{ref}} + 2\phi \approx \sigma^{\text{ref}} + 2 \arg\left[-j I_\ell V_1^*\right] \tag{6.28}$$

where * denotes the complex conjugate, σ^{ref} is the conduction angle reference, and in steady-state $\sigma = \pi - 2\gamma$.

We refer to (6.26)–(6.28) as the reduced model. Please note that it is only the second nonlinear order, but it does pose analytical challenges, as the state and the control input become entangled in (6.28) in forming σ and then passed through the memoryless nonlinearity $C_{\mathit{eff}}(\sigma)$. The reduced model (6.26), (6.28) represents a single-phase TCSC model. The real and negative imaginary parts of (6.26) are also the desired $d - q$ (direct and quadrature) expressions of a three-phase TCSC model when we assume that the current waveforms in the three phases are balanced. In the section on UPFC we explored the effects of phase unbalance.

In the case where the line frequency $\omega_0(t)$ has significant time variation, the Fourier analysis should be performed with respect to the basis functions $e^{jk\theta(t)}$, where $\theta(t)$ is the instantaneous phase of the line current, as suggested in [45]. A complete analysis in this framework is quite elaborate; for our TCSC model a good approximation is to select the conduction angle reference as

$$\sigma^{\text{ref}}(t) = \sigma^{\text{ref}} \frac{\omega_0(t)}{\omega_0^{\text{ref}}} \tag{6.29}$$

which represents the modulation of the conduction angle ("phase correction") under variable fundamental frequency. This correction may sometimes be significant in the SSR analysis, as demonstrated later in the section.

It has been stressed in many studies that the synchronization policy strongly affects the dynamic behavior of the TCSC. Our derivations assume that the firing angle is synchronized with the line current zeros [188]. An alternative is the equidistant firing policy, which is the synchronization scheme used for the eigenvalue analysis based on sampled-data models in the benchmark test for SSR studies with TCSC [202, 203]. In equidistant policy, the firing angle is computed with respect to a periodic reference waveform [189].

6.3.2 Validation of the Dynamic Phasor Model for TCSC

To verify the validity of our phasor models, we compared their behavior with detailed time-domain simulations of the switched TCSC circuit. The models were implemented in Matlab and EMTP. The system parameters were taken from

[174, 194] and corresponds to the TCSC Kayenta Substation installed in Arizona. More details on the simulation setup are described in [188, 189].

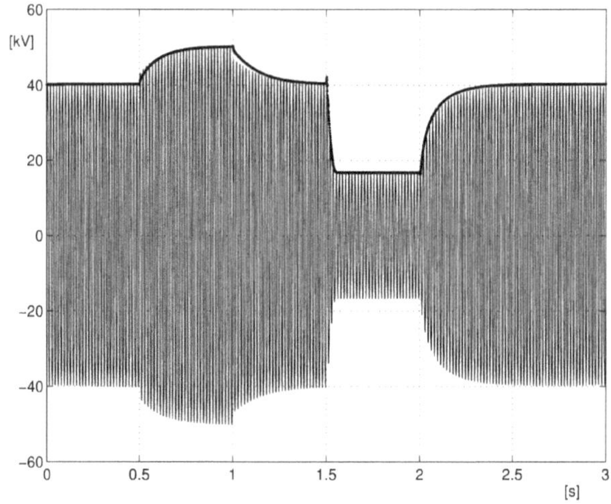

Fig. 6.11 Response of capacitor voltage in circuit of Fig. 6.10 to step changes in line current magnitude and firing angle. The plot starts with the steady-state corresponding to $\mid I_\ell \mid = 1\,\mathsf{kA}$ and $\gamma = 57°$; at $t = 0.5\,\mathsf{sec}$ a line current step change to $\mid I_\ell \mid = 1.25\,\mathsf{kA}$ is introduced, which goes back to $\mid I_\ell \mid = 1\,\mathsf{kA}$ at $t = 1\,\mathsf{sec}$. A firing angle step change from $\gamma = 57°$ to $\gamma = 70°$ and back occurs at $t = 1.5\,\mathsf{sec}$ and $t = 2\,\mathsf{sec}$, respectively.

Figure 6.11 shows the voltage response for changes in the magnitude of the line current and in the firing angle; predictions of the dynamic phasor model are shown as the envelope of the response (amplitude). The plot displays results from the time domain and the full fundamental phasor model and starts with the steady-state corresponding to $\mid I_\ell \mid = 1\,\mathsf{kA}$ and $\gamma = 57°$. At $t = 0.5\,\mathsf{sec}$ we introduce a line current step change to $\mid I_\ell \mid = 1.25\,\mathsf{kA}$, which goes back to $\mid I_\ell \mid = 1\,\mathsf{kA}$ at $t = 1\,\mathsf{sec}$. There follows a firing angle step change from $\gamma = 57°$ to $\gamma = 70°$ and back at $t = 1.5\,\mathsf{sec}$ and $t = 2\,\mathsf{sec}$, respectively. This sequence of perturbations is taken from [5]. The phasor model tracks the envelope of the actual time-domain waveform of Fig. 6.11 very well.

We face a methodological barrier when trying to establish relevance of our nonlinear full and reduced models at frequencies of interest in SSR studies. This barrier stems from the lack of a unified frequency-domain theory for the class of nonlinear phasor models that we have derived. Our validation plan is thus two-pronged – start with small-signal models and then complement their results with selected large-signal simulations. We linearize (6.26) and (6.28) and compare the relevant transfer functions with those obtained from EMTP simulations. For the EMTP simulations, the line was modeled as a three-phase current source, with small perturbations in the real (I_{d1}) and imaginary ($-I_{q1}$) parts of the line current phasor I_ℓ. Measuring the amplitude and phase-shift of the V_{d1} and V_{q1} components of the capacitor voltage v, we obtain quantities that can be compared with predictions of our models (Fig. 6.12).

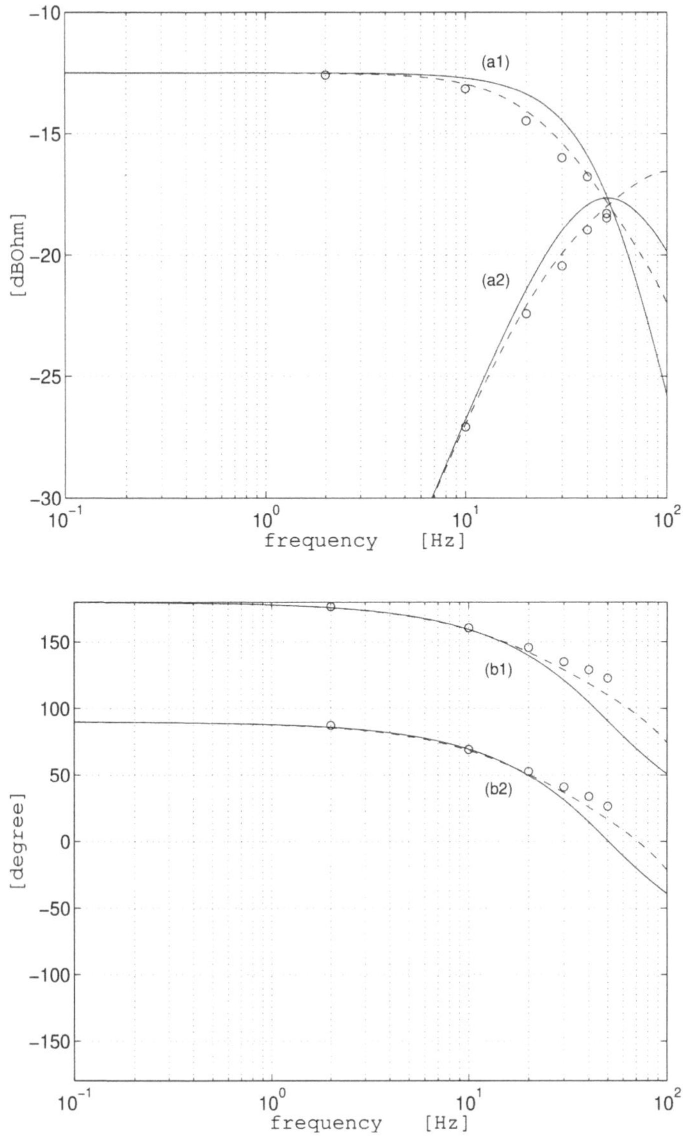

Fig. 6.12 Amplitude modulation of the line current. Top plot: magnitude of V_{d1}/I_{q1} (a1) and V_{q1}/I_{q1} (a2). Bottom plot: phase of V_{d1}/I_{q1} (b1) and V_{q1}/I_{q1} (b2). Symbol "o" refers to results of EMTP simulations, while the continuous lines are from linearizations of the dynamic phasor model: solid line for the reduced model and dashed line for the full model.

In this set of simulations, the turn-on of the thyristors was synchronized to the line current i_ℓ, and open-loop operation was assumed. The operating point is $\sigma^{\text{ref}} = 40°$, $I_{d1}^{\text{ref}} = 0$, and $I_{q1}^{\text{ref}} = 1$. The TCSC parameters correspond to the IEEE first benchmark [202], shown schematically in Fig. 6.13.

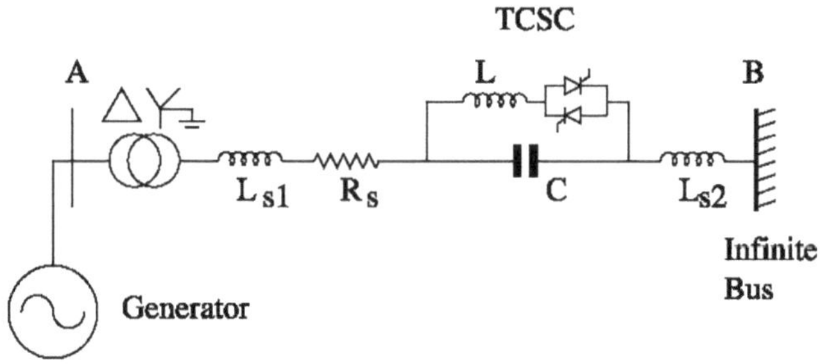

Fig. 6.13 IEEE first benchmark test with TCSC.

When embedding our TCSC model in larger power system models that involve synchronous generators, we have to address a coordinate transformation issue. While the TCSC model is in the synchronous reference frame, the generator models (after Park transformation) are in the rotor frame. Then additional transformation [204] is used to account for the motion of the rotor frame with respect to the synchronous reference frame. The resulting model is nonlinear but time-invariant and can be linearized by standard methods for use in SSR eigenvalue analysis.

The electrical part of the system is shown in Fig. 6.13. The TCSC provides 29% series compensation when the conduction angle is zero. The turbine-generator shaft is modeled as six rotational masses with torsional springs. Recalling the notation from [202], there are five torsional modes — TM1 through TM5 — for the mechanical system, with respective frequencies at approximately 16, 20, 25, 32, and 47 Hz and the swing mode TM0.

In Fig. 6.14 solid lines correspond to the damping of modes TM0 through TM4 with synchronization with respect to the line current zeros. The dashed lines correspond to a TCSC model with the phase correction (6.29). Symbol "o" comes from the sampled-data eigenvalue analysis, which is based on *equidistant* synchronization. The discrepancies between models are attributed to different switching policies.

Next, we focus on the operating range of σ greater than approximately 35° where our analysis of line current-based synchronization suggests instability (note the negative damping for several modes, most notably TM1), but the equidistant synchronization predicts stable operation. We thus perform a large-signal EMTP simulation that repeats the numerical experiment in [202], but with line current-based synchronization. Specifically:

(a) The thyristors are blocked at time 0, and a simultaneous three-phase line-to-ground bus fault is simulated at point A of Fig. 6.13 at 117 msec and cleared after 4.5 cycles.
(b) At 2 sec, the thyristors are commanded to operate with $\sigma^{\mathrm{ref}} = 40°$, synchronized to the zero crossings of the line current.

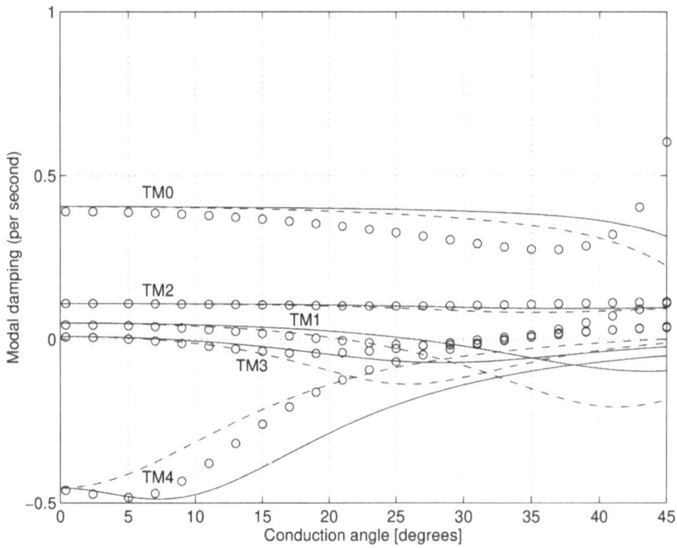

Fig. 6.14 Modal dampings when synchronization is with respect to the *line current zeros*: solid lines for the TCSC model without the phase correction (6.29) and dashed lines with the phase correction. Symbol "o" corresponds to the sampled-data eigenvalue analysis for the *equidistant* synchronization.

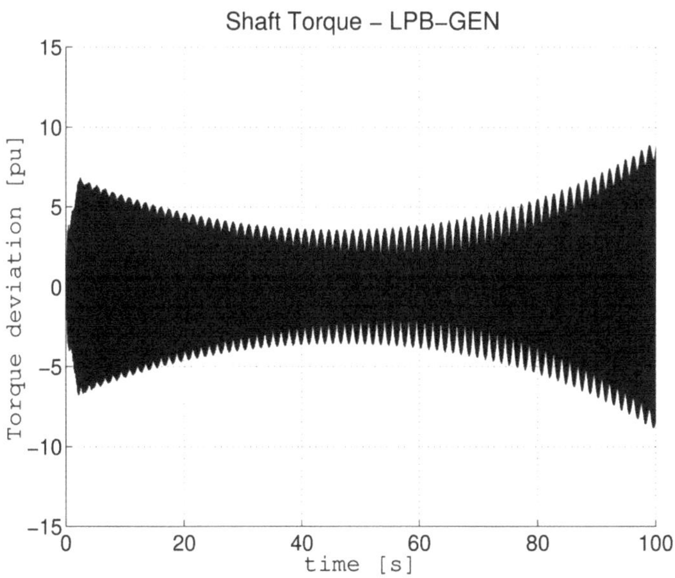

Fig. 6.15 Shaft response from EMTP simulation when synchronization is with respect to the *line current zeros*, showing growing oscillations consistent with the behavior predicted by our eigenvalue analysis results for the line current synchronization.

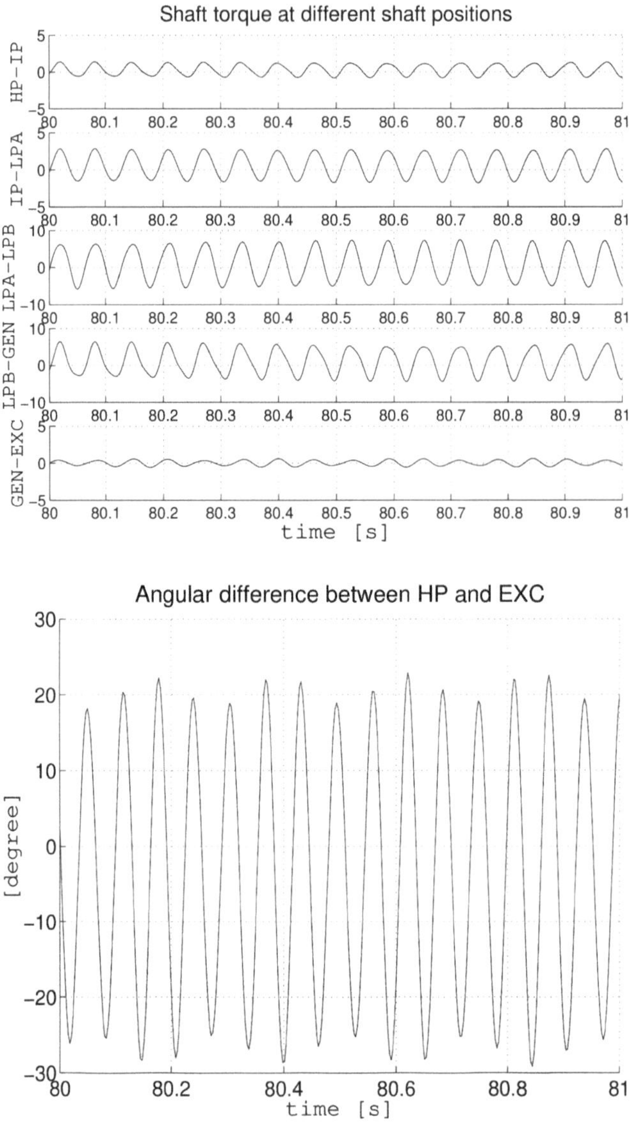

Fig. 6.16 Expanded view (between 80 sec and 81 sec) of torque variations (in pu) at various locations along the shaft (upper panel), and variations in angular difference between the two ends of the shaft (lower panel), obtained from EMTP simulation when synchronization is with respect to the line current zeros.

In this case the instability develops after a considerable time, as seen in Fig. 6.15 which shows the torsional response of the shaft. Note that a simulation that ended at 40 sec would likely (and erroneously) conclude that the system is poorly damped but stable. The growing oscillations evident after 50 sec are consistent with the behavior predicted by our eigenvalue analysis results for line current synchronization, see Fig. 6.14. To further explore details of SSR oscillations, Fig. 6.16 focuses on a one-second time interval (between 80 and 81 sec) of torque variations at various locations along the shaft, as well as the variations in angular difference between the two ends of the shaft. The dominant unstable mode TM1 (16 Hz) is clearly evident, in agreement with predictions in Fig. 6.14 (dashed line, conduction angle $\sigma = 40°$). In actual operation, the angular differences between ends of a shaft of this magnitude would be incompatible with continued operation.

6.4 Nonlinear Control of TCSC

6.4.1 Model Extensions

Control applications of dynamic phasor models are numerous, and more are emerging continuously. Many follow the standard path of designing controllers for nonlinear systems, in which a system model is linearized, tools of linear control synthesis are used to produce a candidate solution, and the verification consists of full (nonlinear) system simulations and possibly experiments. This workflow is a testament to the sophistication and industrial relevance of linear control that, together with clever engineering in system architecture, remains the driving force behind much of energy processing. In this section, however, we review approaches that make nonlinear dynamic phasor models (e.g., of TCSC) compatible with some nonlinear control techniques [218–220].

Recall that the TCSC model in [188] contains a mix of a control and a state-dependent term that is then passed through the static nonlinearity that models the effective capacitance. The new model is obtained starting from (6.26)–(6.28), with alterations that make it a better fit to common formulations in nonlinear control design which are written in terms of real quantities, and often bilinear or linear in terms of the control input. Following [218–220], we rewrite the model as

$$C\frac{d}{dt}V = I_\ell - \mathcal{J}\omega_0 C_{eff}(\beta)V \tag{6.30}$$

where C is the capacitance, ω_0 is the line fundamental frequency, vectors V and I_ℓ are the fundamental dynamic phasors for the capacitor voltage v and the line current i_ℓ, respectively. In this rewriting we use matrix $\mathcal{J} = -\mathcal{J}^\top = \begin{bmatrix} 0 & -1 \\ 1 & 0 \end{bmatrix}$ instead of the complex number j, so that phasor quantities are expressed as real vectors $V = [V_1, V_2]^\top$ and $I_\ell = [I_{\ell 1}, I_{\ell 2}]^\top \in \mathbb{R}^2$ (with real and imaginary parts of the corresponding complex phasor as the entries), and $C_{eff}(\beta)$ is the equivalent capacitance of the circuit in terms of an angle β that equals half of the prevailing conduction angle (i.e., $\beta = \sigma/2$). Thus $\omega_0 C_{eff}(\beta)$ is the *effective* (or apparent) admittance of the TCSC, which in turn is the actual control input.

As before, $C_{eff}(\beta)$ is given by the expression (recall (6.27))

$$C_{eff}(\beta) = C\left[1-\frac{2}{\pi}\left(\mathcal{S}\left[\beta+\frac{\sin(2\beta)}{2}\right]+2\left(\frac{\mathcal{S}}{\eta}\right)^2\left[\tan(\beta)-\eta\tan(\eta\beta)\right]\cos^2\beta\right)\right]^{-1} \tag{6.31}$$

where we have defined $\eta - \frac{1}{\omega_0\sqrt{LC}}$ and $\mathcal{S} = \frac{\eta^2}{\eta^2-1}$. Recall that the transient behavior is captured via the prevailing angle

$$\beta = \beta^{\mathrm{ref}} + 2\phi \ , \quad \phi = \arctan\left(\frac{V_2}{-V_1}\right) \tag{6.32}$$

which fuses the control input β^{ref} and the state-dependent angle ϕ. For control purposes, we are interested in "untangling" the two components; we show that such an approximation amounts to adding a nonlinear term whose role is to add damping, which is inherent to many thyristor circuits [10].

Consider the following approximation:

$$C_{eff}(\beta) = C_{eff}(\beta^{\mathrm{ref}} + 2\phi) = C_{eff}(\beta^{\mathrm{ref}}) + 2\phi \left.\frac{\partial C_{eff}}{\partial\beta}\right|_{\beta^{\mathrm{ref}}}$$

Notice that, in the capacitive region,

$$f(\beta^{\mathrm{ref}}) \stackrel{\mathrm{def}}{=} \left.\frac{\partial C_{eff}}{\partial\beta}\right|_{\beta^{\mathrm{ref}}} < 0 \qquad \text{for all } \beta$$

For $\phi \ll 1$ we can approximate $\phi = \arctan\left(\frac{V_2}{-V_1}\right) \cong -\frac{V_2}{V_1}$. The voltage subsystem equation is now

$$\begin{aligned} C\frac{d}{dt}V &= I_\ell - \mathcal{J}\omega_0 C_{eff}(\beta) \\ &\cong I_\ell - \mathcal{J}\omega_0 C_{eff}(\beta^{\mathrm{ref}})V + 2\mathcal{J}\omega_0\frac{V_2}{V_1}f(\beta^{\mathrm{ref}})V \end{aligned}$$

The third term on the right-hand side can be written as

$$\begin{aligned} 2\mathcal{J}\omega_0\frac{V_2}{V_1}f(\beta^{\mathrm{ref}})V &= 2\omega_0\frac{V_2}{V_1}f(\beta^{\mathrm{ref}})\begin{bmatrix}-V_2\\ V_1\end{bmatrix} \\ &= \underbrace{2\omega_0 f(\beta^{\mathrm{ref}})\begin{bmatrix}-\frac{V_2^2}{V_1^2} & 0\\ 0 & 1\end{bmatrix}}_{\mathcal{R}(V,\,\beta^{\mathrm{ref}})}\begin{bmatrix}V_1\\ V_2\end{bmatrix} \equiv \mathcal{R}(V,\,\beta^{\mathrm{ref}})\,V \end{aligned}$$

Notice from the structure of $\mathcal{R}(V,\ \beta^{\text{ref}})$ that such a nonlinear term is adding damping to the second row, but not to the first row; thus the controller design needs to add damping in the dynamics of V_1.

We observe that $\mathcal{R}(V,\ \beta^{\text{ref}})$ depends on both the control and the state. One possible path for synthesizing control relies on a second step in which $\mathcal{R}(V,\ \beta^{\text{ref}})$ is replaced by its value at the desired equilibrium (which is linear in terms of the state)

$$2\omega_0 f(\beta^{\text{ref}}) \begin{bmatrix} -\frac{V_2^2}{V_1^2} & 0 \\ 0 & 1 \end{bmatrix} V \quad \rightarrow \quad -\underbrace{\begin{bmatrix} 0 & 0 \\ 0 & 2\omega_0 \left| f\left(\overline{\beta^{\text{ref}}}\right)\right| \end{bmatrix}}_{\mathcal{R}\left(\overline{\beta^{\text{ref}}}\right)} V$$

where $\mathcal{R}\left(\overline{\beta^{\text{ref}}}\right) = \text{diag}\{0,\ R\} \geq 0$ is a positive semidefinite matrix, $R = 2\omega_0 \left| \left(\overline{\beta^{\text{ref}}}\right)\right|$, and the constant $\overline{\beta^{\text{ref}}}$ is the equilibrium value of β^{ref}. The system model then becomes

$$C\frac{d}{dt}V = I_\ell - \mathcal{J}\omega_0 C_{\mathit{eff}}(\beta^{\text{ref}})V - \mathcal{R}\left(\overline{\beta^{\text{ref}}}\right)V \tag{6.33}$$

The bilinear structure becomes clear when we define $u \overset{\text{def}}{=} \omega_0 C_{\mathit{eff}}(\beta^{\text{ref}})$, i.e., the effective admittance, as the control input. The *control objective* consists of finding the controller u (and hence an angle β^{ref} through the mapping $u(\beta^{\text{ref}})$) such that regulation of the voltages vector $V = [V_1,\ V_2]^\top$ is guaranteed toward a constant reference $V^{\text{ref}} = [V_1^{\text{ref}},\ 0]^\top$. Notice that the effective admittance u is restricted to take positive values if the TCSC system is to behave as a capacitor. We assume that the line current equals $i_\ell(t) = I_{\ell 2}\sin(\omega_0 t)$, with $I_{\ell 2}$ a positive constant. Thus, $I_\ell = [0\ ,\ -I_{\ell 2}]^T$, implying $V^{\text{ref}} = [\,V_1^{\text{ref}},\ V_2^{\text{ref}}\,]^T = [\,-\frac{I_{\ell 2}}{\omega_0 C_{\text{eff}}(\beta^{\text{ref}})}\ ,\ 0\,]^T$.

In all cases a key control requirement is that the response of the system should be (significantly) faster than the response obtained by operating the system in open loop. Moreover, adaptation can be added to address parametric uncertainties. Four nonlinear controllers stemming from different approaches are described in [218–220]:

(i) An approximate feedback linearization control (FLC) – the stored energy in the capacitor is the linearizing output, but the procedure is made challenging by some parameters that are unknown a priori (hence the term "approximate").

(ii) A sliding surface design – we define a surface on the state space which is a linear combination of the states. We prove that this surface contains the desired equilibrium and that all trajectories starting on this surface converge toward the desired equilibrium point. This design procedure is similar to the sliding mode approach, except that we are not requiring the system to create a sliding regime (possibly by applying high gain).

(iii) Control in transformed coordinates – (6.33) can be transformed by polar coordinates in the form in which the control enters linearly; a nonlinear PI control design is described in [219].

(iv) Immersion and invariance approach (I&I) – the method relies on notions of system *immersion* and *manifold invariance*. It can be used to design stabilizing control laws for nonlinear systems without necessarily requiring the knowledge of a control Lyapunov function.

6.4.2 Simulation Results

We again use a system model with the following parameters: $L = 6.8\,\text{mH}$ and $C = 176.80\,\mu\text{F}$, at a constant frequency $f = 60\,\text{Hz}$. These parameters correspond to the Kayenta substation [178]. The intervals where the apparent admittance u is restricted are given by

$$\begin{aligned} 0.0151 &\le u \le 0.0662 \quad \text{(capacitive region)} \\ -0.2469 &\le u \le -0.0182 \quad \text{(inductive region)} \end{aligned}$$

The line current has been set to $i_\ell = 600\sin(\omega_0 t)$ Amp. The numerical experiment involves the switched circuit model and consists of fixing a reference for the apparent admittance $u^{\text{ref}} = 0.06\,℧$ at the beginning, and then after time $t = 0.2$ sec, we switch to a new reference $u^{\text{ref}} = 0.03\,℧$. This corresponds to a step change in the desired voltage reference V_1^{ref} going from -10 kV to -20 kV.

The responses of the first three controllers proposed, namely (i) approximate FLC, (ii) SMC-like design, and (iii) transformed (polar) coordinates, are quite similar. We show the simulation responses of state V_1 in Fig. 6.17 where it can be observed that the first three proposed controllers guarantee zero steady-state error and a response faster than the open-loop response. The transient responses for the current $i(t)$, voltage $v(t)$ and line current $i_\ell(t)$ are shown in Fig. 6.18.

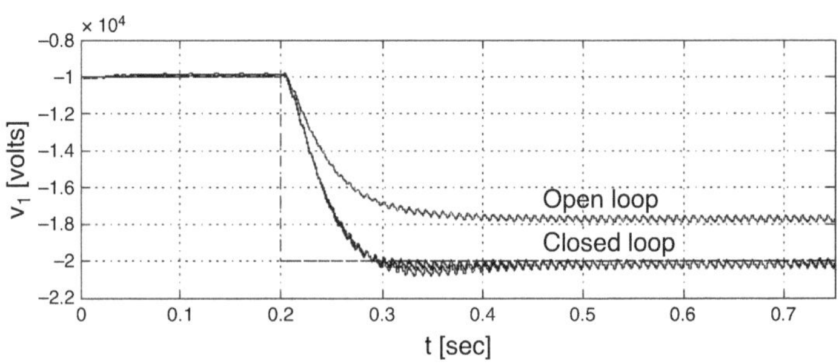

Fig. 6.17 (–) Response of $V_1(t)$ for the system in open loop and under the first three proposed controllers and (- -) reference $V_1^{\text{ref}}(t)$.

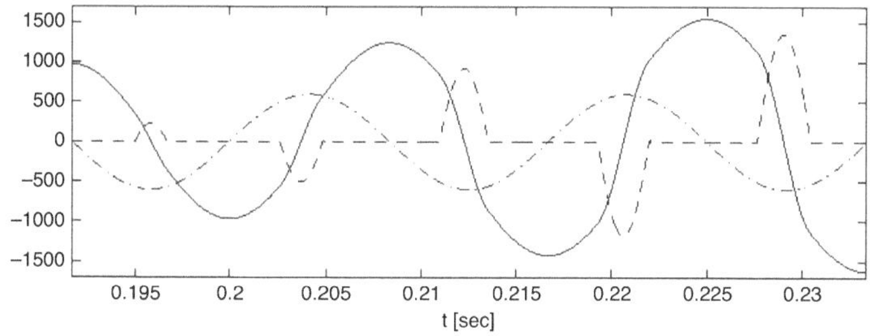

Fig. 6.18 Approximate FLC: (–) scaled voltage $0.1 * v(t)$, (- -) current across the inductor $i(t)$, and (-·) line current $i_\ell(t)$.

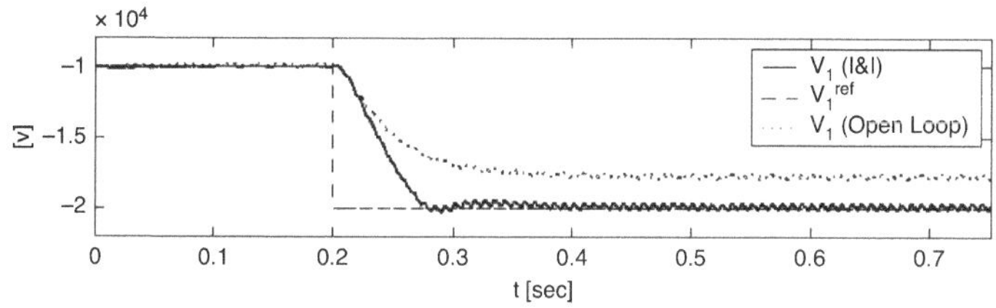

Fig. 6.19 Comparison between responses of phasor voltage V_1 (–) with I&I controller and (··) in open loop, with (--) reference V_1^{ref}.

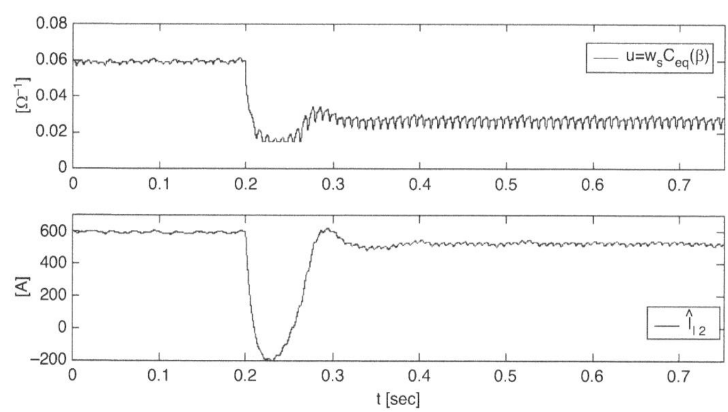

Fig. 6.20 I&I controller: transient response of **(top)** control signal $u(t) = \omega_0 C_{eff}(\beta)$ and **(bottom)** estimate $\hat{I}_{\ell 2}$.

In Fig. 6.19 we present a comparison between the response of voltage V_1 obtained with the proposed I&I controller (in solid line) and the one obtained in open loop (dotted line). We observe that the response of the proposed I&I controller results in a steady-state error that is close to zero.

In Fig. 6.20 we show the control signal $u(t) = \omega_0 C_{eff}(\beta(t))$ (top panel) that results from the application of the I&I controller. This signal has been

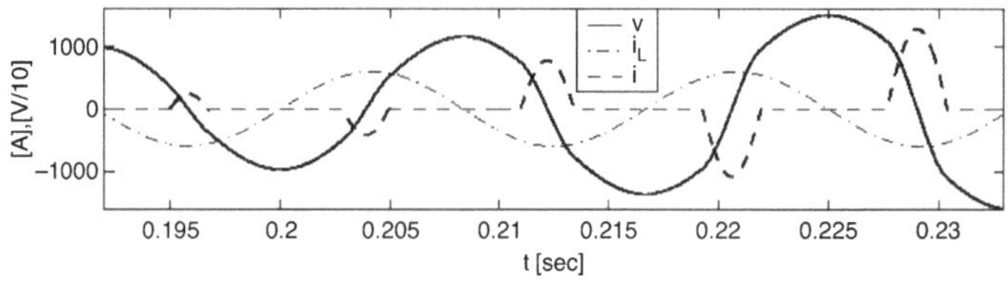

Fig. 6.21 I&I controller: transient responses of (–) voltage $v(t)$ (scaled V/10), (--) current across the inductor $i(t)$, and (-·) line current $i_\ell(t)$.

constrained by the saturation restriction to keep the TCSC within the capacitive operating region, which clearly limits the speed of response. We observe also that the parameter estimate $\hat{I}_{\ell 2}$ (in the bottom panel) converges toward a constant value after a relatively small transient. Finally, the transient responses for current $i(t)$, voltage $v(t)$, and line current $i_\ell(t)$ are shown in Fig. 6.21, where the typical shapes generated by the switched circuit can be observed.

6.5 Further Reading

A Very Abridged History: The analytical tools behind dynamic phasors were introduced in the groundbreaking paper [45]. They share one feature with other useful ideas in engineering in that their historical and intellectual roots reach into many other areas. Among those are ideas from dynamical system averaging [221], frequency-based approaches to oscillations [222], describing functions [43,44], and, closer to home, AC circuit and electrical machine analysis [223–226]. The syntagm "dynamic phasor" was coined by George Verghese in mid-1997 and appeared in titles of [189, 205]; a different orderings of words, such as "phasor ... dynamics," was used before. The extension to dynamic symmetrical components occurred roughly at the same time [213, 227]. A multi-scale extension of dynamic phasors is the development of phasor banks [228].

Recent and Ongoing Developments: The issue of the possible variation of the fundamental component was brought up in the original reference that introduced the dynamic phasors [45], where the authors derive a delay-differential equation to evaluate the time derivative of the phasor. Little advancement seems to have been made in deriving customized approximation; a straightforward but surprisingly accurate option is to simply let ω_0 be a function of time [211, 229]. A number of references addressed common nonlinearities (e.g., magnetic) seen in energy conversion; a key ingredient in evaluating the "right-hand side" of such models (or k-th average) is provided by the describing function. An extensive list of common engineering nonlinearities is covered in [43]. Dynamic phasors have been used to model modern architectures in power electronics such as modular multilevel converters, which play a key role in bringing capabilities of power electronics, such as speed and precision of the response, into power systems [230, 231]. Numerical algorithms that have been specifically tailored for dynamic phasor simulations have been explored in [232]. There is an increasing need to integrate phasor-type simulation tools with pure (or "point-on-wave") time-domain approaches. Dynamic phasors are uniquely suited to fulfill this role [233, 234]. Software tools that are based on dynamic phasors have emerged over the last decade; examples include [235, 236]. A systematic approach to estimate dynamic phasors from a few time-domain samples, with concomitant reduction in the computational effort and storage requirements, was introduced in [77] and further extended as the "sub-cycle dynamic phasors" in [78, 80]. Sub-cycle calculations in practical real-time detection applications suggest that it may be advantageous to calculate at least two sets of phasors to cover both slow and fast transients. The computational cost of such a scheme is still significantly lower than that of a standard full-cycle calculation.

Chapter 7 Electric Machines and Power Systems

In this chapter, we introduce two more classes of applications of dynamic phasors, namely electric machines and power systems. These two application domains are very broad so the presented use cases are far from exhaustive – they are simply meant to illustrate the capabilities of the dynamic phasor modeling approach.

The electric machines section starts with single-phase machines and later considers the polyphase case. The power systems section focuses on unbalanced operation and synchronous generator modeling. As in the previous chapter, we use here exclusively the two-sided form (1.2) of Fourier coefficients.

7.1 Applications to Electric Machines

7.1.1 Single-Phase Electric Machines

In this subsection, we present application of dynamic phasors to modeling of single-phase induction machines. The single-phase induction motor is used very widely, for example, in household and commercial appliances such as refrigerators, washers, driers, fans, and blowers. The analysis problem for this class of machines is made challenging by the lack of rotating magnetic field, so assumptions such as the one about the electromagnetic torque having negligible harmonic content need to be revisited. Our presentation here follows [205].

In this type of motor, a sinusoidal stator current alone produces a pulsating magnetic field and not a revolving one. If, for example, the machine is at standstill and in the absence of an electrical starting mechanism, this excitation will not provide any net torque.

The widely-used "rotating field theory" states that this pulsating wave is equivalent to two magnetic fields (forward and backward) rotating in opposite directions. Now the effectiveness of rotating field theory can be brought to bear on the problem, as each component acts upon the rotor as it would for a balanced polyphase

H. Lev-Ari, A. M. Stanković, *Dynamic Phasors in Energy Processing Systems*,
Power Electronics and Power Systems, https://doi.org/10.1007/978-3-031-50020-6_8

induction motor. The intuitive explanation is that while the two components are equal at standstill, if the motor is made to rotate in one direction (denoted as forward) by additional means (e.g., a capacitor or an additional winding),then the induced currents in the rotor will amplify the forward field and diminish the backward field. The torque components due to the forward- and backward-rotating fields are typically added to obtain the total (average) torque. The simplicity of this approach is perhaps the main reason for its popularity in explaining and deriving equivalent circuits for single-phase induction machines [206–209]. Please note, however, that an electric machine, even with linear magnetics, is a nonlinear system because of current-speed products in the dynamical model. In more practical terms, doubling the voltage input will not exactly double stator current and machine speed. Thus, the rotating field theory in this case applies the superposition principle to a nonlinear system, and we are interested in quantifying the possible error.

Assuming a sinusoidal stator voltage and shorted rotor windings, the following model describes a single-phase induction motor (in a stationary reference frame [210]):

$$\begin{aligned}
\sqrt{2}V\cos(\omega_0 t) &= r_s i_{qs} + \frac{X_s}{\omega_0}\frac{di_{qs}}{dt} + \frac{X_m}{\omega_0}\frac{di_{qr}}{dt} \\
0 &= r_r i_{qr} - \frac{\omega_r}{\omega_0} X_r i_{dr} + \frac{X_r}{\omega_0}\frac{di_{qr}}{dt} + \frac{X_m}{\omega_0}\frac{di_{qs}}{dt} \\
0 &= r_r i_{dr} + \frac{\omega_r}{\omega_0}\left(X_r i_{qr} + X_m i_{qs}\right) + \frac{X_r}{\omega_0}\frac{di_{dr}}{dt} \\
\frac{2H}{\omega_0}\frac{d\omega_r}{dt} &= -T_L - B_f\omega_r + X_m i_{qs} i_{dr}
\end{aligned} \tag{7.1}$$

where i_{qs} represents the stator current, i_{dr} and i_{qr} are the rotor winding currents referenced to the stator, and ω_r is the rotor speed. The model (7.1) can be written in a state-space form to better reveal its structure, including multiplicative nonlinearities: if we define the state $x = [i_{qs}\ i_{qr}\ i_{dr}\ \omega_r]^T$, then the model becomes

$$\begin{aligned}
\dot{x}_1 &= a_{11}x_1 + a_{12}x_2 + a_{13}x_3x_4 + b_1 cos(\omega_0 t) \\
\dot{x}_2 &= a_{21}x_1 + a_{22}x_2 + a_{23}x_3x_4 + b_2 cos(\omega_0 t) \\
\dot{x}_3 &= a_{31}x_1x_4 + a_{32}x_2x_4 + a_{33}x_3 \\
\dot{x}_4 &= a_{41}x_1x_3 + a_{44}x_4 + b_4 T_L
\end{aligned} \tag{7.2}$$

To develop our dynamic phasor model, we will assume a constant load torque T_L and keep first (± 1) harmonic for all currents $x_1 - x_3$ and dc component and second (± 2) harmonic for speed x_4. This set of assumptions can be re-examined and justified via simulations.

Let $X_{1,1} - X_{3,1}$ denote dynamic phasors corresponding to first (fundamental) harmonics of states $x_1 - x_3$, and let $X_{4,0}$ and $X_{4,2}$ denote the zeroth (dc) and second phasor of the speed x_4, respectively. Note that the first subscript identifies the state variable, while the second subscript denotes the harmonic number. We can then rewrite our model (7.2) as

$$\begin{aligned}
\dot{X}_{1,1} &= (a_{11} - j\omega_0)X_{1,1} + a_{12}X_{2,1} + a_{13}X_{3,1}X_{4,0} + a_{13}(X_{3,1})^*X_{4,2} + \frac{b_1}{2} \\
\dot{X}_{2,1} &= a_{21}X_{1,1} + (a_{22} - j\omega_0)X_{2,1} + a_{23}X_{3,1}X_{4,0} + a_{23}(X_{3,1})^*X_{4,2} + \frac{b_2}{2} \\
\dot{X}_{3,1} &= a_{31}X_{1,1}X_{4,0} + a_{31}(X_{1,1})^*X_{4,2} + a_{32}X_{2,1}X_{4,0} \\
&\qquad + a_{31}(X_{2,1})^*X_{4,2} + (a_{33} - j\omega_0)X_{3,1} \\
\dot{X}_{4,0} &= 2a_{41}\Re(X_{1,1}(X_{3,1})^*) + a_{44}X_{4,0} + b_4T_L \\
\dot{X}_{4,2} &= a_{41}X_{1,1}X_{3,1} + (a_{44} - j2\omega_0)X_{4,2}
\end{aligned} \tag{7.3}$$

where we recall that the superscript * denotes complex conjugation. This fifth-order model involves four complex quantities, and it can be replaced by a ninth-order model involving real quantities (since $X_{4,0}$ is real).

We refer the interested reader to [205] for validation of (7.3) and its linearization via simulation of a benchmark model from the literature. Here we focus on a derivation of an equivalent circuit that models the motor in steady-state. To recover the commonly used circuit representations, we introduce a coordinate transformation for the rotor current, viz.,

$$\begin{bmatrix} I_s \\ I_+ \\ I_- \end{bmatrix} = \frac{1}{\sqrt{2}} \begin{bmatrix} 2 & 0 & 0 \\ 0 & 2 & -2j \\ 0 & 2 & 2j \end{bmatrix} \begin{bmatrix} X_{1,1} \\ X_{2,1} \\ X_{3,1} \end{bmatrix} \tag{7.4}$$

and an analogous change of coordinates for the voltage quantities. The resulting steady-state phasor equations are

$$\begin{aligned}
V &= r_sI_s + jX_sI_s + j\frac{X_m}{2}(I_+ + I_-) \\
0 &= \frac{r_r/2}{(1 - \frac{<\omega_r>0}{\omega_0})}I_+ + j\frac{X_r}{2}I_+ + j\frac{X_m}{2}I_s + j\frac{X_r}{2}m_+I_-^* + j\frac{X_m}{2}m_+I_s^* \\
0 &= \frac{r_r/2}{(1 + \frac{<\omega_r>0}{\omega_0})}I_- + j\frac{X_r}{2}I_- + j\frac{X_m}{2}I_s + j\frac{X_r}{2}m_-I_+^* + j\frac{X_m}{2}m_-I_s^*
\end{aligned} \tag{7.5}$$

where

$$m_+ = \frac{\frac{<\omega_r>_2}{\omega_0}}{1 - \frac{<\omega_r>_0}{\omega_0}} \quad \text{and} \quad m_- = \frac{\frac{-<\omega_r>_2}{\omega_0}}{1 + \frac{<\omega_r>_0}{\omega_0}} .$$

This steady-state model has the circuit representation shown in Fig. 7.1 (with $X_r = X_m + X_{lr}$, $X_s = X_m + X_{ls}$).

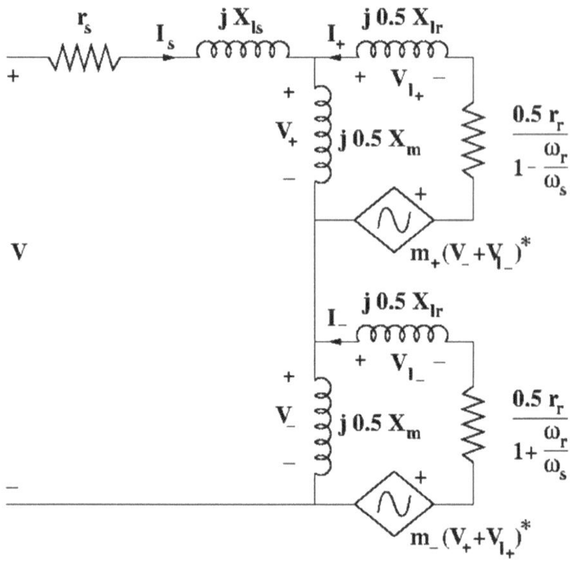

Fig. 7.1 A complete equivalent circuit of a single-phase induction motor in steady-state.

If we neglect the second harmonic in the speed, then model (7.5) simplifies (as $m_+ = m_- = 0$), and the standard equivalent circuit for the single-phase induction machine is recovered. In the traditional interpretation, the I_+ and I_- currents are associated with the forward- and reverse-rotating magnetic fields. This simplified circuit is well known, but our derivation results in a more accurate model, and makes no superposition assumptions involving the forward- and backward-rotating fields. It turns out that the complete equivalent circuit better predicts overall motor efficiency, at least in the example presented in [205]. Please note that further refinements are possible, as more harmonic components can be included. This is particularly useful in studies of harmonics introduced by the mechanical subsystem, for example in the case of position-dependent loads such as some types of pumps [211].

7.1.2 Polyphase Electric Machines

We start by deriving a dynamical phasor model for a three-phase induction machine, possibly with unbalanced supply. Our starting point is the standard model of a squirrel cage induction machine from [212, p.152], expressed in terms of space vectors in the stationary coordinate frame (this model neglects the magnetic saturation and slot harmonics):

$$\begin{aligned}\vec{v}_s &= (r_s + L_s \frac{d}{dt})\vec{i}_s + L_m \frac{d}{dt}\vec{i}_r \\ 0 &= L_m \frac{d}{dt}\vec{i}_s + (r_r + L_r \frac{d}{dt})\vec{i}_r - j\omega_r \frac{P}{2}(L_m \vec{i}_s + L_r \vec{i}_r) \\ J\frac{d}{dt}\omega_r &= \frac{3P}{4} L_m \Im(\vec{i}_s \vec{i}_r^{\,*}) - B\omega_r - T_L \end{aligned} \tag{7.6}$$

where the subscript "s" corresponds to stator quantities, the subscript "r" corresponds to rotor quantities, and $\Im(\cdot)$ denotes the imaginary part of a complex quantity. As before, in electrical equations, nonlinearity appears in the form of speed–current products, while in the mechanical equation it appears in the form of a current product.

Recall that space vectors are a tool for modeling electrical machines and are particularly useful in the case of three-conductor supply arrangement (so that zero-sequence components can be treated separately). Complex space vectors are defined as (note that the phase quantities are assumed to be *real*)

$$\vec{x}(t) \stackrel{\text{def}}{=} \frac{2}{3}\Big(x_a(t) + \alpha x_b(t) + \alpha^* x_c(t)\Big). \tag{7.7}$$

From the mathematical standpoint, these are complex *scalars* and have the potential to encode two-dimensional information, which is important in cylindrical structures such as standard electric machines. Scalings other than $\frac{2}{3}$ are, of course, possible, but this one is useful in three-phase PWM.

One advantage of the space vector concept in electric drive applications is the ease of dealing with rotational coordinate transformations, as they amount to premultiplications with complex exponentials [212]. It can be shown [213] that the frequency-domain representation of a space vector is

$$\vec{x}(\tau) = \frac{2}{\sqrt{3}} \sum_{\ell=-\infty}^{\infty} e^{j\ell\omega_0\tau} X_{p,\ell}(\tau) \tag{7.8}$$

Since the phase quantities are real-valued, we have $X_{p,-\ell} = X_{n,\ell}^*$, for all integer ℓ, and $\vec{x}$ encodes the information about positive and negative sequence quantities (recall that in general $X_{p,\ell} \neq X_{p,-\ell}$); zero-sequence variables have to be treated separately. In the special case of $\ell = \pm 1$ and no zero-sequence component

$$\vec{x}(\tau) = X_{p,1}(\tau)e^{j\omega_0\tau} + X_{n,1}^*(\tau)e^{-j\omega_0\tau}$$

We assume that currents contain $\pm$1-st harmonic ($I_{p,s}$, $I_{p,r}$ and $I_{n,s}^*$, $I_{n,r}^*$, respectively) and that the mechanical speed contains dc (zero-th) component $\Omega_{r,0}$

and 2-nd harmonic $\Omega_{r,2}$ (due to torque ripple). As before, this is a decision that can be motivated by simulations of the time-domain model (7.6). If more current harmonics and consequently more torque and speed harmonics are included, the model accuracy will improve [9], but the model size and complexity will also increase. Recalling that the time-domain multiplication corresponds to discrete convolution of phasors, we obtain

$$\begin{aligned} V_p &= \Big(r_s + j\omega_0 L_s + L_s\frac{d}{dt}\Big)I_{p,s} + \Big(j\omega_0 L_m + L_m\frac{d}{dt}\Big)I_{p,r} \\ 0 &= \Big(j\omega_0 L_m + L_m\frac{d}{dt}\Big)I_{p,s} + \Big[r_r + \Big(j\omega_0 L_r + L_r\frac{d}{dt}\Big)\Big]I_{p,r} \\ &\quad - j\frac{P}{2}\Big[\Omega_{r,0}\big(L_m I_{p,s} + L_r I_{p,r}\big) + \Omega_{r,2}\big(L_m I_{n,s}^* + L_r I_{n,r}^*\big)\Big] \end{aligned} \tag{7.9a}$$

$$\begin{aligned} V_n^* &= \Big(r_s - j\omega_0 L_s + L_s\frac{d}{dt}\Big)I_{n,s}^* + \Big(-j\omega_0 L_m + L_m\frac{d}{dt}\Big)I_{n,r}^* \\ 0 &= \Big(-j\omega_0 L_m + L_m\frac{d}{dt}\Big)I_{n,s}^* + \Big[r_r + \Big(-j\omega_0 L_r + L_r\frac{d}{dt}\Big)\Big]I_{n,r}^* \\ &\quad - j\frac{P}{2}\Big[\Omega_{r,0}(L_m I_{n,s}^* + L_r I_{n,r}^*) + \Omega_{r,2}^*\big(L_m I_{p,s} + L_r I_{p,r}\big)\Big] \end{aligned} \tag{7.9b}$$

$$\begin{aligned} J\frac{d}{dt}\Omega_{r,0} &= \frac{3P}{4}L_m \Im\big(I_{p,s}I_{p,r}^* + I_{n,s}^* I_{n,r}\big) - B\Omega_{r,0} - T_L \\ J\frac{d}{dt}\Omega_{r,2} &= \frac{3P}{j8}L_m\big(I_{p,s}I_{n,r} - I_{n,s}I_{p,r}\big) - \big(B + j2J\omega_0\big)\Omega_{r,2} \end{aligned} \tag{7.9c}$$

This set of nonlinear equations can be cast in state-space form after straightforward, but somewhat tedious algebraic manipulations. Note that while (7.9a)–(7.9c) comprise six differential equations (11 real-valued states) compared to three in (7.6) (five real-valued states), the averaged equations (7.9a)–(7.9c) have a number of useful features. It is a time-invariant model, where inputs and consequently states tend to vary slowly compared to ω_0. When used for simulation in our Matlab implementation (with no attempt to speed up either model), improvements of two to five times were commonly observed. Dynamic phasors achieve "simultaneous demodulation" in that all variables in (7.9a)–(7.9c) are constant ("dc") in a steady-state. This is not achievable by a single reference frame transformation in case of an asymmetrically supplied ac machine.

As in the case of a single-phase induction machine, (7.9a)–(7.9c) in a steady-state yield an equivalent circuit that differs from the conventional result [212]. Again there exists a coupling between the positive and negative sequence subcircuits proportional to $\Omega_{r,2}$, as shown in Fig. 7.2 ("s" denotes the slip, $L_s = L_{ls} + L_m$, $L_r = L_{lr} + L_m$, $m_p = \Omega_{r,2}/(\omega_0 s)$ and $m_n = -\Omega_{r,2}/[\omega_0(2-s)]$.

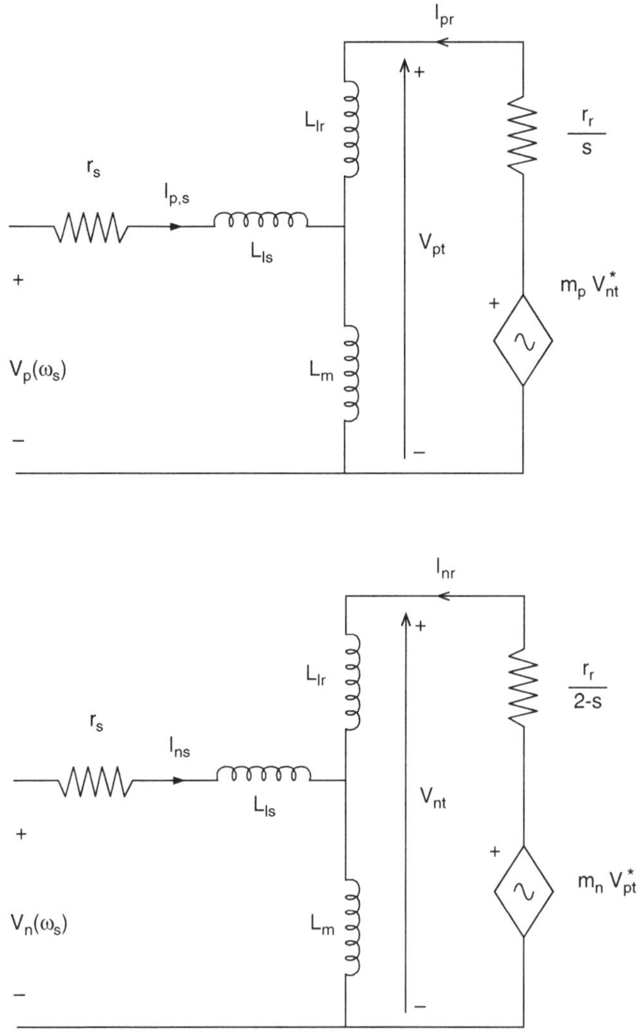

Fig. 7.2 Equivalent circuit of a three-phase induction motor with unbalanced supply.

Next, we focus on the verification of the dynamic phasor model and illustrate the accuracy of (7.9a)–(7.9c), via a hardware experiment: at $t \approx 1.7$ sec the supply voltage suddenly changes from a balanced set to unbalanced set with different magnitudes. The induction motor is a three-phase, isolated Y-connected unit with the following parameter values: $r_s = 2.55\,\Omega$, $r_r = 2.38\,\Omega$, $L_{ls} = L_{lr} = 4.15\,\text{mH}$, $L_m = 82.62\,\text{mH}$, $P = 4$, $J = 4.8e^{-4}\,\text{kg}\cdot\text{m}^2$, $B = 0.5e^{-4}\,\text{N}\cdot\text{m}\cdot\text{sec/rad}$. The supply voltages in experiments are provided by an HP 6834B programmable AC power source; phase voltages are $\approx$10Vrms before the transient that is initiated by a zero crossing of v_c; after the onset of the transient, v_a doubles, v_b remains unchanged, while v_c is set to zero. Actual numerical values of rms phase voltages were $[9.89\ \ 9.89\ \ 9.89]^T \mapsto [23.83\ \ 10.33\ \ 0]^T$.

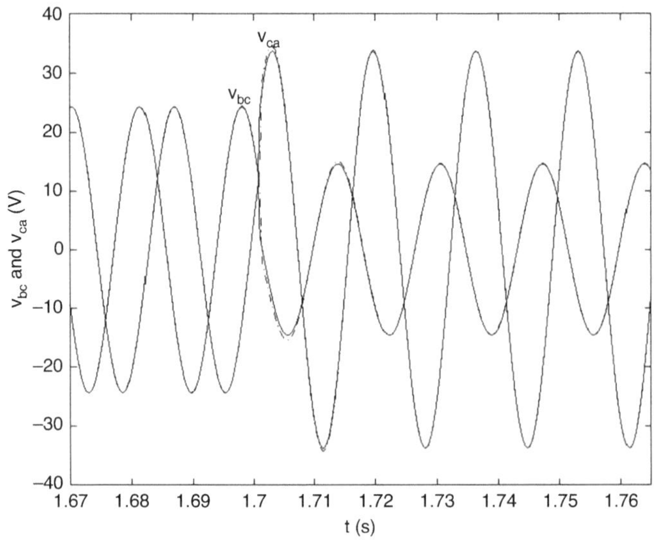

Fig. 7.3 Transient in line voltages following a sudden loss of balance in supply voltages: solid line – experiment; dash-dotted line – simulation.

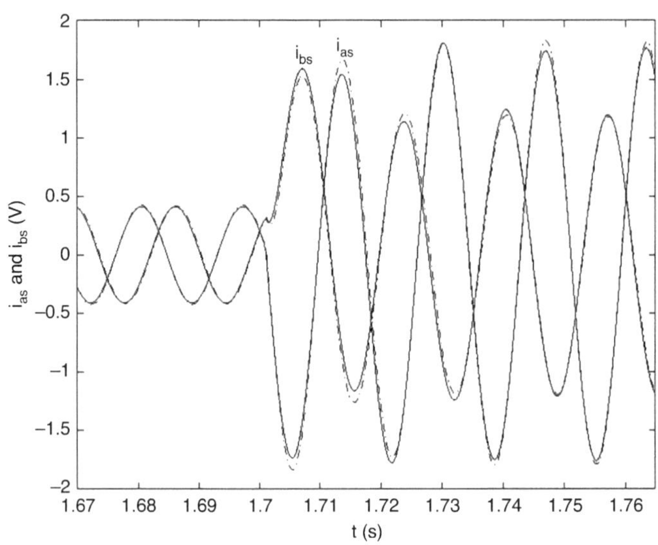

Fig. 7.4 Transients in phase a and b currents following a sudden loss of balance in supply voltages: solid line – experiment; dash-dotted line – simulation.

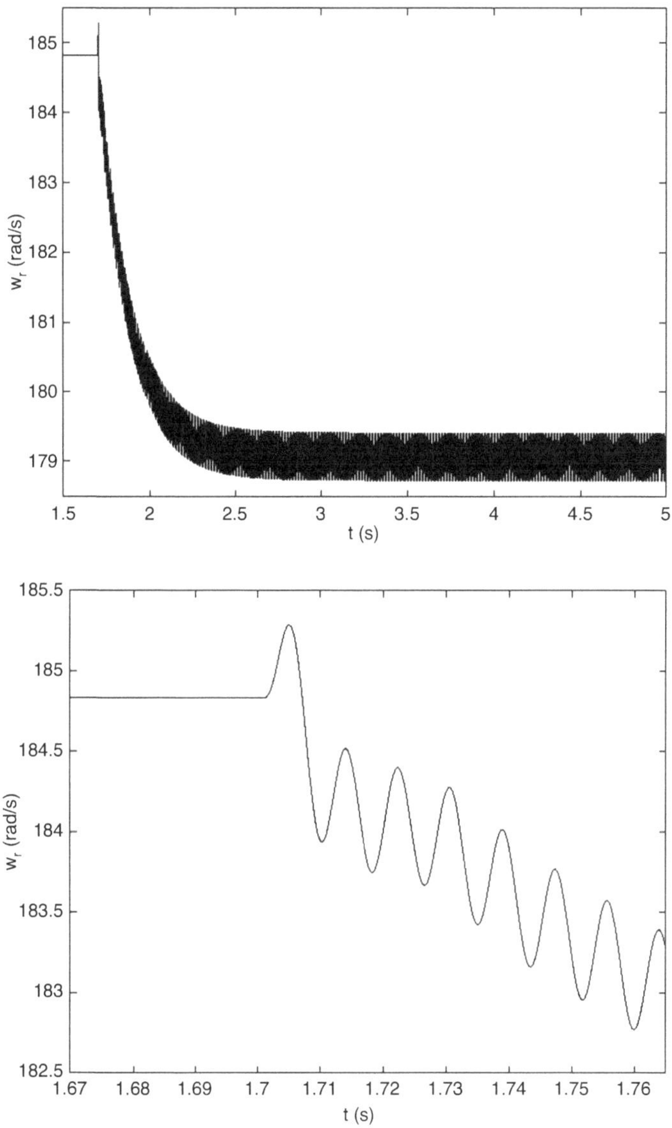

Fig. 7.5 Mechanical speed transient following a sudden loss of balance in supply voltages (top panel); expanded portion with time axis matching the electrical transients (bottom panel).

In Fig. 7.3 we display the input voltages (solid line – *experimental* data, dash-dotted line – simulation), while in Fig. 7.4 we show the currents in phases a and b. It turns out that there is practically no noticeable difference between the predictions of (7.9a)–(7.9c) and of (7.6).

We observe a very good agreement between the experimental data and the predictions of the dynamic phasor model, even though the voltage transient is quite abrupt. The speed transients predicted by (7.9c) and (7.6) are almost identical – in Fig. 7.5 we show simulation results obtained from (7.9c).

Essentially, the same procedure can be applied to other polyphase AC machines; the case of permanent-magnet synchronous machine is described in [213], with very similar results.

7.2 Applications to Power Systems

This section describes applications of dynamic phasors to polyphase AC power systems. We focus on unbalanced operation and faults and make use of dynamic symmetrical components introduced earlier. We present simulations of unbalanced faults involving a three-phase synchronous generator connected to an infinite bus through a transmission line and show that dynamic phasor models provide very accurate descriptions of observed transients. Given our emphasis on faults, we maintain fixed generator excitation for simplicity. Some other application ideas and leads are listed at the end of the section.

Large-power networks are among the largest systems not only by (spatial) size but also in terms of exposure to the environment. This elevates the importance of fault analyses that aim to quantify waveforms of interest, such as currents and voltages, under extraordinary operating conditions. The accuracy of such studies is a key figure of merit, as it guides sizing of components and tuning of protection. Traditionally, available models belong to two broad classes – detailed time-domain models and simple frequency-domain descriptions. Once again, dynamic phasors offer an in-between option, offering a capability to bridge the two domains while preserving their respective strengths.

In existing power systems, still dominated by electromechanical energy sources, fault analyses are most sensitive to variations in generator models [214]. This may change in the near future, as inverter-connected sources, such as renewables, become more widespread. This section thus concentrates on models of synchronous machines, in addition to transmission line models. Our attention is on unbalanced (asymmetrical) faults, as balanced (symmetric) faults present a simpler special case. As in the case of other electric machines covered earlier, dynamic phasors reveal and quantify coupling between speed ripple and currents (or fluxes) in various dynamic sequence components of asynchronous machine. In our simulations, we study a benchmark power system example and compare dynamic phasors with detailed time-domain "EMTP-like" simulations (provided by the package PSCAD) and with standard "quasistationary" models used in transient stability studies (provided by the package EUROSTAG).

Once again, we need to carefully weigh the decision about harmonics to include in our investigation. We assume the presence of the following harmonics (all expressed in the rotor frame that is the standard studies of multi-machine power systems): **(1)** zeroth (DC), which is the only one present in a balanced steady-state and remains a key component in transients; **(2)** $\pm$ 1-st harmonic, which is due to the fault and is present in balanced faults as well (corresponding to DC quantities in the stationary frame); **(3)** $\pm$ 2-nd harmonic, which is due to the asymmetry, as these harmonics are consequences of negative sequence components (especially currents) present during asymmetric faults. In this analysis, we consider linear magnetics for all components, so the nonlinearities of interest are multiplicative. With our assumptions about the frequency content for various states and given our preceding examples of induction and permanent-magnet machines, perhaps the main remaining challenge is clerical, i.e., keeping track of various components. The model of a synchronous generator in 3-phase machine variables is taken from [210, p. 214]. The generator is modeled with one damper winding in the d axis and two damper windings in the q axis. The dynamic phasor model of a synchronous generator from these equations can be written as

$$\begin{aligned}
\frac{1}{\omega_0}\frac{d}{dt} < I >_k &= < L^{-1}(V - RI - \omega_r \frac{dL}{d\theta_r} I) >_k - jk < I >_k \\
2H\frac{d}{dt} < \omega_r >_k &= < T_L >_k + \frac{1}{3} < I_{abcs}^T \frac{dL_{ss}}{d\theta_r} I_{abcs} >_k \\
&\quad - \frac{2}{3} < I_{abcs}^T \frac{dL_{sr}}{d\theta_r} I_{qdr} >_k - jk\,\omega_0\, 2H < \omega_r >_k \\
\frac{1}{\omega_0}\frac{d}{dt} < \delta_r >_k &= < \omega_r - 1 >_k - jk < \delta_r >_k
\end{aligned} \tag{7.10}$$

where

$$V = \left[V_{abcs}^T \ V_{qdr}^T\right]^T, \quad I = \left[-I_{abcs}^T \ I_{qdr}^T\right]^T, \quad R = \mathsf{diag}\Big(\left[r_s \ r_s \ r_s \ r_{kq1} \ r_{kq2} \ r_{fd} \ r_{kd}\right]\Big)$$

and

$$L = \begin{bmatrix} L_{ss} & L_{sr} \\ \frac{2}{3}L_{rs} & L_{rr} \end{bmatrix}$$

Note that $X_{abcs} = [x_{as} \ x_{bs} \ x_{cs}]^T$, $X_{qdr} = [x_{kq1} \ x_{kq2} \ x_{fd} \ x_{kd}]$ and that the frequency content of the inductance matrices is described by the dynamic phasors

$$\begin{aligned}
< L_{ss} >_0 &= \begin{bmatrix} l_{ls} + l_A & -\frac{l_A}{2} & -\frac{l_A}{2} \\ -\frac{l_A}{2} & l_{ls} + l_A & -\frac{l_A}{2} \\ -\frac{l_A}{2} & -\frac{l_A}{2} & l_{ls} + l_A \end{bmatrix} \\
< L_{ss} >_2 &= -\frac{l_B}{2} \begin{bmatrix} 1 & \alpha^* & \alpha \\ \alpha^* & \alpha & 1 \\ \alpha & 1 & \alpha^* \end{bmatrix} e^{j2\delta_r}
\end{aligned}$$

$$
\begin{aligned}
< L_{sr} >_1 &= \frac{1}{2}\begin{bmatrix} l_{mq} & l_{mq} & -jl_{md} & -jl_{md} \\ l_{mq}\alpha^* & l_{mq}\alpha^* & -jl_{md}\alpha^* & -jl_{md}\alpha^* \\ l_{mq}\alpha & l_{mq}\alpha & -jl_{md}\alpha & -jl_{md}\alpha \end{bmatrix} e^{j\delta_r} \\
< L_{rs} >_1 &= < L_{sr} >_1^T \\
< L_{rr} >_0 &= \begin{bmatrix} l_{kq1} & l_{mq} & 0 & 0 \\ l_{mq} & l_{kq2} & 0 & 0 \\ 0 & 0 & l_{fd} & l_{md} \\ 0 & 0 & l_{md} & l_{kd} \end{bmatrix} \\
< \frac{dL_{ss}}{d\theta_r} >_2 &= 2j < L_{ss} >_2 \\
< \frac{dL_{sr}}{d\theta_r} >_1 &= j < L_{sr} >_1 \\
< \frac{dL_{rs}}{d\theta_r} >_1 &= j < L_{sr} >_1
\end{aligned}
$$

where all other harmonics related to inductances are assumed to be zero, and $\alpha = e^{j2\pi/3}$. We refer the interested reader to [215] for details about the dynamic sequence component models of the generator and transmission line. The one additional simplification made there is that the speed has only the DC component in electrical (voltage) equations, which is reasonable for large machines.

We use the numerical example of a synchronous generator connected to an infinite bus from [49, Example 13.4] and consider the case in which a one-phase fault on circuit 2 at point F (Fig. 7.6) cleared by isolating the faulted circuit simultaneously at both ends with a fault-clearing time of 79.3 msec (while assuming a constant field voltage); please see [215] for details. Standard time-domain simulations included for reference were performed by the PSCAD program, while simulations with quasistationary models (e.g., as used in typical transient stability studies) were performed with the package EUROSTAG.

In Figs. 7.7 and 7.8, we show the phase currents and the line-to-neutral voltages at the synchronous generator terminal, respectively, and note an outstanding agreement between the detailed time-domain and dynamic phasor simulations.

A modeling simplification that is commonly used in software tools for electromechanical transients (such as EUROSTAG) relies on the fact that transients associated with the transmission network tend to decay very rapidly. To remove such dynamics from consideration in a logically consistent way, generator stator transients are neglected as well [49]. This simplification is easy to achieve with dynamic phasors by setting the corresponding time derivatives to zero. This is in agreement with the overall model structure, as power systems are routinely described with differential-algebraic equations (DAE). The assumption of per unit value of ω_r being equal to 1 in the stator voltage equations is made to counterbalance the effect of neglecting stator flux linkage transients [49].

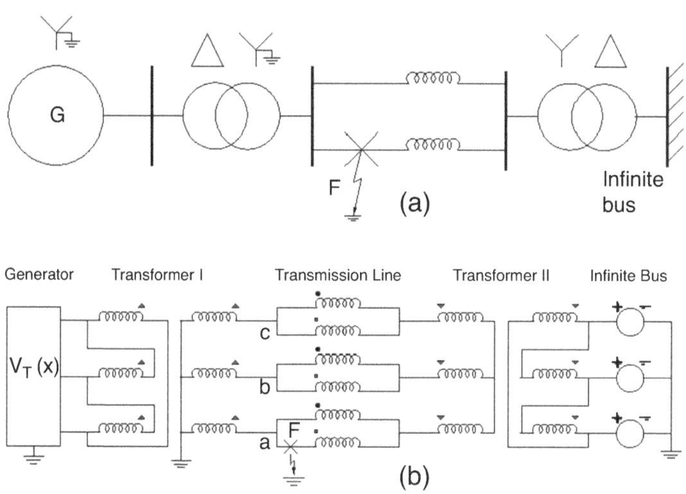

Fig. 7.6 Single machine infinite bus example: (**a**) single-line diagram; (**b**) 3-phase system representation.

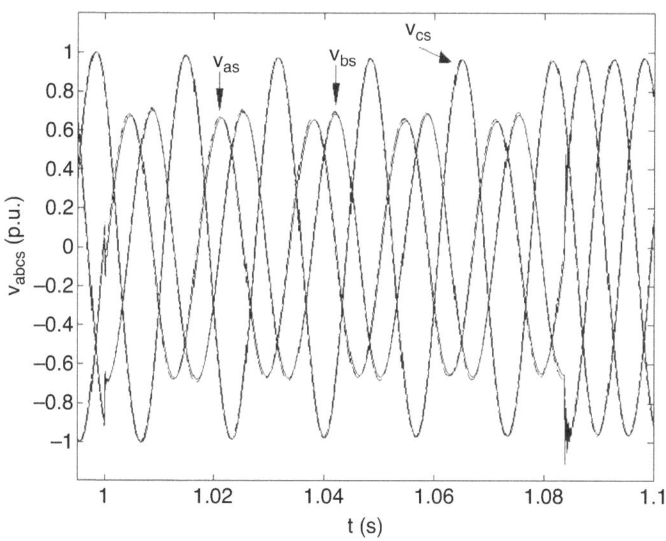

Fig. 7.7 Transients in synchronous machine terminal voltages: solid line – PSCAD; dash-dotted line – phasor dynamics (sequence component and three-phase model).

Next we compare the detailed time-domain models (as implemented, e.g., in PSCAD), the phasor dynamic models with and without network transients, and quasistationary models used in transient stability studies (e.g., EUROSTAG). We look at the rotor angle of the generator – in Fig. 7.9 we show the overall transient following the fault, while in Fig. 7.10 we zoom on the beginning portion of the

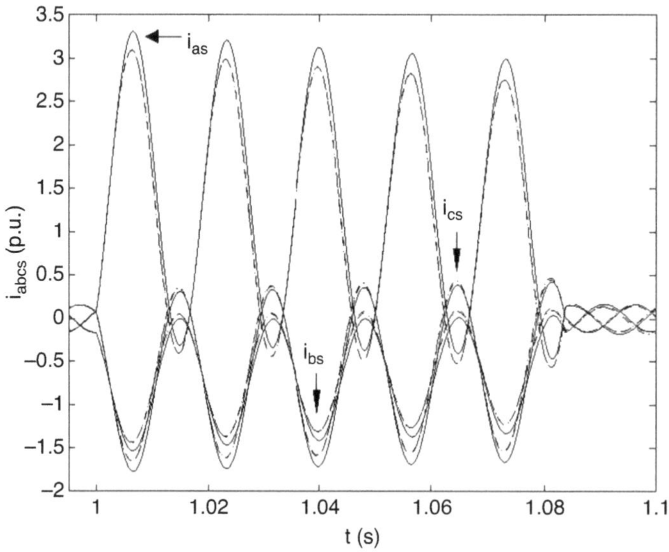

Fig. 7.8 Transients in synchronous machine terminal currents: solid line – PSCAD; dash-dotted line – phasor dynamics (sequence component and three-phase model).

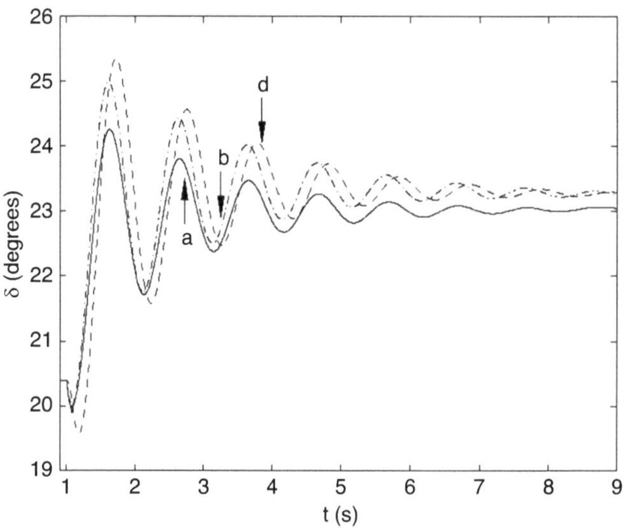

Fig. 7.9 Rotor angle transient: solid line – PSCAD; dash-dotted line – phasor dynamics; dashed line – EUROSTAG.

transient. The overall agreement is quite good, especially since the pointwise differences in time depend on details of fault representation that are not accessible to users of standard programs (e.g., notice the fault duration in trace "d" in Fig. 7.10). Both the detailed time-domain model and the phasor dynamic model with network

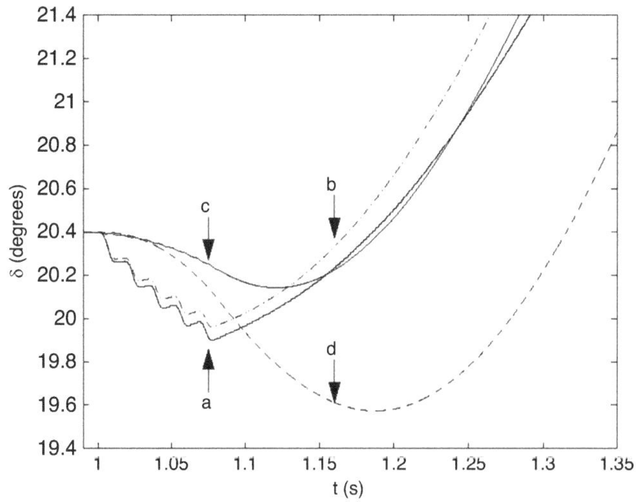

Fig. 7.10 Details of the initial rotor angle transient: solid line (a) – PSCAD; dash-dotted line (b) – phasor dynamics; solid line (c) – phasor dynamics (stator transients neglected); dashed line (d) – EUROSTAG.

transients take into account the non-DC component of the speed variation (which is excluded when network transients are disregarded).

Applications of dynamic phasors that explored some novel domains include [216] that studied ferroresonance and addressed the case of systems with nonlinear magnetics with models in descriptor form (e.g., a singular matrix multiplying derivatives on the left-hand side of state-space models) and the use of dynamic phasor models in calculating bifurcation diagrams. Another direction was explored in [217] that looked into participation factors of states in a limit-cycle behavior. After a dynamic phasor-motivated coordinate transformation, the participation factors are actually calculated for the linearization of the new model at the Hopf bifurcation and then validated through simulations of the full nonlinear model. This procedure identifies the state variables (and indirectly controls) that are key in the inception of oscillatory behavior. This is useful since in time-domain simulation the oscillations appear in all of the state variables.

7.3 Further Reading

There have been many publications in the areas of electric machines and power systems that use dynamic phasors as the key modeling tool. We mention here only a few references that are directly connected with the use cases that are described in this book. At the component level, [211] describes the modeling of electromechanical systems with position-dependent loads (e.g., certain types of compressors) that result in a rich harmonic response. Reference [237] introduces the use of dynamic phasors to model phasor measurement units (PMUs), which are of great relevance for many emerging control and estimation applications. At the system level, [75] analyzes unbalanced polyphase systems with harmonics and the interplay with dynamic symmetrical components. Reference [238] is an example of the use of dynamic phasors in modeling microgrids. This is a particularly effective tool selection because microgrids tend to be faster than typical distribution networks and with potentially complicated harmonic flows. Finally, [239] covers the modeling of line-commutated high-voltage DC systems (HVDC) from a system perspective and describes their embedding into conventional AC networks.

Part III

Applications to Power Flow in Electric Energy Processing Systems

"The miracle of the appropriateness of the language of mathematics for the formulation of the laws of physics is a wonderful gift, which we neither understand nor deserve."

Eugene P. Wigner, 1926

Chapter 8 Power Quality in Steady-State

Ever since AC devices and systems achieved importance and later dominance in production and transmission of electric energy, engineers have been interested in quantifying their performance. In the general (polyphase, nonsinusoidal) case, the problem is made more difficult by possible presence of couplings among phases and of higher harmonics (due to sources and to circuit nonlinearities).

The issue of phase difference ϕ between voltage and current in single-phase single-frequency systems was clarified in the 1880s [240,241], resulting in a simple definition of the power factor ("cos ϕ"). However, this definition does not extend to polyphase multi-frequency systems because of the large number of distinct phase differences involved. An alternative definition of the power factor – the ratio between average (real, active) power P and apparent power S (the largest possible under unchanged circuit conditions) – was introduced by Steinmetz in 1892 [242,243]. This more general definition also admits a geometric interpretation, but only in an appropriately defined Hilbert space of polyphase multi-frequency waveforms.

The early part of the twentieth century has seen a surge of interest in extending the concepts of real and reactive power to polyphase multi-frequency circuits, as evidenced by the specialized symposia organized by AIEE [244] and by numerous publications written in response to the "Romanian Questionnaire" [245–247]. Two key publications from this era are the papers of Budeanu [248] and Fryze [27], each offering a different and profound view of the problem. Budeanu proposed a 3-component decomposition, namely, $S^2 = P^2 + Q_B^2 + D^2$, in which S is (scalar) apparent power, P is real power, Q_B is Budeanu's reactive power, and "cross-terms" obtained when multiplying voltage and current norms are grouped in the distortion power D. Fryze originated a time-domain current decomposition that can be interpreted geometrically in terms of projections and easily extended to transient operation.

H. Lev-Ari, A. M. Stanković, *Dynamic Phasors in Energy Processing Systems*,
Power Electronics and Power Systems, https://doi.org/10.1007/978-3-031-50020-6_9

The apparent power in a *sinusoidal single-phase system* has only two components – the real (a.k.a. active) power P and the reactive power $Q \equiv Q_B$. In this case, the voltage and current waveforms are determined by a single phasor each, viz.,

$$v(t) = \sqrt{2}\,\Re\left\{V e^{\jmath\omega t}\right\} \quad , \quad i(t) = \sqrt{2}\,\Re\left\{I\, e^{\jmath\omega t}\right\}$$

so that $V_{rms} = |V|$, $I_{rms} = |I|$, $S = |V|\,|I|$ (see Sect. 8.1), as well as

$$P = \Re\left\{V I^*\right\} \equiv |V|\,|I| \cos\phi \quad , \quad Q = \Im\left\{V I^*\right\} \equiv |V|\,|I| \sin\phi$$

where V, I are the (complex-valued) voltage and current phasors, respectively, and $\phi \stackrel{\text{def}}{=} \arg\left(V/I\right)$ is the phase-shift of the current relative to the voltage. Consequently, power quality can be described in terms of a single (real) coefficient, the *power factor*

$$\text{PF} \stackrel{\text{def}}{=} \frac{P}{V_{rms}\,I_{rms}} = \frac{\Re\{V I^*\}}{|V|\,|I|} = \cos\phi$$

The apparent power can also be described by the so-called complex apparent power $P + jQ$ [49] because $|P + jQ|^2 = P^2 + Q^2 = S^2$ (see Sect. 8.2).

Following the work of Budeanu and Fryze, many authors have aimed to characterize concepts of reactive and inactive power in the most general (i.e., multi-harmonic and polyphase) case. In this case, the voltage and current waveforms for each individual phase are represented by multiple phasors, as in (1.4a), viz.

$$\begin{aligned} v_p(t) &= V_0^{(p)} + \sum_{k=1}^{\infty} \sqrt{2}\,\Re\left\{V_k^{(p)}\, e^{jk\omega_0 t}\right\} \\ i_p(t) &= I_0^{(p)} + \sum_{k=1}^{\infty} \sqrt{2}\,\Re\left\{I_k^{(p)}\, e^{jk\omega_0 t}\right\} \end{aligned} \qquad 1 \le p \le m \qquad (8.1)$$

where m is the total number of phases. In such a system, there are multiple phase-shifts, given by $\phi_k^{(p)} \stackrel{\text{def}}{=} \arg\left(V_k^{(p)}/I_k^{(p)}\right)$: however, even when all these shifts are zero, we still have, in general, $S^2 > P^2 + Q_B^2$, as explained in Sect. 8.3. Two well-known attempts to address shortcomings of Budeanu's 3-component decomposition were proposed by Shepherd and Zakikhani [249] and Sharon [250]. Both of these (single-phase) 3-component decompositions rely on a modified definition of "reactive power" in a multi-harmonic system: see Sect. 8.5 for additional details.

Subsequent research by Czarnecki resulted in increasing the number of components of the squared apparent power from 3 to 5 [251]. We were able to produce a further refinement, resulting in a 7-component decomposition [252, 253]. A key ingredient in such refinements was the introduction of an *equivalent admittance*

$Y_k^{(p)} \equiv g_k^{(p)} - j b_k^{(p)} \overset{\text{def}}{=} \frac{I_k^{(p)}}{V_k^{(p)}}$ (see also Sect. 8.4). This concept was also central to the (earlier) FBD approach of Depenbrock [254, 255].

Examination of statistical spread within the two sets of parameters defined by the equivalent admittance – the conductances $\{g_k^{(p)}\ ;\ 1 \leq p \leq m\ ,\ 0 \leq k < \infty\}$ and susceptances $\{b_k^{(p)}\ ;\ 1 \leq p \leq m\ ,\ 0 \leq k < \infty\}$ – led us to formulate an alternative compact derivation of our 7-component decomposition of apparent power (see Sect. 8.5). A further increase in the number of components was achieved by utilizing the symmetrical sequence decomposition of P and Q_B (Sect. 8.6), resulting in an 11-component decomposition of the squared apparent power (Sect. 8.7).

Finally, we observe that our 7/11-component decomposition induces a matching decomposition of the current waveform $i(\cdot)$ into a sum of mutually-orthogonal components: see Sect. 8.8 for details. Notice that such multi-component decompositions can no longer be described in terms of complex numbers – which have only two components – or even quaternions, which have one real and three "imaginary" components [256]. Rather, one may need to rely on higher-dimensional conceptual frameworks, such as offered, for instance, by Geometric Algebra: see Sect. 8.9 for further details.

8.1 Hilbert Space of Finite-RMS Periodic Polyphase Waveforms

All time-domain polyphase waveforms $x(t)$ considered in this chapter are periodic, with a fundamental period T. We restrict our discussion to *finite-rms* periodic waveforms, i.e., we assume that

$$\| x(\cdot) \|^2 \overset{\text{def}}{=} \frac{1}{T} \int_T x(s)\, x^\top(s)\, ds \equiv X_{rms}^2 < \infty \tag{8.2}$$

The notation $\int_T$ indicates integration over an interval of length T. We are interested in energy processing systems where this quantity is typically fixed in advance to $1/60$ sec (or $1/50$ sec).

We will use here the conceptual framework introduced in Sects. 1.1 and 1.2. This means that we consider an m-phase system, i.e., a system with $m + 1$ conductors ("wires"), as shown in Fig. 1.1, which we reproduce here as Fig. 8.1. We express all relations in terms of the $1 \times m$ voltage and current vector waveforms

$$\begin{aligned} v(t) &\overset{\text{def}}{=} \begin{bmatrix} v_1(t) & v_2(t) & \ldots & v_m(t) \end{bmatrix} \\ i(t) &\overset{\text{def}}{=} \begin{bmatrix} i_1(t) & i_2(t) & \ldots & i_m(t) \end{bmatrix} \end{aligned} \tag{8.3}$$

where all voltages are referenced to the $(m + 1)$-st ("neutral") conductor, and all currents have reference directions toward the load.

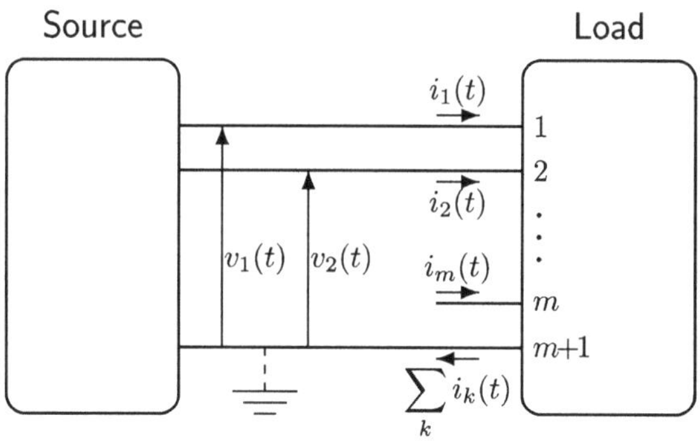

Fig. 8.1 Circuit schematic of an $m+1$-wire polyphase system.

Since we are interested here in introducing fundamental conceptual tools, we consider a simple case of a polyphase system with a common neutral wire, as shown in Fig. 8.1, and we avoid intricacies such as the choice of a voltage measurement reference point, or the possibility of all $m+1$ conductors being of "equal importance." It is possible to map the general case into (8.3) with such aspects included; ways to do so are detailed in [257, 258].

Next, recall that one can view $v(t)$ and $i(t)$, the periodic vector waveforms representing polyphase load voltage and current (nonlinear load, nonsinusoidal voltage source), as elements in the Hilbert space of m-phase, T-periodic waveforms that satisfy the finite-rms constraint (8.2). Such waveforms can be represented by a (one-sided) Fourier series, viz.,

$$x(t) = X_0 + \sum_{k=1}^{\infty} \sqrt{2}\, \Re\left\{X_k\, e^{jk\omega_0 t}\right\} \tag{8.4a}$$

where $X_k = \left[X_k^{(1)}\ X_k^{(2)}\ \ldots\ X_k^{(m)}\right]$ is the polyphase rms phasor (= scaled vector Fourier coefficient) describing the k-th harmonic of a polyphase waveform $x(t)$, given by the expression (1.4c), namely,

$$X_k = \begin{cases} \dfrac{1}{T}\displaystyle\int_T x(s)\, ds & k = 0 \\[2ex] \dfrac{\sqrt{2}}{T}\displaystyle\int_T x(s)\, e^{-jk\omega_0 s}\, ds & k \geq 1 \end{cases} \tag{8.4b}$$

where $\omega_0 \stackrel{\text{def}}{=} 2\pi/T$. The Parseval relation between any two T-periodic polyphase waveforms $x(\cdot)$, $y(\cdot)$ and their rms phasors can be expressed in the form

$$\langle\, x(\cdot)\,,\, y(\cdot)\,\rangle \equiv \frac{1}{T}\int_T x(s)\, y^{\top}(s)\, ds = \sum_{k=0}^{\infty} \Re\left\{\, X_k\, Y_k^H \right\} \tag{8.5a}$$

where the superscript $^\top$ denotes transposition, the superscript ^{H} denotes conjugate transposition, and $\{X_k\,;\ 0 \le k < \infty\}$, $\{Y_k\,;\ 0 \le k < \infty\}$ are the phasor sequences associated with $x(\cdot)$, $y(\cdot)$, respectively.

The Parseval identity (8.5) makes it possible to express the rms voltage and current, as well as the active power P, either in terms of waveforms or in terms of their phasor arrays, as was shown in (1.10b), namely,

$$V_{rms}^2 \stackrel{\text{def}}{=} \|\, v(\cdot)\|^2 = \|\mathbf{V}\|^2 \ , \quad I_{rms}^2 \stackrel{\text{def}}{=} \|\, i(\cdot)\|^2 = \|\,\mathbf{I}\,\|^2$$
$$P \stackrel{\text{def}}{=} \langle\, v(\cdot)\,,\, i(\cdot)\,\rangle = \Re\left\{\mathbf{V}\,\mathbf{I}^H\right\} = V_0\, I_0^\top + \sum_{k=1}^{\infty} \Re\left\{\, V_k\, I_k^H \right\} \tag{8.6}$$

where $\mathbf{V} = \begin{bmatrix} V_0 & V_1 & V_2 & \ldots \end{bmatrix}$, $\mathbf{I} = \begin{bmatrix} I_0 & I_1 & I_2 & \ldots \end{bmatrix}$ are the phasor arrays constructed from the two phasor sequences $\{V_k\,;\ 0 \le k < \infty\}$, $\{I_k\,;\ 0 \le k < \infty\}$ associated with the two polyphase waveforms $v(\cdot)$ and $i(\cdot)$, respectively.

8.2 Power Factor

One common power quality metric is the so-called power factor

$$\text{PF} \stackrel{\text{def}}{=} \frac{P}{S} = \frac{\langle\, v(\cdot)\,,\, i(\cdot)\,\rangle}{\|\, v(\cdot)\|\ \|\, i(\cdot)\|} \equiv \frac{\Re\left\{\mathbf{V}\,\mathbf{I}^H\right\}}{\|\mathbf{V}\|\ \|\,\mathbf{I}\,\|} \tag{8.7}$$

We conclude from the Cauchy–Schwarz inequality for finite-rms periodic polyphase waveforms that $|\langle\, v(\cdot)\,,\, i(\cdot)\,\rangle| \le \|\, v(\cdot)\|\ \|\, i(\cdot)\| \equiv V_{rms}\, I_{rms} \stackrel{\text{def}}{=} S$, so that the magnitude of the power factor is always bounded by unity. In geometric terms, this ratio is the cosine of the angle ψ between the Hilbert waveform-space elements $v(\cdot)$ and $i(\cdot)$ or, equivalently, between the associated phasor arrays $\mathbf{V}$ and $\mathbf{I}$ (Fig. 8.2). This means that the apparent power S is the highest average power delivered to the load, among all loads that have the same rms current $\|\, i(\cdot)\| \equiv \|\,\mathbf{I}\,\|$ (we assume that the voltage across the load remains unchanged as we change the load seen by the source). For a non-ideal load $P < S$, and this mismatch is the target of compensation devices. This relation can also be described in terms of the

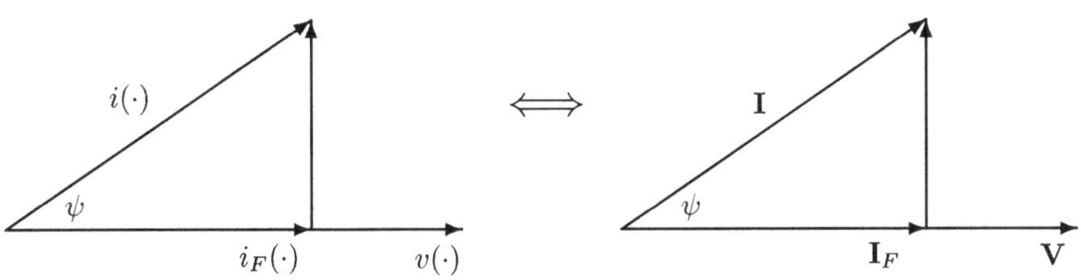

Fig. 8.2 Geometric (Hilbert space) relation between the voltage and current waveforms (and the associated phasor arrays), displaying the waveform angle ψ.

rms (i.e., norm) difference between the current $i(\cdot)$ and the so-called *Fryze current* $i_F(\cdot)$ (see Sect. 8.8 for details).

An alternative (and equivalent) power quality metric is the *power quality gap*

$$S^2 - P^2 \equiv \| v(\cdot)\|^2 \, \| i(\cdot)\|^2 - \left[\langle v(\cdot), i(\cdot) \rangle\right]^2 \equiv \| \mathbf{V} \|^2 \, \| \mathbf{I} \|^2 - \left[\Re\{ \mathbf{V}\mathbf{I}^H \}\right]^2 \tag{8.8}$$

which is always non-negative. This quantity can be expressed explicitly in terms of the individual voltage and current phasors shown in (8.1): see, e.g., [259]. Unfortunately, the resulting representation involves a very large number of terms, each one corresponding to a particular product of a voltage phasor with a current phasor. A more attractive approach is to seek a decomposition of the gap $S^2 - P^2$ into a modest number (say, less than ten) of physically meaningful terms. We develop this approach in detail in the remainder of this chapter, culminating with the introduction of a 7/11-component decomposition of S^2 in Sect. 8.7.

Sinusoidal Single-Phase Waveforms

The phasor-based representation

$$v(t) = \sqrt{2}\,\Re\left\{V e^{j\omega_0 t}\right\} \quad , \qquad i(t) = \sqrt{2}\,\Re\left\{I\, e^{j\omega_0 t}\right\} \tag{8.9a}$$

implies that

$$\| v(\cdot)\| = | V | \quad , \qquad \| i(\cdot)\| = | I |$$

as well as

$$S = \| v(\cdot)\| \, \| i(\cdot)\| = | V | \, | I | \equiv |V I^*|$$

$$P = \langle v(\cdot), i(\cdot) \rangle = \Re\{V I^*\}$$

which is consistent with (8.6). In this simple case, the phasor array $\mathbf{V}$ (resp. $\mathbf{I}$) reduces to the complex scalar phasor V (resp. I), and thus the Hilbert space of phasor arrays reduces to the two-dimensional space of complex numbers (Fig. 8.3). Also, the conjugate transpose reduces in this case to a scalar complex conjugate, which we denote by the superscript *. Consequently, the Hilbert space angle ψ coincides with the phase-shift angle ϕ, viz.,

$$\cos\psi = \frac{P}{S} = \frac{\Re\{V I^*\}}{|V|\,|I|} \equiv \cos\phi \tag{8.9b}$$

where $\phi \overset{\text{def}}{=} \arg\left(V/I\right)$ is the phase-shift between the voltage and the current.

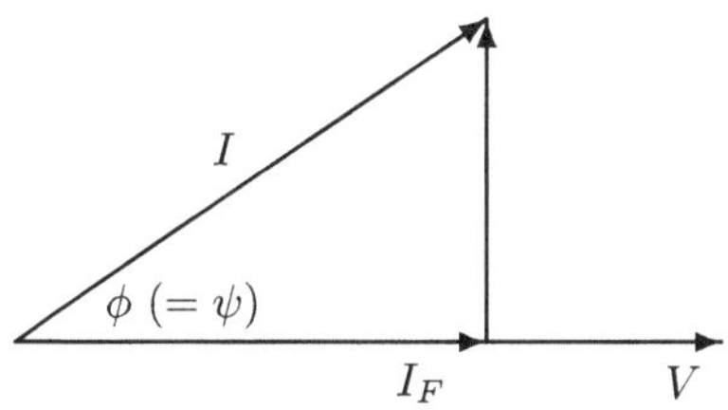

Fig. 8.3 Geometric/phase-shift relation between the scalar phasors V and I.

Similarly, $S^2 - P^2 = \left|VI^*\right|^2 - \left[\Re\{VI^*\}\right]^2 = \left[\Im\{VI^*\}\right]^2$, so that we can interpret the power quality gap $S^2 - P^2$ entirely in terms of the so-called *complex apparent power*

$$S_c \overset{\text{def}}{=} VI^* = P + jQ \tag{8.10a}$$

where $Q \overset{\text{def}}{=} \Im\{VI^*\}$ is known as the *reactive power*. Indeed,

$$S^2 \equiv |V|^2|I|^2 = |S_c|^2 = P^2 + Q^2 \tag{8.10b}$$

so that the power quality gap $S^2 - P^2$ is entirely due to the reactive power Q. The compactness of the representation (8.10) is a direct consequence of the fact that all sinusoidal single-phase waveforms can be represented in a two-dimensional basis, such as $e_1(t) = \sqrt{2}\cos\omega_0 t$, $e_2(t) = \sqrt{2}\sin\omega_0 t$ (see also Appendix E).

Nonsinusoidal Polyphase Waveforms

In contrast, the representation of m-phase waveforms that consist of L harmonics (and no DC) requires $2mL$ basis waveforms, which complicates the interpretation of the power quality gap $S^2 - P^2$ in this more general case.

The Parseval relation (1.10) provides explicit expressions for the rms voltage and current, as well as the active power, so that

$$V_{rms}^2 \equiv \|\mathbf{V}\|^2 = \sum_{k,p} |V_k^{(p)}|^2 \quad , \quad I_{rms}^2 \equiv \|\mathbf{I}\|^2 = \sum_{k,p} |I_k^{(p)}|^2 \tag{8.11a}$$

$$P = \Re\{\mathbf{V}\mathbf{I}^H\} = \sum_{k,p} |V_k^{(p)}|\,|I_k^{(p)}|\cos\phi_k^{(p)} \tag{8.11b}$$

where $V_k^{(p)}$, $I_k^{(p)}$ are the scalar voltage and current phasors associated with the k-th harmonic in the p-th phase, and $\phi_k^{(p)} \overset{\text{def}}{=} \arg\left(I_k^{(p)}/V_k^{(p)}\right)$. Also,

$$\mathbf{V} = \begin{bmatrix} V_0 & V_1 & V_2 & \ldots \end{bmatrix} \quad , \quad \mathbf{I} = \begin{bmatrix} I_0 & I_1 & I_2 & \ldots \end{bmatrix} \tag{8.11c}$$

where

$$V_k = [\,V_k^{(1)} \;\; V_k^{(2)} \;\; \ldots \;\; V_k^{(m)}\,] \quad , \qquad I_k = [\,I_k^{(1)} \;\; I_k^{(2)} \;\; \ldots \;\; I_k^{(m)}\,] \tag{8.11d}$$

are the polyphase voltage and current phasors for the k-th harmonic.

The relations (8.11) allow us to express both the power factor P/S and the power quality gap $S^2 - P^2$ in terms of the scalar parameters $\{(|V_k^{(p)}|, |I_k^{(p)}|, \phi_k^{(p)})\,;\; 0 \le k < \infty, 1 \le p \le m\}$. For instance,

$$\frac{P}{S} = \frac{\sum_{k,p} |V_k^{(p)}|\,|I_k^{(p)}| \cos\phi_k^{(p)}}{\sqrt{\sum_{k,p} |V_k^{(p)}|^2}\;\sqrt{\sum_{k,p} |I_k^{(p)}|^2}}$$

However, such expressions do not appear to identify the causes of power quality imperfections. For instance, the power factor can be strictly less than unity even when $\phi_k^{(p)} = 0$ for all (k, p): in this case

$$\frac{P}{S} = \frac{\sum_{k,p} |V_k^{(p)}|\,|I_k^{(p)}|}{\sqrt{\sum_{k,p} |V_k^{(p)}|^2}\;\sqrt{\sum_{k,p} |I_k^{(p)}|^2}}$$

which, in general, is still less than unity as a consequence of the Cauchy–Schwarz inequality for scalar sequences.

8.3 Budeanu Reactive Power

The first step toward establishing a parsimonious decomposition of the power quality gap was taken by Budeanu [248], who extended the classical notion of reactive power to a more general (single-phase, multi-harmonic) case, by adding up the contributions from all harmonics, namely, $Q_B = \sum_k |V_k|\,|I_k| \sin\phi_k$. This concept was soon extended to polyphase waveforms, by adding the contributions from all phases, viz.,

$$Q_B \stackrel{\text{def}}{=} \Im\left\{\mathbf{V}\mathbf{I}^H\right\} = \sum_{k,p} |\,V_k^{(p)}|\,|I_k^{(p)}|\,\sin\phi_k^{(p)} \tag{8.12}$$

This extended concept was included, for instance, in a proposed set of standard definitions of quantities related to electrical power, assembled and published in 1935 under the sponsorship of the American Institute of Electrical Engineers [247]. Notice that $Q_B = 0$ when $\phi_k^{(p)} = 0$ for all k and p, but the converse does not hold: it is possible to have a non-vanishing Q_B even when all phase-shifts $\phi_k^{(p)}$ vanish.

Furthermore, the power gap $S^2 - P^2$ cannot be attributed solely to the occurrence of the reactive power Q_B. Indeed,

$$P^2 + Q_B^2 = \left|\Re\{\mathbf{V}\,\mathbf{I}^H\}\right|^2 + \left|\Im\{\mathbf{V}\,\mathbf{I}^H\}\right|^2 = \left|\mathbf{V}\,\mathbf{I}^H\right|^2$$

and using a Cauchy–Schwarz inequality for phasor arrays, we conclude that

$$P^2 + Q_B^2 = \left|\mathbf{V}\,\mathbf{I}^H\right|^2 \leq \|\,\mathbf{V}\,\|^2\;\|\,\mathbf{I}\,\|^2 \equiv S^2 \tag{8.13}$$

Budeanu used the term *distortion power* to characterize the residual power quality gap

$$D \stackrel{\text{def}}{=} \sqrt{S^2 - \left(P^2 + Q_B^2\right)} \tag{8.14}$$

Notice that D is a purely arithmetical, unsigned power metric that appears to have no clear physical meaning (i.e., no single cause). We observe that $D = 0$ if and only if (8.13) holds with *equality,* which can happen only when the phasor arrays $\mathbf{I}$ and $\mathbf{V}$ are *collinear*, namely $I_k^{(p)} = Y\,V_k^{(p)}$ for all k and all p, for some complex-valued scalar constant Y. This observation forms the starting point for the introduction in Sect. 8.4 of additional inactive power components that represent three distinct circuit imperfections.

8.4 A Statistical Spread Perspective

The preceding discussion used a Cauchy–Schwarz inequality to demonstrate that Budeanu's distortion power D vanishes only when the phasor arrays $\mathbf{I}$ and $\mathbf{V}$ are collinear. In turn, this implies that the residual power quality gap $D^2 \equiv S^2 - (P^2 + Q_B^2)$ is related to the variation of the *equivalent admittances*

$$Y_k^{(p)} \stackrel{\text{def}}{=} \frac{I_k^{(p)}}{V_k^{(p)}} = g_k^{(p)} - j b_k^{(p)} \tag{8.15a}$$

as a function of the harmonic index k and the phase index p.

When $Y_k^{(p)}$ is independent of both k and p, the phasor arrays $\mathbf{I}$ and $\mathbf{V}$ are collinear, and $D = 0$. Otherwise, a non-vanishing D should be closely related to the variation of the parameter sets

$$\begin{aligned} \mathbf{G} &\stackrel{\text{def}}{=} \{g_k^{(p)}\,;\; 1 \leq p \leq m\,;\; 0 \leq k < \infty\} = \begin{bmatrix} G_0 & G_1 & G_2 & \ldots \end{bmatrix} \\ \mathbf{B} &\stackrel{\text{def}}{=} \{b_k^{(p)}\,;\; 1 \leq p \leq m\,;\; 0 \leq k < \infty\} = \begin{bmatrix} B_0 & B_1 & B_2 & \ldots \end{bmatrix} \end{aligned} \tag{8.15b}$$

where $G_k = \left[g_k^{(1)} \; g_k^{(2)} \; \ldots \; g_k^{(m)}\right]$ and $B_k = \left[b_k^{(1)} \; b_k^{(2)} \; \ldots \; b_k^{(m)}\right]$ are the row vectors representing the k-th harmonic equivalent polyphase conductances and susceptances, respectively. Notice that $\mathbf{G}$ and $\mathbf{B}$ have the same (polyphase) block structure as the phasor arrays $\mathbf{V}$ and $\mathbf{I}$ (recall (8.11c)). Again, it will be convenient to use a vector (i.e., phasor array) notation for the equivalent admittance array, viz.,

$$\mathbf{Y} = \frac{\mathbf{I}}{\mathbf{V}} = \mathbf{G} - j\mathbf{B} \tag{8.15c}$$

where we define the *ratio of two phasor arrays* as element-wise division (same as the array division operation *A./B* in MATLAB). The parameter sets $\mathbf{G}$ and $\mathbf{B}$ serve as a starting point for the construction in Sect. 8.5 of a 5-component decomposition of apparent power, eventually leading to the 7/11-component decomposition of [252, 253, 270].

The statistical spread-based approach that we describe in Sect. 8.5 can be motivated by the observation that the power metrics P and Q_B are directly related to the (weighted) means of the parameter arrays $\mathbf{G}$ and $\mathbf{B}$, namely

$$P = \mu_g V_{rms}^2 \;, \qquad Q_B = \mu_b V_{rms}^2 \tag{8.16a}$$

where μ_g and μ_b are weighted means, viz.,

$$\mu_g \stackrel{\text{def}}{=} \mathcal{M}_w \mathbf{G} \;, \qquad \mu_b \stackrel{\text{def}}{=} \mathcal{M}_w \mathbf{B} \tag{8.16b}$$

We define the *weighted mean* operation $\mathcal{M}_w$, for any given $\mathbf{X} = \left[X_0 \quad X_1 \quad X_2 \quad \ldots \right]$ that consists of $1 \times m$ (polyphase) blocks X_k, as

$$\mathcal{M}_w \mathbf{X} \stackrel{\text{def}}{=} \sum_{k,p} w_k^{(p)} x_k^{(p)} \equiv \mathbf{W}\,\mathbf{X}^T \tag{8.17a}$$

where $\mathbf{W} = \{w_k^{(p)}\}$, and the weights

$$w_k^{(p)} \stackrel{\text{def}}{=} \frac{|V_k^{(p)}|^2}{V_{rms}^2} \tag{8.17b}$$

reflect the relative contributions of the voltage phasors $V_k^{(p)}$ to P and Q_B (see Appendix D). Notice that the weights $\{w_k^{(p)}\}$ are real-valued, non-negative, and add up to unity because $V_{rms}^2 = \sum_{(k,p)} |V_k^{(p)}|^2$. The relations (8.16) follow directly from the observation that $V_k^{(p)}\left[I_k^{(p)}\right]^* = \left|V_k^{(p)}\right|^2\left[Y_k^{(p)}\right]^*$, so that $\mathbf{V}\mathbf{I}^H = V_{rms}^2 \sum_{k,p} w_k^{(p)}\left[Y_k^{(p)}\right]^* \equiv V_{rms}^2 \mathbf{W}\,\mathbf{Y}^H$. In turn, this implies that $P = \Re\left\{\mathbf{V}\mathbf{I}^H\right\} = V_{rms}^2 \mathbf{W}\,\mathbf{G}^T = V_{rms}^2\left(\mathcal{M}_w \mathbf{G}\right)$ and $Q_B = \Im\left\{\mathbf{V}\mathbf{I}^H\right\} =$

$V^2_{rms}\mathbf{W}\,\mathbf{B}^T = V^2_{rms}\left(\mathcal{M}_w\,\mathbf{B}\right)$, where we used the fact that both $\mathbf{G}$ and $\mathbf{B}$ are real-valued.

At this point, we should notice that the equivalent admittance $Y_k^{(p)}$ *cannot be defined* when the component $V_k^{(p)}$ vanishes. Thus, we are led to introduce the index set

$$\Omega_v \stackrel{\text{def}}{=} \{(k,p)\,;\ \frac{|V_k^{(p)}|}{V_{rms}} > \varepsilon\} \tag{8.18}$$

of non-negligible $V_k^{(p)}$ phasors. The value of the threshold ε should be set to keep the relative error $1 - \sum_{(k,p)\in\Omega_v} \frac{|V_k^{(p)}|^2}{V^2_{rms}}$ sufficiently small. In particular, we assume that the DC voltage component V_0 is small enough to be excluded from Ω_v.

We now *redefine* the various phasor arrays ($\mathbf{V}, \mathbf{I}, \mathbf{Y}, \mathbf{G}$, and $\mathbf{B}$) to contain only *in-band* components, i.e., (k,p) combinations that are included in Ω_v: for instance, $\mathbf{V} \stackrel{\text{def}}{=} \{V_k^{(p)}\,;\ (k,p)\in\Omega_v\}$ and $\mathbf{I} \stackrel{\text{def}}{=} \{I_k^{(p)}\,;\ (k,p)\in\Omega_v\}$. Also, we redefine the weights as

$$w_k^{(p)} \stackrel{\text{def}}{=} \frac{|V_k^{(p)}|^2}{\|\mathbf{V}\|^2} \tag{8.19}$$

We will assume from now on that the threshold ε is chosen small enough so that $\|\mathbf{V}\|$ is a very accurate approximation of $V_{rms} \equiv \|v(\cdot)\|$, thus allowing us to substitute one quantity with the other, as needed. However, for the current, we will retain the exact relation

$$I^2_{rms} \equiv \|i(\cdot)\|^2 = \sum_{(k,p)\notin\Omega_v} |I_k^{(p)}|^2 + \|\mathbf{I}\|^2 \tag{8.20a}$$

Consequently,

$$S^2 \stackrel{\text{def}}{=} V^2_{rms}\,I^2_{rms} = S^2_\perp + V^2_{rms}\,\|\mathbf{I}\|^2 \tag{8.20b}$$

where the *out-of-band* component $S_\perp$ captures the apparent power contributed by current components that are excluded from $\mathbf{I}$, viz.,

$$S^2_\perp \stackrel{\text{def}}{=} V^2_{rms} \sum_{(k,p)\notin\Omega_v} |I_k^{(p)}|^2 \tag{8.20c}$$

The 2-component expression (8.20b) can now be refined by splitting the norm $\|\mathbf{I}\|^2$ into several components, as detailed in the following section.

8.5 A 5-Component Decomposition of Apparent Power

A key step in relating the residual power quality gap $S^2 - (P^2 + Q_B^2) - S_\perp^2$ to the variation (a.k.a. "statistical spread") of $\mathbf{G}$ and $\mathbf{B}$ is the *quadrature decomposition* of the (in-band) current norm, viz.,

$$\| \mathbf{I} \|^2 = \sum_{(k,p)\in\Omega_v} |I_k^{(p)}|^2 = \sum_{(k,p)\in\Omega_v} [g_k^{(p)}]^2 |V_k^{(p)}|^2 + \sum_{(k,p)\in\Omega_v} [b_k^{(p)}]^2 |V_k^{(p)}|^2$$

where we used the relation

$$|I_k^{(p)}|^2 = |Y_k^{(p)}|^2 \, |V_k^{(p)}|^2 = \left([g_k^{(p)}]^2 + [b_k^{(p)}]^2\right) |V_k^{(p)}|^2 \qquad \text{for all } (k, p) \in \Omega_v$$

Consequently, the squared current phasor-array norm $\| \mathbf{I} \|^2$ decomposes into *two* components, viz.,

$$\begin{aligned} \| \mathbf{I} \|^2 &= V_{rms}^2 \left(\sum_{(k,p)\in\Omega_v} [g_k^{(p)}]^2 w_k^{(p)} + \sum_{(k,p)\in\Omega_v} [b_k^{(p)}]^2 w_k^{(p)} \right) \\ &= V_{rms}^2 \left(\mathcal{M}_w \, \mathbf{G}^2 + \mathcal{M}_w \, \mathbf{B}^2 \right) \end{aligned} \tag{8.21}$$

where we introduced a shorthand notation for element-wise squaring, viz.,

$$\mathbf{X}^2 \stackrel{\text{def}}{=} \mathbf{X} \circ \mathbf{X} \tag{8.22}$$

Here $\circ$ denotes the Schur–Hadamard product [260] (same as the array multiplication operation $A\,.*\,B$ in MATLAB). Thus, $\mathbf{X}^2$ is an array consisting of the squares of the elements of $\mathbf{X}$.

The decomposition (8.21) of the squared in-band current norm $\| \mathbf{I} \|^2$ involves the weighted means of the *squared* parameter sets $\mathbf{G}^2$ and $\mathbf{B}^2$. A well-known result from statistics (see Appendix D) allows a further decomposition of $\mathcal{M}_w \, \mathbf{G}^2$ and $\mathcal{M}_w \, \mathbf{B}^2$, viz.,

$$\begin{aligned} \mathcal{M}_w \, \mathbf{G}^2 &\equiv \sum_{(k,p)\in\Omega_v} w_k^{(p)} \, [g_k^{(p)}]^2 = \sigma_g^2 + \mu_g^2 \\ \mathcal{M}_w \, \mathbf{B}^2 &\equiv \sum_{(k,p)\in\Omega_v} w_k^{(p)} \, [b_k^{(p)}]^2 = \sigma_b^2 + \mu_b^2 \end{aligned} \tag{8.23a}$$

where

$$\begin{aligned} \sigma_g^2 &\stackrel{\text{def}}{=} \mathcal{M}_w \, \mathbf{G}^2 - (\mathcal{M}_w \, \mathbf{G})^2 = \sum_{(k,p)\in\Omega_v} w_k^{(p)} \, (g_k^{(p)} - \mu_g)^2 \\ \sigma_b^2 &\stackrel{\text{def}}{=} \mathcal{M}_w \, \mathbf{B}^2 - (\mathcal{M}_w \, \mathbf{B})^2 = \sum_{(k,p)\in\Omega_v} w_k^{(p)} \, (b_k^{(p)} - \mu_b)^2 \end{aligned} \tag{8.23b}$$

and we used the definitions (8.16b). Notice that for any (real-valued) array $\mathbf{X}$, the statistical metric $\mathcal{M}_w \mathbf{X}^2 - (\mathcal{M}_w \mathbf{X})^2$ coincides with the *variance* of the set of elements of the array $\mathbf{X}$ (see Appendix D). Consequently, $\sigma_g^2 \geq 0$ and $\sigma_b^2 \geq 0$.

Combining these results with the decompositions (8.20) and (8.21), we obtain a five-component decomposition of apparent power, viz.,

$$S^2 = P^2 + Q_B^2 + \underbrace{N_g{}^2 + N_b{}^2 + S_\perp^2}_{D^2} \tag{8.24a}$$

where $S_\perp$ is defined in (8.20c), and

$$\begin{aligned} P &= V_{rms}^2\, \mu_g \quad , \quad & Q_B &= V_{rms}^2\, \mu_b \\ N_g &\stackrel{\text{def}}{=} V_{rms}^2\, \sigma_g \quad , \quad & N_b &\stackrel{\text{def}}{=} V_{rms}^2\, \sigma_b \end{aligned} \tag{8.24b}$$

It will be convenient to define an "N-metric," which quantifies the effect of uneven spread over phases and harmonics of the equivalent admittance, namely,

$$N \stackrel{\text{def}}{=} \sqrt{N_g^2 + N_b^2} \tag{8.24c}$$

In summary, we observe that Budeanu's distortion power D combines the effects of three distinct circuit-based imperfections:

- The presence of out-of-band current components, caused by load nonlinearity, which gives rise to the apparent power component $S_\perp$.
- The uneven spread over phases and multiple harmonics of the equivalent conductances $\mathbf{G} = \{g_k^{(p)}\}$. This effect is quantified via the component N_g.
- The uneven spread over phases and multiple harmonics of the equivalent susceptances $\mathbf{B} = \{b_k^{(p)}\}$. This effect is quantified via the component N_b.

Notice, of course, that $D^2 = N_g^2 + N_b^2 + S_\perp^2 \equiv N^2 + S_\perp^2$.

Some of the earlier decompositions of apparent power in a *single-phase system* can now be re-interpreted in terms of the 5-component decomposition (8.24). For instance:

- The three-component decomposition $S^2 = S_R^2 + S_X^2 + S_D^2$ of Shepherd and Zakikhani [249] is obtained from (8.24) by defining the aggregate components $S_R = \sqrt{P^2 + N_g^2}$ and $S_X = \sqrt{Q_B^2 + N_b^2}$. Also their "distortion power" component S_D coincides with our "out-of-band" component $S_\perp$.
- The three-component decomposition $S^2 = P^2 + S_c^2 + S_Q^2$ of Sharon [250] partially refines that of Shepherd and Zakikhani by splitting off P from the S_R-component. This means that, using the notation of the 5-component decomposition (8.24), $S_c = \sqrt{N_g^2 + S_\perp^2}$ and $S_Q = \sqrt{Q_B^2 + N_b^2}$.

- The *single-phase* three-component decomposition $S^2 = P^2 + D_s^2 + Q^2$ of Czarnecki [261] corresponds to a slightly different aggregation than the two previous cases, namely $D_s = N_g$ and $Q = \sqrt{Q_B^2 + N_b^2}$. This author also discusses the possibility of an additional component like $S_\perp$, caused by the presence of load-generated current harmonics.

A comparison with Czarnecki's 5-component decomposition for polyphase systems [251, 265], known as *Currents' Physical Components (CPC)*, is provided in Sect. 8.8.

Example 8.1 (Switched resistive load): A single-phase system, consisting of a switched purely-resistive load (with duty factor D), driven by a sinusoidal voltage (see also Sect. L.1.1), as shown in Fig. 8.4.

Since the voltage is purely-sinusoidal and single phase, it follows that N_g and N_b vanish, and $P = \Re\{V_1 I_1^*\}$, $Q_B = \Im\{V_1 I_1^*\}$. The rms voltage phasor is $V_1 = \frac{V_{max}}{\sqrt{2}j}$, where V_{max} is the voltage amplitude, so $V_{rms} = |V_1|$. In contrast, the current consists of multiple harmonics: $I_{rms} = \frac{I_{max}}{\sqrt{2}}\sqrt{D + \frac{1}{\pi}\sin\pi D}$, while the fundamental harmonic current phasor is $I_1 = \frac{I_{max}}{\sqrt{2}j}\left[D + \frac{1}{\pi}\sin\pi D\right]$ (where I_{max} denotes the current amplitude). Notice that $\phi_1 \stackrel{\text{def}}{=} \arg\frac{V_1}{I_1} = 0$, so that $Q_B = |V_1|\,|I_1|\sin\phi_1 = 0$. Nevertheless, $S^2 > P^2$ because of non-vanishing higher current harmonics, all of which belong in the out-of-band component. Consequently,

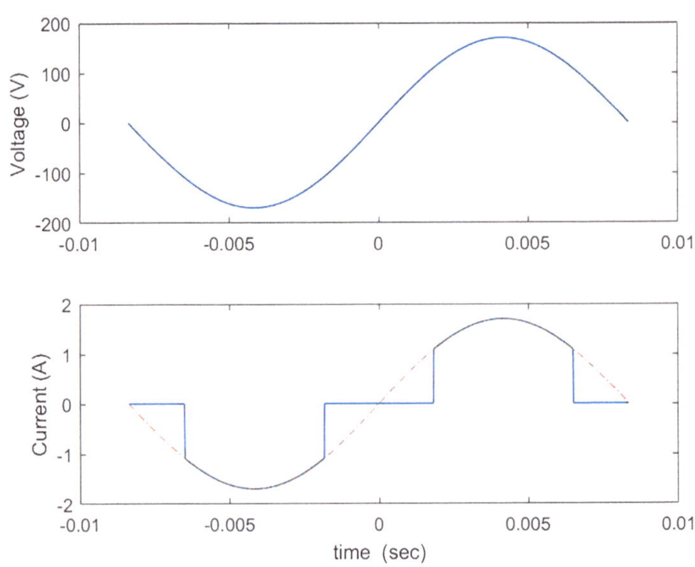

Fig. 8.4 Switched resistive load: voltage (top) and current (bottom, solid line).

$$S = \frac{V_{max}\, I_{max}}{2}\sqrt{D + \frac{\sin \pi D}{\pi}} \quad , \qquad P = \frac{V_{max}\, I_{max}}{2}\left(D + \frac{\sin \pi D}{\pi}\right)$$

so that $S_\perp = \sqrt{S^2 - P^2}$, since Q_B, N_g, and N_b all vanish. In this case, the inactive power is entirely due to the out-of-band current harmonics, rather than to a phase-shift between voltage and current phasors. ■

Example 8.2 (Single-phase purely-reactive load): A single-phase, ideal linear inductor of unit reactance at the fundamental frequency, driven by fundamental and 5-th harmonic voltages, of magnitudes 1 and 0.1, respectively (see also Sect. L.1.2). The corresponding voltage and current phasors are

$$V_1 = 1 \ , \qquad I_1 = -j \ , \qquad V_5 = 0.1 \ , \qquad I_5 = -0.02j$$

so that $g_k = 0$ for all k, and $b_1 = 1$, $b_5 = 0.2$.

In this example $V_{rms}^2 = 1.01$ and $I_{rms}^2 = 1.0004$, so that $S = 1.0052$. Also, the only non-vanishing weights are $w_1 = |V_1|^2/V_{rms}^2 = 0.9901$ and $w_5 = |V_5|^2/V_{rms}^2 = 0.0099$, resulting in

$$\mu_g = 0 \ , \quad \sigma_g = 0 \ , \quad \mu_b = 0.9921 \ , \quad \sigma_b = 0.0792$$

Consequently, the five power components are

$$P = 0 \, , \quad Q_B = 1.0020 \, , \quad N_g = 0 \, , \quad N_b = 0.0800 \, , \quad S_\perp = 0$$

Notice that most of the apparent power is contributed by Q_B, as expected for a purely-reactive load, with a small contribution from N_b, which is caused by the uneven spread of $\{b_k\}$ across harmonics. While the numerical discrepancy between S^2 and $P^2 + Q_B^2$ is small, it is important to note that a non-vanishing distortion power component $D = \sqrt{N_g^2 + N_b^2 + S_\perp^2}$ will exist whenever the load is frequency-dependent, and the driving voltage has higher harmonics. ■

Example 8.3 (RLC load, balanced sinusoidal excitation): A three-phase system, driven by balanced voltages (of unit magnitude) at the fundamental frequency. The unbalanced load consists of a resistor in phase a, an inductor in phase b, and a capacitor in phase c, each drawing a unit magnitude current (see Sect. L.1.3 for additional details). In this example $V_{rms}^2 = 3 = I_{rms}^2$, so that $S = 3$. The three weights $w_1^{(p)} = |V_1^{(p)}|^2/V_{rms}^2$ are all equal to $1/3$, resulting in

$$\mu_g = 1/3 \ , \quad \sigma_g^2 = 2/9 \ , \quad \mu_b = 0 \ , \quad \sigma_b^2 = 2/3$$

Consequently, the five power components are

$$P = 1 \ , \ \ N_g = \sqrt{2} \ , \ \ Q_B = 0 \ , \ \ N_b = \sqrt{6} \ , \ \ S_\perp = 0$$

The inactive power is entirely due to the imbalance of conductances and susceptances: the N-metric contributes 89% of S^2! . Notice that the contributions of the capacitor and inductor to Budeanu's reactive power cancel each other, resulting in $Q_B = 0$. ■

Example 8.4 (Strongly-unbalanced load, balanced sinusoidal excitation): A three phase Δ to Y transformer with 1:1 winding ratio, driven on the primary side by a balanced fundamental harmonic voltage with magnitude 1 (line to neutral). The load is an impedance of $Z = \sqrt{3} + j$, connected between phase a and the neutral terminal on the secondary side (see also Sect. L.1.6). A real-life example of this scenario is provided by an electric train, which loads only a single phase of a three-phase system, and is supplied through such a Δ to Y transformer [265, Illustration 1].[1]

In this example $V_{rms}^2 = 3$ and $I_{rms}^2 = \frac{3}{2}$, so that $S = \frac{3}{\sqrt{2}}$. Also, the three weights $w_1^{(p)} = \left|V_1^{(p)}\right|^2 / V_{rms}^2$ are all equal to $1/3$, so that

$$\mu_g = \sqrt{3}/4 \ , \ \ \sigma_g^2 = 1/8 \ , \ \ \mu_b = 1/4 \ , \ \ \sigma_b^2 = 1/8$$

Consequently, the five power components are

$$P = \frac{3\sqrt{3}}{4} \ , \ \ Q_B = \frac{3}{4} \ , \ \ N_g = \frac{3}{2\sqrt{2}} \ , \ \ N_b = \frac{3}{2\sqrt{2}} \ , \ \ S_\perp = 0$$

Notice that most of the apparent power (i.e., 50% of S^2) is contributed by N_g and N_b, which are caused by the uneven spread of $\{ g_1^{(p)} \}$ and $\{ b_1^{(p)} \}$ across phases, with a smaller contribution from Q_B, caused by the phase-shifts between voltage and current.

8.6 Symmetrical Sequence Components for P and Q_B

Symmetrical sequence components of voltage and current phasors, which were introduced in Sect. 1.3, also provide a decomposition of P and Q_B that leads to a refinement of the 5-component power decomposition (8.24). Following the procedure described in Sect. 1.3, we can express the quantity $\mathbf{V}\mathbf{I}^H$, used to

[1] The same example, but with different parameter values, is presented in [265, Fig. 10].

determine both P and Q_B, in terms of the voltage and current symmetrical sequence components phasors, viz.,

$$\mathbf{V}\mathbf{I}^H \equiv \sum_{k=0}^{\infty} V_k I_k^H = \sum_{k=0}^{\infty} \left[V_{k,0} \;\; V_{k,+} \;\; V_{k,-} \right] \left[I_{k,0} \;\; I_{k,+} \;\; I_{k,-} \right]^H$$

where we used the unitarity property $\mathcal{M}_{sc} \mathcal{M}_{sc}^H = I$. Recall that the subscript "0" denotes the zero-sequence component, while the subscripts "+" and "−" denote the positive and negative sequence components, respectively.

Splitting this expression into contributions from individual sequence components, we obtain additive decompositions for P and Q_B, viz.,

$$P = P_0 + P_+ + P_- \quad , \qquad Q_B = Q_{B,0} + Q_{B,+} + Q_{B,-} \qquad (8.25a)$$

where

$$\begin{aligned} P_0 &= \sum_{k=0}^{\infty} \Re\left\{ V_{k,0} I_{k,0}^H \right\} , & Q_{B,0} &= \sum_{k=0}^{\infty} \Im\left\{ V_{k,0} I_{k,0}^H \right\} \\ P_+ &= \sum_{k=0}^{\infty} \Re\left\{ V_{k,+} I_{k,+}^H \right\} , & Q_{B,+} &= \sum_{k=0}^{\infty} \Im\left\{ V_{k,+} I_{k,+}^H \right\} \\ P_- &= \sum_{k=0}^{\infty} \Re\left\{ V_{k,-} I_{k,-}^H \right\} , & Q_{B,-} &= \sum_{k=0}^{\infty} \Im\left\{ V_{k,-} I_{k,-}^H \right\} \end{aligned} \qquad (8.25b)$$

This means that the 5-component decomposition (8.24) consists, in fact, of nine components, viz.,

$$\begin{aligned} S^2 = S_\perp^2 &+ (P_+ + P_- + P_0)^2 + N_g^2 \\ &+ (Q_{B,+} + Q_{B,-} + Q_{B,0})^2 + N_b^2 \end{aligned} \qquad (8.25c)$$

8.7 A 7/11-Component Decomposition of Apparent Power

The 5/9-component decomposition (8.25c) can be further refined to provide an indication of load imbalance. Recall that a load is considered balanced (across phases) when the equivalent admittance $Y_k^{(p)} \equiv g_k^{(p)} - jb_k^{(p)}$ is independent of the phase index "p." One possible measure of imbalance is provided by the variation of $g_k^{(p)}$ and $b_k^{(p)}$ around their harmonic-dependent means

$$\mu_g(k) \stackrel{\text{def}}{=} \sum_p g_k^{(p)} \frac{w_k^{(p)}}{\sum_{i=1}^m w_k^{(i)}} \quad , \qquad \mu_b(k) \stackrel{\text{def}}{=} \sum_p b_k^{(p)} \frac{w_k^{(p)}}{\sum_{i=1}^m w_k^{(i)}} \qquad (8.26)$$

Notice that, in general, the range of p-values used in the above summations can vary with the harmonic index k, although this is not a common occurrence. In order to streamline our presentation, we will assume in this section that the set Ω_v is "rectangular," i.e., the *k-dependent cross-section* set $\Omega_v(k) \overset{\text{def}}{=} \{p;\ (k,p) \in \Omega_v\}$ coincides, for each choice of k, with the full range $1 \le p \le m$. The non-rectangular case is addressed in Appendix E.

In order to refine the 5-component power decomposition (8.24), we now apply the ANOVA-based decomposition of variance for grouped data (see Appendix D, Sect. D.2) to split σ_g^2, the weighted variance of the elements of the array $\mathbf{G}$, into a sum of two partial variances, namely, $\sigma_g^2 = \sigma_{gu}^2 + \sigma_{gs}^2$. We recall that

$$\sigma_g^2 = \mathcal{M}_w\left(\mathbf{G} - \mu_g\,\mathbf{1}\right)^2 \equiv \sum_{(k,p)\in\Omega_v} w_k^{(p)}\left(g_k^{(p)} - \mu_g\right)^2 \tag{8.27a}$$

where $\mathbf{1}$ denotes an (row) array consisting entirely of unity elements, viz.,

$$\mathbf{1} \overset{\text{def}}{=} \begin{bmatrix} 1 & 1 & 1 & \dots \end{bmatrix} \tag{8.27b}$$

Thus σ_g^2 describes the combined effect of variation across harmonics and phases, while variation across phases alone is captured by

$$\sigma_{gu}^2 \overset{\text{def}}{=} \mathcal{M}_w\left(\mathbf{G} - \mathbf{G}_h\right)^2 \equiv \sum_{(k,p)\in\Omega_v} w_k^{(p)}\left(g_k^{(p)} - \mu_g(k)\right)^2 \tag{8.28a}$$

which is the (weighted) variance of the array $\mathbf{G} - \mathbf{G}_h$, where

$$\mathbf{G}_h \overset{\text{def}}{=} \begin{bmatrix} \boldsymbol{\mu}_g(0) & \boldsymbol{\mu}_g(1) & \boldsymbol{\mu}_g(2) & \dots \end{bmatrix}, \quad \boldsymbol{\mu}_g(k) \overset{\text{def}}{=} \mu_g(k)[\underbrace{1\ 1\ \dots\ 1}_{m}] \tag{8.28b}$$

In the language of Appendix D, σ_{gu}^2 is the *variance within the groups*, where we associate one data group with each harmonic. Thus σ_{gu}^2 represents the effect of phase imbalance, which causes variation of the equivalent conductances $\{g_k^{(p)}\}$ across phases within each individual harmonic. On the other hand, we recognize

$$\sigma_{gs}^2 \overset{\text{def}}{=} \mathcal{M}_w\left(\mathbf{G}_h - \mu_g\,\mathbf{1}\right)^2 \equiv \sum_{(k,p)\in\Omega_v} w_k^{(p)}\left(\mu_g(k) - \mu_g\right)^2 \tag{8.28c}$$

as the *variance between the groups*, thus capturing the effect of harmonic distortion.

Similarly, let

$$\mathbf{B}_h \stackrel{\text{def}}{=} \left[\boldsymbol{\mu}_b(0) \;\; \boldsymbol{\mu}_b(1) \;\; \boldsymbol{\mu}_b(2) \;\; \ldots \;\; \right] \tag{8.29a}$$

where $\boldsymbol{\mu}_b(k) \stackrel{\text{def}}{=} \mu_b(k)[\underbrace{1 \; 1 \; \ldots \; 1}_{m}]$, and define the *variance within the groups* for the elements of $\mathbf{B}$, viz.,

$$\sigma_{bu}^2 \stackrel{\text{def}}{=} \mathcal{M}_w \left(\mathbf{B} - \mathbf{B}_h\right)^2 \equiv \sum_{(k,p)\in\Omega_v} w_k^{(p)} \left(b_k^{(p)} - \mu_b(k)\right)^2 \tag{8.29b}$$

and the *variance between the groups*

$$\sigma_{bs}^2 \stackrel{\text{def}}{=} \mathcal{M}_w \left(\mathbf{B}_h - \mu_b \, \mathbf{1}\right)^2 \equiv \sum_{(k,p)\in\Omega_v} w_k^{(p)} \left(\mu_b(k) - \mu_b\right)^2 \tag{8.29c}$$

so that $\sigma_b^2 = \sigma_{bs}^2 + \sigma_{bu}^2$.

Applying these results to the decomposition (8.24) results in splitting N_g^2 and N_b^2 into a sum of two components each, viz.,

$$N_g^2 = N_{gs}^2 + N_{gu}^2 \quad , \quad N_b^2 = N_{bs}^2 + N_{bu}^2 \tag{8.30a}$$

where

$$\begin{aligned} N_{gs} &\stackrel{\text{def}}{=} V_{rms}^2 \, \sigma_{gs} \quad , \quad & N_{gu} &\stackrel{\text{def}}{=} V_{rms}^2 \, \sigma_{gu} \\ N_{bs} &\stackrel{\text{def}}{=} V_{rms}^2 \, \sigma_{bs} \quad , \quad & N_{bu} &\stackrel{\text{def}}{=} V_{rms}^2 \, \sigma_{bu} \end{aligned} \tag{8.30b}$$

The final result is a decomposition of S^2 into the sum of seven distinct components, viz.,

$$\begin{aligned} S^2 = {} & S_\perp^2 + (P_+ + P_- + P_0)^2 + N_{gs}^2 + N_{gu}^2 \\ & + (Q_{B,+} + Q_{B,-} + Q_{B,0})^2 + N_{bs}^2 + N_{bu}^2 \end{aligned} \tag{8.31}$$

This result is known as a 7/11-component decomposition [253], in view of the (secondary) splitting into symmetrical sequence components of the real power P and Budeanu's reactive power Q_B. A detailed comparison with Czarnecki's 5-component decomposition is provided in Sect. 8.8.

Notice that, by definition, the components $S_\perp$, N_{gs}, N_{gu}, N_{bs}, N_{bu} are always non-negative and thus cannot be network-conservative. Only the real power P

and the reactive power Q_B, as well as their symmetrical sequence components are network-conservative, as we demonstrate in Sect. 9.6 (see also discussion in Appendix I).

Finally, we should mention that an alternative splitting of N_g^2 and N_b^2 can be obtained by swapping the roles of harmonic and phase in the ANOVA-based method, resulting in a "data group" being associated with each individual phase, rather than with each harmonic.

Example 8.5 (RLC load, two harmonics): A three-phase system, driven by balanced fundamental and fifth harmonic voltages, with magnitudes 1 and 0.1, respectively, so that the in-band index set is $\Omega_v = \{(k, p);\ k = 1, 5 \text{ and } p = a, b, c\}$. The unbalanced load consists of a resistor in phase a, an inductor in phase b, and a capacitor in phase c, each drawing a unit magnitude current at the fundamental harmonic (see Sect. L.1.4 for additional details). In this example, $V_{rms}^2 = 3.03$ and $I_{rms}^2 = 3.2604$, so that $S = 3.1431$. The three fundamental-frequency weights $w_1^{(p)} = \left|V_1^{(p)}\right|^2/V_{rms}^2$ are all equal to 0.3300, while the 5-th harmonic weights $w_5^{(p)} = \left|V_5^{(p)}\right|^2/V_{rms}^2$ are all equal to $0.01\, w_1^{(p)} =$ 0.0033. Consequently,

μ_g	$\mu_g(1)$	$\mu_g(5)$	μ_b	$\mu_b(1)$	$\mu_b(5)$
$\frac{1}{3}$	$\frac{1}{3}$	$\frac{1}{3}$	−0.0158	0	−1.6

and

σ_g	σ_{gs}	σ_{gu}	σ_b	σ_{bs}	σ_{bu}
0.4714	0	0.4714	0.8617	0.1584	0.8470

The resulting seven power components are

P	N_{gs}	N_{gu}	Q_B	N_{bs}	N_{bu}	$S_\perp$
1.0100	0	1.4284	−0.0480	0.4800	2.5663	0

Notice that most of the apparent power is contributed by N_{gu} and N_{bu}, which are generated by the uneven spread of $\{g_k^{(p)}\}$ and $\{b_k^{(p)}\}$ across phases, with a smaller contribution from (a) N_{bs} – caused by frequency dependence of the susceptances $\{b_k^{(p)}\}$, and (b) Q_B – caused by the phase-shifts between voltage and current. Also notice that $N_{gs} = 0$ as a result of frequency independence of the conductances $\{g_k^{(p)}\}$. ■

Example 8.6 (Adjustable-speed drive): This lab-generated steady-state case involves a 350-HP industrial adjustable-speed drive (ASD), operating in parallel with a linear unbalanced electrical load (see Sect. L.1.7 for additional details). The seven power components for this case (in kW, kVar, and kVA, respectively) are

P	N_{gs}	N_{gu}	Q_B	N_{bs}	N_{bu}	$S_\perp$
223.3	3.668	7.025	51.55	4.549	7.154	6.078

Also, $V_{rms} = 386$ V and $I_{rms} = 595$ A, so that $S = 229.6$ kVA. Notice that each one of the six inactive power components — N_{gs}, N_{gu}, Q_B, N_{bs}, N_{bu}, and $S_\perp$ — has a non-negligible contribution to the power quality gap $S^2 - P^2$. In particular, N_{gu} and N_{bu} indicate a non-negligible load imbalance, while N_{gs} and N_{bs} imply a (small) harmonic distortion. ■

Example 8.7 (Hydro-Ottawa distribution grid, steady-state): This real-world steady-state case was obtained from Hydro-Ottawa, the largest local distribution company in eastern Ontario, Canada (see Sect. L.1.8 for additional details). The seven power components for this case (in MW, MVar, and MVA, respectively) are

P	N_{gs}	N_{gu}	Q_B	N_{bs}	N_{bu}	$S_\perp$
8.16	0.52	1.31	−1.94	0.82	0.83	0

Also $V_{rms} = 26.3$ kV and $I_{rms} = 326$ A, resulting in $S = 8.57$ MVA and a power factor of 0.95. The negative and zero-sequence components of P and Q_B are negligible, so that $P_+ = P$ and $Q_{B,+} = Q_B$. Notice that each one of the five inactive power components — N_{gs}, N_{gu}, Q_B, N_{bs}, and N_{bu} — has a non-negligible contribution to the power quality gap $S^2 - P^2$. In particular, N_{gu} and N_{bu} indicate a significant load imbalance, while N_{gs} and N_{bs} imply a noticeable harmonic distortion, as is also evident from the current and voltage waveforms shown in Sect. L.1.8. ■

8.8 Induced Orthogonal Current Decompositions

The S^2-decompositions (8.24) and (8.31), which rely on splitting I_{rms}^2 into several distinct components, induce matching decompositions in both the waveform domain and the phasor-array domain. For instance, the out-of-band current component associated with $S_\perp$ is defined as

$$i_\perp(t) \stackrel{\text{def}}{=} \sqrt{2} \sum_{(k,p)\notin\Omega_v} \Re\left\{ I_k^{(p)} e^{jk\omega_0 t} \right\} \eta_p \tag{8.32a}$$

where η_p is a $1 \times m$ unit vector, viz.,

$$\eta_p \stackrel{\text{def}}{=} [\underbrace{0 \ldots 0\ 1}_{p}\ 0 \ldots 0] \quad , \quad 1 \le p \le m \tag{8.32b}$$

This implies that $S_\perp = V_{rms} \| i_\perp(\cdot) \|$ (recall (8.20c)). Also, $i(t) = i_\Omega(t) \oplus i_\perp(t)$, where

$$i_\Omega(t) \stackrel{\text{def}}{=} \sqrt{2} \sum_{(k,p)\in\Omega_v} \Re\left\{ I_k^{(p)} e^{jk\omega_0 t} \right\} \eta_p \tag{8.32c}$$

and $\oplus$ denotes a sum of two mutually-orthogonal waveforms. The orthogonality property $i_\Omega(\cdot) \perp i_\perp(\cdot)$ follows from the fact that all (k-th harmonic, p-th phase) waveforms are orthogonal to all (ℓ-th harmonic, q-th phase) waveforms, as long as either $k \neq \ell$ or $p \neq q$, as demonstrated in detail in Sect. 1.2.2.

3-Component Decomposition

Next, we observe that the voltage–current relation $I_k^{(p)} = \left(g_k^{(p)} - jb_k^{(p)}\right) V_k^{(p)}$, which was first introduced in (8.15a), induces a decomposition of $i_\Omega(t)$ into two mutually-orthogonal *quadrature components*. Starting from the expression (8.32c), namely,

$$i_\Omega(t) = \sqrt{2} \sum_{(k,p)\in\Omega_v} \Re\left\{ \left(g_k^{(p)} - jb_k^{(p)}\right) V_k^{(p)} e^{jk\omega_0 t} \right\} \eta_p$$

we conclude that

$$i_\Omega(t) = \underbrace{\sum_{(k,p)\in\Omega_v} g_k^{(p)} v_k^{(p)}(t)}_{i_\mathcal{V}(t)} + \underbrace{\sum_{(k,p)\in\Omega_v} b_k^{(p)} \left[\mathcal{H} v_k^{(p)}(t)\right]}_{i_\Upsilon(t)} \stackrel{\text{def}}{=} i_\mathcal{V}(t) \oplus i_\Upsilon(t) \tag{8.33a}$$

where

$$\begin{aligned} v_k^{(p)}(t) &\stackrel{\text{def}}{=} \sqrt{2}\, \Re\left\{ V_k^{(p)} e^{jk\omega_0 t} \right\} \eta_p \\ \mathcal{H}\, v_k^{(p)}(t) &\stackrel{\text{def}}{=} \sqrt{2}\, \Re\left\{ \left(-j\, V_k^{(p)}\right) e^{jk\omega_0 t} \right\} \eta_p \end{aligned} \tag{8.33b}$$

Notice that $v_k^{(p)}(t)$ are special polyphase waveforms, each one representing the contribution of a single phase.

We recognize $\mathcal{H}$ as the well-known *Hilbert transform* [82, 253, 267]: when applied to a periodic waveform, it imparts a 90° phase-shift to every harmonic, namely

$$\mathcal{H}\left\{\sqrt{2}\sum_k \Re\left\{X_k\, e^{jk\omega_0 t}\right\}\right\} = \sqrt{2}\sum_k \Re\left\{(-jX_k)e^{jk\omega_0 t}\right\}$$

The two waveforms $v_k^{(p)}(\cdot)$ and $\mathcal{H}\, v_k^{(p)}(\cdot)$ belong in the 2-dimensional subspace of all (k-th harmonic, p-th phase) waveforms, as explained in Appendix E. It follows from (1.12b) that

$$\langle\, \mathcal{H}\, v_k^{(p)}(\cdot)\,,\; v_k^{(p)}(\cdot)\,\rangle = \Re\left\{[-jV_k^{(p)}][V_k^{(p)}]^*\right\} = 0$$

which means that these two waveforms form an orthogonal (voltage-derived) basis for this 2-dimensional subspace.

In fact, we demonstrate in Sect. E.2 that $\left\{v_k^{(p)}(\cdot)\,,\, \mathcal{H}\, v_k^{(p)}(\cdot)\,;\; (k,p) \in \Omega_v\right\}$ are all mutually-orthogonal waveforms, i.e., $\langle\, v_k^{(p)}(\cdot)\,,\; v_\ell^{(q)}(\cdot)\,\rangle = \delta_{k,l}\;\delta_{p,q}\,\left|V_k^{(p)}\right|^2$ and $\langle\, \mathcal{H}\, v_k^{(p)}(\cdot)\,,\; \mathcal{H}\, v_\ell^{(q)}(\cdot)\,\rangle = \delta_{k,l}\;\delta_{p,q}\,\left|V_k^{(p)}\right|^2$, as well as $\langle\, v_k^{(p)}(\cdot)\,,\; \mathcal{H}\, v_\ell^{(q)}(\cdot)\,\rangle = 0$ for all k, ℓ, p, q. This establishes the orthogonality property $i_{\mathcal{V}}(\cdot) \perp i_{\Upsilon}(\cdot)$ in (8.33), which can be used, for instance, to obtain the norm identity

$$\|\, i_\Omega(\cdot)\|^2 = \sum_{(k,p)\in\Omega_v} [g_k^{(p)}]^2 |V_k^{(p)}|^2 + \sum_{(k,p)\in\Omega_v} [b_k^{(p)}]^2 |V_k^{(p)}|^2$$

thus providing an alternative derivation of (8.21).

Combining the decomposition of $i(t)$ with that of $i_\Omega(t)$ results in a 3-component orthogonal current decomposition

$$i(t) = i_{\mathcal{V}}(t) \oplus i_{\Upsilon}(t) \oplus i_{\perp}(t) \tag{8.34}$$

The same argument that was used to establish the orthogonality property $i_\Omega(\cdot) \perp i_\perp(\cdot)$ also implies that $i_{\mathcal{V}}(\cdot)$ and $i_{\Upsilon}(\cdot)$ are both orthogonal to $i_\perp(\cdot)$. In view of the expressions (8.33) for $i_{\mathcal{V}}(\cdot)$ and $i_{\Upsilon}(\cdot)$, we conclude that the Hilbert space of all T-periodic m-phase waveforms consists of three mutually-orthogonal subspaces:

- The *voltage-aligned subspace* $\mathcal{V} \overset{\text{def}}{=} span\left\{v_k^{(p)}(\cdot)\,;\; (k,p) \in \Omega_v\right\}$
- The *quadrature subspace* $\Upsilon \overset{\text{def}}{=} span\left\{\mathcal{H}\, v_k^{(p)}(\cdot)\,;\; (k,p) \in \Omega_v\right\}$
- The *out-of-band subspace*, consisting of all harmonics (and phases) excluded from Ω_v

Finally, we observe that mutual orthogonality of the three current components in (8.34) implies a Pythagorean identity, namely, $\| i(\cdot)\|^2 = \| i_{\mathcal{V}}(\cdot)\|^2 + \| i_{\Upsilon}(\cdot)\|^2 + \| i_{\perp}(\cdot)\|^2$. Multiplying this result by V_{rms}^2 generates a 3-component decomposition of S^2, viz.,

$$S^2 = S_{\mathcal{V}}^2 + S_{\Upsilon}^2 + S_{\perp}^2$$

where (recall (8.21))

$$S_{\mathcal{V}} \stackrel{\text{def}}{=} V_{rms}\| i_{\mathcal{V}}(\cdot)\| \equiv V_{rms}^2\sqrt{\mathcal{M}_w\,\mathbf{G}^2}$$

$$S_{\Upsilon} \stackrel{\text{def}}{=} V_{rms}\| i_{\Upsilon}(\cdot)\| \equiv V_{rms}^2\sqrt{\mathcal{M}_w\,\mathbf{B}^2}$$

When $S_{\perp}$ is negligible, this coincides with the 2-component decomposition of S^2 proposed by Shepherd and Zakikhani [249].

7-Component Decomposition

Our next step is to split each one of the current waveforms $i_{\mathcal{V}}(\cdot)$ and $i_{\Upsilon}(\cdot)$ into a sum of three mutually-orthogonal components, which will transform (8.34) into an orthogonal 7-component current decomposition. We show in Appendix E that both $i_{\mathcal{V}}(\cdot)$ and $i_{\Upsilon}(\cdot)$ can be expressed as a sum of 3 mutually-orthogonal components (see Theorem E.1), viz.,

$$i(t) = \underbrace{i_F(t) \oplus i_{gs}(t) \oplus i_{gu}(t)}_{i_{\mathcal{V}}(t)} \oplus \underbrace{i_B(t) \oplus i_{bs}(t) \oplus i_{bu}(t)}_{i_{\Upsilon}(t)} \oplus i_{\perp}(t) \qquad (8.35a)$$

where

$$i_F(t) \stackrel{\text{def}}{=} \mu_g \sum_{(k,p)\in\Omega_v} v_k^{(p)}(t) \quad , \quad i_B(t) \stackrel{\text{def}}{=} \mu_b \sum_{(k,p)\in\Omega_v} \mathcal{H}\, v_k^{(p)}(t) \qquad (8.35b)$$

and μ_g , μ_b are as defined in (8.16). Also,

$$i_{gs}(t) \stackrel{\text{def}}{=} \sum_{(k,p)\in\Omega_v} \big(\mu_g(k) - \mu_g\big)\, v_k^{(p)}(t) \quad , \quad i_{gu}(t) = \sum_{(k,p)\in\Omega_v} \big(g_k^{(p)} - \mu_g(k)\big)\, v_k^{(p)}(t) \qquad (8.35c)$$

$$i_{bs}(t) \stackrel{\text{def}}{=} \sum_{(k,p)\in\Omega_v} \big(\mu_b(k) - \mu_b\big)\, \mathcal{H}\, v_k^{(p)}(t) \quad , \quad i_{bu}(t) = \sum_{(k,p)\in\Omega_v} \big(b_k^{(p)} - \mu_b(k)\big)\, \mathcal{H}\, v_k^{(p)}(t)$$

$$\mu_g(k) \stackrel{\text{def}}{=} \frac{\sum_p g_k^{(p)} w_k^{(p)}}{\sum_p w_k^{(p)}} \quad , \quad \mu_b(k) \stackrel{\text{def}}{=} \frac{\sum_p b_k^{(p)} w_k^{(p)}}{\sum_p w_k^{(p)}} \qquad (8.35d)$$

and $i_\perp(t)$ is as defined in (8.32a). The Corollary to Theorem E.1 shows that these relations result in a 7-component decomposition of apparent power, viz.,

$$S^2 = \left(\mu_g^2 + \sigma_{gs}^2 + \sigma_{gu}^2 + \mu_b^2 + \sigma_{bs}^2 + \sigma_{bu}^2\right) V_{rms}^4 + S_\perp^2$$

which coincides with (8.30), (8.31).

The current component $i_F(t)$ is known in the literature as the *Fryze current* associated with the current waveform $i(t)$. It is, in fact, the orthogonal projection of the current waveform on the one-dimensional subspace defined by the voltage waveform (see, e.g., [29]), as shown in Fig. 8.2, namely,

$$i_F(t) = \frac{\langle\, i(\cdot)\,,\; v(\cdot)\,\rangle}{\|\, v(\cdot)\|^2}\; v(t) \equiv \frac{P}{V_{rms}^2}\; v(t) = \mu_g\; v(t)$$

where we used the relation (8.16). Similarly, the so-called *Budeanu current* $i_B(t)$ can be interpreted as the orthogonal projection of the current waveform $i(t)$ on the subspace defined by $\mathcal{H}v(t)$, the *Hilbert-transformed voltage waveform*, viz.,

$$i_B(t) = \frac{\langle\, i(\cdot)\,,\; \mathcal{H}v(\cdot)\,\rangle}{\|\, \mathcal{H}v(\cdot)\|^2}\; \mathcal{H}v(t) \equiv \frac{Q_B}{V_{rms}^2}\; \mathcal{H}v(t) = \mu_b\; \mathcal{H}v(t)$$

where, again, we used the relation (8.16). These geometric interpretations of the Fryze and Budeanu current components provide a convenient starting point for a re-derivation of such current decompositions via a geometric algebra approach, which we present in Sect. 8.9.

Since all seven current components in (8.35a) are orthogonal to each other, eliminating any single component via a suitable compensation scheme results in reduction of I_{rms}, and thus reduces supply line losses. In addition, eliminating any component *except* $i_F(t)$ leaves the real power P unchanged. In particular, elimination of the $i_B(t)$ component results in $Q_B = 0$, while still preserving real power and reducing I_{rms}. Notice, however, that $Q_B = 0$ can be achieved by a variety of distinct compensation schemes: some of these can result in *increased* I_{rms} [262, 263]. This unfortunate feature of Budeanu's reactive power concept was the subject of repeated criticism throughout the past decades, which ultimately resulted to its removal from the latest revision of IEEE Standard 1459 [264].

A direct comparison of the 7-component current decomposition (8.35) with Czarnecki's *Currents' Physical Components (CPC)* theory [265] reveals that the latter is a slightly-aggregated version of (8.35), as detailed in Table 8.1. The main differences between these two orthogonal current decompositions are:

- The CPC component $i_u(t)$ combines the contributions, attributed to phase imbalance, of both $i_{gu}(t)$ and $i_{bu}(t)$, namely $i_u(t) \equiv i_{gu}(t) + i_{bu}(t)$.
- The CPC component $i_r(t)$ aggregates the Budeanu component $i_B(t)$ with the quadrature-spread component $i_{bs}(t)$, i.e., $i_r(t) \equiv i_B(t) + i_{bs}(t)$.

In a *single-phase system*, the CPC component $i_u(t)$ vanishes, resulting in the 4-component orthogonal current decomposition presented in [261, 266].

Table 8.1 Comparing the 7-component orthogonal current decomposition (8.35) with Czarnecki's 5-component decomposition (CPC)

Czarnecki's 3-phase CPC [251, 265]	i_a	i_s	i_r	i_u	i_g
7-component decomposition (8.35)	i_F	i_{gs}	$i_B \oplus i_{bs}$	$i_{gu} \oplus i_{bu}$	$i_\perp$

5-Component Decomposition

A 5-component orthogonal current decomposition, which matches the power decomposition (8.24), can be obtained from (8.35) by aggregating the "spread" and "imbalance" currents, viz.,

$$i_g(t) \stackrel{\text{def}}{=} i_{gs}(t) + i_{gu}(t) = \sum_{(k,p)\in\Omega_v} \left(g_k^{(p)} - \mu_g\right) v_k^{(p)}(t)$$

$$i_b(t) \stackrel{\text{def}}{=} i_{bs}(t) + i_{bu}(t) = \sum_{(k,p)\in\Omega_v} \left(b_k^{(p)} - \mu_b\right) \mathcal{H}\, v_k^{(p)}(t)$$

so that

$$\| i_g(\cdot)\| = V_{rms}\, \sigma_g \quad , \quad \| i_b(\cdot)\| = V_{rms}\, \sigma_b \tag{8.36a}$$

The resulting 5-component orthogonal current decomposition is

$$i(t) = i_F(t) \oplus i_g(t) \oplus i_B(t) \oplus i_b(t) \oplus i_\perp(t) \tag{8.37}$$

The associated Pythagorean identity

$$\| i(\cdot)\|^2 = \| i_F(\cdot)\|^2 + \| i_g(\cdot)\|^2 + \| i_B(\cdot)\|^2 + \| i_b(\cdot)\|^2 + \| i_\perp(\cdot)\|^2$$

when multiplied by V_{rms}^2, recovers the apparent power decomposition (8.24).

8.9 Multivector Representation of Power Components

The development of 5-component and 7-component decompositions in the preceding sections was based on two closely-related sets of ideas:

- In Sect. 8.5, we used the equivalent conductances and susceptances to decompose the squared-rms current $I_{rms}^2 \equiv \sum_{(k,p)} \left|I_k^{(p)}\right|^2 \equiv \| i(\cdot)\|^2$ into five components, which was followed in Sect. 8.7 by a refinement into seven components.

- In Sect. 8.8, we introduced matching decompositions of the current waveform $i(\cdot)$ itself, expressed as sums of mutually-orthogonal components.

Both approaches can be combined into a single conceptual framework using the tools of *Geometric Algebra.* A tutorial introduction to geometric algebra is provided in Appendix F.

A geometric algebra (also known as a Clifford algebra) is a higher-level space generated by introducing a new kind of product between elements of a given Euclidean space (i.e., "vectors") [256, 268, 269]. For instance, the geometric algebra generated by R^N, the Euclidean space of real-valued N-tuples, is itself a Euclidean space, but of dimension 2^N. Its elements are called *multivectors,* and each multivector consists of $N+1$ distinct grade components: grade-0 components are (real) scalars, grade-1 components are (standard) vectors, grade-2 components are called *bivectors,* grade-r components are called r-*vectors,* and so on.

Here we shall consider the geometric algebra generated by the *voltage-aligned subspace* $\mathcal{V} \stackrel{\text{def}}{=} span\left\{v_k^{(p)}(\cdot)\,;\ (k,p) \in \Omega_v\right\}$, which was introduced in Sect. 8.8. In order to simplify our presentation, we assume in the sequel that the current $i(\cdot)$ has no out-of-band components and that the voltage $v(\cdot)$ is band-limited, so that $\Omega_v = \left\{(k,p)\,;\ 0 \le k \le L\,,\ 1 \le p \le m\right\}$ for some positive integer L. This means that $\mathcal{V}$ is, in fact, a Euclidean space, and its (finite) dimension is $N = m(L+1)$.

The resulting geometric algebra is a direct sum of $N+1$ mutually-orthogonal subspaces, one for each grade r in the range $0 \le r \le N$: the grade-1 subspace coincides with the generating Euclidean space. We will use here upper case letters $(P, Q, I_{rms}, \ldots)$ to indicate scalars and lower case boldface letters $(\mathbf{x}, \mathbf{y}, \ldots)$ to indicate vectors (i.e., waveforms). We will use upper case "blackboard" letters, such as $\mathbb{S}$ or $\mathbb{Q}$, to indicate multivectors.

Our first step toward connecting waveforms with power metrics is to introduce an (partial) apparent power multivector that is the *geometric product*[2] of the voltage $\mathbf{v} \equiv v(\cdot)$ and the voltage-aligned current component $\mathbf{i}_{\mathcal{V}} \equiv i_{\mathcal{V}}(\cdot)$, viz.,

$$\mathbb{S}_{\mathcal{V}} \stackrel{\text{def}}{=} \mathbf{v}\,\mathbf{i}_{\mathcal{V}} = \underbrace{\langle\, \mathbf{v}\,,\, \mathbf{i}_{\mathcal{V}} \,\rangle}_{\text{grade-0}} + \underbrace{\mathbf{v} \wedge \mathbf{i}_{\mathcal{V}}}_{\text{grade-2}} \equiv P \oplus \mathbb{S}_g \tag{8.38a}$$

[2] See Appendix F for a discussion of this concept.

The grade-0 component of this multivector is the scalar

$$\langle \mathbf{v}, \mathbf{i}_{\mathcal{V}} \rangle \equiv \langle\, v(\cdot)\,,\, i_{\mathcal{V}}(\cdot)\,\rangle = \sum_{(k,p)\in\Omega_v} \Re\left\{V_k^{(p)}[\,g_k^{(p)} V_k^{(p)}]^*\right\} = \sum_{(k,p)\in\Omega_v} \Re\left\{V_k^{(p)}[\,I_k^{(p)}]^*\right\}$$

which we recognize as the real power P. The grade-2 component of $\mathbb{S}_{\mathcal{V}}$ is the bivector

$$\mathbb{S}_g \overset{\text{def}}{=} \mathbf{v} \wedge \mathbf{i}_{\mathcal{V}} = \mathbf{v} \wedge \mathbf{i}_g$$

where we deduce the second equality from the properties of the wedge product (see Sect. F.1) and the fact that (a) $\mathbf{i}_{\mathcal{V}} = \mathbf{i}_F + \mathbf{i}_g$, and (b) $\mathbf{i}_F$ is collinear with $\mathbf{v}$, while $\mathbf{i}_g$ is orthogonal to $\mathbf{v}$. It now follows from the special norm identity (F.7) that

$$\|\,\mathbb{S}_g\,\| = \|\,\mathbf{v}\,\|\,\|\,\mathbf{i}_g\,\| = N_g = V_{rms}^2\,\sigma_g \tag{8.38b}$$

where we used (8.36a) to obtain the last (i.e., 3rd) equality. Finally, $\mathbb{S}_{\mathcal{V}} = P \oplus \mathbb{S}_g$ results in the norm identity (recall (F.8c))

$$\|\,\mathbb{S}_{\mathcal{V}}\,\|^2 \equiv \|\,\mathbf{v}\,\|^2\,\|\,\mathbf{i}_{\mathcal{V}}\,\|^2 = P^2 + N_g^2 \tag{8.38c}$$

which is consistent with the two-term split of $\mathcal{M}_w\,\mathbf{G}^2$ shown in (8.23a).

The grade structure of a geometric algebra is symmetric, so that a grade-r subspace and a grade-$(N-r)$ subspace always have the same dimension. Consequently, we can set up a one-to-one correspondence between such pairs of equal dimension spaces. One convenient way of doing so is via the *duality relation* $\mathbb{A} \leftrightarrow \mathbb{A}\,\mathcal{J}$, where the so-called *pseudo-scalar* $\mathcal{J}$ is the geometric product of a complete set of orthonormal basis vectors of the generating Euclidean space, viz., $\mathcal{J} \overset{\text{def}}{=} \mathbf{e}_1\mathbf{e}_2 \ldots \mathbf{e}_N$. For instance, the dual of a scalar α is the pseudo-scalar $\alpha\,\mathcal{J}$, the dual of a vector $\mathbf{x}$ is the so-called pseudo-vector (an $(N-1)$-vector) $\mathbf{x}\,\mathcal{J}$, and so on. Dual multivectors have the same norm, viz., $\|\mathbb{A}\,\mathcal{J}\| = \|\mathbb{A}\|$ for every multivector $\mathbb{A}$. The pseudo-scalar $\mathcal{J}$ is independent of the choice of basis, up to a change of sign.

Our next step is to introduce an (partial) apparent power multivector $\mathbb{S}_{\Upsilon}$ that is a function of the quadrature current component $i_{\Upsilon}(\cdot)$ and the voltage $\mathbf{v}$. However, since $i_{\Upsilon}(\cdot) \notin \mathcal{V}$, the geometric product of this current with $\mathbf{v}$ is not defined. This challenge can be overcome in either one of two alternative ways:

- Redefine the generating Euclidean space as $\mathcal{V} \oplus \Upsilon$ and construct the corresponding geometric algebra, which will have a dimension of 2^{2N}.

- Retain the geometric algebra of dimension 2^N generated by the Euclidean space $\mathcal{V}$ alone, and associate a *pseudo-vector* with the quadrature current $i_\Upsilon(\cdot)$, using the duality relation of geometric algebra.

We describe here the second (more compact) approach, originally introduced in [269]. Since every pseudo-vector can be expressed as $\mathbf{x}\,\mathcal{J}$ for some 1-vector $\mathbf{x} \in \mathcal{V}$, the overall current multivector is defined as

$$\mathbf{i} \stackrel{\text{def}}{=} \underbrace{\mathbf{i}_\mathcal{V}}_{\text{grade-1}} \oplus \underbrace{\mathbf{i}_\Upsilon\,\mathcal{J}}_{\text{grade-(N-1)}} \tag{8.39a}$$

where (compare with (8.33))

$$\mathbf{i}_\Upsilon \stackrel{\text{def}}{=} \mathcal{H}^{-1}\{i_\Upsilon(t)\} - \sum_{(k,p)\in\Omega_v} b_k^{(p)} v_k^{(p)}(t) \tag{8.39b}$$

Notice that $\mathbf{i}_\Upsilon \in \mathcal{V}$, so that $\mathbf{i}_\Upsilon\,\mathcal{J}$ is indeed a pseudo-vector (i.e., a grade-$(N-1)$ multivector). Consequently, the two current components in (8.39a) are mutually-orthogonal, and $\|\,\mathbf{i}\,\|^2 = \|\,\mathbf{i}_\mathcal{V}\,\|^2 + \|\,\mathbf{i}_\Upsilon\,\|^2 = \sum_{(k,p)\in\Omega_v}\big([g_k^{(p)}]^2 + [b_k^{(p)}]^2\big)\big|V_k^{(p)}\big|^2$, which matches (8.21). The apparent power multivector that corresponds to the pseudo-vector $\mathbf{i}_\Upsilon\,\mathcal{J}$ is the pseudo-bivector $\mathbb{S}_\Upsilon\,\mathcal{J}$, where

$$\mathbb{S}_\Upsilon \stackrel{\text{def}}{=} \mathbf{v}\,\mathbf{i}_\Upsilon = \underbrace{\langle\,\mathbf{v}\,,\,\mathbf{i}_\Upsilon\,\rangle}_{\text{grade-0}} + \underbrace{\mathbf{v}\wedge\mathbf{i}_\Upsilon}_{\text{grade-2}} \equiv Q_B \oplus \mathbb{S}_b \tag{8.40a}$$

The grade-0 component of this multivector is the scalar

$$\begin{aligned}\langle\,\mathbf{v}\,,\,\mathbf{i}_\Upsilon\,\rangle \equiv \langle\,v(\cdot)\,,\,\mathcal{H}^{-1}i_\Upsilon(\cdot)\,\rangle &= \sum_{(k,p)\in\Omega_v} \Re\left\{V_k^{(p)}[\,b_k^{(p)}V_k^{(p)}\,]^*\right\}\\ &= \sum_{(k,p)\in\Omega_v} \Im\left\{V_k^{(p)}[\,I_k^{(p)}\,]^*\right\}\end{aligned}$$

which we recognize as the Budeanu reactive power Q_B. The grade-2 component of $\mathbb{S}_\Upsilon$ is the bivector

$$\mathbb{S}_b \stackrel{\text{def}}{=} \mathbf{v}\wedge\mathbf{i}_\Upsilon = \mathbf{v}\wedge\mathbf{i}_b$$

where $\mathbf{i}_b$ denotes the part of $\mathbf{i}_\Upsilon$ that is orthogonal to $\mathbf{v}$. In fact,

$$\mathbf{i}_b = \mathcal{H}^{-1}\,i_b(t) = \sum_{(k,p)\in\Omega_v} \big(b_k^{(p)} - \mu_b\big)\,v_k^{(p)}(t)$$

because the property $i_b(\cdot) \perp v(\cdot)$ implies that $\mathbf{i}_b \perp \mathbf{v}$ as well.[3] Consequently, $\| \mathbf{i}_b \| = \| i_b(\cdot) \| = V_{rms}\, \sigma_b$. The special norm identity (F.7) now implies that

$$\| \mathbb{S}_b \| \;=\; \| \mathbf{v} \| \, \| \mathbf{i}_b \| \;=\; N_b \;=\; V_{rms}^2 \, \sigma_b \tag{8.40b}$$

where again we used (8.36a) to obtain the last (i.e., 3rd) equality. Thus, $\mathbb{S}_\Upsilon = Q_B \oplus \mathbb{S}_b$: this results in the norm identity (recall (F.8c))

$$\| \mathbb{S}_\Upsilon \|^2 \;\equiv\; \| \mathbf{v} \|^2 \, \| \mathbf{i}_\Upsilon \|^2 \;=\; Q_B^2 + N_b^2 \tag{8.40c}$$

which is consistent with the two-term split of $\mathcal{M}_w \, \mathbf{B}^2$ shown in (8.23a).

In summary, the apparent power multivector

$$\mathbb{S} \;\stackrel{\text{def}}{=}\; \mathbb{S}_\mathcal{V} \oplus \left(\mathbb{S}_\Upsilon \, \mathcal{J}\right) \;=\; \underbrace{P}_{\text{grade-0}} \oplus \underbrace{\mathbb{S}_g}_{\text{grade-2}} \oplus \underbrace{Q_B \, \mathcal{J}}_{\text{grade-N}} \oplus \underbrace{\mathbb{S}_b \, \mathcal{J}}_{\text{grade-(N-2)}} \tag{8.41a}$$

results in the Pythagorean norm identity

$$\| \mathbb{S} \|^2 \;=\; P^2 + \| \mathbb{S}_g \|^2 + Q_B^2 + \| \mathbb{S}_b \|^2 \;\equiv\; P^2 + N_g^2 + Q_B^2 + N_b^2 \tag{8.41b}$$

which matches the 5-component decomposition of apparent power (8.24) under the assumption that $S_\perp = 0$. Notice that each one of the multivectors $\mathbb{S}_g$ and $\mathbb{S}_b \, \mathcal{J}$ belongs in a subspace of dimension $N(N-1)/2$. In particular, in a sinusoidal single-phase system $N = 3$, so that $\mathbb{S}_g$ and $\mathbb{S}_b \, \mathcal{J}$ consist of three coordinate-based components each. In this case $P + \mathbb{S}_g$ is indistinguishable from a *quaternion* [256], and similarly for $Q_B \, \mathcal{J} + \mathbb{S}_b \, \mathcal{J}$.

[3] Recall that $\langle \, \mathcal{H} x(\cdot) \, , \, \mathcal{H} y(\cdot) \, \rangle = \langle \, x(\cdot) \, , \, y(\cdot) \, \rangle$ for any $x(\cdot)$ and $y(\cdot)$.

Chapter 9 Power Flow in the Presence of Transients

The importance of AC devices and systems in all areas of electric energy generation, transmission, and conversion naturally raises questions about characterization of their operating regimes. The emerging electric energy systems are accommodating many more generating units, as renewable sources tend to be smaller in rating and much more volatile than the conventional dispatchable units. At the same time, loads are becoming electronics-driven and increasingly software-defined, leading to richer and adjustable responses to changes in operation. Consequently, transients have been increasing both in frequency and significance. With the persistent increase in control bandwidths brought about by power electronics, it has become important to develop fast-varying metrics that can capture both transient and steady-state behavior.

We rely on (one-sided, rms) dynamic phasors, as defined in Chap. 2 via a local Hilbert space, to generate time-dependent versions of the steady-state decompositions that were presented in Chap. 8. In particular, the 7/11-component decomposition of apparent power still applies, but all of its components are time-varying in the general case when the voltage and/or current waveforms contain transient (i.e., non-periodic) components [270]. This observation also applies to the rms voltage and current, apparent power, as well as the real power $P(t)$, and its symmetrical sequence components, as detailed in Sect. 9.1.

A dynamic $7/11$-component decomposition provides accurate information about the presence of a transient, its time of onset, and its duration. In particular, the negative sequence components $P_{-}(t)$ and $Q_{-}(t)$, as well as the combined N-metric, i.e., $N(t) = \sqrt{N_{gs}^2(t) + N_{gu}^2(t) + N_{bs}^2(t) + N_{bu}^2(t)}$, can serve as sensitive indicators of such transient characteristics. A typical example[1] is shown in Fig. 9.1: notice that all three metrics provide a clear indication of the transient interval. The same metrics can also be used to generate a library of transient signatures,

[1] See Sect. L.2.2 in Appendix L for a detailed description of this ("wind-farm") dataset.

H. Lev-Ari, A. M. Stanković, *Dynamic Phasors in Energy Processing Systems*,
Power Electronics and Power Systems, https://doi.org/10.1007/978-3-031-50020-6_10

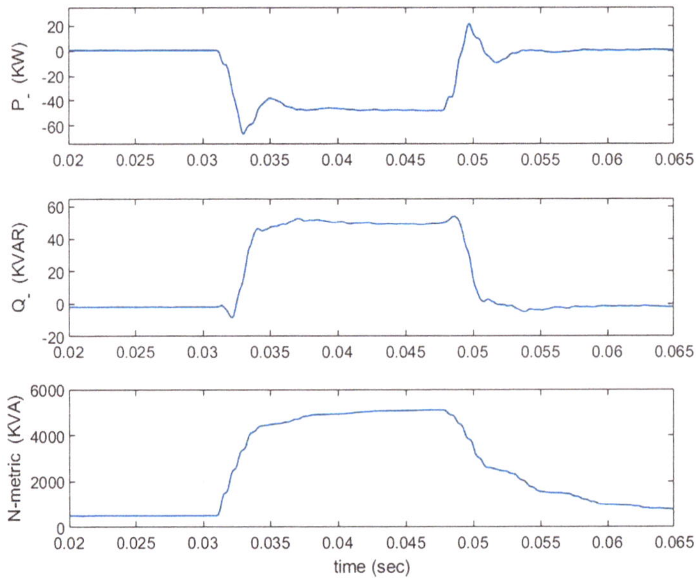

Fig. 9.1 Transient detection capabilities of dynamic power metrics: P_{-} (top), Q_{-} (middle), and N-metric (bottom), for the wind-farm dataset. The steady-state level of real power in this example is $P \approx 5\,\text{MW}$.

in support of real-time fault classification. During steady-state operation all components of this dynamic decomposition become time-invariant, fully coinciding with the constant-phasor-based characterization presented in Chap. 8. Thus, a dynamic 7/11-component decomposition can be used to analyze power quality in both steady-state and in the presence of varying operating conditions. For instance, collecting statistics of selected dynamic power metrics over time intervals of interest can provide information about variations of programmable loads and equipment aging. This evidence, in turn, is valuable input to power system planning, operation, and control.

An alternative characterization of local behavior, with applications in protection, control, and monitoring, can be obtained via the so-called instantaneous metrics proposed by Akagi [271,272]. However, the geometry of the unconstrained problem (unbalanced, with harmonics, not exclusively 3-phase) requires more information than what can be obtained from a single instantaneous (vector) sample. A discussion of several shortcomings of the original definition of instantaneous reactive power is given in [273]. Moreover, in the case of polyphase systems with more than three phases, the definition of instantaneous reactive power is not unique [274, 275]. We demonstrate in Sect. 9.2 that metrics derived from the dynamic 7/11-component decomposition of apparent power are much more accurate than their instantaneous counterparts (albeit at a higher implementation cost) both in steady-state and in the presence of transients.

In Sect. 9.3 we show that power metrics based on *sub-cycle* dynamic phasors can bridge the gap between the purely-instantaneous (single-sample) power metrics of [272] and the 7/11-component dynamic decomposition, which relies on a full-cycle (sliding-window) record of voltage and current waveforms. Thus, the approach we present in Sect. 9.3 provides a flexible trade-off between computational cost, which increases with the number of time-domain samples used, and the attendant improvement in quality of the resulting dynamic power metrics. Additional examples that illustrate the capabilities of sub-cycle power metrics can be found in [78, 80].

When operating in steady-state, the design parameters should be adjusted for accuracy of the sub-cycle dynamic phasors and of the derived sub-cycle power metrics, using their full-cycle counterparts as a benchmark for comparison, as detailed in Sect. 9.4.1. Similar considerations apply in the presence of slow (in-band) transients: examples are provided in Sect. 9.4.2. In contrast, full-cycle metrics cannot serve as a benchmark in the presence of a short wide-band transient: the primary concern in this case becomes accuracy of the estimates of transient duration and time of onset, as explained in Sect. 9.4.3.

Thus, in practical real-time implementations it may be advantageous to calculate in parallel several sets of sub-cycle metrics, each one corresponding to a different choice of the design parameters $\mathcal{S}$ and ξ, to cover the possibility of steady-state operation being interrupted by occasional slow or fast transients (see Sect. 9.4.4). Given the low cost of each individual sub-cycle scheme, the overall computational effort of such a multi-objective implementation can still be significantly below that of evaluating a single set of full-cycle power quality metrics.

Finally, Sect. 9.5 provides a comparison of sub-cycle power metrics with the "instantaneous reactive power" (IRP) metric proposed in [271, 272] and the derivative-based IRP metrics introduced in [276–278]. It turns out that even the most rudimentary sub-cycle scheme, which uses only two (polyphase) voltage and current samples, and thus can estimate only the fundamental dynamic phasor, outperforms its IRP competitors in both steady-state operation and in the presence of transients. While implementation cost of a 2-sample sub-cycle scheme is comparable to that of the various IRP approaches, the averaging built into the sub-cycle waveform-to-phasor map results in a considerable reduction of steady-state fluctuation. In addition, the sub-cycle metrics $P_{-}(t)$, $Q_{-}(t)$ and $N(t)$, which are not available in any IRP implementation, provide estimates of transient duration and time of onset that are much more accurate than those derived from IRP metrics [291].

9.1 Full-Cycle Dynamic 7/11-Component Power Decomposition

We introduced in Sect. 2.1 an explicitly time-variant Hilbert space framework for real-valued signals with finite local power. We now apply this framework to construct dynamic power decompositions from given polyphase voltage and current waveforms. The resulting power decomposition consists of time-dependent components, evaluated from the polyphase waveform segment $\{x(s);\ t - T < s \leq t\}$. The numerical calculation of all the components in our 7/11-component decomposition of apparent power can be conveniently carried out in terms of the (local) dynamic phasors we introduced in Sect. 2.1.

First, we observe that the finite-rms constraint for polyphase waveforms (i.e., real-valued $1 \times m$ vector processes) is

$$\frac{1}{T}\int_{t-T}^{t} x(s)\, x^{\top}(s)\, ds \;<\; \infty \qquad \text{for all } t$$

and the local (= sliding-window) inner-product in this space is

$$\langle\, x(\cdot)\,,\; y(\cdot)\,\rangle\,(t) \;\stackrel{\text{def}}{=}\; \frac{1}{T}\int_{t-T}^{t} x(s)\, y^{\top}(s)\, ds \tag{9.1}$$

The corresponding sliding-window (one-sided, rms) Fourier-series representation of a polyphase waveform $x(\cdot)$ is (recall (2.2))

$$x(s) = X_0(t) \;+\; \sqrt{2}\sum_{k=1}^{\infty} \Re\left\{X_k(t)\, e^{jk\omega_0 s}\right\}\,, \quad t - T \;<\; s \;\le t \tag{9.2a}$$

and

$$X_k(t) \;=\; \begin{cases} \dfrac{1}{T}\displaystyle\int_{t-T}^{t} x(s)\, ds & k = 0 \\[2ex] \dfrac{\sqrt{2}}{T}\displaystyle\int_{t-T}^{t} x(s)\, e^{-jk\omega_0 s}\, ds & k \ge 1 \end{cases} \tag{9.2b}$$

where $\omega_0 = {}^{2\pi}/_{T}$, and T is the duration of a cycle (or the period in a steady-state). The polyphase dynamic rms phasor $X_k(t)$ is a function of the window location parameter t, except when $x(\cdot)$ is a periodic signal with period T: in this case $X_k(t)$ is independent of t and coincides with the standard (one-sided, rms, polyphase) Fourier phasor.

The corresponding Parseval identity for polyphase waveforms is (recall (2.3))

$$\langle\, x(\cdot)\,,\; y(\cdot)\,\rangle\,(t) \;=\; X_0(t)\, Y_0^{\top}(t) \;+\; \sum_{k=1}^{\infty} \Re\left\{\, X_k(t)\, Y_k^{H}(t)\right\} \;=\; \Re\left\{\mathbf{X}(t)\,\mathbf{Y}^{H}(t)\right\} \tag{9.3a}$$

where we used the dynamic (time-dependent) *phasor arrays*

$$\begin{aligned} \mathbf{X}(t) &\;\stackrel{\text{def}}{=}\; \left[\, X_0(t) \quad X_1(t) \quad X_2(t) \quad \ldots \;\right] \\ \mathbf{Y}(t) &\;\stackrel{\text{def}}{=}\; \left[\, Y_0(t) \quad Y_1(t) \quad Y_2(t) \quad \ldots \;\right] \end{aligned} \tag{9.3b}$$

Recall that the superscript ^{H} denotes conjugate transposition.

The identity (9.3) defines an isometry between the local Hilbert space of polyphase waveforms on the t-dependent interval $(t - T \quad t]$, and the Hilbert space of time-dependent polyphase phasor arrays associated with these waveforms. This means that all inner-products can be evaluated via appropriate dynamic phasors. In particular, the dynamic real power

$$P(t) \;=\; \frac{1}{T}\int_{t-T}^{t} v(s)\, i^{\top}(s)\, ds \;=\; \langle\, v(\cdot)\,,\, i(\cdot)\,\rangle(t) \tag{9.4a}$$

is an inner-product of the voltage and current waveform segments within the interval $(t - T \quad t]$. Thus, $P(t)$ can be expressed in terms of the corresponding dynamic phasor arrays, viz.,

$$P(t) \;=\; \Re\{\mathbf{V}(t)\,\mathbf{I}^{H}(t)\} \;\equiv\; \sum_k \Re\{V_k(t)\, I_k^{H}(t)\} \tag{9.4b}$$

which is a dynamic version of the steady-state relation (8.6).

In order to obtain a dynamic counterpart of the 7/11-component decomposition (8.31) we need first to evaluate the local Fourier phasor arrays $\mathbf{I}(t)$, $\mathbf{V}(t)$, and subsequently the local equivalent linear load admittance $\mathbf{Y}(t) \equiv \mathbf{G}(t) - j\mathbf{B}(t) \stackrel{\text{def}}{=} \frac{\mathbf{I}(t)}{\mathbf{V}(t)}$. We can then use the same expressions as in the steady-state case, but with the appropriate dynamic (= sliding-window) quantities. For instance V_{rms}^2 and I_{rms}^2 are replaced with their time-variant versions

$$\begin{aligned} V_{rms}^2(t) &\stackrel{\text{def}}{=} \|\, v(\cdot)\|^2(t) \;=\; \|\,\mathbf{V}(t)\,\|^2 \;\equiv\; \sum_k \|\, V_k(t)\,\|^2 \\ I_{rms}^2(t) &\stackrel{\text{def}}{=} \|\, i(\cdot)\|^2(t) \;=\; \|\,\mathbf{I}(t)\,\|^2 \;\equiv\; \sum_k \|\, I_k(t)\,\|^2 \end{aligned} \tag{9.5a}$$

where $\| \quad \|$ denotes the Euclidean norm of a (row) vector. The resulting dynamic apparent power is

$$S(t) \;\stackrel{\text{def}}{=}\; V_{rms}(t)\, I_{rms}(t) \tag{9.5b}$$

Also, the real power component P is replaced by its dynamic counterpart, viz.,

$$P(t) \;=\; \Re\{\mathbf{V}(t)\,\mathbf{I}^{H}(t)\} \;\equiv\; \sum_k \Re\{V_k(t)\, I_k^{H}(t)\} \;=\; \mu_g(t)\, V_{rms}^2(t) \tag{9.5c}$$

where

$$\mu_g(t) \;=\; \frac{1}{V_{rms}^2(t)}\; \mathcal{M}_w\, \mathbf{G}(t) \tag{9.5d}$$

and similarly for $Q_B(t)$ and $\mu_b(t)$. Here the elements of the time-variant "weight vector" $\mathbf{W}(t) = \{w_k^{(p)}(t)\}$ reflect the relative contributions of the dynamic voltage phasors $V_k^{(p)}(t)$, viz.,

$$w_k^{(p)}(t) \stackrel{\text{def}}{=} \frac{|V_k^{(p)}(t)|^2}{V_{rms}^2(t)} \tag{9.5e}$$

so that the averaging operator $\mathcal{M}_w$ is time-variant as well. Again, notice that the weights $w_k^{(p)}(t)$ are real-valued, non-negative, and add up to unity for every t.

All other quantities used in evaluating the 7/11-component decomposition are also determined from the waveform segment $\{x(s);\ \ t - T\ < s\ \le t\}$ and thus are all time-variant. This means that the static decomposition (8.31) is now replaced by a dynamic counterpart, viz.,

$$\begin{aligned} S^2(t) \;=\; & S_\perp^2(t) + \left(P_+(t) + P_-(t) + P_0(t)\right)^2 + N_{gs}^2(t) + N_{gu}^2(t) \\ & + \left(Q_{B,+}(t) + Q_{B,-}(t) + Q_{B,0}(t)\right)^2 + N_{bs}^2(t) + N_{bu}^2(t) \end{aligned} \tag{9.6}$$

Only the symmetrical sequence components of $P(t)$ and $Q_B(t)$ are *network-conservative*, in the sense defined in Sect. 9.6 and examined in detail in Appendix I. The remaining five components are non-negative by definition and thus can never satisfy the network-conservativity property.

The following example illustrates the utility of our *dynamic* 7/11-component power decomposition in an industrial setting (the paper-mill dataset, which is described in detail in Sect. L.2.4). Since $S_\perp(t)$ and the zero-sequence components $P_0(t)$ and $Q_{B,0}(t)$ are negligible in this case, we show in Fig. 9.2 the remaining eight components of $S(t)$, as well as the combined dynamic real and reactive power metrics $P(t) = P_+(t) + P_-(t)$ and $Q_B(t) = Q_{B,+}(t) + Q_{B,-}(t)$. Notice that (a) the steady-state interval $\{t \le 0.18 \text{ sec}\}$ exhibits significant *harmonic distortion*, as indicated by the considerable values of N_{gs} and N_{bs} in that interval; and (b) the *unbalanced transient* causes a large increase in the values of N_{gu} and N_{bu} within the transient interval $\{0.18 \text{ sec} \le t \le 0.30 \text{ sec}\}$. Further analysis of this example is provided in Sect. 9.4.2.

9.2 Purely-Instantaneous Power Metrics

The instantaneous (real) power in a polyphase system is $p(t) = v(t)\, i^\top(t)$, and the instantaneous voltage and current norms are $v(t)\, v^\top(t)$ and $i(t)\, i^\top(t)$, respectively, where the polyphase voltage and current are as defined in (1.6), viz.,

$$v(t) = \begin{bmatrix} v_1(t) & v_2(t) & \ldots & v_m(t) \end{bmatrix} \quad , \qquad i(t) = \begin{bmatrix} i_1(t) & i_2(t) & \ldots & i_m(t) \end{bmatrix}$$

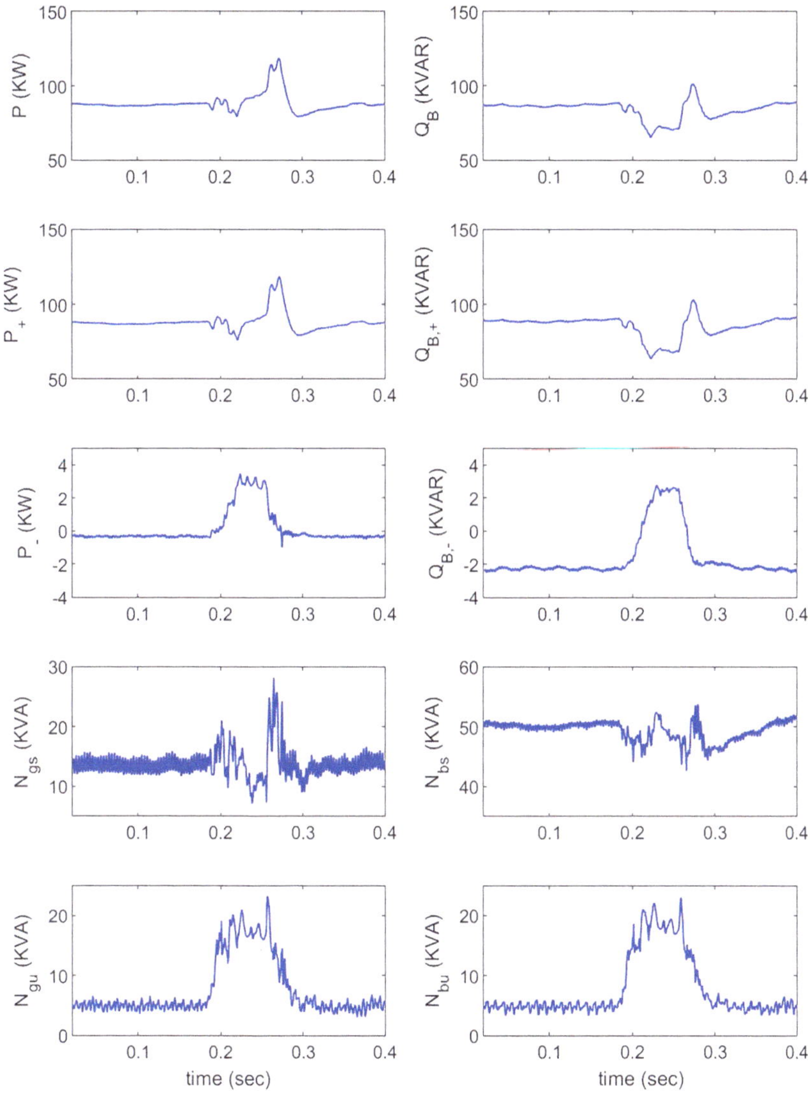

Fig. 9.2 Dynamic power quality metrics (paper-mill example): non-vanishing components of (9.6).

The Cauchy-Schwarz inequality for real (row) vectors implies that $|p(t)| \leq s(t)$, where

$$s(t) \stackrel{\text{def}}{=} \sqrt{\left[v(t)\, v^\top(t)\right]\left[i(t)\, i^\top(t)\right]} \tag{9.7a}$$

is known as the *the instantaneous apparent power* [271, 272]. The concepts of instantaneous real and apparent power focus explicitly on polyphase systems, since for a single-phase system $s(t) = |p(t)|$. It is well-known (see, e.g., [259, 272] and our Appendix C) that the non-negative difference $s^2(t) - p^2(t)$ can be written as a sum of squares, namely,

$$s^2(t) - p^2(t) = \sum_{r<\ell} q_{r\ell}^2(t) = \tfrac{1}{2} \parallel \boldsymbol{Q}(t) \parallel_F^2 \tag{9.7b}$$

where $\| \cdot \|_F$ is the matrix Frobenius norm, and the "instantaneous reactive power" terms $q_{r\ell}(t)$ are the (strictly above diagonal) elements of the skew-symmetric matrix

$$\boldsymbol{Q}(t) \stackrel{\text{def}}{=} i^\top(t)v(t) - v^\top(t)i(t) \tag{9.7c}$$

For instance, in a three-phase system described by

$$v(t) = \begin{bmatrix} v_a(t) & v_b(t) & v_c(t) \end{bmatrix} , \quad i(t) = \begin{bmatrix} i_a(t) & i_b(t) & i_c(t) \end{bmatrix} \tag{9.8}$$

there are three instantaneous reactive power quantities, viz., $q_{ab}(t)$, $q_{ac}(t)$, and $q_{bc}(t)$. In view of the expression (9.7b), some authors [279] proposed interpreting the row vector $\begin{bmatrix} q_{bc}(t) & q_{ac}(t) & q_{ab}(t) \end{bmatrix}$ as a vector (cross) product between the polyphase current vector $i(t)$ and the polyphase voltage vector $v(t)$. However, this interpretation does not carry over to polyphase systems with more than 3 phases: in the general case, one has to define instantaneous inactive power components in terms of the skew-symmetric matrix (a.k.a. "tensor") $\boldsymbol{Q}(t)$, which involves the *outer product* of the instantaneous polyphase voltage and current [280–283]. A compact unified geometric framework for describing both the active and inactive instantaneous power components, which relies on a *Geometric Algebra* construction, is presented in Sect. 9.2.2 below.

9.2.1 The Clarke Transform in 3-Phase Systems

The number of non-vanishing instantaneous reactive power quantities $(q_{r\ell}(t))$ in a three-phase system can be reduced by using, for instance, the Clarke transform [272, 284–286], viz.,

$$\begin{gathered} \begin{bmatrix} v_\alpha(t) & v_\beta(t) & v_0(t) \end{bmatrix} = \begin{bmatrix} v_a(t) & v_b(t) & v_c(t) \end{bmatrix} \mathcal{M} \\ \begin{bmatrix} i_\alpha(t) & i_\beta(t) & i_0(t) \end{bmatrix} = \begin{bmatrix} i_a(t) & i_b(t) & i_c(t) \end{bmatrix} \mathcal{M} \end{gathered} \tag{9.9a}$$

where $\mathcal{M}$ is the 3×3 orthogonal matrix

$$\mathcal{M} = \frac{1}{\sqrt{3}} \begin{pmatrix} \sqrt{2} & 0 & 1 \\ -\frac{\sqrt{2}}{2} & \sqrt{\frac{3}{2}} & 1 \\ -\frac{\sqrt{2}}{2} & -\sqrt{\frac{3}{2}} & 1 \end{pmatrix} \stackrel{\text{def}}{=} \mathcal{M}_{Clarke} \tag{9.9b}$$

The associated Clarke-transformed $\boldsymbol{Q}(\cdot)$ matrix is

$$\boldsymbol{Q}_{Clarke}(t) \stackrel{\text{def}}{=} \mathcal{M}^\top \boldsymbol{Q}(t)\, \mathcal{M} = \begin{pmatrix} 0 & q_{\alpha\beta}(t) & q_{\alpha 0}(t) \\ -q_{\alpha\beta}(t) & 0 & q_{\beta 0}(t) \\ -q_{\alpha 0}(t) & -q_{\beta 0}(t) & 0 \end{pmatrix} \tag{9.10}$$

where $q_{\alpha\beta}(t) = i_\alpha(t)\, v_\beta(t) - v_\alpha(t)\, i_\beta(t)$, and similarly for $q_{\alpha 0}(t)$ and $q_{\beta 0}(t)$. Notice that since $\mathcal{M}\,\mathcal{M}^\top = I$, we obtain the identity

$$\begin{aligned} \| \boldsymbol{Q}_{Clarke}(t) \|_F^2 &= \| \mathcal{M}^\top \boldsymbol{Q}(t)\, \mathcal{M} \|_F^2 = \operatorname{tr}\{\mathcal{M}^\top \boldsymbol{Q}(t)\, \mathcal{M}\, \mathcal{M}^\top \boldsymbol{Q}^\top(t)\, \mathcal{M}\} \\ &= \| \boldsymbol{Q}(t) \|_F^2 \end{aligned}$$

where we used the properties $\| A \|_F^2 \equiv \operatorname{tr}\{A A^H\}$ and $\operatorname{tr}\{A\,B\} = \operatorname{tr}\{B\,A\}$. Thus, the gap $s^2(t) - p^2(t)$ can be expressed also in terms of the elements of $\boldsymbol{Q}_{Clarke}(t)$, rather than in terms of the original $\boldsymbol{Q}(t)$ (see Sect. H.1 for a detailed discussion).

The instantaneous real power $p(t) \stackrel{\text{def}}{=} v(t)\, i^\top(t)$ can also be expressed in terms of the Clarke voltage and current components, viz.,

$$p(t) = v(t)\, i^\top(t) \equiv \begin{bmatrix} v_\alpha(t) & v_\beta(t) & v_0(t) \end{bmatrix} \mathcal{M}^{-1} \mathcal{M}^{-\top} \begin{bmatrix} i_\alpha(t) & i_\beta(t) & i_0(t) \end{bmatrix}^\top$$

Using the fact that $\mathcal{M}\,\mathcal{M}^\top = I$ we obtain the well-known expression

$$p(t) = v_\alpha(t)\, i_\alpha(t) + v_\beta(t)\, i_\beta(t) + v_0(t)\, i_0(t) \tag{9.11}$$

The *cycle-averaged* value of $p(t)$ is, of course, the (full-cycle) dynamic real power $P(t)$, because $\frac{1}{T}\int_{t-T}^{t} p(s)\, ds = P(t)$ (recall (9.4a)). One can also establish a number of useful relations between the *cycle-averaged* values of the elements of the matrix $\boldsymbol{Q}_{Clarke}(t)$, and the symmetrical sequence components of the voltage and current (see, e.g., [274]). The most interesting of these relations is $\frac{1}{T}\int_{t-T}^{t} q_{\alpha\beta}(s)\, ds = Q_{B,+}(t) - Q_{B,-}(t)$, which holds both in steady-state operation (see [272]) and during transients, even when the voltage and current zero-sequence components do not vanish. Additional relations are reviewed in Appendix H.

In the special case when $i_0(t) = 0 = v_0(t)$ in a three-phase system (e.g., in a 3-wire system using the virtual star point as a voltage reference), we find that $q_{\alpha 0}(t) = 0 = q_{\beta 0}(t)$, so that the only non-vanishing instantaneous reactive power quantity is $q_{\alpha\beta}(t)$. Consequently,

$$Q_{Clarke}(t) = \begin{pmatrix} 0 & q_{\alpha\beta}(t) & 0 \\ -q_{\alpha\beta}(t) & 0 & 0 \\ 0 & 0 & 0 \end{pmatrix}$$

and $s^2(t) - p^2(t) = q^2_{\alpha\beta}(t)$. Thus, in this case the instantaneous power quality gap $s^2(t) - p^2(t)$ consists entirely of the (square of) a single signed "instantaneous reactive power" quantity, namely $q_{\alpha\beta}(t)$. The *negative of this quantity*, namely $-q_{\alpha\beta}(t)$, is known as "Akagi reactive power" (see, e.g., [272]).

Finally, the so-called complex Park theory (see, e.g., [286]) exploits the fact that the assumption $i_0(t) = 0 = v_0(t)$ results in a two-dimensional representation of the instantaneous voltage and current in a three-phase system. This makes it possible to define the complex-valued *Park vector* voltage and current, namely,

$$v_{\mathsf{pv}}(t) \stackrel{\text{def}}{=} v_\alpha(t) + j\ v_\beta(t) \quad , \quad i_{\mathsf{pv}}(t) \stackrel{\text{def}}{=} i_\alpha(t) + j\ i_\beta(t) \tag{9.12a}$$

which results in the identities

$$|v_{\mathsf{pv}}(t)|^2 = v^2_\alpha(t) + v^2_\beta(t) \equiv \|v(t)\|^2 \quad , \quad |i_{\mathsf{pv}}(t)|^2 = i^2_\alpha(t) + i^2_\beta(t) \equiv \|i(t)\|^2$$

Also notice that

$$v_{\mathsf{pv}}(t)\, i^*_{\mathsf{pv}}(t) = \big[\underbrace{v_\alpha(t)\, i_\alpha(t) + v_\beta(t)\, i_\beta(t)}_{p(t)}\big] + j\,\big[\underbrace{i_\alpha(t)\, v_\beta(t) - v_\alpha(t)\, i_\beta(t)}_{q_{\alpha\beta}(t)}\big] \tag{9.12b}$$

which recovers the fundamental relation $s^2(t) = p^2(t) + q^2_{\alpha\beta}(t)$, since

$$s^2(t) \equiv \|v(t)\|^2\, \|i(t)\|^2 = |v_{\mathsf{pv}}(t)|^2\, |i_{\mathsf{pv}}(t)|^2 = p^2(t) + q^2_{\alpha\beta}(t)$$

These relations can be extended to polyphase systems with any number of phases by exploiting the multi-dimensionality of the space generated via Geometric Algebra.

9.2.2 Geometric Algebra Approach to Instantaneous Power Quantities

When the number of phases exceeds three (i.e., $m > 3$) there are always multiple instantaneous reactive power quantities, even if the zero-sequence voltage and current vanish. For instance, in a six-phase system there are 15 such quantities in general, and making $i_0(t) = 0 = v_0(t)$ only reduces their number to 10. Thus, while the instantaneous voltage and current can always be viewed as elements in the Euclidean space of real-valued $1 \times m$ vectors, a conceptually-compact interpretation of the set of $m(m-1)/2$ distinct elements of $\boldsymbol{Q}(t)$ requires a higher-dimensional space.

One such compact – and very elegant – interpretation is provided by the framework of *geometric algebra* [256] (see also Appendix F). This framework is representation-free and thus extends readily to any polyphase system and any choice of coordinates, such as symmetrical sequence components. Starting with the Euclidean space of $1 \times m$ vectors, geometric algebra introduces higher-dimensional entities, called *multivectors*, constructed via operations such as the *wedge product* and the *geometric product.* Since our entire discussion involves a single (fixed) time-instant, and for clarity of presentation, we suppress in the remainder of this section the time dependence of $v(t)$ and $i(t)$ and represent them instead using a lower-case boldface notation, namely $\boldsymbol{v}$ and $\boldsymbol{i}$. Also, we can choose any convenient orthonormal basis $\mathcal{B} = \{\boldsymbol{b}_r\}$ for the space of real-valued $1 \times m$ vectors and use it to provide a coordinate-based representation of polyphase waveforms, e.g., $\boldsymbol{v} = \sum_r v_r \boldsymbol{b}_r$, $\boldsymbol{i} = \sum_r i_r \boldsymbol{b}_r$.

We use uppercase "blackboard" boldface to denote *bivectors* (= grade 2 multivectors). For instance, we can define a bivector counterpart of (9.7c), viz.,

$$\mathbb{Q} \stackrel{\text{def}}{=} \boldsymbol{i} \wedge \boldsymbol{v} = \sum_{r<\ell} q_{r\ell}\, \boldsymbol{b}_r \boldsymbol{b}_\ell \equiv \frac{1}{2}\begin{bmatrix}\boldsymbol{b}_1 & \boldsymbol{b}_2 & \dots & \boldsymbol{b}_m\end{bmatrix} \boldsymbol{Q}_{\mathcal{B}} \begin{bmatrix}\boldsymbol{b}_1 & \boldsymbol{b}_2 & \dots & \boldsymbol{b}_m\end{bmatrix}^{\top}$$

where $\wedge$ denotes the wedge product, $\boldsymbol{Q}_{\mathcal{B}}$ is the representation of $\mathbb{Q}$ in the orthonormal basis $\mathcal{B}$, and $q_{r\ell} \stackrel{\text{def}}{=} i_r v_\ell - i_\ell v_r$. In particular, the matrix $\boldsymbol{Q}(t)$ used in (9.7) is the representation of $\mathbb{Q}$ in the natural (unit-vector) basis $\{\boldsymbol{e}_r ;\ 1 \le r \le m\}$ of R^m, the space of real-valued $1 \times m$ vectors. Similarly, the matrix $\boldsymbol{Q}_{Clarke}(t)$ of (9.10) is obtained as a representation of $\mathbb{Q}$ in the orthonormal Clarke basis

$$\boldsymbol{b}_1 = \frac{1}{\sqrt{3}}\begin{bmatrix}\sqrt{2} & -\frac{\sqrt{2}}{2} & -\frac{\sqrt{2}}{2}\end{bmatrix}$$

$$\boldsymbol{b}_2 = \frac{1}{\sqrt{3}}\begin{bmatrix}0 & \frac{\sqrt{3}}{2} & -\frac{\sqrt{3}}{2}\end{bmatrix}$$

$$\boldsymbol{b}_3 = \frac{1}{\sqrt{3}}\begin{bmatrix}1 & 1 & 1\end{bmatrix}$$

In geometric algebra one can associate norms with bivectors (and, in general with multivectors of any grade), resulting in expressions such as

$$\| \mathbb{Q} \|^2 \equiv \| \boldsymbol{i} \wedge \boldsymbol{v} \|^2 = \sum_{r<\ell} q_{r\ell}^2$$

which recovers (9.7b) when $\{\boldsymbol{b}_r\}$ are the standard unit vectors. Notice that while the individual $\{q_{r\ell}\}$ depend on the choice of basis, the sum $\sum_{r<\ell} q_{r\ell}^2$ is independent of this choice. We can also define an apparent power multivector, viz.,

$$\mathbb{S} = p(t) + \mathbb{Q}$$

which is the m-phase generalization of (9.12b) [274, 275]. When $m = 3$, with non-vanishing $v_0(t)$ and $i_0(t)$, this expression for $\mathbb{S}$ is, essentially, a *quaternion*, consisting of a scalar component, and a three-element bivector (see Appendix F for additional details).

9.2.3 Purely-Instantaneous Compensation

The power gap $s^2(t) - p^2(t)$ can be eliminated by *instantaneous compensation*: a shunt compensator is used to inject a current $i_x(t)$ so that the resulting compensated load current $i_a(t) \stackrel{\text{def}}{=} i(t) - i_x(t)$ is given by

$$i_a(t) = \frac{p(t)}{v(t)\, v^\top(t)}\, v(t) \tag{9.13a}$$

where we recall that $p(t) = v(t)\, i^\top(t)$ is the instantaneous power. Notice that the denominator $v(t)\, v^\top(t)$ is strictly positive as long as $v(t) \neq 0$. We now observe that

$$i_a(t)\, i^\top(t) = \frac{p^2(t)}{v(t)\, v^\top(t)} = i_a(t)\, i_a^\top(t) \tag{9.13b}$$

and, consequently,

$$i_a(t)\, i_x^\top(t) \equiv i_a(t) \left[i(t) - i_a(t)\right]^\top = 0 \tag{9.13c}$$

In addition,

$$i_x(t)\, i_x^\top(t) \equiv \left[i(t) - i_a(t)\right] \left[i(t) - i_a(t)\right]^\top = i(t)\, i^\top(t) - i_a(t)\, i_a^\top(t)$$

This leads to a direct relation between the compensating current $i_x(t)$ and the power gap $s^2(t) - p^2(t)$, viz.,

$$i_x(t)\, i_x^\top(t) = \frac{s^2(t) - p^2(t)}{v(t)\, v^\top(t)} = \sum_{r<\ell} \frac{q_{r\ell}^2(t)}{v(t)\, v^\top(t)} \tag{9.14}$$

where we used the identity (9.7b). Clearly, instantaneous compensation achieves $q_{r\ell}(t) = 0$ for all r, ℓ because $i_x(t) = 0$ for a compensated load. Equivalently, we observe that since instantaneous compensation reduces the load current from $i(t)$ to $i_a(t)$, the corresponding (compensated) $\boldsymbol{Q}$-matrix *vanishes*, viz.,

$$\begin{aligned}
\text{compensated} \quad \boldsymbol{Q}(t) &= i_a^\top(t)\, v(t) - v^\top(t)\, i_a(t) \\
&= \left[\frac{p(t)}{v(t)\, v^\top(t)}\, v(t)\right]^\top v(t) \; - \; v^\top(t)\left[\frac{p(t)}{v(t)\, v^\top(t)}\, v(t)\right] \\
&= \frac{p(t)}{v(t)\, v^\top(t)}\left[v^\top(t)\, v(t) - v^\top(t)\, v(t)\right] = 0
\end{aligned}$$

which, again, confirms the fact that instantaneous compensation reduces $q_{r\ell}(t)$ to zero and thus annihilates the power gap $s^2(t) - p^2(t)$.

We present in Appendix H several bounds on the instantaneous power gap $s^2(t) - p^2(t)$, expressed in terms of the cycle-averaged "compensated-power" metric

$$Q_x^2(t) \stackrel{\text{def}}{=} V_{rms}^2(t)\left[\frac{1}{T}\int_{t-T}^{t} \sum_{r<\ell} \left(\frac{q_{r\ell}^2(s)}{v(s)\, v^\top(s)}\right) ds\right] \tag{9.15}$$

which quantifies the portion of the full-cycle dynamic apparent power $S(t)$ that can be compensated without memory, i.e., by using only instantaneous operations to reduce $\sum_{r<\ell} q_{r\ell}^2(t)$ to zero.

9.2.4 Fluctuations of Purely-Instantaneous Metrics

The purely-instantaneous power metrics exhibit strong fluctuations (as compared with full-cycle dynamic metrics) in both steady-state operation and during transients. We use the paper-mill dataset (see Sect. L.2.4 for a detailed description) to compare in Fig. 9.3 the purely-instantaneous power metrics $p(t)$ and $q_{\alpha\beta}(t)$ with their full-cycle "equivalents" $P(t)$ and $Q_B(t)$.

The highly "noisy" nature of purely-instantaneous power metrics necessitates using filtering in most real-life applications [287–290] (see also discussion in Sect. 9.5), which increases both implementation cost and group-delay. Moreover, such filtering invalidates the claim of "instantaneous processing," since digital implementations of "filtered instantaneous" schemes require memory (or values of relevant quantities from previous time instances), like many other dynamic frameworks. We introduce in the following section an alternative approach that offers a significant reduction in the level of fluctuation, but at a fraction of the cost of either the "filtered instantaneous" approach or the full-cycle approach.

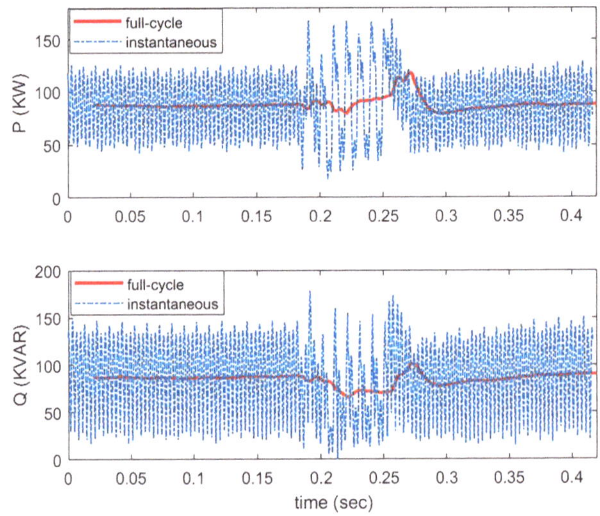

Fig. 9.3 Fluctuations in paper-mill example: real power (top) and reactive power (bottom).

9.3 Sub-cycle Dynamic Power Quality Metrics

Sub-cycle dynamic phasors, which were introduced in Sect. 3.3, can be used to construct sub-cycle counterparts of the full-cycle dynamic power quality metrics of Sect. 9.1, via a sliding-window version of the statistics-based approach that was presented in Sects. 8.5, 8.6, and 8.7. It should be noticed that sub-cycle metrics are derived from *sub-cycle phasors* and thus have different numerical values than their full-cycle counterparts, which rely on full-cycle phasors. Since sub-cycle metrics use a controllable number of selected time-domain samples, they bridge the gap between the dynamic full-cycle power metrics of Sect. 9.1, and the purely-instantaneous power metrics of Sect. 9.2 [77, 78, 80]. In particular, sub-cycle power quality metrics offer a flexible trade-off between computational cost, which increases with the number of samples used, and the accuracy of the resulting dynamic power quality metrics.

In order to obtain a sub-cycle counterpart of the dynamic decomposition (9.6), for a given pair $\{v(t), i(t)\}$ of polyphase voltage and current waveforms, we need first to evaluate the *sub-cycle dynamic phasor arrays* (of size $1 \times mL$), viz.,

$$\widehat{\mathbf{V}}(t) = \left[\, \widehat{V}_{k_1}(t) \;\; \widehat{V}_{k_2}(t) \;\; \ldots \;\; \widehat{V}_{k_L}(t) \,\right], \quad \widehat{V}_k(t) = \left[\, \widehat{V}_k^{(1)}(t) \;\; \ldots \;\; \widehat{V}_k^{(m)}(t) \,\right] \tag{9.16a}$$

$$\widehat{\mathbf{I}}(t) = \left[\, \widehat{I}_{k_1}(t) \;\; \widehat{I}_{k_2}(t) \;\; \ldots \;\; \widehat{I}_{k_L}(t) \,\right], \quad \widehat{I}_k(t) = \left[\, \widehat{I}_k^{(1)}(t) \;\; \ldots \;\; \widehat{I}_k^{(m)}(t) \,\right]$$

where the polyphase sub-cycle voltage phasors $\left\{\widehat{V}_{k_i}(t)\,;\;\; 1 \le i \le L\right\}$ and current phasors $\left\{\widehat{I}_{k_i}(t)\,;\;\; 1 \le i \le L\right\}$ are determined via the waveform-to-phasor map (3.30a), namely,

$$\begin{pmatrix} \widehat{V}_{k_1}(t)\, e^{jk_1\omega_0 t} \\ \widehat{V}_{k_2}(t)\, e^{jk_2\omega_0 t} \\ \vdots \\ \widehat{V}_{k_L}(t)\, e^{jk_L\omega_0 t} \end{pmatrix} = \frac{1}{\sqrt{L}} \begin{bmatrix} \mathbf{I}_L & \mathbf{0}_L \end{bmatrix} \mathcal{W}^{-1} \begin{pmatrix} v(t) \\ v(t-\Delta_1) \\ \vdots \\ v(t-\Delta_{M-1}) \end{pmatrix}$$

$$\begin{pmatrix} \widehat{I}_{k_1}(t)\, e^{jk_1\omega_0 t} \\ \widehat{I}_{k_2}(t)\, e^{jk_2\omega_0 t} \\ \vdots \\ \widehat{I}_{k_L}(t)\, e^{jk_L\omega_0 t} \end{pmatrix} = \frac{1}{\sqrt{L}} \begin{bmatrix} \mathbf{I}_L & \mathbf{0}_L \end{bmatrix} \mathcal{W}^{-1} \begin{pmatrix} i(t) \\ i(t-\Delta_1) \\ \vdots \\ i(t-\Delta_{M-1}) \end{pmatrix} \tag{9.16b}$$

As we observed in Sects. 3.2–3.3, this waveform-to-phasor map relies on two sets of design parameters: the set of harmonic indices $\boldsymbol{\mathcal{S}} \stackrel{\text{def}}{=} \{k_1, k_2, \ldots, k_L\}$, and the set of shifts $\boldsymbol{\Delta} \stackrel{\text{def}}{=} \{\Delta_1, \Delta_2, \ldots, \Delta_{M-1}\}$. We consider here only the *uniformly-shifted* case, so that $\Delta_i = i\,(\xi T)$, where the normalized inter-sample shift ξ satisfies the *sub-cycle fractionality constraint* $0 < \xi \le \frac{1}{2M}$.

We demonstrate in Sect. 9.4 that the control parameters of sub-cycle power quality metrics can be selected to optimize their performance for several distinct categories of power system waveforms. We also discuss the trade-off between cost, steady-state error and transient response. In contrast, both full-cycle metrics and purely-instantaneous metrics have no control parameters, and thus cannot be customized at all. The former tend to "smear-out" such transients [78], while the latter exhibit excessive fluctuations (even in steady-state) [77].

9.3.1 Fundamental Sub-cycle Power Metrics

Sub-cycle versions of the fundamental metrics (i.e., real power and rms voltage and current) are obtained by replacing the full-cycle phasors in (9.5) by their sub-cycle counterparts, namely,

$$P(t) \stackrel{\text{def}}{=} \Re\left\{\widehat{\mathbf{V}}(t)\,\widehat{\mathbf{I}}^H(t)\right\} \equiv \sum_{k\in\boldsymbol{\mathcal{S}}} \Re\left\{\widehat{V}_k(t)\,\widehat{I}_k^{\,H}(t)\right\} \tag{9.17a}$$

and

$$V_{rms}^2(t) \stackrel{\text{def}}{=} \widehat{\mathbf{V}}(t)\,\widehat{\mathbf{V}}^H(t) \equiv \sum_{k\in\mathcal{S}} \widehat{V}_k(t)\,\widehat{V}_k^H(t)$$
$$I_{rms}^2(t) \stackrel{\text{def}}{=} \widehat{\mathbf{I}}(t)\,\widehat{\mathbf{I}}^H(t) \equiv \sum_{k\in\mathcal{S}} \widehat{I}_k(t)\,\widehat{I}_k^H(t) \tag{9.17b}$$

where $\widehat{\mathbf{V}}(t)$ and $\widehat{\mathbf{I}}(t)$ are the sub-cycle dynamic phasor arrays defined in (9.16). Similarly, the sub-cycle dynamic apparent power is defined as

$$S(t) \stackrel{\text{def}}{=} V_{rms}(t)\,I_{rms}(t) \tag{9.17c}$$

To complete the analogy with full-cycle definitions, the sub-cycle version of Budeanu's reactive power is defined as

$$Q(t) \stackrel{\text{def}}{=} \Im\{\widehat{\mathbf{V}}(t)\,\widehat{\mathbf{I}}^H(t)\} \equiv \sum_{k\in\mathcal{S}} \Im\left\{\widehat{V}_k(t)\,\widehat{I}_k^H(t)\right\} \tag{9.17d}$$

We now observe that these fundamental sub-cycle dynamic metrics still satisfy the inequality

$$P^2(t)+Q^2(t) = \left|\,\Re\{\widehat{\mathbf{V}}(t)\,\widehat{\mathbf{I}}^H(t)\}\,\right|^2 + \left|\,\Im\{\widehat{\mathbf{V}}(t)\,\widehat{\mathbf{I}}^H(t)\}\,\right|^2 \equiv \left|\,\widehat{\mathbf{V}}(t)\,\widehat{\mathbf{I}}^H(t)\,\right|^2$$
$$\leq \left(\widehat{\mathbf{V}}(t)\,\widehat{\mathbf{V}}^H(t)\right)\left(\widehat{\mathbf{I}}(t)\,\widehat{\mathbf{I}}^H(t)\right) \equiv S^2(t)$$

which relies on a Cauchy-Schwarz inequality for complex-valued (row) vectors.

9.3.2 Sub-cycle Averaging

In the special case when $\xi = \frac{1}{2M}$, the square matrix $\mathcal{W}$ of (9.16) is orthogonal (recall Theorem G.3), which establishes a *Parseval-type relation* between the time-domain waveform samples and the associated sub-cycle phasor arrays $\widehat{\mathbf{V}}(t)$ and $\widehat{\mathbf{I}}(t)$, as shown in Sect. G.4, namely,

$$P(t) \stackrel{\text{def}}{=} \sum_{k\in\mathcal{S}} \Re\{\,\widehat{V}_k(t)\,\widehat{I}_k^H(t)\,\} = \frac{1}{M}\sum_{i=0}^{M-1} p(t-\Delta_i) \tag{9.18a}$$

where $p(t) \stackrel{\text{def}}{=} v(t)\,i^\top(t)$ is the instantaneous power. In other words, in this special case the phasor-based sub-cycle power $P(t)$ can also be determined by averaging the instantaneous power samples $\{\,p(t-\Delta_i);\;\; 0\leq i\leq M-1\,\}$. Similar Parseval-type relations hold for $V_{rms}(t)$ and $I_{rms}(t)$, viz.,

$$V_{rms}^2(t) \stackrel{\text{def}}{=} \sum_{k\in\mathcal{S}} \widehat{V}_k(t)\, \widehat{V}_k^H(t) = \frac{1}{M}\sum_{i=0}^{M-1} v(t-\Delta_i)\, v^\top(t-\Delta_i)$$

$$I_{rms}^2(t) \stackrel{\text{def}}{=} \sum_{k\in\mathcal{S}} \widehat{I}_k(t)\, \widehat{I}_k^H(t) = \frac{1}{M}\sum_{i=0}^{M-1} i(t-\Delta_i)\, i^\top(t-\Delta_i) \tag{9.18b}$$

The sub-cycle reactive power $Q(t)$ also has a time-domain equivalent, albeit a more complicated one (see Sect. G.5).

The excessive fluctuation of the instantaneous power metrics $s(t)$, $p(t)$, and $q(t)$, which were introduced in Sect. 9.2, can be significantly attenuated by suitable temporal averaging [77]. In particular, the *sub-cycle* metrics $P(t)$, $V_{rms}^2(t)$, and $I_{rms}^2(t)$ exhibit significantly reduced fluctuations as compared with their instantaneous counterparts (see, e.g., the examples in Sect. 9.5).

To illustrate the effect of sub-cycle averaging (using $\xi = \frac{1}{2M}$), let us consider the purely sinusoidal (steady-state) polyphase case, namely

$$v(t) = \sqrt{2}\,\Re\left\{V_1\, e^{j\omega_0 t}\right\}\ , \qquad i(t) = \sqrt{2}\,\Re\left\{I_1\, e^{j\omega_0 t}\right\}$$

where V_1 and I_1 are constant polyphase phasors, i.e., row vectors of length m. In this case the product $p(t) \equiv v(t)\, i^\top(t)$ consists of a DC component and a significant 2nd harmonic, and the same holds true also for $v(t)\, v^\top(t)$ and $i(t)\, i^\top(t)$, viz.,

$$v(t)\, i^\top(t) = \Re\left\{V_1\, I_1^H\right\} + \Re\left\{V_1\, I_1^T\, e^{2j\omega_0 t}\right\}$$

$$v(t)\, v^\top(t) = V_1\, V_1^H + \Re\left\{V_1\, V_1^T\, e^{2j\omega_0 t}\right\}$$

$$i(t)\, i^\top(t) = I_1\, I_1^H + \Re\left\{I_1\, I_1^T\, e^{2j\omega_0 t}\right\}$$

The second harmonic components in these expressions can be completely eliminated by selecting $M = 2$ and $\Delta_1 = \frac{T}{4}$ in (9.18): in this case

$$\frac{1}{2}\left[e^{2j\omega_0 t} + e^{2j\omega_0 (t-\frac{T}{4})}\right] = \frac{1}{2}\left[1 + e^{-j\pi}\right] e^{2j\omega_0 t} = 0$$

so the resulting sub-cycle $P(t)$, $V_{rms}^2(t)$, and $I_{rms}^2(t)$ have only a DC component.

However, in the more general periodic (i.e., multi-harmonic) case, sub-cycle averaging cannot completely suppress the fluctuations in the sub-cycle metrics $P(t)$,

$V^2_{rms}(t)$, and $I^2_{rms}(t)$. Still the *average over the sliding-window* $[t - T \quad t]$ of each one of these time-variant quantities coincides with the corresponding classical (steady-state) value. For instance,

$$\frac{1}{T}\int_{t-T}^{t} P(s)\,ds = \sum_{i=0}^{M-1}\left[\frac{1}{T}\int_{t-T}^{t} p(s-\Delta_i)\,ds\right] = P_{av}$$

and similarly for $V^2_{rms}(t)$, $I^2_{rms}(t)$. Notice that the sliding-window average is independent of both the number of samples and their locations.

On the other hand, the *level of fluctuation* of sub-cycle metrics around their sliding-window averages depends strongly on the presence of *spurious harmonics* (recall (3.37)). While the choice $\mathcal{S} = \{1\}$ is sufficient in the sinusoidal case, a larger harmonic index set $\mathcal{S}$ is needed to achieve satisfactory suppression of fluctuations in the multi-harmonic case: Sects. 9.4.1 and 9.4.2 provide several examples that illustrate this point.

9.3.3 5-Component Decomposition of Sub-cycle Apparent Power

The sub-cycle gap $S^2(t) - \left[P^2(t) + Q^2(t)\right]$ can be decomposed, similar to the full-cycle case, into several sub-cycle dynamic power quality components. The statistics-based technique of Sect. 8.5 can now be applied to decompose this gap into a sum of (squares of) several components that describe the combined effect of harmonic distortion and load imbalance among phases. We use sub-cycle phasors to construct a dynamic sub-cycle counterpart of the full-cycle 5/9-component decomposition (8.25c), which involves the "un-split" metrics $N_g(t)$ and $N_b(t)$: in other words, we treat $\widehat{\mathbf{V}}(t)$ and $\widehat{\mathbf{I}}(t)$ as singly-indexed phasor arrays, ignoring the distinction between load imbalance and harmonic distortion.

The time-dependent equivalent admittance array

$$\mathbf{Y}(t) = \frac{\widehat{\mathbf{I}}(t)}{\widehat{\mathbf{V}}(t)} \equiv \widehat{\mathbf{G}}(t) - j\widehat{\mathbf{B}}(t) \tag{9.19a}$$

obtained by element-wise division of $\widehat{\mathbf{I}}(t)$ by $\widehat{\mathbf{V}}(t)$ serves as the starting point for a statistics-based 5-component decomposition. As in the full-cycle case, the gap $S^2(t) - \left[P^2(t) + Q^2(t)\right]$ is directly related to the fact that the elements of $\widehat{\mathbf{G}}(t) - j\widehat{\mathbf{B}}(t)$ are not necessarily all equal to each other. To capture the variation (across phases and harmonics) of the elements of $\widehat{\mathbf{G}}(t)$ and $\widehat{\mathbf{B}}(t)$, we use the (time-dependent) weight vector

$$\mathbf{w} \stackrel{\text{def}}{=} \frac{1}{V^2_{rms}(t)}\left[\mathsf{abs}\left(\widehat{\mathbf{V}}(t)\right)\right]^2 \tag{9.19b}$$

where abs( ) denotes an element-wise absolute-value operation (same as in MATLAB).

Consequently, the weighted mean $\mathcal{M}_w$ is a time-dependent operator, since it uses a set of time-varying weights. We can now define, as in Sect. 8.5,

$$\mu_g(t) \stackrel{\text{def}}{=} \mathcal{M}_w\,\widehat{\mathbf{G}}(t) = \frac{P(t)}{V_{rms}^2(t)} \;,\quad \mu_b(t) \stackrel{\text{def}}{=} \mathcal{M}_w\,\widehat{\mathbf{B}}(t) = \frac{Q(t)}{V_{rms}^2(t)}$$

$$\sigma_g^2(t) \stackrel{\text{def}}{=} \mathcal{M}_w\left(\widehat{\mathbf{G}}(t) - \mu_g(t)\mathbf{1}\right)^2 \;,\quad \sigma_b^2(t) \stackrel{\text{def}}{=} \mathcal{M}_w\left(\widehat{\mathbf{B}}(t) - \mu_b(t)\mathbf{1}\right)^2 \tag{9.20a}$$

so that

$$\mathcal{M}_w\,\widehat{\mathbf{G}}^2(t) = \mu_g^2(t) + \sigma_g^2(t) \;,\quad \mathcal{M}_w\,\widehat{\mathbf{B}}^2(t) = \mu_b^2(t) + \sigma_b^2(t)$$

The definition of the equivalent admittance $\widehat{\mathbf{G}}(t) - j\widehat{\mathbf{B}}(t)$ implies that

$$\sum_{k\in\boldsymbol{\mathcal{S}}} \|\,\widehat{I}_k(t)\,\|^2 = V_{rms}^2(t)\left[\mathcal{M}_w\,\widehat{\mathbf{G}}^2(t) + \mathcal{M}_w\,\widehat{\mathbf{B}}^2(t)\right]$$

and, consequently,

$$\begin{aligned} S^2(t) &= V_{rms}^4(t)\left[\mu_g^2(t) + \sigma_g^2(t) + \mu_b^2(t) + \sigma_b^2(t)\right] + S_\perp^2(t) \\ &= P^2(t) + N_g^2(t) + Q^2(t) + N_b^2(t) + S_\perp^2(t) \end{aligned} \tag{9.20b}$$

where we define

$$N_g(t) \stackrel{\text{def}}{=} \sigma_g(t)\,V_{rms}^2(t) \;,\quad N_b(t) \stackrel{\text{def}}{=} \sigma_b(t)\,V_{rms}^2(t) \tag{9.20c}$$

As an illustration, we compare in Fig. 9.4 the full-cycle and sub-cycle versions of the fundamental dynamic power quality metrics $P(t)$ and $Q(t)$ for the paper-mill example (Sect. L.2.4).

9.3.4 Symmetrical Sequence Components for $P(t)$ and $Q(t)$

Following the procedure described in Sects. 2.3 and 8.6, we can use the sub-cycle voltage and current phasors to determine the symmetrical sequence components of the sub-cycle $P(t)$ and $Q(t)$. First, the symmetrical sequence components of the sub-cycle dynamic phasors $\{\widehat{V}_k(t)\,;\; k \in \boldsymbol{\mathcal{S}}\}$ and $\{\widehat{I}_k(t)\,;\; k \in \boldsymbol{\mathcal{S}}\}$ are obtained in a 3-phase system by applying the unitary transformation (1.13b), namely,

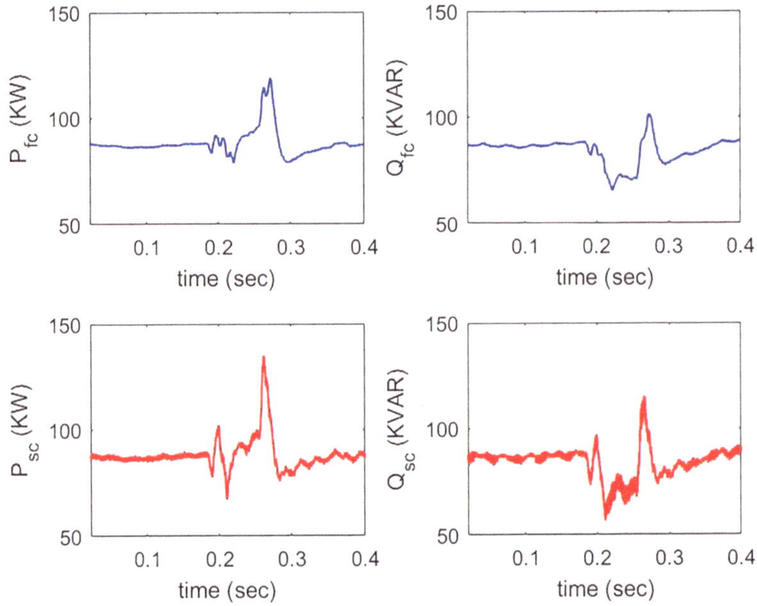

Fig. 9.4 Fundamental power metrics (paper-mill example): full-cycle (top), and sub-cycle with $\mathcal{S} = \{1, 3, 5, 7\}$ and $\xi = \frac{1}{16}$ (bottom).

$$\begin{aligned} \begin{bmatrix} \widehat{V}_{k,0}(t) & \widehat{V}_{k,+}(t) & \widehat{V}_{k,-}(t) \end{bmatrix} &\stackrel{\text{def}}{=} \widehat{V}_k(t)\, \mathcal{M}_{sc} \\ \begin{bmatrix} \widehat{I}_{k,0}(t) & \widehat{I}_{k,+}(t) & \widehat{I}_{k,-}(t) \end{bmatrix} &\stackrel{\text{def}}{=} \widehat{I}_k(t)\, \mathcal{M}_{sc} \end{aligned} \tag{9.21}$$

where the 3×3 unitary matrix $\mathcal{M}_{sc}$ is defined in (1.13c). Recall that the subscript "0" denotes the zero-sequence component, while the subscripts "+" and "−" denote the positive and negative sequence components, respectively.

Next, the quantity $\widehat{V}_k(t)\, \widehat{I}_k^H(t)$, used to determine both $P(t)$ and $Q(t)$, is expressed in terms of the voltage and current symmetrical sequence phasors, viz.,

$$\widehat{V}_k(t)\, \widehat{I}_k^H(t) = \begin{bmatrix} \widehat{V}_{k,0}(t) & \widehat{V}_{k,+}(t) & \widehat{V}_{k,-}(t) \end{bmatrix} \begin{bmatrix} \widehat{I}_{k,0}(t) & \widehat{I}_{k,+}(t) & \widehat{I}_{k,-}(t) \end{bmatrix}^H$$

where we used the unitarity property $\mathcal{M}_{sc}\, \mathcal{M}_{sc}^H = I$. As in Sect. 8.6, this identity induces additive decompositions for $P(t)$ and $Q(t)$, viz.,

$$\begin{aligned} P(t) &= P_0(t) + P_+(t) + P_-(t) \\ Q(t) &= Q_0(t) + Q_+(t) + Q_-(t) \end{aligned} \tag{9.22a}$$

where

$$
\begin{aligned}
P_0(t) &= \sum_{k \in \mathcal{S}} \Re\left\{\widehat{V}_{k,0}(t)\, \widehat{I}_{k,0}^{H}(t)\right\} \\
P_+(t) &= \sum_{k \in \mathcal{S}} \Re\left\{\widehat{V}_{k,+}(t)\, \widehat{I}_{k,+}^{H}(t)\right\} \\
P_-(t) &= \sum_{k \in \mathcal{S}} \Re\left\{\widehat{V}_{k,-}(t)\, \widehat{I}_{k,-}^{H}(t)\right\}
\end{aligned}
\tag{9.22b}
$$

and

$$
\begin{aligned}
Q_0(t) &= \sum_{k \in \mathcal{S}} \Im\left\{\widehat{V}_{k,0}(t)\, \widehat{I}_{k,0}^{H}(t)\right\} \\
Q_+(t) &= \sum_{k \in \mathcal{S}} \Im\left\{\widehat{V}_{k,+}(t)\, \widehat{I}_{k,+}^{H}(t)\right\} \\
Q_-(t) &= \sum_{k \in \mathcal{S}} \Im\left\{\widehat{V}_{k,-}(t)\, \widehat{I}_{k,-}^{H}(t)\right\}
\end{aligned}
\tag{9.22c}
$$

This means that the 5-component decomposition (9.20b) consists, in fact, of nine components, viz.,

$$
\begin{aligned}
S^2(t) = S_\perp^2(t) &+ \Big(P_+(t) + P_-(t) + P_0(t)\Big)^2 + N_g^2(t) \\
&+ \Big(Q_+(t) + Q_-(t) + Q_0(t)\Big)^2 + N_{\bar{b}}^2(t)
\end{aligned}
\tag{9.23}
$$

which has the same structure as the steady-state decomposition (8.25c), but different numerical values.

The positive and negative sequence components of $P(t)$ and $Q(t)$ for the paper-mill example (Sect. L.2.4) are shown in Fig. 9.5: the zero-sequence components are negligible in this case. Notice that both $P_-(t)$ and $Q_-(t)$ provide an indication of the event duration (between 0.18 and 0.27 sec). Similar observations were made in Sect. 9.1 about the full-cycle counterparts of these negative sequence components (see Sect. 9.4.2 for a continued discussion of this example).

9.3.5 7/11-Component Decomposition of Sub-cycle Apparent Power

The sub-cycle 5/9-component apparent power decomposition (9.23) can be further refined to provide an indication of load imbalance, thus producing a sub-cycle

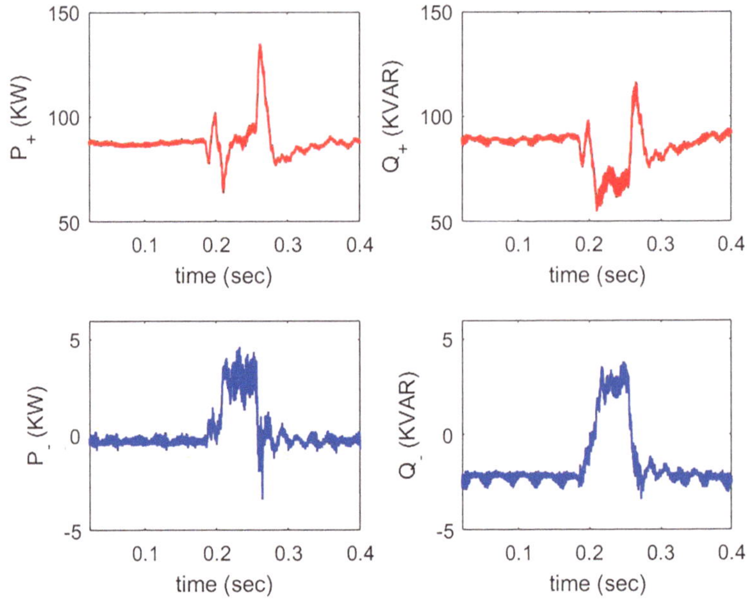

Fig. 9.5 Sub-cycle symmetrical sequence components of $P(t)$ and $Q(t)$: paper-mill example, with $\mathcal{S}=\{1,3,5,7\}$ and $\xi = \frac{1}{16}$.

counterpart of the full-cycle decomposition (9.6), namely,

$$\begin{aligned} S^2(t) \;=\; & S_{\perp}^2(t) + \big(P_+(t) + P_-(t) + P_0(t)\big)^2 + N_{gs}^2(t) + N_{gu}^2(t) \\ & + \big(Q_+(t) + Q_-(t) + Q_0(t)\big)^2 + N_{bs}^2(t) + N_{bu}^2(t) \end{aligned} \tag{9.24}$$

This is accomplished by applying an ANOVA-based decomposition of variance for grouped data, as detailed in Sect. 8.7, to the time-dependent sub-cycle phasor arrays $\widehat{\mathbf{V}}(t)$ and $\widehat{\mathbf{I}}(t)$ that were defined in (9.16). Again, it should be noticed that the components of this decomposition are derived from sub-cycle dynamic phasors and thus have different numerical values than their full-cycle counterparts in (9.6).

Another important difference between the sub-cycle and full-cycle power decomposition schemes is the number of harmonics involved in assembling the dynamic phasor arrays $\widehat{\mathbf{V}}(t)$ and $\widehat{\mathbf{I}}(t)$. The sub-cycle versions of these arrays are typically much shorter than their full-cycle counterparts, which impacts the quality of the resulting sub-cycle power quality metrics. While using $M = 8$ (or, even, $M = 4$) in a sub-cycle scheme is sufficient to obtain a reasonable approximation of $P(t)$ and $Q(t)$, as demonstrated in Sects. 9.4.1 and 9.4.2, a much higher number of waveform samples, say $M \geq 16$, is needed to obtain reliable sub-cycle estimates of the refined metrics $N_{gs}(t)$, $N_{gu}(t)$, $N_{bs}(t)$, and $N_{bu}(t)$. In contrast, it turns out that accurate estimation of the duration and onset time of a wide-band transient requires using a small number of samples, say $M = 4$ or less, as detailed in Sect. 9.4.3. These observations suggest that the number of waveform samples used by a sub-cycle scheme should be selected based on application-specific considerations.

9.4 Customizing the Response of Sub-cycle Power Metrics

Both the steady-state behavior and the transient response of sub-cycle dynamic power quality metrics depend on the harmonic index set $\mathcal{S}$ and the normalized inter-sample shift ξ. In this section we use real-life examples to examine the impact of these design parameters on the behavior of sub-cycle power metrics in three distinct scenarios: (a) steady-state operation, (b) slowly-varying harmonic content (a.k.a. "in-band transient"), and (c) fast wide-band transient. Additional examples can be found in [78, 80].

When operating in steady-state, the design parameters should be adjusted for accuracy of the sub-cycle dynamic phasors and of the derived sub-cycle power metrics, using their full-cycle counterparts as a benchmark for comparison. However, full-cycle metrics cannot serve as a benchmark in the presence of a short wide-band transient: the primary concern in this case becomes accuracy of the estimates of transient duration and time of onset. Thus, in practical real-time implementations it may be advantageous to calculate in parallel several sets of sub-cycle metrics, corresponding to different choices of $\mathcal{S}$ and ξ, to cover the possibility of steady-state operation, as well as both slow and fast transients. Given the low cost of each individual sub-cycle scheme, the overall computational effort of such a multi-objective implementation can still be significantly below that of evaluating a single set of full-cycle power quality metrics (see Table 9.1).

Table 9.1 Cost (per phase, per sample) comparison for full-cycle vs. sub-cycle

	Evaluation of $\mathbf{V}, \mathbf{I}$	Evaluation of P, Q	Evaluation of N_g, N_b	Total
Full-cycle ($N = 128$)	3584	256	640	4480
Full-cycle ($N = 256$)	8192	512	1280	9984
Sub-cycle ($M = 2$)	8	4	10	22
Sub-cycle ($M = 4$)	32	8	20	60
Sub-cycle ($M = 8$)	128	16	40	184
Sub-cycle ($M = 16$)	512	32	80	624

In order to facilitate performance comparisons we will use explicit notation to distinguish sub-cycle dynamic power quality metrics from their full-cycle counterparts. For instance, we will use the notation $P_{sc}(t)$ for the sub-cycle real power and $P_{fc}(t)$ for the full-cycle equivalent, and similarly for other power quality metrics.

9.4.1 Steady-State Operation

We have observed in Sect. 3.3 that, for periodic waveforms, reasonably accurate sub-cycle phasors are obtained by using the special choice $\xi = \xi_{orth}(M) \equiv \frac{1}{2M}$, and a somewhat higher value of M. Consequently, the same observation applies also to the sub-cycle dynamic power metrics. For instance, consider a real-life periodic dataset (see Sect. L.1.8 for a detailed description of this example, obtained from Hydro-Ottawa). Figure 9.6 demonstrates the close resemblance between the full-cycle dynamic power $P_{fc}(t)$ and its sub-cycle counterpart $P_{sc}(t)$. The latter exhibits a minor fluctuation, caused by spurious harmonics, which can be reduced by increasing the value of M from 4 to 8. Notice that the average of $P_{sc}(t)$ over the cycle-long window $(t - T \quad t]$ matches exactly the (constant) $P_{fc}(t)$.

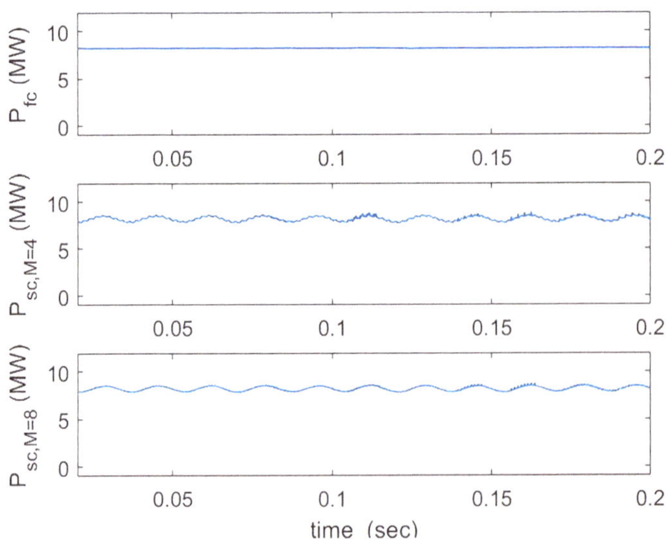

Fig. 9.6 Steady-state example (steady-state Hydro-Ottawa, real power): full-cycle (top), sub-cycle with $\mathcal{S}=\{1, 5\}$ (middle), and sub-cycle with $\mathcal{S}=\{1, 3, 5, 7\}$ (bottom).

Table 9.1, which slightly extends the one presented in [80], compares the cost of a full-cycle scheme using $N = 128$ or 256 samples/cycle (typical number in modern PMUs), with a sub-cycle scheme that uses a sparse subset of $M = 2, 4, 8$

or 16 waveform samples. Notice the impressive reduction in computational cost achieved with sub-cycle dynamic power metrics, as compared with their full-cycle counterparts. Also notice that in the Hydro-Ottawa example the most attractive cost-performance trade-off is achieved with $M = 4$: this choice reduces the cost by a factor of three (as compared with $M = 8$), while causing only a minor increase in fluctuation.

Again, it should be pointed out that while using $M = 4$ in a sub-cycle scheme is sufficient to obtain a reasonable approximation of $P(t)$ and $Q(t)$, a much higher value, say $M \geq 16$, is needed to obtain reliable sub-cycle estimates of the refined metrics $N_{gs}(t)$, $N_{gu}(t)$, $N_{bs}(t)$, and $N_{bu}(t)$. This suggests two different customizations of the sub-cycle scheme for application in steady-state monitoring of power quality, namely:

- **Panoptic implementation:** uses $M = 16$ (or higher) to generate the complete 7/11-component decomposition (9.24), at about 14% of the cost of a typical ($N = 128$) full-cycle implementation.
- **Compact implementation:** uses $M = 8$ (or even $M = 4$) to generate the 5-component decomposition (9.20b), or the 5/9-component decomposition (9.23), at about 4% of the cost of a typical ($N = 128$) full-cycle implementation.

Notice that the relatively small number of harmonics used in a compact implementation precludes a reliable split of N_g into N_{gs} and N_{gu} (and of N_b into N_{bs} and N_{bu}): this is why a reduction in the number of samples used necessitates an attendant reduction in the number of apparent power components generated from these samples. Moreover, the small number of sub-cycle harmonics also rules out the possibility of a non-vanishing out-of-band component $S_{\perp}$.

9.4.2 Tracking of In-Band Transients

Based on our observations in Sect. 3.3 about in-band transients, we should expect to achieve good accuracy of sub-cycle metrics by using the same design choices – i.e., a compact implementation using $M = 4, 6$ or 8, and $\xi = \frac{1}{2M}$ – that work well in a steady-state scenario with limited monitoring objectives, such as tracking $P(t)$, $Q(t)$ and their symmetrical sequence components. To illustrate this point we examine again the paper-mill dataset (described in Sect. L.2.4), which contains a 5-cycle-long unbalanced fault. The results shown in Fig. 9.7 are indeed consistent with our prediction: the sub-cycle dynamic power $P_{sc}(t)$ matches it full-cycle counterpart $P_{fc}(t)$ in the steady-state interval preceding the fault (i.e., $t \leq 0.18$ sec) as well as in the recovery interval following the fault ($t > 0.3$ sec), while providing a reasonable approximation within the fault interval. The level of fluctuation caused by spurious harmonics is somewhat reduced when we increase the number of waveform samples from $M = 4$ to $M = 8$.

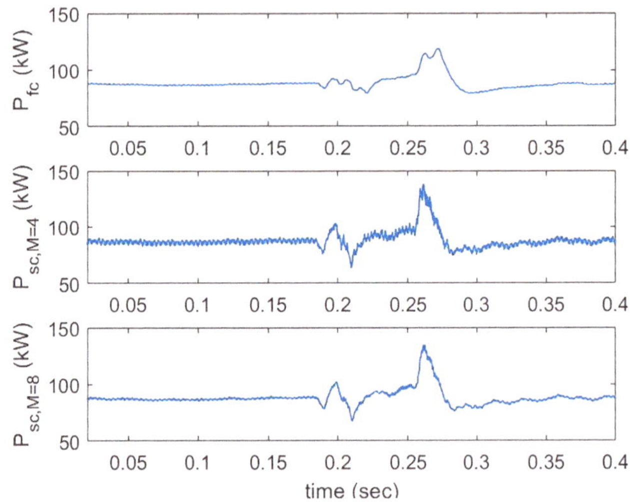

Fig. 9.7 In-band-transient example (Paper-mill, real power): full-cycle (top), sub-cycle with $\boldsymbol{\mathcal{S}}=\{1,5\}$, $\xi=\frac{1}{8}$ (middle), and sub-cycle with $\boldsymbol{\mathcal{S}}=\{1,3,5,7\}$, $\xi=\frac{1}{16}$ (bottom).

In Sect. 3.3 we suggested using transient-generated metrics as indicators of the occurrence and duration of a transient. Since dynamic power quality metrics combine information from multiple phasors, they are natural candidates for the role of a transient-generated metric. To be specific, the components $N_g(t)$ and $N_b(t)$ tend to increase in the presence of harmonic distortion, while $P_-(t)$ and $Q_-(t)$ are affected mainly by phase imbalance. Thus, any transient that involves increased harmonic distortion and/or declining phase imbalance should be clearly visible in at least one of these metrics.

The paper-mill example exhibits considerable steady-state values for both N_g and N_b (on the same order of magnitude as the real power P) in the interval that precedes the onset of the fault, as seen in Fig. 9.2. In contrast, the negative sequence power component $P_-(t)$ is negligible in steady-state, which makes it suitable to serve as a transient-generated metric for this example: both the full-cycle and the sub-cycle version of $P_-(t)$ provide a clear and accurate indication of the presence and duration of the transient (Fig. 9.8), and similarly for $Q_-(t)$.

In summary, sub-cycle dynamic power quality metrics offer an attractive cost-effective alternative to their full-cycle counterparts, both in steady-state and in the presence of (slow) in-band transients: according to Table 9.1, the cost of a compact $(M=8)$ sub-cycle implementation is only 4% of the cost of a typical $(N=128)$ full-cycle implementation.

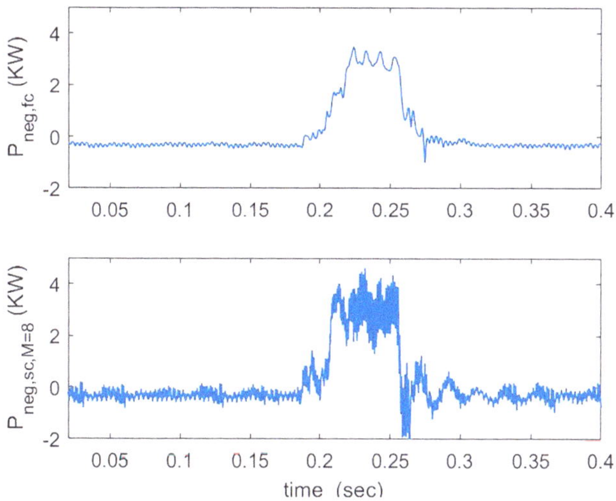

Fig. 9.8 In-band-transient example (Paper-mill, negative sequence power): full-cycle (top), and sub-cycle with $\boldsymbol{\mathcal{S}}=\{1,3,5,7\}$, $\xi = \frac{1}{16}$ (bottom).

9.4.3 Response to Wide-Band Transients

In contrast to the steady-state and in-band-transient cases, the behavior of the sub-cycle dynamic power $P_{sc}(t)$ in the presence of a wide-band transient can differ radically from that of its full-cycle counterpart $P_{fc}(t)$. Both power metrics can be interpreted as processed versions of the instantaneous power $p(t) \equiv v(t)\, i^{\top}(t)$: thus $P_{fc}(t)$ is the full-cycle average of $p(t)$ over the t-dependent interval $(t - T \quad T]$ (recall (9.4a)), while $P_{sc}(t)$, with $\xi = \frac{1}{2M}$, is the average of M samples of $p(t)$ selected from the same interval (recall (9.18a)). Wide-band transients have a negligible impact on $P_{fc}(t)$ because full-cycle averaging of $p(t)$ acts as a narrow-band lowpass filter (see Fig. 9.9, top), suppressing all higher frequency components. On the other hand, the frequency response of sub-cycle sample-averaging (see Fig. 9.9, bottom) allows some wide-band higher frequency content to show up in $P_{sc}(t)$.

To illustrate this effect, we consider the wind-farm dataset (see Sect. L.2.2 for a detailed description), which contains a cycle-long, wide-band transient caused by switching on a bank of capacitors. Again, the sub-cycle dynamic power $P_{sc}(t)$ matches its full-cycle counterpart $P_{fc}(t)$ (Fig. 9.10) in the steady-state interval preceding the onset of the transient (i.e., $t \le 0.031$ sec). However, these two power metrics differ greatly within the transient interval (i.e., in $0.031\,\text{sec} \le t \le 0.048\,\text{sec}$): the sub-cycle $P_{sc}(t)$ exhibits a sizeable transient component, comparable to its steady-state component, while the transient component of the full-cycle $P_{fc}(t)$ is barely perceptible (notice the difference in the vertical scales used in the middle and bottom plots of Fig. 9.10). This is not surprising, given the

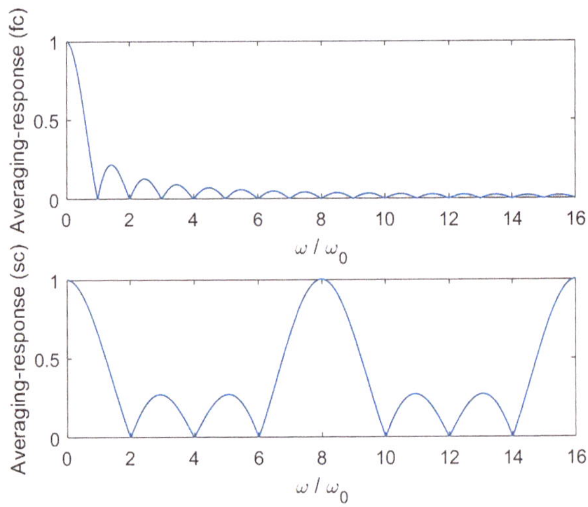

Fig. 9.9 Averaging response: full-cycle (top); and sub-cycle with $\mathcal{S}=\{1, 5\}$, $\xi = \frac{1}{8}$ (bottom).

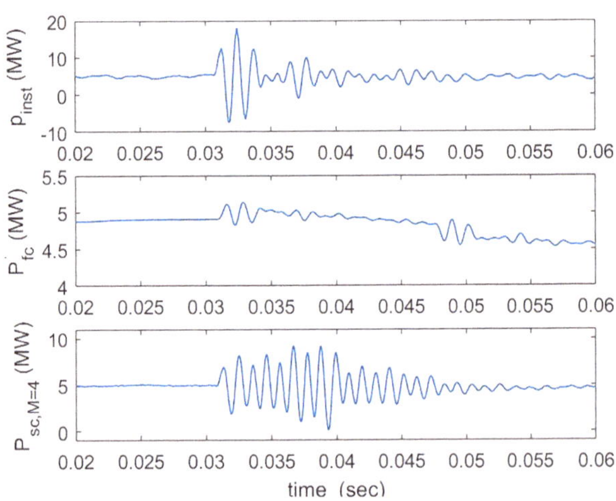

Fig. 9.10 Wide-band transient example (wind-farm, real power): instantaneous (top), full-cycle (middle), and sub-cycle with $\mathcal{S}=\{1, 5\}$, $\xi = \frac{1}{8}$ (bottom).

frequency ($\approx$ 950 Hz) of the transient in the instantaneous power $p(t)$ (see Fig. 9.10, top), and the narrow bandwidth of the full-cycle averaging response (recall Fig. 9.9, top).

Following the introduction of a two-criterion approach in Sect. 3.3, we now turn to identify a metric that can provide accurate information about the occurrence and duration of a wide-band transient. As before, we consider the use of N_g, N_b or P_-, Q_- for this task. In the wind-farm example the best "transient-generated"

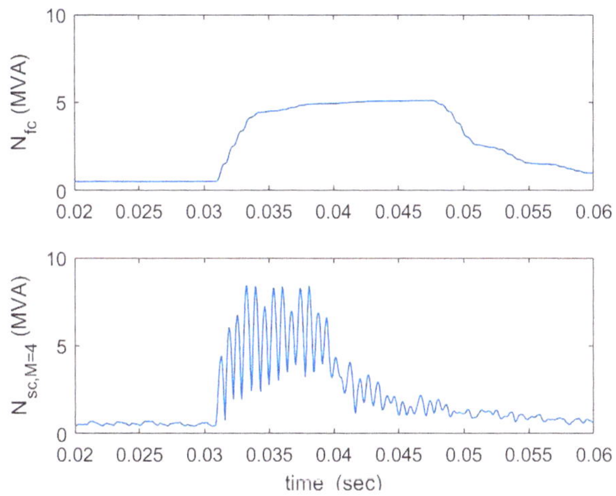

Fig. 9.11 Wide-band transient example (wind-farm, N-metric): full-cycle (top), and sub-cycle with $\boldsymbol{S} = \{1, 5\}$, $\xi = \frac{1}{8}$ (bottom).

metric is provided by $N_g(t)$ and/or $N_b(t)$. We present in Fig. 9.11 the *combined N-metric*

$$N(t) \stackrel{\text{def}}{=} \sqrt{N_g^2(t) + N_b^2(t)} \tag{9.25}$$

which can be viewed as the sub-cycle equivalent of Budeanu's "distortion power."

Notice that the full-cycle $N_{fc}(t)$ tends to exaggerate the duration of the transient interval. In contrast, the sub-cycle $N_{sc}(t)$ identifies two distinct transient sub-intervals:

- A strong transient in $0.031\,\text{sec} \le t \le 0.040\,\text{sec}$
- A decaying weaker transient in $0.040\,\text{sec} \le t \le 0.048\,\text{sec}$

This conclusion is consistent with the shape of the instantaneous power transient component (recall Fig. 9.10, top).

In order to highlight the advantages of using sub-cycle metrics for accurate estimation of the transient's moment of onset and interval of duration, we now consider a synthetic dataset (see Sect. L.2.1 for a detailed description), for which we have exact prior knowledge of such timing information. In this example, a higher frequency wide-band transient is generated by injecting a decaying 15-th harmonic component, limited to the short time interval $0.03\,\text{sec} \le t \le 0.035\,\text{sec}$. Notice that the transient interval is clearly visible in the instantaneous power waveform (Fig. 9.12, top). Again, the sub-cycle dynamic power $P_{sc}(t)$ matches perfectly its full-cycle counterpart $P_{fc}(t)$ (Fig. 9.12, middle and bottom) in the steady-state interval preceding the onset of the transient (i.e., $t \le 0.030\,\text{sec}$). However, these two power metrics differ greatly in the transient interval $0.030\,\text{sec} \le t \le 0.035\,\text{sec}$, and

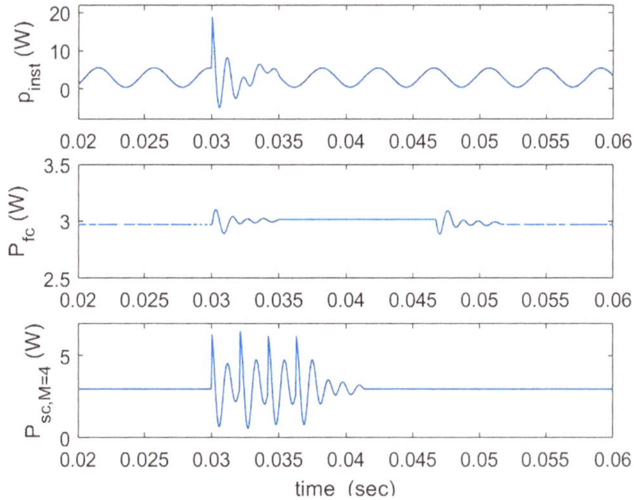

Fig. 9.12 Wide-band higher frequency transient example (synthetic-transient, real power): instantaneous (top), full-cycle (middle), and sub-cycle with $\boldsymbol{\mathcal{S}}=\{1,5\}$, $\xi = \frac{1}{8}$ (bottom).

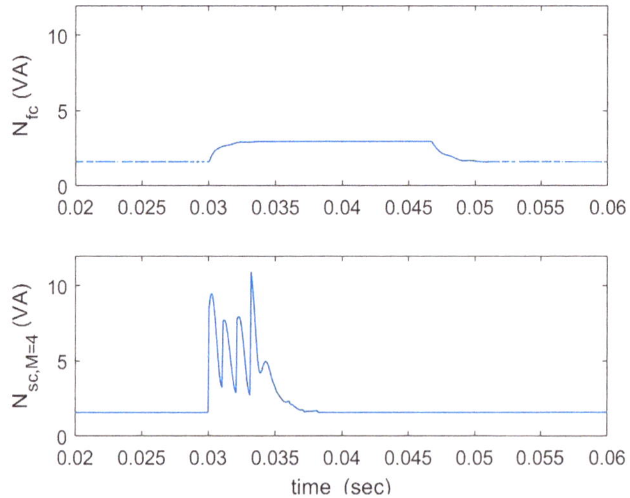

Fig. 9.13 Wide-band higher frequency transient example (synthetic-transient, N-metric): full-cycle (top), and sub-cycle with $\boldsymbol{\mathcal{S}}=\{1,5\}$, $\xi = \frac{1}{16}$ (bottom).

even beyond it: the sub-cycle power $P_{sc}(t)$ exhibits a sizeable transient, comparable to its steady-state component, while the transient component of the full-cycle $P_{fc}(t)$ is barely perceptible (notice the difference in the vertical scales used in the middle and bottom plots of Fig. 9.12). In addition, notice that the full-cycle power $P_{fc}(t)$ contains a *second transient*, starting at $t \approx 0.046$ sec, which is caused by the cycle-long averaging used to generate this full-cycle metric.

As before, we conclude that the real power metrics $P_{fc}(t)$, $P_{sc}(t)$ can provide only limited information about the transient, which motivates us to consider the use of N_g, N_b, P_-, or Q_- as indicators of transient occurrence and duration. In this synthetic-transient example the voltage and current are balanced, so that the negative sequence power component is negligible, thus making $N_g(t)$, $N_b(t)$ better choices for the role of a "transient-generated" metric. We present in Fig. 9.13 the combined N-metric, which was defined in (9.25). Notice that the full-cycle $N_{fc}(t)$ greatly exaggerates the duration of the transient. In contrast, the sub-cycle $N_{sc}(t)$, obtained with $\xi = \frac{1}{16}$, identifies correctly the onset and duration of the transient interval.

9.4.4 Parallel Multi-objective Sub-cycle Implementation

The empirical study presented in this section reveals that no single sub-cycle scheme can capture the key characteristics of the full range of power system waveforms, from steady-state, through (slow) in-band transients, to (fast) wide-band transients. Our results suggest that in practical real-time implementations it may be advantageous to calculate at least three sets of sub-cycle metrics, corresponding to three different choices of $\mathcal{S}$ and ξ, to cover the need for monitoring steady-state power quality, as well as capturing both slow and fast transients. Given the very low cost of each individual sub-cycle scheme, the overall computational effort of such a multi-objective approach is still significantly below that of evaluating a single set of full-cycle power quality metrics (recall Table 9.1).

To be specific, we propose to implement in parallel the following three sub-cycle modules:

(a) **P-module:** a panoptic (i.e., all-inclusive) steady-state module, using $\mathcal{S} = \{1, 3, 5, 7, 9, 11, 13, 17\}$ and $\xi = 1/32$ to evaluate all power quality metrics specified by the 7/11-component decomposition (9.24).
(b) **C-module:** a compact module for either a steady-state or slow-transient scenario, using $\mathcal{S} = \{1, 3, 5, 7\}$ and $\xi = 1/16$ to evaluate the selected metrics $P(t)$, $Q(t)$, $P_-(t)$, $Q_-(t)$ and $N(t) = \sqrt{N_g^2(t) + N_b^2(t)}$.
(c) **A-module:** an agile fast-transient module, using $\mathcal{S} = \{1, 5\}$ and $\xi = 1/16$ to generate a more detailed (i.e., wider bandwidth) version of the transient-sensitive metrics $P_-(t)$, $Q_-(t)$, and $N(t)$.

Notice that implementation of the proposed modules imposes a constraint on the data-sampling scheme: the number of samples per cycle has to be a multiple of 32, such as 128, 160, 192, 224, or 256 samples per cycle.

We recall that non-negligible triplen harmonics can occur during transients, as demonstrated in Figs. 3.13 and 4.2, as well as in steady-state operation of a distribution grid that contains multiple power inverters [81]. This observation, combined with the $\mathcal{S}$-validity constraints described in Sect. G.6, motivates inclusion of such harmonics in the proposed index sets $\mathcal{S}$ for the P-module and C-module.

In summary, sub-cycle dynamic power quality metrics offer a cost-effective alternative to their full-cycle counterparts. In particular, they can provide accurate information about of the time of onset and the duration of both fast (wide-band) and slow (in-band) transients, while still (very nearly) matching the performance of full-cycle metrics in steady-state operation. In contrast, full-cycle metrics tend to exaggerate the duration of such transients. Notice that the combined cost of the proposed three-module sub-cycle scheme, according to Table 9.1, is still less than 20% of the cost of a typical ($N = 128$) full-cycle implementation.

9.5 2-Sample Sub-cycle: An Alternative to Instantaneous Metrics

A 2-sample sub-cycle scheme for evaluating dynamic real and reactive power (with $\xi = \frac{1}{4}$) has, essentially, the same implementation cost as the purely-instantaneous (Akagi) approach but offers numerous advantages:

- The sub-cycle approach can be used with any number of phases, including single-phase systems. In contrast, most purely-instantaneous metrics are defined only for 3-phase systems.
- The level of fluctuation caused by higher frequency components, both in steady-state and in transient operation, is significantly lower for the 2-sample sub-cycle power metrics $P(t)$ and $Q(t)$, as compared with that of the purely-instantaneous $p(t)$ and $q_{\alpha\beta}(t)$.
- Practical applications of $q_{\alpha\beta}(t)$ to compensation and active filter control require filtering to suppress the unwanted fluctuating component [287–290], which increases both implementation cost and group-delay.
- The sub-cycle scheme generates several additional dynamic power metrics, namely $N_g(t)$, $N_b(t)$, as well as the symmetrical sequence components of $P(t)$ and $Q(t)$, at a modest increase in computational cost. Some of these metrics can be used to detect onset of faults, and accurately determine their duration.

In addition, sub-cycle power metrics can be customized, by selecting the value of the control parameter ξ, so as to meet the requirements of specific applications [78]. In this section we analyze the properties of 2-sample sub-cycle dynamic power metrics and demonstrate their advantages, as compared with the purely-instantaneous approach.

When $M = 2$ the only possible choice for a primary harmonic index set is $\mathcal{S} = \{1\}$. This means that all sub-cycle power metrics are defined in terms of the voltage phasor $\widehat{V}_1(t)$ and current phasor $\widehat{I}_1(t)$. Each one of these is generated from two waveform samples, separated by $\Delta \equiv \xi T$, via the waveform-to-phasor map (3.30a), viz.,

$$\widehat{X}_1(t)\, e^{j\omega_0 t} = \begin{bmatrix} 1 & 0 \end{bmatrix} \mathcal{W}^{-1} \begin{pmatrix} x(t) \\ x(t-\xi T) \end{pmatrix}, \quad \text{where } \mathcal{W} = \frac{1}{\sqrt{2}} \begin{pmatrix} 1 & 1 \\ e^{-j2\pi\xi} & e^{j2\pi\xi} \end{pmatrix}$$

so that

$$\widehat{H}_1(s) = \frac{1}{j\sqrt{2}\,\sin 2\pi\xi} \left[e^{j2\pi\xi} - e^{-s\xi T} \right] \tag{9.26a}$$

This results in explicit expressions for the fundamental harmonic sub-cycle voltage and current phasors (recall (9.16)), viz.,

$$\begin{aligned} \widehat{V}_1(t)\, e^{j\omega_0 t} &= \frac{1}{j\sqrt{2}\sin 2\pi\xi} \left[e^{j2\pi\xi}\, v(t) - v(t-\xi T) \right] \\ \widehat{I}_1(t)\, e^{j\omega_0 t} &= \frac{1}{j\sqrt{2}\sin 2\pi\xi} \left[e^{j2\pi\xi}\, i(t) - i(t-\xi T) \right] \end{aligned} \tag{9.26b}$$

where $v(t)$, $i(t)$ are the (polyphase) voltage and current waveforms, respectively. Notice that since these expressions involve a single inter-sample shift ($\Delta = \xi T$) the 2-sample sub-cycle scheme is always *uniformly-shifted.* In particular, making the special choice $\xi = \frac{1}{2M} = \frac{1}{4}$ reduces the dynamic phasor expressions (9.26b) to

$$\begin{aligned} \widehat{V}_1(t)\, e^{j\omega_0 t} &= \tfrac{1}{\sqrt{2}} \left[v(t) + j\, v(t - \tfrac{T}{4}) \right] \\ \widehat{I}_1(t)\, e^{j\omega_0 t} &= \tfrac{1}{\sqrt{2}} \left[i(t) + j\, i(t - \tfrac{T}{4}) \right] \end{aligned} \qquad \left(\xi = \tfrac{1}{4}\right) \tag{9.26c}$$

which requires no multiplications, except for scaling by $1/\sqrt{2}$.

9.5.1 2-Sample Sub-cycle Real and Reactive Power

All 2-sample sub-cycle power metrics can be expressed in terms of the voltage phasor $\widehat{V}_1(t)$ and current phasor $\widehat{I}_1(t)$. In particular (recall (9.17))

$$V_{rms}^2(t) = \left\| \widehat{V}_1(t) \right\|^2, \quad I_{rms}^2(t) = \left\| \widehat{I}_1(t) \right\|^2, \quad S(t) = V_{rms}(t)\, I_{rms}(t)$$

and

$$P(t) = \Re\left\{ \widehat{V}_1(t)\, \widehat{I}_1^H(t) \right\}, \quad Q(t) = \Im\left\{ \widehat{V}_1(t)\, \widehat{I}_1^H(t) \right\}$$

The 5/9-component decomposition (9.23) still holds, so that we have at our disposal a considerable number of dynamic power quality metrics, which can be used to characterize both steady-state and transient behavior. This should be contrasted with the fact that the purely-instantaneous scheme, which we described in Sect. 9.2, consists of only two power metrics, $p(t)$ and $q_{\alpha\beta}(t)$.

In general, the 2-sample sub-cycle real and reactive power metrics $P(t)$ and $Q(t)$ exhibit a much lower level of fluctuation in steady-state operation than their purely-instantaneous counterparts.

The difference in the level of steady-state fluctuation is evident, in particular, when the polyphase waveforms $v(t)$ and $i(t)$ are both *sinusoidal*, namely

$$v(t) = \sqrt{2}\,\Re\left\{V_1 e^{j\omega_0 t}\right\}, \quad i(t) = \sqrt{2}\,\Re\left\{I_1 e^{j\omega_0 t}\right\}$$

In this case we obtain $\widehat{V}_1(t) = V_1$ and $\widehat{I}_1(t) = I_1$, for any choice of ξ, so that $S(t) = \|V_1\|\,\|I_1\|$, and

$$P(t) = \Re\left\{V_1\, I_1^H\right\}, \quad Q(t) = \Im\left\{V_1\, I_1^H\right\}$$

which are the same as the results obtained using the standard (i.e., full-cycle) approach. In contrast, the *purely-instantaneous* real power $p(t)$ is time-varying, with a potentially sizeable 2-nd harmonic component, viz.,

$$p(t) = \Re\left\{V_1\, I_1^H\right\} + \Re\left\{V_1\, I_1^\top e^{2j\omega_0 t}\right\}$$

Similar results hold for the purely-instantaneous reactive power $q_{\alpha\beta}(t)$. Notice that sinusoidal 3-phase voltage and current generate sinusoidal Clarke components, viz.,

$$\begin{aligned}
v_\alpha(t) &= \sqrt{2}\,\Re\left\{V_\alpha e^{j\omega_0 t}\right\}, & i_\alpha(t) &= \sqrt{2}\,\Re\left\{I_\alpha e^{j\omega_0 t}\right\} \\
v_\beta(t) &= \sqrt{2}\,\Re\left\{V_\beta e^{j\omega_0 t}\right\}, & i_\beta(t) &= \sqrt{2}\,\Re\left\{I_\beta e^{j\omega_0 t}\right\} \\
v_0(t) &= \sqrt{2}\,\Re\left\{V_0 e^{j\omega_0 t}\right\}, & i_0(t) &= \sqrt{2}\,\Re\left\{I_0 e^{j\omega_0 t}\right\}
\end{aligned}$$

where

$$\left[V_\alpha\ V_\beta\ V_0\right] = V_1\, \mathcal{M}_{Clarke}, \quad \left[I_\alpha\ I_\beta\ I_0\right] = I_1\, \mathcal{M}_{Clarke}$$

and the matrix $\mathcal{M}_{Clarke}$ is given by (9.9b). Thus, assuming that zero-sequence components vanish (i.e., $V_0 = 0 = I_0$), the purely-instantaneous reactive power is given by the expression

$$q_{\alpha\beta}(t) = \Re\left\{V_\beta I_\alpha^* - V_\alpha I_\beta^*\right\} + \Re\left\{(V_\beta I_\alpha - V_\alpha I_\beta)\, e^{2j\omega_0 t}\right\}$$

(continued)

We observe that, similar to $p(t)$, the purely-instantaneous reactive power $q_{\alpha\beta}(t)$ consists of two components: (i) a DC component and (ii) a 2nd-harmonic component. The 2nd-harmonic component of both $q_{\alpha\beta}(t)$ and $p(t)$ vanishes when the current and voltage are perfectly balanced.

To illustrate this fundamental difference between 2-sample sub-cycle metrics and purely-instantaneous metrics, consider again the synthetic RLC load example with multiple harmonics (described in detail in Sect. L.1.5). When the source voltage is sinusoidal, the sub-cycle metrics are constant (and coincide with their full-cycle counterparts), while the instantaneous metrics exhibit a large fluctuating component. Adding a (spurious) 7-th harmonic voltage component, with amplitude that is 5% of the fundamental, results in further distortion of the instantaneous metrics but only contributes a mild fluctuating component to the 2-sample sub-cycle $P(t)$ and $Q(t)$ (Fig. 9.14). This is, of course, a direct consequence of the (sample-based) averaging operation that generates $P(t)$ from $p(t)$ (recall (9.18a)).

Similar results are obtained for transient waveforms that involve non-negligible harmonic distortion, such as the paper-mill example (described in Sect. L.2.4), or the transient-Hydro-Ottawa example (described in Sect. L.2.3). Both Figs. 9.15 and 9.16 exhibit a large fluctuating component in the instantaneous power waveform, which is generated by the multiple harmonics present in the current waveforms in both the fault interval and the steady-state interval that precedes it. In contrast, the sub-cycle

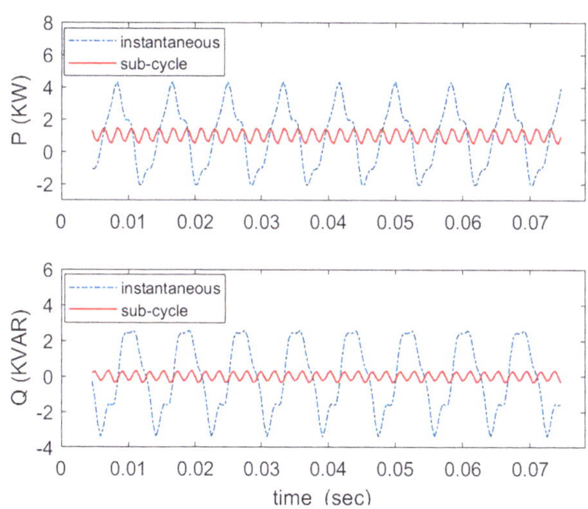

Fig. 9.14 Synthetic steady-state example (RLC load): real power (top), and reactive power (bottom). Dash-dot line used for instantaneous metrics, and solid line used for sub-cycle metrics (evaluated with $M = 2, \xi = \frac{1}{4}$).

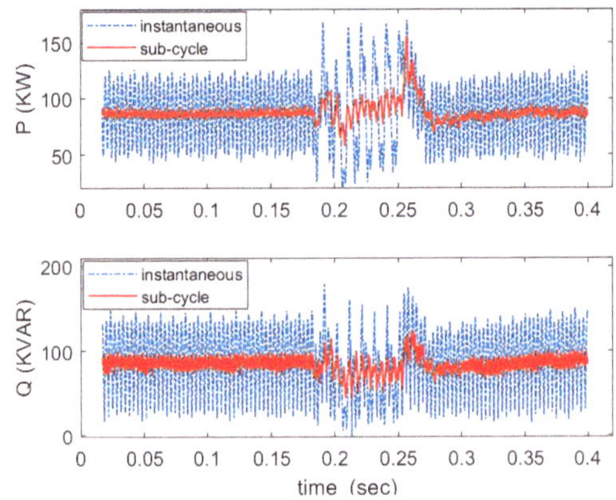

Fig. 9.15 Real-life transient example (paper-mill): real power (top) and reactive power (bottom). Dash-dot line used for instantaneous metrics, and solid line used for sub-cycle metrics (obtained with $M = 2, \xi = \frac{1}{4}$).

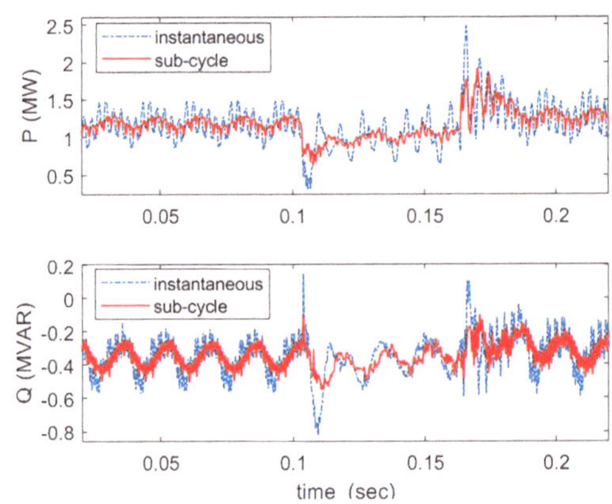

Fig. 9.16 Real-life transient example (transient-Hydro-Ottawa): real power (top), and reactive power (bottom). Dash-dot line used for instantaneous metrics, and solid line used for sub-cycle metrics (obtained with $M = 2, \xi = \frac{1}{4}$).

metrics exhibit a relatively mild fluctuation in both examples. Additional examples that illustrate the same advantage can be found in [291].

We conclude that, in general, the 2-sample sub-cycle scheme offers a very significant improvement in the quality of dynamic power metrics, as compared with the purely-instantaneous metrics of [271, 272], at a comparable computational cost. Of course, further improvement in accuracy can be achieved by increasing the number of samples used (albeit at a higher cost), as we have demonstrated in Sect. 9.4.

Table 9.2 Cost (per sample) comparison of instantaneous vs. 2-sample sub-cycle

	Evaluating real power		Evaluating reactive power		Combined cost	
	mult.	add.	mult.	add.	mult.	add.
Purely-instantaneous	3	2	12	7	15	9
Sub-cycle ($M = 2, \xi = 1/4$)	7	4	7	4	14	8

In terms of implementation cost in a 3-phase system, we observe that evaluation of the instantaneous $p(t)$ and $q_{\alpha\beta}(t)$ requires 15 real multiplications and 9 real additions (Table 9.2). In contrast, evaluation of the 2-sample sub-cycle $P(t)$ and $Q(t)$ for $\xi = \frac{1}{4}$ requires a comparable 14 real multiplications and 8 real additions but produces much cleaner results with a significantly reduced fluctuating component (see, e.g., Figs. 9.14, 9.15, and 9.16).

Moreover, the sub-cycle scheme provides additional power quality metrics — $N_g(t)$ and $N_b(t)$, as well as the symmetrical sequence components of $P(t)$ and $Q(t)$ — that are not available in the purely-instantaneous framework. For instance, the sub-cycle negative sequence power component $P_-(t)$ often provides accurate information about the duration and time of onset of a transient (see Fig. 9.17), and similarly for $Q_-(t)$. Additional demonstration of the accuracy of these event-sensitive metrics can be found in [291].

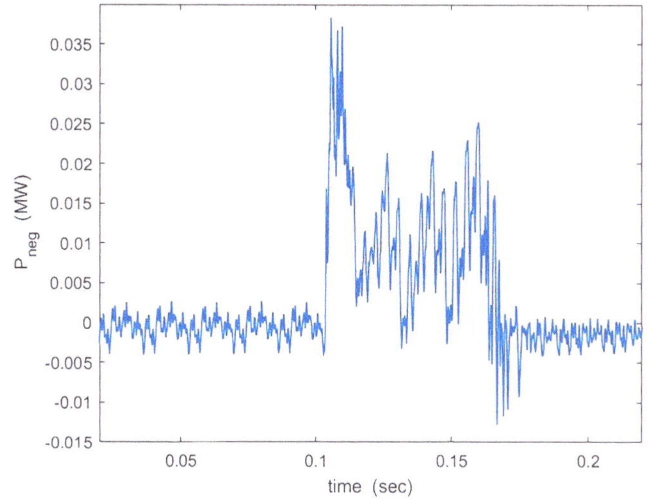

Fig. 9.17 Sub-cycle negative sequence power in the transient-Hydro-Ottawa example (with $M = 2, \xi = \frac{1}{4}$).

9.5.2 Analytical Comparison with Purely-Instantaneous Metrics

Any linear combination of the (polyphase) waveform samples $x(t)$ and $x(t-\Delta)$ can be equivalently described as a combination of the present sample $x(t)$ and the "discrete-derivative" $\frac{1}{\Delta}\big[x(t)-x(t-\Delta)\big]$, also known as a divided-difference (see, e.g., [292]). For instance, we can rewrite the waveform-to-phasor map (9.26) in the form

$$\widehat{H}_1(s) \;=\; \frac{1}{\sqrt{2}}\left(\frac{2\pi\xi}{\sin 2\pi\xi}\right)\left[\left(\frac{\sin\pi\xi}{\pi\xi}\right)e^{j\pi\xi} - j\,D_\xi\,(s)\right] \tag{9.27a}$$

where D_ξ is a *normalized* divided-difference operator, viz.,

$$D_\xi\,(s) \;\stackrel{\text{def}}{=}\; \frac{1-e^{-s\Delta}}{\omega_0\Delta} \;\equiv\; \frac{1-e^{-(s/\omega_0)\,2\pi\xi}}{2\pi\xi} \tag{9.27b}$$

In phasor terms this means that, for any polyphase waveform $x(\cdot)$,

$$\widehat{X}_1(t)\,e^{j\omega_0 t} \;=\; \frac{1}{\sqrt{2}}\left(\frac{2\pi\xi}{\sin 2\pi\xi}\right)\left[\left(\frac{\sin\pi\xi}{\pi\xi}\,e^{j\pi\xi}\right)x(t) - j\Big(D_\xi\,x(t)\Big)\right]$$

For small values of ξ the $\widehat{H}_1(s)$ expression reduces to a simpler, more intuitive, form. Using a first-order approximation, we obtain

$$\widehat{H}_1(s) \;\approx\; \frac{1}{\sqrt{2}}\Big[1 - j\,D_\xi\,(s)\Big] \tag{9.28a}$$

or, equivalently,

$$\begin{aligned}\widehat{V}_1(t)e^{j\omega_0 t} \;&\approx\; \frac{1}{\sqrt{2}}\Big[v(t) - j\,D_\xi\,v(t-\xi T)\Big]\\ \widehat{I}_1(t)e^{j\omega_0 t} \;&\approx\; \frac{1}{\sqrt{2}}\Big[i(t) - j\,D_\xi\,i(t-\xi T)\Big]\end{aligned} \tag{9.28b}$$

The corresponding sub-cycle real and reactive power can be approximated, for small values of ξ, by the expressions

$$\begin{aligned}P(t) \;&\approx\; \frac{1}{2}\Big[p(t) + \Big(D_\xi\,v(t)\Big)\Big(D_\xi\,i(t)\Big)^{\top}\Big]\\ Q(t) \;&\approx\; \frac{1}{2}\Big[v(t)\Big(D_\xi\,i(t)\Big)^{\top} - \Big(D_\xi\,v(t)\Big)\,i^{\top}(t)\Big]\end{aligned} \tag{9.29}$$

Similarly,

$$
\begin{aligned}
V_{rms}^2(t) &\approx \frac{1}{2}\Big[\,\|v(t)\|^2 + \|D_\xi\, v(t)\|^2\Big] \\
I_{rms}^2(t) &\approx \frac{1}{2}\Big[\,\|i(t)\|^2 + \|D_\xi\, i(t)\|^2\Big]
\end{aligned}
\tag{9.30}
$$

Finally, when $\xi \downarrow 0$, the operator D_ξ converges to a "normalized" derivative, viz.,

$$
D_0\, x(t) \stackrel{\text{def}}{=} \lim_{\xi\to 0} D_\xi\, x(t) = \left(\frac{1}{\omega_0}\frac{d}{dt}\right) x(t)
$$

for any waveform $x(t)$. In this case, the approximations (9.28)–(9.30) become *exact*, and they involve both the *instantaneous voltage and current*, as well as the *derivatives* of these waveforms.

In particular, when $\xi = 0$ we have

$$
Q(t)\Bigg|_{\xi=0} = \frac{1}{2}\left[v(t)\left(\frac{1}{\omega_0}\frac{d}{dt}\,i(t)\right)^{\top} - \left(\frac{1}{\omega_0}\frac{d}{dt}\,v(t)\right)i^{\top}(t)\right] \stackrel{\text{def}}{=} \breve{Q}_0(t)
\tag{9.31}
$$

which fully coincides with the "instantaneous derivative reactive power" $q^{\beta}(t)$ of [277]. It also coincides (up to scaling by $1/\omega_0$) with the proposed "reactive power rate" $p_{react,2}(t)$ of [276] and, consequently, shares its properties, namely,

- In a sinusoidal system $\breve{Q}_0(t) = \Im\{V_1 I_1^H\}$. In fact, this property holds for any value of ξ, as we observed earlier.
- $\breve{Q}_0(t)$ is positive for inductors, negative for capacitors, and zero for resistors.
- The cycle-averaged value of $\breve{Q}_0(t)$ is related to energy storage in inductors and capacitors [277].
- $\breve{Q}_0(t)$ is network-conservative: see Sect. 9.6 below for a definition and detailed discussion of this property, which also holds for any value of ξ.

Notice that the scaling by $\frac{1}{\omega_0}$ endows $\breve{Q}_0(t)$ with the correct dimensionality, namely *Volt · Ampere*.

However, the reliance of (9.31) on current and voltage derivatives results in very high sensitivity to the presence of higher harmonics. To illustrate this effect we evaluate the 2-sample sub-cycle $Q(t)\Big|_{\xi=0}$ for the paper-mill dataset (described in Sect. L.2.4) and present it in Fig. 9.18: the top plot in this figure shows this metric, while the bottom plot is identical to the bottom plot of Fig. 9.15, which displays the 2-sample sub-cycle $Q(t)$ (for $\xi = \frac{1}{4}$) and the purely-instantaneous $q_{\alpha\beta}(t)$. Notice that the amplitude of fluctuation of the derivative-based reactive power

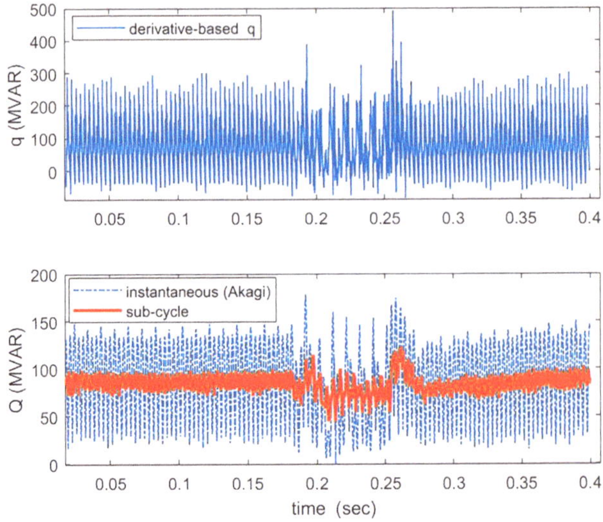

Fig. 9.18 Steady-state and transient fluctuation of dynamic reactive power (paper-mill example): derivative-based reactive power (top), sub-cycle (with $M = 2, \xi = \frac{1}{4}$), and purely-instantaneous reactive power (bottom).

$Q(t)\Big|_{\xi=0} \equiv q^{\beta}(t) \equiv \frac{1}{\omega_0} p_{react,2}(t)$ is approximately 15 times larger than that of the sub-cycle $Q(t)\Big|_{\xi=1/4}$, and 3 times larger than that of the purely-instantaneous $q_{\alpha\beta}(t)$. Similar results are obtained for the transient-Hydro-Ottawa dataset.

In summary, a 2-sample sub-cycle scheme (with $\xi = \frac{1}{4}$) outperforms the purely-instantaneous ("negative-Akagi") reactive power $q_{\alpha\beta}(t)$, and the "generalized instantaneous" reactive power metrics of [276, 277], in both steady-state and in transient operation. It has low implementation cost, and it offers several additional dynamic power metrics, which can be used to detect onset of faults and accurately determine their duration, at a modest increase in computational cost.

9.6 Tellegen's Theorem and Network-Conservativity

Tellegen's theorem, in its most basic form (see Sect. I.1), states that the instantaneous power $p(t)$ is *network-conservative,* namely, the summation of $p(t)$ over all branches of any electrical network adds up to zero, for every individual time-instant t. The theorem relies only on Kirchhoff's laws: it holds for any set of branch voltages that satisfy Kirchhoff's voltage law (KVL) on a given network, and any set of branch currents that satisfy Kirchhoff's current law (KCL) on the same network. Consequently, the theorem remains valid even if we use an operator to transform all branch voltages, provided that the transformed voltages satisfy KVL. Similarly, the theorem still holds if we apply another operator to transform all branch currents, provided that the transformed currents satisfy KCL. In summary, given any two network-invariant[2] linear operators $\mathcal{A}$ and $\mathcal{B}$, the product $\left[\mathcal{A}\, v(t)\right]\left[\mathcal{B}\ i(t)\right]^{\top}$ is network-conservative, provided that the original branch voltages $v(t)$ satisfy KVL and the original branch currents $i(t)$ satisfy KCL, for every time-instant t (see also the discussion in Sect. I.1).

For instance, starting with the set of actual three-phase branch voltages $v(t) = [v_a(t)\ \ v_b(t)\ \ v_c(t)]$ (which, of course, satisfy KVL), we observe that the Clarke-transformed branch voltages $v_\alpha(t)$ and $v_\beta(t)$ also satisfy KVL, since they are (memoryless) linear combinations of the branch phase voltages (recall (9.9)). Similarly, the Clarke-transformed branch currents $i_\alpha(t)$ and $i_\beta(t)$ satisfy KCL, since they are (memoryless) linear combinations of the actual branch phase currents $[i_a(t)\ \ i_b(t)\ \ i_c(t)]$ (which satisfy KCL). Thus, each one of the four products $\{v_\alpha(t)\, i_\alpha(t),\, v_\alpha(t)\, i_\beta(t),\, v_\beta(t)\, i_\alpha(t),\, v_\beta(t)\, i_\beta(t)\}$ is network-conservative, for every choice of t. We conclude that the "negative-Akagi" instantaneous reactive power $q_{\alpha\beta}(t) = v_\beta(t)\, i_\alpha(t) - v_\alpha(t)\, i_\beta(t)$, constructed under the assumption that the zero-sequence voltage and zero-sequence current vanish for each network branch (recall (9.10)), is also network-conservative for every time-instant t.

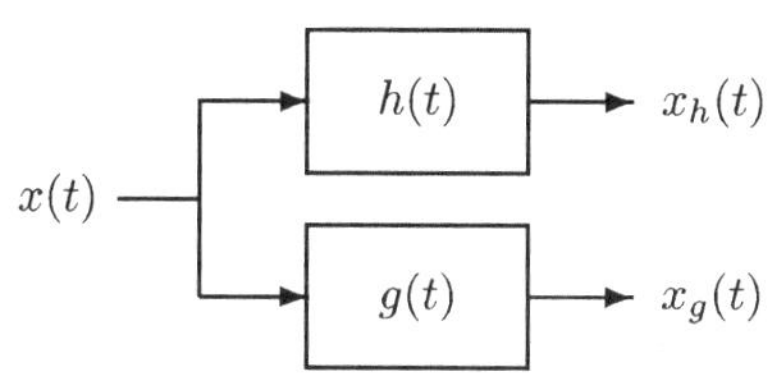

Fig. 9.19 Generation of LTI-filtered waveforms.

This result can be extended to obtain other definitions of network-conservative instantaneous reactive power (such as the ones proposed in [276–278]), using single-input/single-output LTI filters to generate transformed branch voltage and branch current waveforms that still satisfy Kirchhoff's laws (see discussion in Sect. I.2). Applying two different filters, as in Fig. 9.19, to each one of the branch voltage and current waveforms, we can define a generalized network-conservative "instantaneous reactive power" $\breve{q}(t)$, viz.,

$$\breve{q}(t) \ \stackrel{\text{def}}{=}\ \frac{1}{2}\left[\, v_h(t)\, i_g^{\top}(t) - v_g(t)\, i_h^{\top}(t) \,\right] \tag{9.32a}$$

[2] An operator on a function space (a.k.a. waveform space or signal space) is called network-invariant if its definition does not depend on the network location of the waveform to which this operator is applied.

In a sinusoidal system $\breve{q}(t) = \left[-\Im\{H(j\omega_0)\, G^*(j\omega_0)\}\right] \Im\{V_1 I_1^H\}$, which motivates imposing the scaling constraint

$$\Im\{H(j\omega_0)\, G^*(j\omega_0)\} \;=\; -1 \tag{9.32b}$$

Notice that the expression (9.32a) implicitly assumes that $v_h(t)$, $v_g(t)$ are voltages (i.e., measured in *Volts*), and $i_h(t)$, $i_g(t)$ are currents (measured in *Amperes*). This imposes additional constraints on the scaling of the impulse responses $h(\cdot)$ and $g(\cdot)$: the corresponding transfer functions $H(s) = \mathcal{L}\{h(t)\}$ and $G(s) = \mathcal{L}\{g(t)\}$ must be pure gains (i.e., dimensionless).

We have already observed that the "negative-Akagi" instantaneous reactive power $q_{\alpha\beta}(t)$ has the form (9.32a). Also, the expression (9.31) for the sub-cycle reactive power $\breve{Q}_0(t)$ (with $M = 2$ and $\xi \to 0$) is obtained from this generalized expression by setting $h(t) = \delta(t)$ (i.e., no filtering), and $g(t) = \frac{1}{\omega_0}\dot{\delta}(t)$, which corresponds to a (scaled) derivative operation. Similarly, the proposed "reactive power rate" $p_{react,2}(t)$ of [276], which has the dimension of *Volt* $\cdot$ *Ampere/sec*, corresponds to the choice $h(t) = \delta(t)$, and $g(t) = \dot{\delta}(t)$.

The high level of fluctuation of such instantaneous reactive power metrics (recall Fig. 9.18) can be reduced by suitable lowpass filtering. For instance, we can apply a cycle-long sliding-window average to $\breve{q}(t)$, viz.,

$$\breve{Q}(t) \;\stackrel{\text{def}}{=}\; \frac{1}{T}\int_{t-T}^{t} \breve{q}(s)\, ds$$

In particular, cycle averaging of $q_{\alpha\beta}(t)$ generates the difference $Q_{B,+}(t) - Q_{B,-}(t)$, both in steady-state operation and during transients. A detailed discussion of the properties of such cycle-averaged metrics is provided in Appendix H.

Two generalized definitions of "instantaneous reactive power," limited to the case of periodic (non-sinusoidal) single-phase voltage and current waveforms, are used in [277, 278] to construct a decomposition of apparent power that forms the basis for proposed compensation strategies of active and hybrid power filters. One definition (denoted by $q^{\beta}(t)$) intended for control of *fast compensators,* is of the form (9.32a) with $h(t) = \delta(t)$ and $g(t) = \frac{1}{\omega_0}\dot{\delta}(t)$ (i.e., the same as in our expression (9.31)). A controller based on this "instantaneous reactive power" metric will have a fast response, due to the fact that the waveform filter $h(t)$ enhances high-frequency content. The other definition of reactive power in [277, 278] (denoted by $q^{\alpha}(t)$), intended for control of *slow compensators,* is also of the form (9.32a), but with the choice $h(t) = \omega_0\, u(t) \circledast \left[\delta(t) - u_T(t)\right]$ and $g(t) = \delta(t)$, where $\circledast$ denotes linear convolution, $u(t)$ is the unit-step function, and $u_T(t)$ is a cycle-long rectangular pulse, viz.,

$$u_T(t) \;\stackrel{\text{def}}{=}\; \begin{cases} \frac{1}{T} & 0 \le t \le T \\ 0 & \text{else} \end{cases}$$

Controllers based on this generalized "instantaneous reactive power" metric will have a slow response, due to the lowpass nature of the waveform filter $h(t)$ used to evaluate this metric. The cycle-averaged version of this reactive power metric is related to energy storage in inductors and capacitors: additional details are provided in Sect. H.5.

We should observe, however, that designing compensators or controllers with the objective of making $\breve{q}(t)$ vanish may lead to unexpected (and undesired) results, because $\breve{q}(t) = 0$ can be achieved with a variety of non-ideal loads. One sufficient condition for achieving $\breve{q}(t) = 0$ is that the (compensated) load satisfies $i(t) = v(t)\,\boldsymbol{G}$, for some *symmetric* real-valued $m \times m$ matrix $\boldsymbol{G}$. Indeed, this polyphase current-voltage relation implies that

$$i_h(t) \;\equiv\; h(t) \circledast i(t) \;=\; h(t) \circledast \big(v(t)\,\boldsymbol{G}\big) \;=\; \big(h(t) \circledast v(t)\big)\,\boldsymbol{G} \;\equiv\; v_h(t)\,\boldsymbol{G}$$

Similarly, $i_g(t) = v_g(t)\,\boldsymbol{G}$, so that

$$2\,\breve{q}(t) = v_h(t)\,\big(v_g(t)\,\boldsymbol{G}\big)^{\top} - v_g(t)\,\big(v_h(t)\,\boldsymbol{G}\big)^{\top} = v_h(t)\,\boldsymbol{G}\,v_g^{\top}(t) - v_g(t)\,\boldsymbol{G}\,v_h^{\top}(t) = 0$$

where we used the symmetry property $\boldsymbol{G} = \boldsymbol{G}^{\top}$. This means that $\breve{q}(t)$ may vanish even if the load is not ideal: we recall that an ideal polyphase load (i.e., one that is purely resistive and balanced) must satisfy the relation $i(t) = g\,v(t)$, where g is a *positive real scalar.*

Since Tellegen's theorem holds also for voltage and current waveforms generated via filter-banks (see Sect. I.3), it applies to the full-cycle dynamic phasor product $V_k(t)\,I_k^H(t)$, used to determine dynamic real and reactive power. To be specific, recall that the *passband* full-cycle dynamic voltage phasor $V_k(t)\,e^{jk\omega_0 t}$ is obtained by passing the (polyphase) voltage waveform $v(t)$ through the single-input/single-output LTI filter $H_k(s)$ (Fig. 2.1), and the passband full-cycle dynamic current phasor $I_k(t)\,e^{jk\omega_0 t}$ is obtained by passing the current waveform $i(t)$ through the same filter. Consequently, the voltage-current phasor product

$$\Big[\,V_k(t)\,e^{jk\omega_0 t}\Big]\Big[\,I_k(t)\,e^{jk\omega_0 t}\Big]^H \;\equiv\; V_k(t)\,I_k^H(t)$$

is network-conservative, for each individual harmonic k. We conclude that the full-cycle dynamic real power $P(t) \stackrel{\text{def}}{=} \sum_k \Re\{V_k(t)\,I_k^H(t)\}$ and the full-cycle dynamic reactive power $Q_B(t) \stackrel{\text{def}}{=} \sum_k \Im\{V_k(t)\,I_k^H(t)\}$ are both network-conservative, for each time-instant t.

Similar observations hold for the sub-cycle scheme. The *passband* sub-cycle dynamic voltage phasor $\widehat{V}_k(t)\,e^{jk\omega_0 t}$ is obtained by passing the (polyphase) voltage waveform $v(t)$ through the single-input/single-output LTI filter $\widehat{H}_k(s)$ (Fig. 3.7), and the passband sub-cycle dynamic current phasor $\widehat{I}_k(t)\,e^{jk\omega_0 t}$ is obtained by passing the current waveform $i(t)$ through the same filter. Consequently, the voltage-current phasor product $\widehat{V}_k(t)\,\widehat{I}_k^H(t)$ is network-conservative, for each individual harmonic $k \in \boldsymbol{\mathcal{S}}$. Again, we conclude that the sub-cycle dynamic real power $P_{sc}(t) \stackrel{\text{def}}{=} \sum_{k\in\boldsymbol{\mathcal{S}}} \Re\left\{\widehat{V}_k(t)\,\widehat{I}_k^H(t)\right\}$ and the sub-cycle dynamic reactive power $Q_{sc}(t) \stackrel{\text{def}}{=} \sum_{k\in\boldsymbol{\mathcal{S}}} \Im\left\{\widehat{V}_k(t)\,\widehat{I}_k^H(t)\right\}$ are both network-conservative, for each time-instant t.

Appendices

Appendix A: Linear System with Periodic Excitation

We consider the linear state-space model

$$\frac{d}{dt}\, x(t) \;=\; A\, x(t) + B\, u(t) \quad , \qquad t_0 \le t < t_1 \tag{A.1a}$$

with the initial condition $x(t_0)$. We assume that A is a *stability matrix*, namely

$$\Re\left\{\lambda_i(A)\right\} \;<\; 0 \tag{A.1b}$$

and the input signal

$$u(t) \;=\; U_0 + \sqrt{2}\, \sum_{k=1}^{\infty} \Re\left\{U_k\, e^{jk\omega_0(t-t_0)}\right\} \quad , \qquad t_0 \le t < t_1 \tag{A.1c}$$

is periodic with period T, so that the fundamental frequency is $\omega_0 = \frac{2\pi}{T}$. We assume throughout this Appendix that $t_1 - t_0 \ge T$. Notice that in the special case $t_1 = t_0 + T$ periodicity has no meaning, so that (A.1c) is a Fourier-series representation of $u(t)$ in the cycle-long interval $t_1 \le t \le t_0 + T$.

A.1 Time-Domain Solution

The solution of the differential equation (A.1) consists of a steady-state (periodic) component

H. Lev-Ari, A. M. Stanković, *Dynamic Phasors in Energy Processing Systems*,
Power Electronics and Power Systems, https://doi.org/10.1007/978-3-031-50020-6_11

$$x_{ss}(t) = X_0 + \sqrt{2} \sum_{k=1}^{\infty} \Re\left\{X_k\, e^{jk\omega_0(t-t_0)}\right\} \tag{A.2a}$$

as well as a transient component $x_{tr}(t)$ that depends on the initial condition $x(t_0)$. Since $x_{ss}(t)$ satisfies the differential equation (A.1), we conclude that $jk\omega_0 X_k = AX_k + BU_k$ or, equivalently,

$$X_k = -(A - jk\omega_0 I)^{-1} BU_k \tag{A.2b}$$

This observation is consistent with the fact that the frequency response of the map from $u(\cdot)$ to $x(\cdot)$, at the harmonic frequency $k\omega_0$, is $(sI - A)^{-1}B\Big|_{s=jk\omega_0}$. Notice that since $\Re\left\{\lambda_i(A)\right\} < 0$, then $\Re\left\{\lambda_i(A - jk\omega_0 I)\right\} < 0$ as well, so that the matrix $(A - jk\omega_0 I)$ is non-singular for every value of the harmonic index k. Also notice that the steady-state component is independent of the initial condition $x(t_0)$.

The transient component $x_{tr}(t) \stackrel{\text{def}}{=} x(t) - x_{ss}(t)$ satisfies the *homogeneous* differential equation

$$\frac{d}{dt}\, x_{tr}(t) = A\, x_{tr}(t) \tag{A.3a}$$

which is obtained by subtracting $\frac{d}{dt}\, x_{ss}(t) = Ax_{ss}(t) + Bu(t)$ from the original equation (A.1a). This homogeneous equation admits the well-known solution

$$x_{tr}(t) = e^{A(t-t_0)} x_{tr}(t_0) \tag{A.3b}$$

where

$$x_{tr}(t_0) \stackrel{\text{def}}{=} x(t_0) - x_{ss}(t_0) \tag{A.3c}$$

and (recall (A.2))

$$x_{ss}(t_0) = X_0 + \sqrt{2} \sum_{k=1}^{\infty} \Re\left\{X_k\right\} \tag{A.3d}$$

Notice that since $x_{ss}(t_0)$ is completely determined by the phasors $\{U_k\,;\ k \geq 0\}$, Eq. (A.3c) establishes a one-to-one correspondence between $x(t_0)$ and $x_{tr}(t_0)$. Similarly, we also have a one-to-one correspondence between the initial condition $x(t_0)$ and the terminal condition $x_{tr}(t_1)$: since $x_{tr}(t_1) = e^{A(t_1-t_0)} x_{tr}(t_0)$, it follows that

$$x_{tr}(t_1) = e^{A(t_1-t_0)}\Big[x(t_0) - x_{ss}(t_0)\Big] \tag{A.4a}$$

Consequently, we can express the terminal state $x(t_1)$ directly in terms of $x_{tr}(t_1)$ and the phasors $\{U_k\,;\ k \geq 0\}$, viz.,

$$\begin{aligned} x(t_1) &= x_{tr}(t_1) + x_{ss}(t_1) \\ &= x_{tr}(t_1) - A^{-1}BU_0 - \sqrt{2}\sum_{k=1}^{\infty} \Re\left\{(A - jk\omega_0 I)^{-1} BU_k\, e^{jk\omega_0(t_1 - t_0)}\right\} \end{aligned} \tag{A.4b}$$

A.2 Phasor-Domain Solution

We can associate dynamic phasors with any waveform, as discussed in Sect. 2.1. Since the dynamic phasor definition (2.2b) involves a waveform segment of length T, the phasor sets $\{X_k(t)\}$ and $\{U_k(t)\}$ that are associated, respectively, with $x(\cdot)$ and $u(\cdot)$ can be defined only for $t \geq t_0 + T$. While $U_k(t) = U_k$ (for $t \geq t_0 + T$), due to the periodicity of the input $u(t)$ (recall (A.1c)), each dynamic phasor $X_k(t)$ contains a transient component, viz.,

$$X_k(t) \;=\; X_k \;+\; \xi_k(t)\ , \qquad t \geq t_0 + T \tag{A.5}$$

where $\xi_k(t)$ represents the transient waveform component $x_{tr}(t)$ and X_k represents the steady-state component $x_{ss}(t)$ (recall (A.2)). According to the definition (2.2), $\xi_k(t)$ is determined from $x_{tr}(t)$ via

$$\begin{aligned} \xi_k(t) &\stackrel{\text{def}}{=} \frac{a_k}{T}\int_{t-T}^{t} x_{tr}(s)\, e^{-jk\omega_0 s}\, ds \\ &= \left[e^{-jk\omega_0 t_0}\, \frac{a_k}{T}\int_{t-T}^{t} e^{(A - jk\omega_0 I)(s - t_0)}\, ds \right] x_{tr}(t_0) \end{aligned}$$

where $a_0 = 1$ and $a_k = \sqrt{2}$ for all $k \geq 1$. The identity $\int_{t_1}^{t_2} e^{Ms}\, ds = M^{-1}\big(e^{Mt_2} - e^{Mt_1}\big)$ holds for any non-singular matrix M, so that

$$\xi_k(t) \;=\; e^{(A - jk\omega_0 I)(t - t_0)}\, \widehat{\xi}_k(t_0) \tag{A.6a}$$

where

$$\widehat{\xi}_k(t_0) \;\stackrel{\text{def}}{=}\; a_k\, (A - jk\omega_0 I)^{-1}\, \frac{I - e^{-AT}}{T}\, e^{-jk\omega_0 t_0}\, x_{tr}(t_0) \tag{A.6b}$$

Notice that the same expression for $\xi_k(t)$ would be obtained by propagating the dynamic equation

$$\frac{d}{dt}\,\xi_k(t) \;=\; (A \;-\; jk\omega_0 I)\;\xi_k(t)$$

starting at $t = t_0$ with the initial condition $\xi_k(t_0) = \widehat{\xi}_k(t_0)$.

Equivalently, we can express $\xi_k(t)$ in terms of the terminal condition $x_{tr}(t_1)$, viz.,

$$\xi_k(t) \;=\; e^{(A-jk\omega_0 I)(t-t_1)}\,\widehat{\xi}_k(t_1) \tag{A.7a}$$

where

$$\begin{aligned}\widehat{\xi}_k(t_1) \;&\stackrel{\text{def}}{=}\; e^{(A-jk\omega_0 I)(t_1-t_0)}\,\widehat{\xi}_k(t_0)\\ &=\; a_k\,(A - jk\omega_0 I)^{-1}\,\frac{I - e^{-AT}}{T}\;e^{-jk\omega_0 t_1}\,x_{tr}(t_1)\end{aligned} \tag{A.7b}$$

In particular, we can use this result to obtain a compact expression for the *terminal* value of the phasor $X_k(\cdot)$, viz.,

$$\begin{aligned}X_k(t_1) &= X_k + \widehat{\xi}_k(t_1)\\ &= (A - jk\omega_0 I)^{-1}\left\{a_k\,\frac{I - e^{-AT}}{T}\;e^{-jk\omega_0 t_1}\,x_{tr}(t_1) \;-\; B\,U_k\right\}\end{aligned} \tag{A.8}$$

where we used (A.7b) for $\widehat{\xi}_k(t_1)$ and (A.2b) for X_k. These results form the basis for our EP initialization approach (see Sect. 2.5).

Appendix B: Duration and Bandwidth of Window Functions

A *window function* is a real-valued continuous-time waveform that is symmetric to the origin and achieves its global maximum at the origin, and its (real-valued and symmetric) Fourier transform satisfies the same two properties. Most windows are non-negative, but this property is usually not shared by their Fourier transforms. Selection of a suitable window function is a key step in construction of a *time–frequency representation* (TFR), as detailed in Chaps. 4 and 5. In contrast, non-symmetric (one-sided) windows are commonly used in applications that rely on causal processing, such as adaptive filtering and fast detection of fault-generated transients. However, since the literature on time–frequency representation is solely concerned with symmetric two-sided windows, we will focus in this Appendix exclusively on properties of such waveforms.

The primary objective of using a particular window type in a TFR application is to control *leakage between subbands* of the TFR: such leakage depends on the intrinsic trade-off between mainlobe bandwidth and sidelobes attenuation level.

A number of specialized window types – such as Hamming, Kaiser and Dolph–Chebyshev – use a constrained optimization formulation to derive desirable window choices. A rich collection of window function types can be found, for instance, in [293].

In addition to controlling inter-subband leakage, the process of window design also involves time-domain scaling, which determines the effective duration and bandwidth of the window. A popular standard-deviation-based metric that quantifies these concepts is presented in Sect. B.3. Unfortunately, this metric assigns an infinite bandwidth to most commonly used window functions. Consequently, practitioners prefer to describe bandwidth in terms of a prescribed-drop level (such as, say, a 3 dB drop): such metrics are examined in Sect. B.4. Moreover, prescribed-drop metrics are explicitly related to the notions of frequency-resolution and time-resolution of a window of interest, as detailed in Sects. B.5–B.8.

B.1 Commonly-Used Two-Sided Window Functions

Although all the different measures of duration and bandwidth we introduce in the sequel are not affected by amplitude scaling of the window function, we will maintain here the same unity DC gain convention we used in Sect. 4.1, namely $H(0) \equiv \int_{-\infty}^{\infty} h(t)\, dt = 1$.

Gaussian Window

A two-sided Gaussian window is given by the expression

$$h(t) = \frac{1}{\sigma\sqrt{2\pi}}\, e^{-\left(t^2/2\sigma^2\right)} \tag{B.1a}$$

where the parameter σ determines the trade-off between time- and frequency-resolution. Its Fourier transform is

$$H(j2\pi f) = e^{-2\pi^2\sigma^2 f^2} \tag{B.1b}$$

Both the window and its Fourier transform are real-valued, non-negative, symmetric to the origin, and have a (single) global maximum at the origin.

Triangular Window

A triangular window (also known as a Bartlett window) of length T_h is given by the expression

$$h(t) = \begin{cases} \frac{2}{T_h}\left(1 + \frac{2}{T_h}\,t\right) & -\frac{T_h}{2} \le t \le 0 \\ \frac{2}{T_h}\left(1 - \frac{2}{T_h}\,t\right) & 0 \le t \le \frac{T_h}{2} \\ 0 & \text{else} \end{cases} \tag{B.2a}$$

and its Fourier transform is

$$H(j2\pi f) = \left(\frac{\sin \pi f\, T_h/2}{\pi f\, T_h/2}\right)^2 \tag{B.2b}$$

Both the window and its Fourier transform are real-valued, symmetric to the origin, and have a global maximum at the origin. Notice that both $h(t)$ and $H(j2\pi f)$ are non-negative, but $H(j2\pi f)$ has multiple local maxima.

Rectangular Window

A rectangular window of length T_h is given by the expression

$$h(t) = \begin{cases} \dfrac{1}{T_h} & |t| \le \frac{T_h}{2} \\ 0 & \text{else} \end{cases} \tag{B.3a}$$

and its Fourier transform is

$$H(j2\pi f) = \frac{\sin \pi f\, T_h}{\pi f\, T_h} \tag{B.3b}$$

Both the window and its Fourier transform are real-valued, symmetric to the origin, and have a global maximum at the origin. Notice that $h(t)$ is non-negative, but $H(j2\pi f)$ has both positive and negative values, and multiple extremum points.

Kaiser Window

A Kaiser window (also known as a Kaiser–Bessel window) of length T_h is a generalization of the rectangular window. It is given by the expression

$$h(t) = \begin{cases} \left(\dfrac{\beta}{T_h\,\sinh\beta}\right) I_0\left(\beta\sqrt{1-(2\,t/\,T_h)^2}\right) & |t| \le \frac{T_h}{2} \\ 0 & \text{else} \end{cases} \tag{B.4a}$$

where $I_0(\cdot)$ is the zero-order modified Bessel function of the first kind, and the non-negative parameter β controls the trade-off between sidelobe leakage and mainlobe bandwidth. The corresponding Fourier transform is

$$H(j2\pi f) = \frac{\sinh\left(\beta\sqrt{1-(\pi f T_h/\beta)^2}\right)}{\left(\sinh\beta\right)\sqrt{1-(\pi f T_h/\beta)^2}} \tag{B.4b}$$

The rectangular window is obtained by setting $\beta = 0$: its Fourier transform has a narrow mainlobe, but very prominent sidelobes. Increasing the value of β suppresses the sidelobes at the expense of a modest increase in the bandwidth of the mainlobe. Notice that $h(t)$ is discontinuous at $t = \pm\frac{T_h}{2}$ for every value of the control parameter β. Both the window function and its Fourier transform are real-valued, symmetric to the origin, and have a global maximum at the origin. However, $H(j2\pi f)$ has both positive and negative values, as well as multiple extremum points, unlike $h(t)$, which is non-negative and has a single maximum.

Hann Window

A Hann window (also known as von Hann, Hanning, or raised-cosine window) of length T_h is given by the expression

$$h(t) = \begin{cases} \frac{1}{T_h}\left[1 + \cos\left(\frac{2\pi t}{T_h}\right)\right] \equiv \frac{2}{T_h}\cos^2\left(\frac{\pi t}{T_h}\right) & |t| \le \frac{T_h}{2} \\ 0 & \text{else} \end{cases} \tag{B.5a}$$

and its Fourier transform is

$$H(j2\pi f) = \frac{\sin\pi f T_h}{\pi f T_h} + \frac{1}{2}\,\frac{\sin\pi(f T_h - 1)}{\pi(f T_h - 1)} + \frac{1}{2}\,\frac{\sin\pi(f T_h + 1)}{\pi(f T_h + 1)} \tag{B.5b}$$

Both the window and its Fourier transform are real-valued, symmetric to the origin, and have a global maximum at the origin. Notice that $h(t)$ is non-negative, but $H(j2\pi f)$ has both positive and negative values, and multiple extremum points.

Hamming Window

A Hamming window of length T_h is given by the expression

$$h(t) = \begin{cases} \frac{1}{T_h}\left[1 + \frac{21}{25}\cos\left(\frac{2\pi t}{T_h}\right)\right] & |t| \le \frac{T_h}{2} \\ 0 & \text{else} \end{cases} \tag{B.6a}$$

and its Fourier transform is

$$H(j2\pi f) = \frac{\sin \pi f T_h}{\pi f T_h} + \frac{21}{50}\frac{\sin \pi(f T_h - 1)}{\pi(f T_h - 1)} + \frac{21}{50}\frac{\sin \pi(f T_h + 1)}{\pi(f T_h + 1)} \tag{B.6b}$$

Both the window and its Fourier transform are real-valued, symmetric to the origin, and have a global maximum at the origin. Notice that $h(t)$ is non-negative, but $H(j2\pi f)$ has both positive and negative values, and multiple extremum points.

B.2 One-Sided Windows

One-sided windows are essential in causal (online) processing of transient waveforms because they provide the capability for fast detection of transient onset. We present in this section a very brief introduction to this topic. However, since the literature on time–frequency representation is concerned solely with symmetric two-sided windows, we will focus in the remainder of this Appendix on the properties of two-sided windows.

Finite-Length Windows

Every (two-sided) finite-length window has a one-sided version, obtained by applying a delay of $T_h/2$, where T_h is the length of the window. This includes in particular the rectangular, triangular, Kaiser, Hann, and Hamming windows.

One-Sided Exponential Window

An exponential window (also known as exponential forgetting window) is given by the expression

$$h(t) = \begin{cases} \frac{1}{\tau}\, e^{-t/\tau} & |t| \geq 0 \\ 0 & \text{else} \end{cases} \tag{B.7a}$$

where the time constant τ determines the trade-off between time- and frequency-resolution. The Fourier transform of this window function is

$$H(j2\pi f) = \frac{1}{1 + j2\pi\tau f} \tag{B.7b}$$

The window function $h(t)$ is real-valued and non-negative. Both the window and the magnitude of its Fourier transform have a unique (global) maximum at the origin.

B.3 Standard-Deviation (SD) Metrics

As stated earlier, we will focus in the remainder of this Appendix on the properties of *two-sided windows*. Since every two-sided window and its Fourier transform are even (i.e., symmetric) functions, we can define measures of width in terms of the standard deviations of $|h(t)|^2$ and $|H(j2\pi f)|^2$, viz.,

$$\Delta_t^{(SD)} \stackrel{\text{def}}{=} 2\sqrt{\frac{\int_{-\infty}^{\infty} t^2\,|h(t)|^2\,dt}{\int_{-\infty}^{\infty} |h(t)|^2\,dt}}$$
$$\Delta_f^{(SD)} \stackrel{\text{def}}{=} 2\sqrt{\frac{\int_{-\infty}^{\infty} f^2\,|H(j2\pi f)|^2\,df}{\int_{-\infty}^{\infty} |H(j2\pi f)|^2\,df}} \tag{B.8}$$

namely, width is defined as twice the corresponding standard deviation. Notice that both measures use the same denominator because $\int_{-\infty}^{\infty} |h(t)|^2 dt = \int_{-\infty}^{\infty} |H(j2\pi f)|^2 df$ as a result of the Parseval identity.

Also, since the Fourier transform of $\frac{dh(t)}{dt}$ is $(j2\pi f)X(j2\pi f)$, we can use the alternative expression

$$\Delta_f^{(SD)} = \frac{1}{\pi}\sqrt{\frac{\int_{-\infty}^{\infty} \left|\frac{dh(t)}{dt}\right|^2 dt}{\int_{-\infty}^{\infty} |h(t)|^2\,dt}} \tag{B.9}$$

to evaluate the effective bandwidth $\Delta_f^{(SD)}$. Notice that this expression implies that $\Delta_f^{(SD)} = \infty$ when the derivative $\frac{dh(t)}{dt}$ is not square-integrable over the interval $-\infty < t < \infty$. In particular, this occurs when the window function $h(t)$ contains a discontinuity, as in the case of the rectangular and Kaiser windows, as well as the one-sided exponential window.

An important characteristic of each window function is its *duration–bandwidth product*, namely

$$\mathcal{K}^{(SD)} \stackrel{\text{def}}{=} \Delta_t^{(SD)}\,\Delta_f^{(SD)} \tag{B.10}$$

For each type of window, this product is a dimensionless constant: thus different types of windows can be compared in terms of the duration–bandwidth product. In particular, the Gaussian window has often been considered "optimal" because for this window $\mathcal{K}^{(SD)}$ is as small as possible [33, 86]. However, this property may change when other measures of width are considered (see, e.g., Sect. B.4).

For the Gaussian window, we can use the expressions (B.1) to obtain

$$\Delta_t^{(SD)} = \sqrt{2}\,\sigma \ , \quad \Delta_f^{(SD)} = \frac{1}{\sqrt{2}\,\pi\,\sigma} \quad \Longrightarrow \quad \mathcal{K}^{(SD)} = \frac{1}{\pi} = 0.318$$

The well-known "uncertainty principle" in signal processing states that for any given window function

$$\mathcal{K}^{(SD)} \geq \frac{1}{\pi} \ (= 0.318) \tag{B.11}$$

with the lower bound being achieved (only) by the Gaussian window [33]. For instance, for the triangular window, we can use the expressions (B.2) to obtain

$$\Delta_t^{(SD)} = \frac{T_h}{\sqrt{10}} \ , \quad \Delta_f^{(SD)} = \frac{2\sqrt{3}}{\pi}\, T_h{}^{-1} \quad \Longrightarrow \quad \mathcal{K}^{(SD)} = \frac{2\sqrt{3}}{\pi\sqrt{10}} = 0.349$$

where we used the alternative expression (B.9) to determine the effective bandwidth $\Delta_f^{(SD)}$. However, for the rectangular window we obtain from (B.3) the results

$$\Delta_t^{(SD)} = \frac{T_h}{\sqrt{3}} \ , \quad \Delta_f^{(SD)} = \infty \quad \Longrightarrow \quad \mathcal{K}^{(SD)} = \infty$$

For other types of windows, we may need to rely on numerical evaluation to determine the duration $\Delta_t^{(SD)}$, bandwidth $\Delta_f^{(SD)}$, and their product $\mathcal{K}^{(SD)}$. The resulting values are summarized in Table B.1.

Notice that the "infinite-bandwidth" property is shared by all windows that contain a discontinuity, such as the rectangular window, the Kaiser window, the Hamming window, and the one-sided exponential window: the derivative $\frac{dh(t)}{dt}$ of such window functions contains Dirac delta functions located at the points of discontinuity, so that evaluating $\Delta_f^{(SD)}$ via (B.9) results in an infinite value. Since this property of SD-based metrics conflicts with our intuitive concept of bandwidth, which is based on commonly used measures of bandwidth (such as a 3 dB drop),

Table B.1 Comparison of standard-deviation duration–bandwidth products

Window function	$\Delta_t^{(SD)}$	$\Delta_f^{(SD)}$	$\mathcal{K}^{(SD)}$
Gaussian	$1.414\ \sigma$	$0.225\ \sigma^{-1}$	0.318
Triangular	$0.316\ T_h$	$1.103\ T_h{}^{-1}$	0.349
Rectangular	$0.577\ T_h$	∞	∞
Kaiser ($\beta = 1$)	$0.542\ T_h$	∞	∞
Kaiser ($\beta = 6$)	$0.283\ T_h$	∞	∞
Hann	$0.283\ T_h$	$1.155\ T_h{}^{-1}$	0.327
Hamming	$0.308\ T_h$	∞	∞

we are motivated to introduce in the following section alternative metrics of window duration and bandwidth. We demonstrate in Sects. B.5–B.7 that drop-based metrics are also directly related to common definitions of *time-resolution* and *frequency-resolution*.

B.4 Prescribed-Drop (PD) Metrics

As an alternative to standard-deviation-based metrics, consider the definition of width in terms of a prescribed drop (say 3 dB, or even 10 dB or 20 dB) from the peak value at the origin. Since every two-sided window and its Fourier transform are even (i.e., symmetric) functions, we define

$$\Delta_t^{(PD)} \stackrel{\text{def}}{=} 2\,t_{cr} \quad , \quad \Delta_f^{(PD)} \stackrel{\text{def}}{=} 2\,f_{cr} \tag{B.12a}$$

where $t_{cr} > 0$ and $f_{cr} > 0$ are determined by solving the prescribed-drop equations

$$20\log_{10}\left|\frac{h(t_{cr})}{h(0)}\right| = -A \quad , \quad 20\log_{10}\left|\frac{H(j2\pi f_{cr})}{H(0)}\right| = -A \tag{B.12b}$$

and A is the chosen value (in dB) for the prescribed drop.

As an example, for the two-sided Gaussian window, we need to solve the equations

$$20\log_{10}\left(e^{-\left(t_{cr}^2/2\sigma^2\right)}\right) = -A \quad , \quad 20\log_{10}\left(e^{-2\pi^2\sigma^2 f_{cr}^2}\right) = -A$$

which results in

$$\Delta_t^{(PD)} = \left[\sqrt{\frac{2A}{5}\ln 10}\,\right]\sigma \ ,$$

$$\Delta_f^{(PD)} = \left[\frac{1}{\pi}\sqrt{\frac{A}{10}\ln 10}\,\right]\sigma^{-1} \quad \Longrightarrow \quad \mathcal{K}^{(PD)} = \frac{\ln 10}{5\pi}\,A$$

For other types of windows, we may need to use numerical evaluation to determine the duration $\Delta_t^{(PD)}$, bandwidth $\Delta_f^{(PD)}$, and their product $\mathcal{K}^{(PD)}$. The resulting values are summarized in Tables B.2, B.3, and B.4.

As evident from Tables B.2, B.3, B.4, and B.5, the PD duration–bandwidth product $\mathcal{K}^{(PD)}$ is comparable for most windows, which means that the choice of a specific window appears to be of only a minor concern in most applications: in particular, notice that the Gaussian window is not necessarily optimal in the PD

Table B.2 Comparison of prescribed-drop (3 dB) duration–bandwidth products

Window function	$\Delta_t^{(PD\star 3)}$	$\Delta_f^{(PD\star 3)}$	$\mathcal{K}^{(PD\star 3)}$
Gaussian	$1.662\ \sigma$	$0.265\ \sigma^{-1}$	0.440
Triangular	$0.292\ T_h$	$1.286\ T_h^{-1}$	0.376
Rectangular	T_h	$0.891\ T_h^{-1}$	0.891
Kaiser ($\beta = 1$)	T_h	$0.917\ T_h^{-1}$	0.917
Kaiser ($\beta = 6$)	$0.350\ T_h$	$1.401\ T_h^{-1}$	0.490
Hann	$0.364\ T_h$	$1.439\ T_h^{-1}$	0.523
Hamming	$0.380\ T_h$	$1.291\ T_h^{-1}$	0.491

Table B.3 Comparison of prescribed-drop (10 dB) duration–bandwidth products

Window function	$\Delta_t^{(PD\star 10)}$	$\Delta_f^{(PD\star 10)}$	$\mathcal{K}^{(PD\star 10)}$
Gaussian	$3.035\ \sigma$	$0.483\ \sigma^{-1}$	1.466
Triangular	$0.684\ T_h$	$2.241\ T_h^{-1}$	1.532
Rectangular	T_h	$1.483\ T_h^{-1}$	1.483
Kaiser ($\beta = 1$)	T_h	$1.534\ T_h^{-1}$	1.534
Kaiser ($\beta = 6$)	$0.620\ T_h$	$2.470\ T_h^{-1}$	1.532
Hann	$0.620\ T_h$	$2.515\ T_h^{-1}$	1.559
Hamming	$0.670\ T_h$	$2.273\ T_h^{-1}$	1.523

Table B.4 Comparison of prescribed-drop (20 dB) duration–bandwidth products

Window function	$\Delta_t^{(PD\star 20)}$	$\Delta_f^{(PD\star 20)}$	$\mathcal{K}^{(PD\star 20)}$
Gaussian	$4.292\ \sigma$	$0.683\ \sigma^{-1}$	2.932
Triangular	$0.900\ T_h$	$2.967\ T_h^{-1}$	2.670
Rectangular	T_h	$5.367\ T_h^{-1}$	5.367
Kaiser ($\beta = 1$)	T_h	$5.227\ T_h^{-1}$	5.227
Kaiser ($\beta = 6$)	$0.830\ T_h$	$3.304\ T_h^{-1}$	2.742
Hann	$0.795\ T_h$	$3.298\ T_h^{-1}$	2.622
Hamming	$0.920\ T_h$	$3.043\ T_h^{-1}$	2.800

Table B.5 Comparison of prescribed-drop (40 dB) duration–bandwidth products

Window function	$\Delta_t^{(PD\star 40)}$	$\Delta_f^{(PD\star 40)}$	$\mathcal{K}^{(PD\star 40)}$
Gaussian	$6.070\ \sigma$	$0.966\ \sigma^{-1}$	5.863
Triangular	$0.990\ T_h$	$10.733\ T_h^{-1}$	10.626
Rectangular	T_h	$62.911\ T_h^{-1}$	62.911
Kaiser ($\beta = 1$)	T_h	$52.865\ T_h^{-1}$	52.865
Kaiser ($\beta = 6$)	T_h	$4.087\ T_h^{-1}$	4.087
Hann	$0.936\ T_h$	$5.590\ T_h^{-1}$	5.233
Hamming	T_h	$3.769\ T_h^{-1}$	3.769

metric. We also notice that products of the form $\Delta_f^{(PD\star A)}\,\Delta_t^{(PD\star B)}$ are always constant for a given window type and vary only mildly between windows. This fact is useful, for instance, in quantifying the trade-off between time- and frequency-resolution (see Sect. B.8) and in capturing the subband-overlap ratio (see Sect. 4.5).

A comparison of Table B.2 with Table B.1 reveals that standard-deviation (SD) metrics are comparable to the 3 dB PD metrics. This makes SD metrics inappropriate for designing time- and frequency-discretization schemes: such schemes are based on a Nyquist-type criterion (see Sect. B.9), which relies on a strictly band-limited (or strictly time-limited) property. Using SD metrics of window length results in underestimates of duration, leading to significant aliasing caused by insufficient density of the resulting frequency-discretization scheme. Similarly, using SD metrics of bandwidth (for the Gaussian, triangular, and Hann window types) results in insufficient density of the time-discretization scheme. In addition, for windows with a discontinuity, the SD metrics are useless both for time-discretization and for determining frequency-resolution (see Sect. B.5 for details). In summary, one should use exclusively PD metrics, with an appropriate prescribed drop, in all time- and frequency-discretization applications.

B.5 Frequency-Resolution of the STFT vs. Window Bandwidth

Given a waveform $x(t)$, we can associate with it a bi-variate time–frequency representation (TFR), as detailed in Sect. 4.1, namely,

$$X(t, f) \;=\; \int_{-\infty}^{\infty} h(t-\tau)\, x(\tau)\, e^{-j2\pi f\tau}\, d\tau \tag{B.13}$$

Notice that, for a fixed time-instant t_0, the time–frequency representation $X(t_0, f)$ depends only on the frequency variable f, and this relation can be interpreted as the Fourier transform of a product, viz.,

$$X(t_0, f) \;=\; \mathcal{F}_f\Big\{w_0(\cdot)\, x(\cdot)\Big\} \;\equiv\; W_0(j2\pi f) \circledast X(j2\pi f)$$

where

$$w_0(t) \stackrel{\text{def}}{=} h(t_0 - t) \qquad \Longrightarrow \qquad W_0(j2\pi f) = H(-j2\pi f)\, e^{-j2\pi f t_0}$$

and

$$\mathcal{F}_\nu\Big\{x(\cdot)\Big\} \stackrel{\text{def}}{=} \int_{-\infty}^{\infty} x(t)\, e^{-j2\pi\nu t}\, dt \;\equiv\; X(j2\pi\nu)$$

The frequency-domain convolution of $X(j2\pi f)$ with $W_0(j2\pi f)$ results in "smearing out" the frequency spectrum, which degrades frequency-resolution.

A quantitative characterization of frequency-resolution in the spectrum of a finite length (i.e., windowed) signal record is commonly obtained via a simple scenario consisting of two steady-state sinusoids, viz.,

$$x(t) = 2\,\Re\left\{e^{j2\pi f_1 t} + Ce^{j2\pi f_2 t}\right\} \tag{B.14a}$$

where, without loss of generality, we may assume that $f_1 < f_2$ and $|C| \leq 1$. Thus,

$$X(j2\pi f) = \delta(f - f_1) + C\,\delta(f - f_2) + \delta(f + f_1) + C^*\delta(f + f_2)$$

and, consequently,

$$X(t_0, f) = \underbrace{W_0\big(j2\pi(f - f_1)\big) + C\,W_0\big(j2\pi(f - f_2)\big)}_{\text{positive frequency peaks}} + \underbrace{W_0\big(j2\pi(f + f_1)\big) + C^*W_0\big(j2\pi(f + f_2)\big)}_{\text{negative frequency peaks}} \tag{B.14b}$$

Plotting the TFR magnitude $|X(t_0, f)|$ as a function of the frequency f results, in general, in a plot with two peaks, for $f \geq 0$ (Fig. B.1). The solid trace in this plot corresponds to the magnitude $|X(t_0, f)|$, while the two dashed-line traces

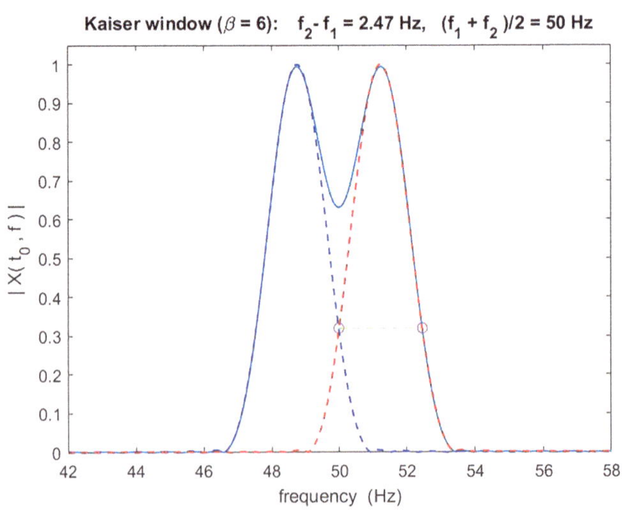

Fig. B.1 Well-resolved spectrum with minimal leakage: $|X(t_0, f)|$ for $T_h = 1, |C| = 1$ (solid line), and its two single-frequency components (dashed lines).

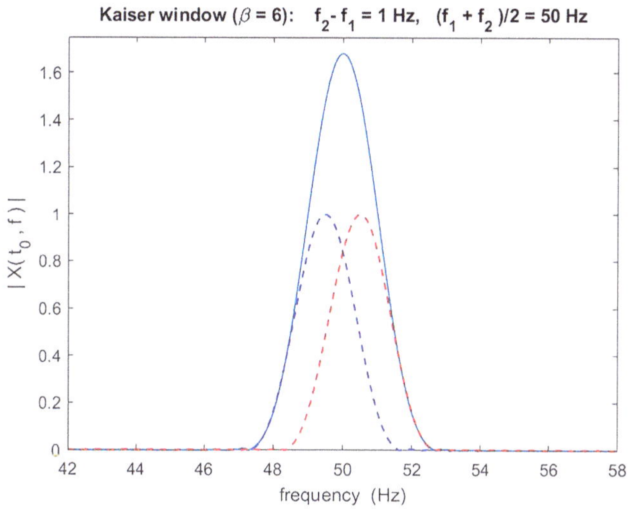

Fig. B.2 Unresolved spectrum (merged peaks): $|X(t_0, f)|$ for $T_h = 1$, $|C| = 1$ (solid line), and its two single-frequency components (dashed lines).

depict the magnitudes of the two individual single-frequency components, namely $|W_0(j2\pi(f - f_1)) + W_0(j2\pi(f + f_1)|$ and $|C\, W_0(j2\pi(f - f_2)) + C^* W_0(j2\pi(f + f_2))|$.

The exact shape of this frequency spectrum depends on the frequency difference $|f_2 - f_1|$ between the two peak locations, as well as the choice of window and the complex amplitude C. When the frequency difference is not too small, the spectrum will display two peaks (as in Fig. B.1). In this "well-resolved" case, we are able to reach the correct conclusion about the number of distinct sinusoids in the waveform $x(t)$ and accurately estimate their frequencies: reducing the frequency difference $|f_2 - f_1|$ causes these two peaks to merge into a single one, so that the two distinct frequencies can no longer be resolved (Fig. B.2). We will use the two-sinusoid scenario to define *frequency-resolution* of the STFT as the value of the frequency difference $|f_2 - f_1|$ for which (a) $|X(t_0, f)|$ displays two peaks (as in Fig. B.1), and (b) the dip between these two peaks has a specified *canonical value*.

In order to simplify our analysis, we will now assume that the frequencies f_1 and f_2 are not too close to the origin: more precisely, we assume that

$$2 \min(f_1, f_2) \gg \Delta_f^{(PD\star 20)}$$

This allows us to ignore the effect of the "mirror image" terms $W_0(j2\pi(f + f_1))$ and $W_0(j2\pi(f + f_2))$ (i.e., the "negative frequency peaks" in (B.14b)) within the frequency interval $f_1 \le f \le f_2$, resulting in the very reasonable approximation

$$X(t_0, f) \approx W_0\big(j2\pi(f - f_1)\big) + C\,W_0\big(j2\pi(f - f_2)\big) \qquad \text{for} \quad f_1 \le f \le f_2$$

In general, the combination of the two single-frequency components in this expression generates peaks at $f \approx f_1$ and $f \approx f_2$ in the plot of $|X(t_0, f)|$, with a "dip" between the peaks, as in Fig. B.1. The depth of this dip is minimized when $C = e^{2j\psi}$, where $\psi \stackrel{\text{def}}{=} \arg\{W_0\big(j\pi(f_2 - f_1)\big)\}$: in this case the two peaks have the same height, and

$$\left|X(t_0, \frac{f_1 + f_2}{2})\right| \approx 2\left|W_0(j2\pi f)\right| \Bigg|_{f=|f_2-f_1|/2}$$

so that

$$20\log_{10}\left(\left|X\left(t_0, \frac{f_1 + f_2}{2}\right)\right|\right) \approx 6 + 20\log_{10}\left(\left|W_0(j2\pi f)\right| \Bigg|_{f=|f_2-f_1|/2}\right) \tag{B.14c}$$

In other words, the midpoint of the dip is 6 dB above the point of intersection between the two single-frequency components (indicated by the left circle in Fig. B.1). Notice that the frequency difference between the two points indicated by circles in Fig. B.1 is exactly $f_2 - f_1$. This observation holds for any choice of window and any values of f_1 and f_2 as seen, for instance, in Fig. B.3, obtained with a rectangular window (= Kaiser window with $\beta \to 0$).

If we choose 4 dB (= the value used in Fig. B.1) as the *canonical-dip* (CD) value, then, according to (B.14c), the resulting frequency-resolution metric

$$F_{res}^{(CD\star 4)} \stackrel{\text{def}}{=} |f_2 - f_1|\Big|_{\text{canonical dip}} = 4\,\text{dB} \tag{B.15a}$$

is characterized by the relation

$$20\log_{10}\left(\left|W_0(j2\pi f)\right| \Bigg|_{f=F_{res}/2}\right) \approx \underbrace{20\log_{10}\left(\left|X(t_0, \frac{f_1 + f_2}{2})\right|\right)}_{-4\,\text{dB}} - 6$$

$$= -10\,\text{dB}$$

or, equivalently,

$$\text{Frequency-resolution}\Bigg|_{\text{canonical dip}=4\,\text{dB}} \equiv F_{res}^{(CD\star 4)} = \Delta_f^{(PD\star 10)} \tag{B.15b}$$

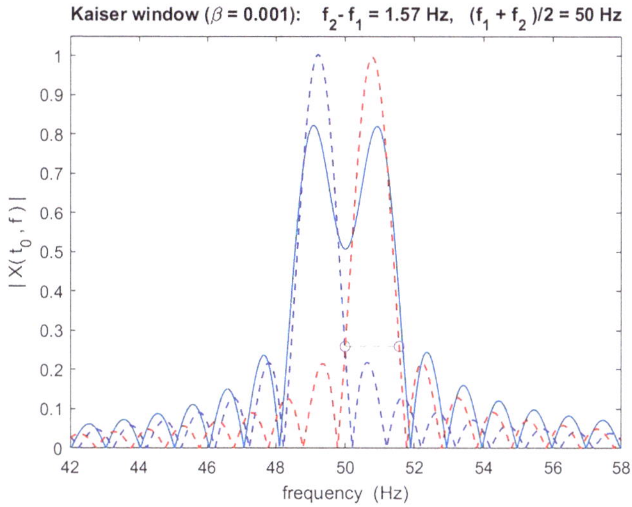

Fig. B.3 Well-resolved spectrum with strong sidelobe leakage: $|X(t_0, f)|$ for $T_h = 1$, $|C| = 1$ (solid line), and its two single-frequency components (blue and red dashed lines).

Thus, this two-sinusoid scenario establishes a direct relation between frequency-resolution and a prescribed-drop (PD) window bandwidth. In particular, the values of $\Delta_f^{(PD\star 10)}$ listed in Table B.3 for the Kaiser and rectangular windows are in full agreement with the $F_{res}^{(CD\star 4)}$ values obtained from Figs. B.1 and B.3, respectively.

The explicit relation (B.15b) provides yet another convincing argument in favor of using prescribed-drop (PD) bandwidth metrics. Notice that it is impossible to establish a similar relation using a standard-deviation (SD) metric. In particular, $\Delta_f^{(SD)} = \infty$ for the Kaiser window (and many others), even though the frequency interval needed for good resolution when using such windows is far from being infinite.

B.6 Frequency-Resolution of a Discretized STFT (D-STFT)

Digital implementation of an STFT involves discretization of both time and frequency, i.e., evaluation of $X(t\,;\,f_c)$ on a time–frequency grid $\left\{(t\,;\,f_c)\,;\; t = nT_s,\, f_c = k\frac{F_s}{L}\right\}$, where n, k are integers, $T_s = \frac{1}{F_s}$ is the time-sampling interval, and L is an integer-valued design parameter that determines the density of the frequency grid. The resulting time–frequency representation (TFR) is known as a D-STFT: for any given discrete-time-instant $t = t_0 \equiv n_0 T_s$, it consists of frequency-sampled values, as shown in Fig. B.4. In general, the two peaks in this figure (and the dip between them) need not fall exactly on a frequency-grid point.

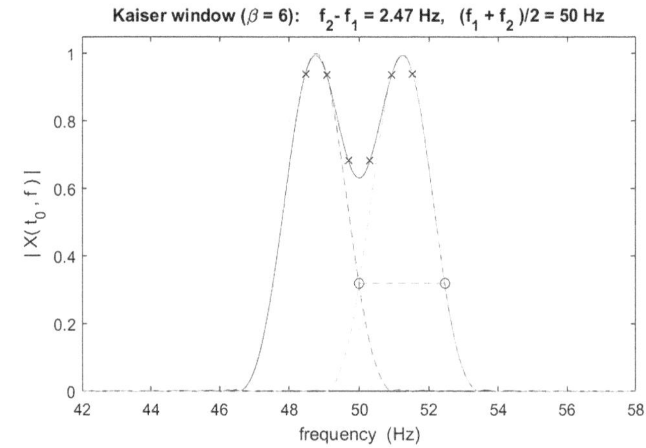

Fig. B.4 Well-resolved spectrum with minimal leakage: $|X(t_0, f)|$ for $T_h = 1, |C| = 1$ (solid line), and its two single-frequency components (dashed lines): frequency-grid points near the local extrema are indicated by "x".

In a worst-case scenario, each one of these three local extrema falls exactly halfway between two adjacent grid points, as shown in the figure. This results in degraded frequency-resolution, caused by a decreased peak height, as well as a reduced dip depth between the peaks. To be specific, the frequency-resolution expression (B.15b) is replaced by the somewhat higher value

$$\text{Frequency-resolution}\Big|_{\text{canonical dip}=4\,\text{dB}} = \Delta_f^{(PD\star(10+2\epsilon))} \tag{B.16}$$

where ϵ denotes the reduction (in dB) of the discretized peak level, and we assume that the discretized minimum point between the two peaks increases, approximately, by $\epsilon\,\text{dB}$ as well. This means that in order to avoid loss of frequency-resolution, we need to choose a value of L such that ϵ is small enough to satisfy the relation

$$\Delta_f^{(PD\star(10+2\epsilon))} \approx \Delta_f^{(PD\star 10)}$$

For instance, $\Delta_f^{(PD\star 10)} = 2.47\,T_h^{-1}$ for a Kaiser ($\beta = 6$) window, while the "perturbed" bandwidth $\Delta_f^{(PD\star 11)} = 2.58\,T_h^{-1}$ is only 4.4% higher. This suggests that $\epsilon = 0.5\,\text{dB}$ is an acceptable value, since the associated relative change in frequency-resolution is sufficiently small.

The next step is to determine a value of the design parameter L that generates a worst-case peak reduction of $\epsilon = 0.5\,\text{dB}$. We notice that, according to Fig. B.4, the spacing between two adjacent frequency-grid points is exactly the same as the $\epsilon\,\text{dB}$ bandwidth of the peak (= bandwidth of window function). In our example,

Table B.6 Value of the coefficient $C_L = \left[T_h \, \Delta_f^{(PD\star 0.5)}\right]^{-1}$

Window function	C_L
Rectangular	2.66
Kaiser ($\beta = 1$)	2.62
Triangular	1.87
Kaiser ($\beta = 6$)	1.73
Hann	1.67
Hamming	1.87

this means that we need to choose L so that $\frac{F_s}{L} \approx \Delta_f^{(PD\star 0.5)}$ or, equivalently,

$$L = \text{ceil}\left\{C_L \, \frac{T_h}{T_s}\right\} \tag{B.17a}$$

where

$$C_L = \left[T_h \, \Delta_f^{(PD\star 0.5)}\right]^{-1} \tag{B.17b}$$

is a window-dependent constant. Table B.6 lists the value of this coefficient for several standard window choices. Notice that setting $L = \text{ceil}\left\{3 \, \frac{T_h}{T_s}\right\}$ meets the accuracy requirement (B.17a) for all listed windows, while the smaller value $L = \text{ceil}\left\{2 \, \frac{T_h}{T_s}\right\}$ is sufficient to meet this requirement for the commonly used Hamming or Kaiser windows.

B.7 Time-Resolution of the STFT vs. Window Duration

In contrast to the frequency–bandwidth/resolution relation (B.15b), we do not have a similar direct relation between time-duration and time-resolution. Most sources treat either the actual window length (T_h) or the effective window duration (Δ_t) as the actual measure of time-resolution. Instead, we propose here a definition that provides a precise measure for time-resolution in a simple scenario: a single (complex) sinusoid, with a short "zero-amplitude" gap, viz.,

$$x(t) = \begin{cases} 0 & |t| < \frac{\Delta}{2} \\ C \, e^{j2\pi f_0 t} & |t| \geq \frac{\Delta}{2} \end{cases} \tag{B.18}$$

where, without loss of generality, we assume that $|C| = 1$. Far away from the gap (i.e., when $|t| \gg \Delta$), there is no overlap between the window and the waveform gap: consequently, the TFR magnitude $|X(t, f_0)|$ is simply that of a single complex sinusoid: a time-invariant constant, equal to the amplitude $|C| \equiv 1$. Indeed, in this case $\int_{t-\frac{\Delta}{2}}^{t+\frac{\Delta}{2}} h(\theta)\, d\theta \ll 1$, so that (recall (B.13))

$$X(t, f_0) = \int_{-\infty}^{\infty} h(t-\tau) \left[x(\tau)\, e^{-j2\pi f_0 \tau} \right] d\tau = C \left[1 - \int_{-\frac{\Delta}{2}}^{\frac{\Delta}{2}} h(t-\tau)\, d\tau \right]$$

$$= C \left[1 - \int_{t-\frac{\Delta}{2}}^{t+\frac{\Delta}{2}} h(\theta)\, d\theta \right] \approx C$$

where we used the fact that $H(0) \equiv \int_{-\infty}^{\infty} h(\tau)\, d\tau = 1$ for all windows. We conclude that $|X(t, f_0)| \approx 1$ for all t sufficiently away from the waveform gap (Fig. B.5).

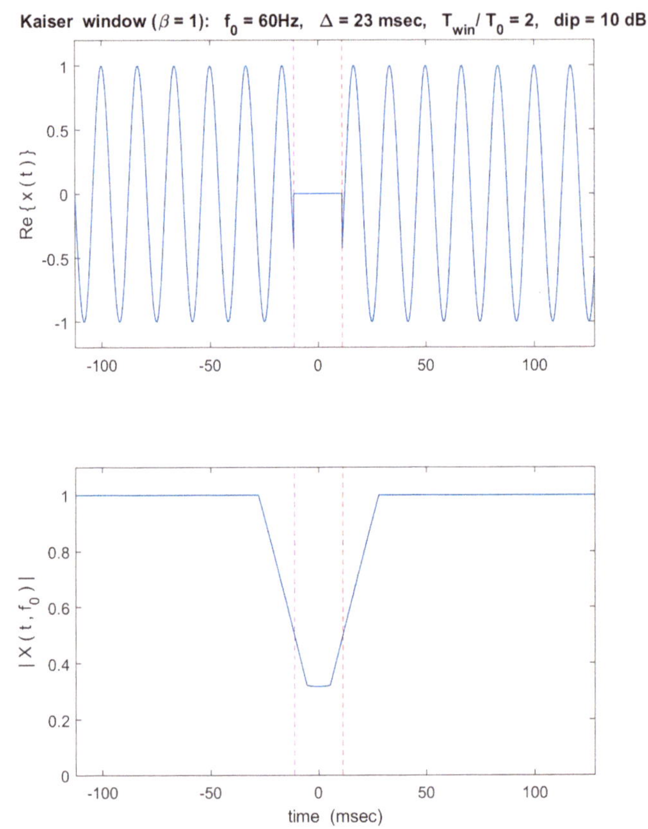

Fig. B.5 Time–frequency representation (bottom) of a narrow-gapped complex sinusoidal 60 Hz waveform (top), obtained with a two-cycle wide Kaiser ($\beta = 1$) window. Dashed lines indicate the gap.

On the other hand, the TFR magnitude $|X(t, f_0)|$ displays a dip in the near vicinity of the waveform gap (see Fig. B.5, bottom plot). The shape and depth of this dip depend on the choice of window: in particular, we notice that the gap in $|X(t, f_0)|$ is wider than the one in $x(t)$. We will use this scenario to define *time-resolution* of the STFT as the width Δ of the waveform gap for which the midpoint dip $|X(0, f_0)|$ has a specified *canonical value.*

The level of the TFR, at the midpoint of the gap, is obtained from the expression

$$\begin{aligned} X(0, f_0) &= \int_{-\infty}^{\infty} h(-\tau)\, x(\tau)\, e^{-j2\pi f_0 \tau}\, d\tau \\ &= C\left[\int_{-\infty}^{\infty} h(-\tau)\, d\tau - \int_{-\frac{\Delta}{2}}^{\frac{\Delta}{2}} h(-\tau)\, d\tau\right] \\ &= C\left[1 - \int_{-\frac{\Delta}{2}}^{\frac{\Delta}{2}} h(\tau)\, d\tau\right] \end{aligned}$$

so that

$$|X(0, f_0)| = 1 - \int_{-\Delta/2}^{\Delta/2} h(\tau)\, d\tau \stackrel{\text{def}}{=} \rho(\Delta) \tag{B.19}$$

For instance,

$$\rho(\Delta) = \begin{cases} 1 - \frac{\Delta}{T_h} & \text{Rectangular window} \\ \left(1 - \frac{\Delta}{T_h}\right)^2 & \text{Triangular window} \\ 1 - \frac{\Delta}{T_h} - \frac{1}{\pi} \sin\pi\left(\frac{\Delta}{T_h}\right) & \text{Hann window} \\ 1 - \frac{\Delta}{T_h} - \frac{0.84}{\pi} \sin\pi\left(\frac{\Delta}{T_h}\right) & \text{Hamming window} \end{cases}$$

where $\Delta \leq T_h$. A canonical dip of A dB defines a time-resolution metric $T_{res}^{(CD\star A)}$ via the equation

$$-A = 20 \log_{10} \rho\left(T_{res}^{(CD\star A)}\right) \tag{B.20}$$

In general, $\rho(\Delta)$ is a monotone decreasing function of Δ (see, e.g., Fig. B.6).

Evaluating this function for any window of choice, we can determine the time-resolution for several canonical-dip levels (Table B.7) and compare it with the various measures of time-duration we introduced earlier in this Appendix. Notice that the ratios $\dfrac{T_{res}^{(CD\star A)}}{\Delta_t^{(PD\star B)}}$ are almost independent of the window type for all useful A and B values. For instance, $T_{res}^{(CD\star 3)} \approx \frac{1}{4}\,\Delta_t^{(PD\star 10)}$, $T_{res}^{(CD\star 10)} \approx \frac{2}{3}\,\Delta_t^{(PD\star 10)}$ and $T_{res}^{(CD\star 20)} \approx \Delta_t^{(PD\star 10)}$ for all windows. This means that time-duration of any window can be used to accurately *estimate* its time-resolution.

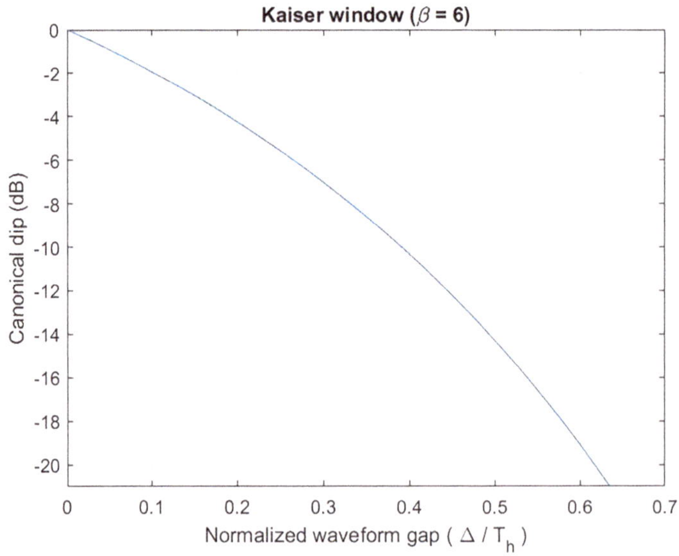

Fig. B.6 The dip function $\rho(\Delta)$ for a Kaiser window (with $\beta = 6$).

Table B.7 Prescribed-dip time-resolution vs. measures of time-duration

Window function	Δ_t / T_h				T_{res} / T_h		
	SD	3 dB	10 dB	20 dB	3 dB	10 dB	20 dB
Rectangular	0.58	1	1	1	0.29	0.68	0.90
Kaiser ($\beta = 1$)	0.54	1	1	1	0.27	0.66	0.88
Triangular	0.32	0.29	0.68	0.90	0.16	0.44	0.68
Kaiser ($\beta = 6$)	0.28	0.35	0.62	0.83	0.15	0.39	0.62
Hann	0.28	0.36	0.62	0.80	0.15	0.39	0.60
Hamming	0.31	0.38	0.67	0.92	0.16	0.42	0.67

Selected prescribed-dip ratios

Window function	$\frac{T_{res}^{(CD\star 3)}}{\Delta_t^{(PD\star 10)}}$	$\frac{T_{res}^{(CD\star 10)}}{\Delta_t^{(PD\star 10)}}$	$\frac{T_{res}^{(CD\star 10)}}{\Delta_t^{(PD\star 20)}}$	$\frac{T_{res}^{(CD\star 20)}}{\Delta_t^{(PD\star 20)}}$	$\frac{T_{res}^{(CD\star 20)}}{\Delta_t^{(PD\star 10)}}$
Rectangular	0.29	0.68	0.68	0.90	0.90
Kaiser ($\beta = 1$)	0.27	0.66	0.66	0.88	0.88
Triangular	0.24	0.65	0.49	0.76	1.00
Kaiser ($\beta = 6$)	0.24	0.63	0.47	0.75	1.00
Hann	0.24	0.63	0.49	0.75	0.97
Hamming	0.24	0.63	0.46	0.73	1.00

B.8 Trade-Off Between Time- and Frequency-Resolution

Combining the results in the last two sections, we find that

$$\begin{aligned}\mathcal{K}_{res}(A,B) &\stackrel{\text{def}}{=} F_{res}^{(CD\star(A-6))}\,T_{res}^{(CD\star B)}\\ &= \Delta_f^{(PD\star A)}\Delta_t^{(PD\star A)}\,\frac{T_{res}^{(CD\star B)}}{\Delta_t^{(PD\star A)}}\\ &= \mathcal{K}^{(PD\star A)}\,\frac{T_{res}^{(CD\star B)}}{\Delta_t^{(PD\star A)}}\end{aligned}$$

For instance,

$$\mathcal{K}_{res}(10,10) \stackrel{\text{def}}{=} F_{res}^{(CD\star 4)}\,T_{res}^{(CD\star 10)} = \mathcal{K}^{(PD\star 10)}\,\frac{T_{res}^{(CD\star 10)}}{\Delta_t^{(PD\star 10)}} \approx 1$$

for all window types, while $\mathcal{K}_{res}(10,20) \stackrel{\text{def}}{=} F_{res}^{(CD\star 4)}\,T_{res}^{(CD\star 20)} \approx 1.5$ for all windows except rectangular and low-β Kaiser window (see Table B.8).

Table B.8 Product of time- and frequency-resolution

Window function	$\mathcal{K}^{(PD\star 10)}$	$\frac{T_{res}^{(CD\star 10)}}{\Delta_t^{(PD\star 10)}}$	$\frac{T_{res}^{(CD\star 20)}}{\Delta_t^{(PD\star 10)}}$	$\mathcal{K}_{res}(10,10)$	$\mathcal{K}_{res}(10,20)$
Rectangular	1.48	0.68	0.90	1.01	1.33
Kaiser ($\beta = 1$)	1.53	0.66	0.88	1.01	1.35
Triangular	1.53	0.65	1.00	0.99	1.53
Kaiser ($\beta = 6$)	1.53	0.63	1.00	0.96	1.53
Hann	1.56	0.63	0.97	0.98	1.51
Hamming	1.52	0.63	1.00	0.96	1.52

B.9 Frequency-Sampling of Time-Limited Waveforms

The celebrated Nyquist–Shannon–Kotelnikov Theorem[1] [32, 33] states that every perfectly band-limited continuous-time waveform can be *recovered without error* from its samples on a sufficiently dense discretized time grid. Replacing the roles of "time" and "frequency," and relying on the duality property of the continuous-time Fourier transform (CTFT), results in a *dual Nyquist theorem*, which states that the CTFT of every *perfectly time-limited* continuous-time waveform can be recovered without error from its samples on a sufficiently dense discretized frequency grid.

[1] Often called simply the "Nyquist theorem."

Theorem B.1 (Dual Nyquist): *Let* $x(t)$ *be a waveform that vanishes outside the time interval* $\left|t\right| < \frac{T_x}{2}$, *(i.e., it is "perfectly time-limited"), and let*

$$X(j2\pi f) \;=\; \int_{-\infty}^{\infty} x(t)\, e^{-j2\pi f t}\, dt \;=\; \int_{-\frac{T}{2}}^{\frac{T_x}{2}} x(t)\, e^{-j2\pi f t}\, dt \tag{B.21}$$

denote its Fourier transform. Also, let

$$X_k \;\stackrel{\text{def}}{=}\; f_0\, X(j2\pi f)\Big|_{f=kf_0} \qquad -\infty < k < \infty \tag{B.22a}$$

denote the (scaled) samples of $X(j2\pi f)$. *If the frequency-sampling interval* f_0 *satisfies the dual Nyquist condition*

$$\frac{1}{f_0} \;\geq\; T_x \tag{B.22b}$$

then $X(j2\pi f)$ *admits a representation in terms of its samples, viz.,*

$$X(j2\pi f) \;=\; \sum_{k=-\infty}^{\infty} X_k\, G\big[j2\pi(f - kf_0)\big] \tag{B.22c}$$

where

$$G(j2\pi f) \;=\; \frac{\sin \pi f T_x}{\pi f} \tag{B.22d}$$

Proof: This result follows, of course, directly from the Nyquist theorem about time sampling of band-limited waveforms: it is obtained by switching around the roles of "time" and "frequency" in the original result. However, one can also establish this result more directly by considering a periodic extension of the time-limited waveform, as we now demonstrate.

First, consider the "zero-padded" waveform segment $\{x(t)\,;\; \left|t\right| \leq \frac{T_0}{2}\}$, where $T_0 \geq T_x$, and extend it periodically, with period T_0. The *Fourier coefficients* of this periodic waveform are given by the expression (for $-\infty < k < \infty$)

$$\begin{aligned} X_k \;&=\; \frac{1}{T_0} \int_{-\frac{T_0}{2}}^{\frac{T_0}{2}} x(t)\, e^{j2\pi k f_0 t}\, dt \;=\; f_0 \int_{-\frac{T_x}{2}}^{\frac{T_x}{2}} x(t)\, e^{j2\pi k f_0 t}\, dt \\ &\equiv\; f_0\, X(j2\pi f)\Big|_{f=kf_0} \end{aligned}$$

where $f_0 \overset{\text{def}}{=} \frac{1}{T_0}$ is the fundamental frequency of the periodically extended waveform. Notice that, by construction, $\frac{1}{f_0} \geq T_x$.

Next, we use the Fourier-series representation of the periodically extended waveform to deduce that

$$x(t) = \sum_{k=-\infty}^{\infty} X_k\, e^{j2\pi k f_0 t} \qquad \text{for } \left|t\right| \leq \frac{T_0}{2}$$

Using this expression in (B.21), and relying on the fact that $T_x \leq T_0$, we obtain

$$\begin{aligned} X(j2\pi f) &= \int_{-\frac{T_x}{2}}^{\frac{T_x}{2}} \left(\sum_{k=-\infty}^{\infty} X_k\, e^{j2\pi k f_0 t} \right) e^{-j2\pi f t}\, dt \\ &= \sum_{k=-\infty}^{\infty} X_k \left[\int_{-\frac{T_x}{2}}^{\frac{T_x}{2}} e^{-j2\pi(f-kf_0)t}\, dt \right] \\ &= \sum_{k=-\infty}^{\infty} X_k\, G\Big[j2\pi(f - kf_0) \Big] \end{aligned}$$

where

$$G(j2\pi f) = \int_{-\frac{T_x}{2}}^{\frac{T_x}{2}} e^{-j2\pi f t}\, dt = \frac{\sin \pi f T_x}{\pi f}$$

This completes the proof of the theorem. ■

It is evident from this proof that the original time-limited waveform $x(t)$ need not be centered at the origin. Delaying the time-limited waveform by Δ seconds results in adding a phase-shift of $2\pi f \Delta$ radians to both X_k and $X(j2\pi f)$, so that the representation (B.22c) still holds.

Appendix C: Lagrange Identity

Theorem C.1 (Lagrange Identity): *For any two (complex-valued) row vectors $X = \{x_i\}$ and $Y = \{y_i\}$ of the same length N, we have*

$$\left\| X \right\|^2 \left\| Y \right\|^2 - \left| X Y^H \right|^2 = \tfrac{1}{2} \left\| X^H Y - Y^H X \right\|_F^2 \equiv \tfrac{1}{2} \sum_{i=1}^{N} \sum_{j=1}^{N} \left| x_i^* y_j - y_i^* x_j \right|^2 \tag{C.1}$$

where $\| \ \|_F$ denotes the Frobenius norm of a matrix.

Proof: Define the square matrix $Q \stackrel{\text{def}}{=} X^H Y - Y^H X$, and observe that

$$\begin{aligned}\| Q \|_F^2 &= \operatorname{tr}\{Q\,Q^H\} \equiv \operatorname{tr}\left\{(X^H Y - Y^H X)\,(Y^H X - X^H Y)\right\} \\ &= \operatorname{tr}\left\{X^H \|Y\|^2 X + Y^H \|X\|^2 Y - X^H (Y X^H) Y - Y^H (X Y^H) X\right\} \\ &= 2\,\|X\|^2 \|Y\|^2 - (X Y^H)^2 - (Y X^H)^2\end{aligned}$$

where we used the identity $\operatorname{tr}\{AB\} = \operatorname{tr}\{BA\}$, which holds for any two matrices A, B of appropriate size. Next, notice that for any complex scalar $\gamma = \alpha + j\beta$ we have the identity

$$\Re\{\gamma^2\} = \alpha^2 + \beta^2 \equiv |\gamma|^2$$

and, since the scalar $Y X^H$ is the complex conjugate of $X Y^H$, we conclude that

$$(X Y^H)^2 + (Y X^H)^2 = 2\Re\left\{(X Y^H)^2\right\} = 2\,|X Y^H|^2$$

so that

$$\| Q \|_F^2 = 2\,\|X\|^2 \|Y\|^2 - 2\,|X Y^H|^2$$

which completes the derivation of the Lagrange identity (C.1). ■

Corollary: *When X and Y are real-valued, the identity reduces to*

$$\|X\|^2 \|Y\|^2 - (X Y^\top)^2 = \tfrac{1}{2}\left\|X^\top Y - Y^\top X\right\|_F^2$$

■

As an example, consider the (full-cycle, time-invariant) phasor arrays $\mathbf{V}$ and $\mathbf{I}$ of Sect. 8.1. Now, $\|\mathbf{V}\|^2 \|\mathbf{I}\|^2 = S^2$ and $|\mathbf{V}\mathbf{I}^H|^2 = P^2 + Q_B^2$, so that

$$S^2 - (P^2 + Q_B^2) = \tfrac{1}{2}\,\|\mathbf{V}^H\mathbf{I} - \mathbf{I}^H\mathbf{V}\|_F^2$$

The same identity holds also in the dynamic case, for both full-cycle and sub-cycle power quality metrics.

Appendix D: Variance Components for Grouped Data

D.1 Mean and Variance of 1-D Arrays

Given a real-valued *data vector* $\mathbf{X} = \begin{bmatrix} x_1 & x_2 & \dots & x_N \end{bmatrix}$, and a *weight vector* $\mathbf{W} = \begin{bmatrix} w_1 & w_2 & \dots & w_N \end{bmatrix}$ such that $w_i \geq 0$ and $\sum_i w_i = 1$, we can define a *weighted mean* operation $\mathcal{M}_w$, viz.,

$$\mathcal{M}_w \mathbf{X} \stackrel{\text{def}}{=} \sum_{i=1}^{N} w_i \, x_i \equiv \mu_x \tag{D.1}$$

Next, we introduce the *weighted variance* of the elements of $\mathbf{X}$, viz.,

$$\sigma_x^2 \stackrel{\text{def}}{=} \sum_{i=1}^{N} w_i \left(x_i - \mu_x\right)^2 = \mathcal{M}_w \mathbf{X}^2 - \left(\mathcal{M}_w \mathbf{X}\right)^2 \tag{D.2a}$$

The second equality in (D.2a) is obtained by observing that

$$\sum_{i=1}^{N} w_i \left(x_i - \mu_x\right)^2 = \sum_{i=0}^{N} w_i \left(x_i^2 - 2\, x_i\, \mu_x + \mu_x^2\right) = \sum_{i=1}^{N} w_i \, x_i^2 - \mu_x^2$$

Notice that we introduced here a shorthand notation for element-wise squaring, viz.,

$$\mathbf{X}^2 \stackrel{\text{def}}{=} \mathbf{X} \circ \mathbf{X} \tag{D.2b}$$

where $\circ$ denotes the Schur–Hadamard product [260] (same as the array multiplication operation $A\,.\!*\,B$ in MATLAB). Thus, $\mathbf{X}^2$ is a vector consisting of the squared elements of $\mathbf{X}$. We observe that (D.2a) provides a decomposition of $\mathcal{M}_w \mathbf{X}^2$, viz.,

$$\mathcal{M}_w \mathbf{X}^2 = \sigma_x^2 + \mu_x^2 \tag{D.2c}$$

D.2 Mean and Variance of 2-D Arrays

Given a real-valued *data matrix* $\mathbf{X} = \left\{x_{ij}\,,\ 1 \leq i \leq M\ ,\ 1 \leq j \leq N\right\}$, and a *weight matrix* $\mathbf{W} = \left\{w_{ij}\,,\ 1 \leq i \leq M\ ,\ 1 \leq j \leq N\right\}$ such that $w_{ij} \geq 0$ and $\sum_{i,j} w_{ij} = 1$, we can extend the *weighted mean* operation $\mathcal{M}_w$ to data matrices, viz.,

$$\mathcal{M}_w \mathbf{X} \stackrel{\text{def}}{=} \sum_{i=1}^{M} \sum_{j=1}^{N} w_{ij}\, x_{ij} \equiv \mu_x \tag{D.3a}$$

The *weighted variance* of the elements of $\mathbf{X}$ is similarly extended to matrices, viz.,

$$\sigma_x^2 \stackrel{\text{def}}{=} \sum_{i=1}^{M}\sum_{j=1}^{N} w_{ij}\left(x_{ij}-\mu_x\right)^2 = \mathcal{M}_w\,\mathbf{X}^2 - \left(\mathcal{M}_w\,\mathbf{X}\right)^2 \qquad \text{(D.3b)}$$

and we still have the decomposition

$$\mathcal{M}_w\,\mathbf{X}^2 = \sigma_x^2 + \mu_x^2 \qquad \text{(D.3c)}$$

2-D arrays are sometimes obtained as a result of comparing several distinct populations (a.k.a. "groups"), such as when carrying out an *Analysis of Variance* (ANOVA) [294]. To be specific, we will assume in the sequel that groups are represented by *rows of the data matrix*: each row of $\mathbf{X}$ represents a distinct group.

One way to compare the rows of the data array $\mathbf{X}$ is by evaluating the weighted mean and variance of each individual row, viz.,

$$\mu_x(i) \stackrel{\text{def}}{=} \frac{\sum_{j=1}^{N} w_{ij}\,x_{ij}}{\sum_{j=1}^{N} w_{ij}}\ , \qquad \sigma_x^2(i) \stackrel{\text{def}}{=} \frac{\sum_{j=1}^{N} w_{ij}\left(x_{ij}-\mu_x(i)\right)^2}{\sum_{j=1}^{N} w_{ij}} \qquad \text{(D.4)}$$

Notice that the weight associated with the i-th row is $\sum_j w_{ij}$. The weighted mean of the row-specific variances $\sigma_x^2(i)$ over the set of rows, i.e., $\sum_i \left[\sum_j w_{ij}\right]\sigma_x^2(i)$ generates an aggregated measure of row variation, viz.,

$$\sigma_{x,\text{within}}^2 \stackrel{\text{def}}{=} \sum_{i=1}^{M}\sum_{j=1}^{N} w_{ij}\Big(x_{ij}-\mu_x(i)\Big)^2 \qquad \text{(D.5)}$$

known as the *variance within groups* of the data array $\mathbf{X}$. The relation between this statistic and the array variance σ_x^2 is highlighted in the following theorem.

Theorem D.1 (Group-Oriented Variance Decomposition): *The variance* σ_x^2 *of the data array* $\mathbf{X}$ *is the sum of the variance within groups and the variance between groups of this array, viz.,*

$$\sigma_x^2 = \sigma_{x,\text{within}}^2 + \sigma_{x,\text{between}}^2 \qquad \text{(D.6a)}$$

where $\sigma_{x,\text{within}}^2$ *is as defined in (D.5), and the* variance between groups *of the data array* $\mathbf{X}$ *is defined as the weighted average variation of the row-specific means* $\mu_x(i)$ *around the array mean* μ_x*, viz.,*

$$\sigma_{x,\text{between}}^2 \stackrel{\text{def}}{=} \sum_{i=1}^{M}\sum_{j=1}^{N} w_{ij}\Big(\mu_x(i)-\mu_x\Big)^2 \qquad \text{(D.6b)}$$

Proof: Starting from the definition (D.3b), we observe that $\left(x_{ij} - \mu_x\right) = \left[x_{ij} - \mu_x(i)\right] + \left[\mu_x(i) - \mu_x\right]$, and thus

$$\sigma_x^2 = \sigma_{x,\text{within}}^2 + \sigma_{x,\text{between}}^2 + 2\sum_{i=1}^{M}\sum_{j=1}^{N} w_{ij}\left[x_{ij} - \mu_x(i)\right]\left[\mu_x(i) - \mu_x\right]$$

Next, we observe that the last term on the right-hand side vanishes, viz.,

$$\begin{aligned}&\sum_{i=1}^{M}\sum_{j=1}^{N} w_{ij}\left[x_{ij} - \mu_x(i)\right]\left[\mu_x(i) - \mu_x\right]\\ &\quad = \sum_{i=1}^{M}\left[\mu_x(i) - \mu_x\right]\left(\sum_{j=1}^{N} w_{ij}\left[x_{ij} - \mu_x(i)\right]\right) = 0\end{aligned}$$

because

$$\sum_{j=1}^{N} w_{ij}\left[x_{ij} - \mu_x(i)\right] = \sum_{j=1}^{N} w_{ij}\, x_{ij} - \mu_x(i)\left[\sum_{j=1}^{N} w_{ij}\right] = 0$$

as a result of the definition (D.4) of the row-specific means $\mu_x(i)$. This completes the proof of the theorem. ■

Corollary: *The weighted square mean* $\mathcal{M}_w\,\mathbf{X}^2$ *can be decomposed into a sum of three distinct components, viz.,*

$$\mathcal{M}_w\,\mathbf{X}^2 = \mu_x^2 + \sigma_{x,\text{within}}^2 + \sigma_{x,\text{between}}^2 \tag{D.7}$$

■

Appendix E: Orthogonal Decomposition of Periodic Polyphase Current Waveforms

E.1 Harmonic-Phase Representation of Finite-Power Periodic Polyphase Waveforms

As shown in Sect. 1.2.2, every T-periodic polyphase waveform $x(t)$ can be represented as a sum of mutually orthogonal elementary components, namely (recall (1.11c)),

$$x(t) = \sum_{p=1}^{m} x_0^{(p)}(t) + \sum_{k=1}^{\infty}\sum_{p=1}^{m} x_k^{(p)}(t) \tag{E.1a}$$

where

$$x_k^{(p)}(t) \;=\; \begin{cases} X_0^{(p)}\,\eta_p & k=0 \\ \sqrt{2}\,\Re\left\{X_k^{(p)}\,e^{jk\omega_0 t}\right\}\eta_p & k\geq 1 \end{cases} \tag{E.1b}$$

and

$$\eta_p \;\stackrel{\text{def}}{=}\; [\,\underbrace{0\,\ldots\,0\;1}_{p}\;0\,\ldots\,0\,] \tag{E.1c}$$

Each component $x_k^{(p)}(\cdot)$ is a polyphase waveform and thus an element in the Hilbert space of finite-power periodic polyphase waveforms that was introduced in Sect. 1.2. According to (1.12b)

$$\langle\, x_k^{(p)}(\cdot)\,,\; y_\ell^{(q)}(\cdot)\,\rangle \;=\; \delta_{k,\ell}\;\delta_{p,q}\;\;\Re\left\{X_k^{(p)}\left[Y_k^{(p)}\right]^*\right\}$$

for any two finite-power periodic polyphase waveforms $x(\cdot)$ and $y(\cdot)$. This means that:

- The subspace of (k-th harmonic/p-th phase) components is orthogonal to the (ℓ-th harmonic/q-th phase)-subspace, except when $k=\ell$ and $p=q$.
- Every finite-power periodic polyphase waveform can be expressed as a sum of such (k-th harmonic/p-th phase) components.

In other words, the Hilbert space of finite-power periodic polyphase waveforms is an *orthogonal direct sum*[2] of all (k-th harmonic/p-th phase) subspaces.

E.2 Voltage-Derived Orthogonal Basis

The set of all (k-th harmonic/p-th phase) waveforms, such as $x_k^{(p)}(\cdot)$, is a 2-dimensional subspace: for any given k and p, all such waveforms can be expressed, for instance, as linear combinations of $\big(\cos k\omega_0 t\big)\eta_p$ and $\big(\sin k\omega_0 t\big)\eta_p$. Alternatively, one can use the (k-th harmonic/p-th phase) components of a given polyphase waveform, such as load voltage, to construct an equivalent basis for this 2-dimensional subspace. For instance, the quadrature decomposition (8.33) suggests the choice $\{v_k^{(p)}(\cdot), \mathcal{H}\,v_k^{(p)}(\cdot)\}$ as a basis for the (k-th harmonic/p-th phase) subspace. According to (1.12b),

$$\langle\, v_k^{(p)}(\cdot)\,,\; \mathcal{H}\,v_k^{(p)}(\cdot)\,\rangle = \Re\left\{V_k^{(p)}\left(-\,j\,V_k^{(p)}\right)^*\right\} = 0$$

[2] See Sect. J.5 for a definition of this concept.

which demonstrates that this is, in fact, an orthogonal basis. Furthermore, the identity (1.12b) also implies that, for all k, ℓ, p, q,

$$\langle\, v_k^{(p)}(\cdot)\,,\; v_\ell^{(q)}(\cdot)\,\rangle \;=\; \delta_{k,\ell}\;\delta_{p,q}\,\left|V_k^{(p)}\right|^2 \;=\; \langle\, \mathcal{H}\,v_k^{(p)}(\cdot)\,,\; \mathcal{H}\,v_\ell^{(q)}(\cdot)\,\rangle \tag{E.2a}$$

as well as

$$\langle\, v_k^{(p)}(\cdot)\,,\; \mathcal{H}\,v_\ell^{(q)}(\cdot)\,\rangle \;=\; 0 \tag{E.2b}$$

We have pointed out in Sect. 8.8 that the space of all T-periodic m-phase waveforms can be expressed as a direct sum of three mutually orthogonal subspaces:

- The *voltage-aligned subspace* $\mathcal{V} \stackrel{\text{def}}{=} \text{span}\,\{v_k^{(p)}(\cdot)\,;\;(k,p)\in\Omega_v\}$
- The *quadrature subspace* $\Upsilon \stackrel{\text{def}}{=} \text{span}\,\{\mathcal{H}\,v_k^{(p)}(\cdot)\,;\;(k,p)\in\Omega_v\}$
- The *out-of-band subspace*, consisting of all harmonics (and phases) excluded from Ω_v

The preceding discussion shows that $\left\{\left(v_k^{(p)}(\cdot)\,,\;\mathcal{H}\,v_k^{(p)}(\cdot)\right);\;(k,p)\in\Omega_v\right\}$ is an orthogonal basis of the *in-band subspace* $\mathcal{V}\oplus\Upsilon$.

E.3 7-Component Orthogonal Current Decomposition

The following theorem establishes the decomposition of the current $i(\cdot)$ into a sum of seven mutually orthogonal components. Notice that in the general (i.e., non-rectangular) case, the *cross-section*

$$\Omega_v(k) \;\stackrel{\text{def}}{=}\; \left\{\,p\,;\;(k,p)\in\Omega_v\right\} \tag{E.3}$$

may be a k-dependent proper subset of the full range $1\le p\le m$.

Theorem E.1 (7-Component Current Decomposition): *The current $i(t)$ can be decomposed into a sum of 7 mutually orthogonal components, viz.,*

$$i(t) \;=\; \underbrace{i_F(t)\oplus i_{gs}(t)\oplus i_{gu}(t)}_{i_{\mathcal{V}}(t)} \;\oplus\; \underbrace{i_B(t)\oplus i_{bs}(t)\oplus i_{bu}(t)}_{i_{\Upsilon}(t)} \;\oplus\; i_\perp(t) \tag{E.4a}$$

where

$$i_F(t) \;\stackrel{\text{def}}{=}\; \mu_g \sum_{(k,p)\in\Omega_v} v_k^{(p)}(t) \quad,\quad i_B(t) \;\stackrel{\text{def}}{=}\; \mu_b \sum_{(k,p)\in\Omega_v} \mathcal{H}\,v_k^{(p)}(t) \tag{E.4b}$$

and μ_g , μ_b , *are as defined in (8.16). Furthermore,*

$$i_{gs}(t) \stackrel{\text{def}}{=} \sum_{(k,p)\in\Omega_v} \left(\mu_g(k) - \mu_g\right) v_k^{(p)}(t) \quad , \quad i_{gu}(t) = \sum_{(k,p)\in\Omega_v} \left(g_k^{(p)} - \mu_g(k)\right) v_k^{(p)}(t) \tag{E.4c}$$

$$i_{bs}(t) \stackrel{\text{def}}{=} \sum_{(k,p)\in\Omega_v} \left(\mu_b(k) - \mu_b\right) \mathcal{H}\, v_k^{(p)}(t) \ , \quad i_{bu}(t) = \sum_{(k,p)\in\Omega_v} \left(b_k^{(p)} - \mu_b(k)\right) \mathcal{H}\, v_k^{(p)}(t)$$

$$\mu_g(k) \stackrel{\text{def}}{=} \frac{\sum_{p\in\Omega_v(k)} g_k^{(p)} w_k^{(p)}}{\sum_{p\in\Omega_v(k)} w_k^{(p)}} \quad , \quad \mu_b(k) \stackrel{\text{def}}{=} \frac{\sum_{p\in\Omega_v(k)} b_k^{(p)} w_k^{(p)}}{\sum_{p\in\Omega_v(k)} w_k^{(p)}} \tag{E.4d}$$

Also, the k-dependent cross-section $\Omega_v(k)$ *is defined in (E.3), and* $i_\perp(t)$ *is as defined in (8.32a).*

Proof*:* The definitions (E.4b–d), combined with the original definition (8.33) of $i_{\mathcal{V}}(t)$ and $i_\Upsilon(t)$, imply the identities

$$i_{\mathcal{V}}(t) = i_F(t) + i_{gs}(t) + i_{gu}(t) \quad , \quad i_\Upsilon(t) = i_B(t) + i_{bs}(t) + i_{bu}(t) \tag{E.5}$$

by direct inspection. Thus we only need to verify the mutual orthogonality of the various current components in (E.4a).

Next, we observe that the three current components $i_F(t), i_{gs}(t), i_{gu}(t)$ belong in the voltage-aligned subspace $\mathcal{V} = span\ \{v_k^{(p)}(\cdot);\ (k,p) \in \Omega_v\}$ and thus are automatically orthogonal to $i_B(t),\ i_{bs}(t),\ i_{bu}(t)$, which are all in the quadrature subspace $\Upsilon = span\ \{\mathcal{H}\, v_k^{(p)}(\cdot);\ (k,p) \in \Omega_v\}$. Since both $\mathcal{V}$ and Υ are orthogonal to the out-of-band subspace, we conclude that the six current components of (E.5) are all orthogonal to the out-of-band current $i_\perp(\cdot)$.

Our next step is to confirm that $i_F(\cdot) \perp i_{gs}(\cdot) \perp i_{gu}(\cdot)$. First, notice that

$$\sum_{p\in\Omega_v(k)} \left(g_k^{(p)} - \mu_g(k)\right) w_k^{(p)} = \sum_{p\in\Omega_v(k)} g_k^{(p)} w_k^{(p)} - \mu_g(k) \sum_{p\in\Omega_v(k)} w_k^{(p)} = 0 \tag{E.6}$$

where we used the definition (E.4d) of $\mu_g(k)$. Now, using the definitions (E.4) of $i_F(t)$ and $i_{gu}(t)$, and the fact that $\|\, v_k^{(p)}(\cdot)\|^2 = \left|V_k^{(p)}\right|^2 = \|\mathbf{V}\|^2\, w_k^{(p)}$ (recall (8.19)), we conclude that

$$\frac{\langle\, i_F(\cdot)\, ,\, i_{gu}(\cdot)\, \rangle}{\mu_g\, \|\mathbf{V}\|^2} = \sum_{(k,p)\in\Omega_v} \left(g_k^{(p)} - \mu_g(k)\right) w_k^{(p)} = 0$$

where we used (E.6). Similarly,

$$\frac{\langle\, i_{gs}(\cdot)\,,\, i_{gu}(\cdot)\,\rangle}{\|\mathbf{V}\|^2} = \sum_{(k,p)\in\Omega_v} \Big(\mu_g(k)-\mu_g\Big)\Big(g_k^{(p)}-\mu_g(k)\Big)\, w_k^{(p)}$$

$$= \sum_k \Big(\mu_g(k)-\mu_g\Big)\Bigg[\sum_{p\in\Omega_v(k)} \Big(g_k^{(p)}-\mu_g(k)\Big)\, w_k^{(p)}\Bigg] = 0$$

where, again, we relied on (E.6). Finally,

$$\frac{\langle\, i_F(\cdot)\,,\, i_{gs}(\cdot)\,\rangle}{\mu_g\,\|\mathbf{V}\|^2} = \sum_{(k,p)\in\Omega_v} \Big(\mu_g(k)-\mu_g\Big) w_k^{(p)}$$

$$= \sum_k \Bigg[\Big(\mu_g(k)-\mu_g\Big) \sum_{p\in\Omega_v(k)} w_k^{(p)}\Bigg]$$

$$= \sum_k \Bigg[\sum_{p\in\Omega_v(k)} g_k^{(p)} w_k^{(p)} - \mu_g \sum_{p\in\Omega_v(k)} w_k^{(p)}\Bigg]$$

$$= \sum_{(k,p)\in\Omega_v} \Big(g_k^{(p)}-\mu_g\Big) w_k^{(p)} = \mu_g - \mu_g = 0$$

which completes the proof that $i_F(\cdot) \perp i_{gs}(\cdot) \perp i_{gu}(\cdot)$. A similar (dual) argument, using $b_k^{(p)}$ instead of $g_k^{(p)}$, confirms that $i_B(\cdot) \perp i_{bs}(\cdot) \perp i_{bu}(\cdot)$ as well, which completes the proof of this theorem. ■

Corollary: *The Pythagorean identity associated with (E.4a) is*

$$\|\, i(\cdot)\|^2 = \|\, i_F(\cdot)\|^2 + \|\, i_{gs}(\cdot)\|^2 + \|\, i_{gu}(\cdot)\|^2 + \|\, i_B(\cdot)\|^2 + \|\, i_{bs}(\cdot)\|^2 + \|\, i_{bu}(\cdot)\|^2 + \|\, i_\perp(\cdot)\|^2$$

A direct calculation shows that

$$\|\, i_F(\cdot)\| = \mu_g \big\|\mathbf{V}\big\| \;,\quad \|\, i_{gs}(\cdot)\| = \sigma_{gs} \big\|\mathbf{V}\big\| \;,\quad \|\, i_{gu}(\cdot)\| = \sigma_{gu} \big\|\mathbf{V}\big\|$$

$$\|\, i_B(\cdot)\| = \mu_b \big\|\mathbf{V}\big\| \;,\quad \|\, i_{bs}(\cdot)\| = \sigma_{bs} \big\|\mathbf{V}\big\| \;,\quad \|\, i_{bu}(\cdot)\| = \sigma_{bu} \big\|\mathbf{V}\big\|$$

Under the assumption that the index set Ω_v *was defined with a sufficiently small threshold* ε*, we can replace* $\big\|\mathbf{V}\big\|$ *in these expressions by* V_{rms}*, resulting in the identity*

$$\frac{S^2}{V_{rms}^2} \equiv I_{rms}^2 = \Big(\mu_g^2 + \sigma_{gs}^2 + \sigma_{gu}^2 + \mu_b^2 + \sigma_{bs}^2 + \sigma_{bu}^2\Big)\, V_{rms}^2 + \frac{S_\perp^2}{V_{rms}^2}$$

Multiplying this result by V_{rms}^2 *recovers the expressions (8.30) and (8.31) for the various 7-component power metrics.* ■

Appendix F: Geometric Algebra

A geometric algebra – also known as Clifford algebra [256, 268] – is a higher-level space generated by introducing a new kind of product between elements of a given Euclidean space (i.e., "vectors"). The geometric algebra generated by R^N, the Euclidean space of real-valued N-tuples is itself a Euclidean space, but of dimension 2^N. Its elements are called multivectors, and each multivector consists of $N+1$ distinct grade components: grade-0 components are scalars, grade-1 components are (standard) vectors, grade-2 components are called bivectors, grade-r components are called r-vectors, and so on.

We provide here a brief overview of the concept of multivectors, their grade components, and their norms. In particular, we show that each multivector can be interpreted as a direct sum of mutually orthogonal grade components, which leads to a Pythagorean property for the multivector norm.

F.1 Multivectors and Their Products

Multivectors are generated by (repeated) application of the geometric product to the elements (i.e., "vectors") of the generating Euclidean space, such as R^N. The result is neither a scalar nor a vector, but rather a generalization of the concept of a vector. The geometric product is associative, distributive (but not commutative), and has the key property that every vector $\mathbf{x}$ in the given Euclidean space satisfies

$$\mathbf{x}^2 = \langle\, \mathbf{x}\,,\, \mathbf{x}\,\rangle \equiv \|\,\mathbf{x}\|^2 \tag{F.1a}$$

where $\mathbf{x}^2$ indicates a geometric product of $\mathbf{x}$ by itself, and $\langle\,\cdot\,,\,\cdot\,\rangle$ is the inner-product associated with the Euclidean space (e.g., (1.8a) for periodic polyphase waveforms). However, $\mathbf{xy} \neq \langle\, \mathbf{x}\,,\, \mathbf{y}\,\rangle$ when $\mathbf{x}, \mathbf{y}$ are two distinct (i.e., non-collinear) vectors, and the difference

$$\mathbf{xy} - \langle\, \mathbf{x}\,,\, \mathbf{y}\,\rangle \stackrel{\text{def}}{=} \mathbf{x} \wedge \mathbf{y} \tag{F.1b}$$

is known as a *wedge product*. Notice that $\langle\, \mathbf{x}\,,\, \mathbf{y}\,\rangle = 0$ if and only if $\mathbf{x}$ is *orthogonal* to $\mathbf{y}$, while $\mathbf{x} \wedge \mathbf{y} = 0$ if and only if $\mathbf{x}$ is *collinear* with $\mathbf{y}$. The wedge product is always anti-symmetric, namely $\mathbf{y} \wedge \mathbf{x} = -\mathbf{x} \wedge \mathbf{y}$ for every two vectors $\mathbf{x}, \mathbf{y}$. As a result

$$\mathbf{y\,x} = -\mathbf{x\,y} \quad \Longleftrightarrow \quad \langle\, \mathbf{x}\,,\, \mathbf{y}\,\rangle = 0 \tag{F.1c}$$

because $\mathbf{x\,y} = \mathbf{x} \wedge \mathbf{y}$ whenever $\mathbf{x}$ and $\mathbf{y}$ are *mutually orthogonal*.

We will use lower case Greek letters $(\alpha, \beta, \ldots)$ to indicate scalars and lower case boldface letters $(\mathbf{x}, \mathbf{y}, \ldots)$ to indicate vectors. We will use upper case "blackboard" letters $(\mathbb{S}, \mathbb{Q}, \ldots)$ to indicate multivectors. The different components of a multivector are distinguished by their *grade*. Scalars are grade-zero multivectors (or 0-vectors for short), vectors are grade-one multivectors (= 1-vectors), and products like $\mathbf{x} \wedge \mathbf{y}$ are grade-two multivectors (= 2-vectors or bivectors). The r-vector component of a multivector $\mathbb{A}$ is denoted by $\langle \mathbb{A} \rangle_r$.

F.2 Basis r-Vectors

Consider an orthonormal basis $\{\mathbf{e}_i \,;\; 1 \le i \le N\}$ of some N-dimensional Euclidean space. The preceding discussion of the properties of the geometric and wedge products implies that

$$\mathbf{e}_i{}^2 = 1 \quad \text{for all } i \tag{F.2a}$$

Also, since $\langle \mathbf{e}_i , \mathbf{e}_j \rangle = 0$ for all $i \neq j$, it follows from (F.1b) that the geometric product of two such basis vectors satisfies the relation $\mathbf{e}_i \, \mathbf{e}_j = \mathbf{e}_i \wedge \mathbf{e}_j$, and consequently

$$\mathbf{e}_i \, \mathbf{e}_j = -\, \mathbf{e}_j \, \mathbf{e}_i \quad \text{for all } i \neq j \tag{F.2b}$$

In summary, the set $\{\mathbf{e}_i \, \mathbf{e}_j \,;\; i < j, 1 \le i, j \le N\}$ forms an orthonormal basis (in the sense defined in Sect. F.4) for the space of all 2-vectors (a.k.a. bivectors). Similarly, products of r elements of the orthonormal basis $\{\mathbf{e}_i \,;\; 1 \le i \le N\}$ form a basis for the space of all r-vectors.

F.3 Geometric Algebra of Periodic Waveforms

As a simple illustration of these concepts, consider the case of sinusoidal voltage and current in a single-phase system, which are related to their single-phase (rms) phasors V, I via (8.9a), namely

$$v(t) = \sqrt{2}\, \Re\left\{V e^{j\omega_0 t}\right\} \quad , \quad i(t) = \sqrt{2}\, \Re\left\{I \, e^{j\omega_0 t}\right\}$$

Both belong to the subspace of single-phase sinusoidal waveforms that has dimension 2, so that each such waveform can be represented by a linear combination of two orthonormal basis elements $\mathbf{e}_1$ and $\mathbf{e}_2$, for instance $\mathbf{e}_1 = \sqrt{2}\cos\omega_0 t$, $\mathbf{e}_2 = \sqrt{2}\sin\omega_0 t$. Thus, for any choice of the two orthonormal basis waveforms $\mathbf{e}_1$ and $\mathbf{e}_2$, we have the representations

$$\begin{aligned} v(t) &\equiv \mathbf{v} = \alpha_1 \mathbf{e}_1 + \alpha_2 \mathbf{e}_2 \\ i(t) &\equiv \mathbf{i} = \beta_1 \mathbf{e}_1 + \beta_2 \mathbf{e}_2 \end{aligned}$$

Recall that we use upper case "blackboard" letters to indicate multivectors, such as the apparent power multivector $\mathbb{S}$, which can be defined as the geometric product of $\mathbf{v}$ and $\mathbf{i}$, viz.,

$$\begin{aligned}\mathbb{S} &\stackrel{\text{def}}{=} \mathbf{v}\,\mathbf{i} = (\alpha_1\mathbf{e}_1 + \alpha_2\mathbf{e}_2)\,(\beta_1\mathbf{e}_1 + \beta_2\mathbf{e}_2) \\ &= \alpha_1\beta_1\,\mathbf{e}_1^2 + \alpha_1\beta_2\,\mathbf{e}_1\mathbf{e}_2 + \alpha_2\beta_1\,\mathbf{e}_2\mathbf{e}_2 + \alpha_2\beta_2\,\mathbf{e}_2^2\end{aligned}$$

where we used the distributivity property of the geometric product, as well as the fact that scalars commute with all multivectors [256]. Since ${\mathbf{e}_1}^2 = \|\mathbf{e}_1\|^2 = 1$ and ${\mathbf{e}_2}^2 = \|\mathbf{e}_2\|^2 = 1$, as well as $\mathbf{e}_1\,\mathbf{e}_2 = -\,\mathbf{e}_2\,\mathbf{e}_1$ (recall (F.2)), this reduces to

$$\mathbb{S} = \underbrace{\alpha_1\beta_1 + \alpha_2\beta_2}_{\langle\,\mathbf{v}\,,\mathbf{i}\,\rangle} + \underbrace{(\alpha_1\beta_2 - \alpha_2\beta_1)\,\mathbf{e}_1\mathbf{e}_2}_{\mathbf{v}\wedge\mathbf{i}}$$

We recognize

$$\alpha_1\beta_1 + \alpha_2\beta_2 \equiv \langle\,\mathbf{v}\,,\,\mathbf{i}\,\rangle = P$$

as the real (average) power, which is independent of the choice of basis. Similarly, the wedge product $\mathbf{v}\wedge\mathbf{i} = (\alpha_1\beta_2 - \alpha_2\beta_1)\,\mathbf{e}_1\mathbf{e}_2$ is also independent of the choice of basis. The wedge product is clearly neither a scalar nor a vector, but a new entity called a *bivector*. Recall that when the generating Euclidean space is spanned by a finite orthonormal set, say $\{\mathbf{e}_i\,;\ 1 \le i \le N\}$, bivectors can always be represented as linear combinations of the basis bivectors $\left\{\mathbb{E}_{i,j}\,;\ \ i < j,\ 1 \le i,\,j, \le N\right\}$, where $\mathbb{E}_{i,j} \stackrel{\text{def}}{=} \mathbf{e}_i\,\mathbf{e}_j$. In our example, the generating Euclidean space is two-dimensional ($N = 2$), so there is only one such term in the multivector $\mathbb{S}$, corresponding to the basis bivector $\mathbf{e}_1\mathbf{e}_2$.

Recall that the grade-r component of a multivector $\mathbb{A}$ is denoted by $\langle\,\mathbb{A}\,\rangle_r$. We notice that the apparent power multivector $\mathbb{S} = \mathbf{v}\,\mathbf{i}$ in our example consists of two distinct grades, viz.,

$$\langle\,\mathbb{S}\,\rangle_0 = \langle\,\mathbf{v}\,,\,\mathbf{i}\,\rangle = \alpha_1\beta_1 + \alpha_2\beta_2 \equiv P$$

and

$$\langle\,\mathbb{S}\,\rangle_2 = \mathbf{v}\wedge\mathbf{i} = (\alpha_1\beta_2 - \alpha_2\beta_1)\,\mathbf{e}_1\mathbf{e}_2$$

If we use the simple choice of basis waveforms

$$\mathbf{e}_1 = \sqrt{2}\cos\omega_0 t \quad , \qquad \mathbf{e}_2 = \sqrt{2}\sin\omega_0 t \tag{F.3}$$

then

$$\alpha_1 = \Re\{V\},\ \ \alpha_2 = -\Im\{V\},\ \ \beta_1 = \Re\{I\},\ \ \beta_2 = -\Im\{I\}$$

and $P = \Re\{VI^*\}$, while

$$\alpha_1\beta_2 - \alpha_2\beta_1 = \Im\{VI^*\} \stackrel{\text{def}}{=} Q_B$$

which we recognize as Budeanu's reactive power [248, 253]. The superscript * denotes complex conjugation.

An alternative, voltage-derived, choice of basis waveforms is

$$\mathbf{e}_1 = \sqrt{2}\,\Re\left\{\frac{V}{|V|}\,e^{j\omega_0 t}\right\} \;\equiv\; \frac{v(t)}{|V|} \tag{F.4a}$$

$$\mathbf{e}_2 = \sqrt{2}\,\Im\left\{\frac{V}{|V|}\,e^{j\omega_0 t}\right\} \;\equiv\; \sqrt{2}\,\Re\left\{-j\,\frac{V}{|V|}\,e^{j\omega_0 t}\right\} \tag{F.4b}$$

which results in $v(t) \equiv \mathbf{v} = |V|\,\mathbf{e}_1$. Also

$$\begin{aligned} i(t) &= \sqrt{2}\,\Re\{I e^{j\omega_0 t}\} = \sqrt{2}\,\Re\left\{\frac{IV^*}{|V|}\cdot\frac{V}{|V|}\,e^{j\omega_0 t}\right\} \\ &= \sqrt{2}\,\Re\left\{\frac{P - jQ_B}{|V|}\cdot\frac{V}{|V|}\,e^{j\omega_0 t}\right\} \end{aligned}$$

so that $\mathbf{i} = \frac{P}{|V|}\,\mathbf{e}_1 + \frac{Q_B}{|V|}\,\mathbf{e}_2$. Now

$$\mathbb{S} \;\equiv\; \mathbf{v}\,\mathbf{i} \;=\; P + Q_B\,\mathbf{e}_1\mathbf{e}_2 \tag{F.5}$$

which is the same as the result obtained with the previous choice of basis waveforms. Voltage-derived basis waveforms play a key role in our construction of the apparent power multivector for polyphase (and polyharmonic) voltage and current (Sect. 8.9).

Notice that the relation $\mathbf{e}_2 = \mathcal{H}\{\mathbf{e}_1\}$ holds for both the simple basis (F.3) and the voltage-derived basis (F.4), where $\mathcal{H}\{\,\}$ denotes the *Hilbert transform* [32,267] of a given waveform. For periodic waveforms, this linear transform is completely determined by

$$\mathcal{H}\{e^{jk\omega_0 t}\} = -j(\operatorname{sgn} k)e^{jk\omega_0 t}$$

so that, for $k \geq 1$,

$$\mathcal{H}\{\cos k\omega_0 t\} = \sin k\omega_0 t, \qquad \mathcal{H}\{\sin k\omega_0 t\} = -\cos k\omega_0 t$$

Two key properties of the Hilbert transform (for periodic waveforms with no DC component) are $\|\mathcal{H}\{x\}\| = \|x\|$ and $\langle\, \mathcal{H}\{x\}\,,\, x\,\rangle = 0$, so that $\{\mathbf{e}_1, \mathbf{e}_2\}$ is indeed an orthonormal basis because $\langle\, \mathbf{e}_1\,,\, \mathbf{e}_2\,\rangle = 0$ and $\|\mathbf{e}_1\| = 1 = \|\mathbf{e}_2\|$ in both (F.3) and (F.4).

F.4 Multivector Norm and Scalar Product

The *norm* of a multivector $\mathbb{A}$ is defined as

$$\|\mathbb{A}\|^2 \stackrel{\text{def}}{=} \left\langle \mathbb{A}\mathbb{A}^{\dagger} \right\rangle_0 \tag{F.6a}$$

where the superscript $(\cdot)^{\dagger}$ denotes a reversal operation, viz.,

$$\left(\mathbf{x}_1\mathbf{x}_2 \ \ldots \ \mathbf{x}_{p-1}\mathbf{x}_p\right)^{\dagger} \stackrel{\text{def}}{=} \mathbf{x}_p \, \mathbf{x}_{p-1} \ \ldots \ \mathbf{x}_2 \, \mathbf{x}_1 \tag{F.6b}$$

As a result, the multivector norm of a 1-vector $\mathbf{x}$ coincides with the original Euclidean norm $\|\mathbf{x}\| = \sqrt{\langle \, \mathbf{x} \, , \, \mathbf{x} \, \rangle}$, and the norm of a product of vectors is a product of the individual vector norms, viz.,

$$\|\mathbf{x}_1\mathbf{x}_2 \cdots \mathbf{x}_p\| = \|\mathbf{x}_1\| \, \|\mathbf{x}_2\| \cdots \|\mathbf{x}_p\| \tag{F.6c}$$

In particular, $\|\mathbb{S}\| \equiv \|\mathbf{v}\,\mathbf{i}\| = \|\mathbf{v}\| \, \|\mathbf{i}\|$, which coincides with the usual definition of apparent power as the unsigned scalar $\|\mathbf{v}\| \, \|\mathbf{i}\|$.

Recall that $\mathbf{x}\,\mathbf{y} = \mathbf{x} \wedge \mathbf{y}$ when $\mathbf{x}$ and $\mathbf{y}$ are mutually orthogonal. This implies the special multivector norm identity

$$\|\, \mathbf{x} \wedge \mathbf{y} \,\| = \|\, \mathbf{x}\,\mathbf{y} \,\| = \|\, \mathbf{x} \,\| \, \|\, \mathbf{y} \,\| \qquad \text{when} \ \ \mathbf{x} \perp \mathbf{y} \tag{F.7}$$

Since, in general, any vector $\mathbf{y}$ can be expressed as a sum of a part $\mathbf{y}_{\|}$ that is collinear with $\mathbf{x}$, and a part $\mathbf{y}_{\perp}$ that is orthogonal to $\mathbf{x}$, we conclude that $\|\, \mathbf{x} \wedge \mathbf{y} \,\| = \|\, \mathbf{x} \,\| \, \|\, \mathbf{y}_{\perp} \,\|$ holds for any $\mathbf{x}$ and $\mathbf{y}$.

The squared norm of any multivector is additive over grades, viz.,

$$\|\mathbb{A}\|^2 = \sum_{r=0}^{N} \|\langle \, \mathbb{A} \, \rangle_r\|^2$$

so that in our sinusoidal single-phase example $\mathbb{S} = P + Q_B \, \mathbf{e}_1\mathbf{e}_2$ results in the Pythagorean relation

$$\|\mathbb{S}\|^2 = \|\mathbf{v}\|^2 \, \|\mathbf{i}\|^2 = P^2 + Q_B^2$$

because $\|\mathbf{e}_1\mathbf{e}_2\| = \|\mathbf{e}_1\| \, \|\mathbf{e}_2\| = 1$. Thus the definition of $\mathbb{S}$ provides us with the correct magnitude $\|\mathbf{v}\| \, \|\mathbf{i}\|$ and the correct (Pythagorean) relation to the magnitudes of the two power components P and Q_B.

The linear subspaces associated with distinct grades are, in fact, Euclidean spaces with respect to the so-called *scalar product*

$$\langle \, \mathbb{A} \, , \, \mathbb{B} \, \rangle \stackrel{\text{def}}{=} \left\langle \mathbb{A} \, \mathbb{B}^{\dagger} \right\rangle_0 \tag{F.8a}$$

Thus a geometric algebra can also be interpreted as a *direct sum* of mutually orthogonal Euclidean spaces. One obvious outcome of the definition (F.8a) is additivity by grade, viz.,

$$\langle \mathbb{A}, \mathbb{B} \rangle = \sum_{r=0}^{N} \langle \langle \mathbb{A} \rangle_r, \langle \mathbb{B} \rangle_r \rangle \tag{F.8b}$$

and, in particular, the Pythagorean property

$$\langle \mathbb{A}, \mathbb{A} \rangle = \|\mathbb{A}\|^2 = \sum_{r=0}^{N} \|\langle \mathbb{A} \rangle_r\|^2 \tag{F.8c}$$

which we used in our sinusoidal single-phase example to obtain the relation $\|\mathbb{S}\|^2 = P^2 + Q_B^2$. Notice that when $\mathbb{A}$, $\mathbb{B}$ are 1-vectors, their scalar product coincides with the conventional inner-product.

F.5 Grade Structure and Duality

In general, multivectors generated by an N-dimensional Euclidean space (over real scalars) can have grades ranging from zero to N. The set of all r-vectors, for any fixed r in the range $[0 \;\; N]$, is a linear space over the field of real scalars, and its dimension is $\binom{N}{r} = \frac{N!}{r!\,(N-r)!}$. Thus, the subspace of all (real) scalars has dimension 1, the subspace of all 1-vectors has dimension N (it coincides with the generating Euclidean space), the space of all 2-vectors has dimension $\frac{N(N-1)}{2}$, and so on. Recall that the geometric product mixes up grades so that, for instance, the geometric product of two 1-vectors, such as the apparent power multivector $\mathbb{S} = \mathbf{v}\,\mathbf{i}$, always consists of two grades, a scalar, and a 2-vector.

The grade structure of a geometric algebra is symmetric, so that a grade-r subspace and a grade-$(N - r)$ subspace always have the same dimension. Consequently, we can set up a one-to-one correspondence between such pairs of equal dimension spaces. One convenient way of doing so is via the *duality relation* $\mathbb{A} \leftrightarrow \mathbb{A}\,\mathcal{J}$, where the so-called pseudo-scalar $\mathcal{J}$ is the geometric product of a complete set of orthonormal basis vectors, viz., $\mathcal{J} \overset{\text{def}}{=} \mathbf{e}_1\mathbf{e}_2 \ldots \mathbf{e}_N$. For instance the dual of a scalar α is the pseudo-scalar $\alpha\,\mathcal{J}$, the dual of a vector $\mathbf{x}$ is the so-called pseudo-vector (an $(N - 1)$-vector) $\mathbf{x}\,\mathcal{J}$, and so on. Dual multivectors have the same norm, viz., $\|\mathbb{A}\,\mathcal{J}\| = \|\mathbb{A}\|$ for every multivector $\mathbb{A}$. The pseudo-scalar $\mathcal{J}$ is independent of the choice of basis, up to a sign change. Notice that $\mathcal{J}^2 = (-1)^{N(N-1)/2}$: this follows directly from the two properties $\mathbf{e}_i{}^2 = 1$ and $\mathbf{e}_j\,\mathbf{e}_i = -\mathbf{e}_i\,\mathbf{e}_j$, which hold for any two orthonormal basis vectors $\mathbf{e}_i$ and $\mathbf{e}_j$.

In particular, when $N = 2$, as in our sinusoidal single-phase example, the entire geometric algebra consists of 3 grades only: scalars (grade 0), vectors (grade 1), and pseudo-scalars (grade 2). Notice that in this case $\mathbb{S} = P + Q_B\,\mathcal{J}$ shares many properties with the classical complex apparent power $S_c \overset{\text{def}}{=} P + j\,Q_B$: in addition to $\|\mathbb{S}\|^2 = P^2 + Q_B^2 = |S_c|^2$, we also have $\mathbb{S}^\dagger = P - Q_B\,\mathcal{J}$ (similar to $S_c^* = P - j\,Q_B$) and $\mathcal{J}^2 = (\mathbf{e}_1\mathbf{e}_2)(\mathbf{e}_1\mathbf{e}_2) = -1$, similar to $j^2 = -1$. However, the role of $\mathcal{J}$ as a duality transformation generalizes to geometric algebras generated by Euclidean spaces of arbitrary dimensions, while the use of the complex apparent power is limited to the case $N = 2$, i.e., to single-phase sinusoidal waveforms.

The multiplicity of grades for higher-dimensional Euclidean spaces allows us to construct apparent power multivectors with multiple physically significant components, one component per grade (see Sect. 8.9). Our approach relies on quadrature, duality, and grade separation, which allow us to recover *all five components* of (8.24).

Appendix G: A Least-Squares Approach to Dynamic Phasor Estimation

We study here the sliding-window polyphase linear least-squares phasor estimation problem (3.16), namely

$$\min_{\{\widehat{X}_k\,;\;k\,\in\,\boldsymbol{\mathcal{S}}\}} \sum_{i=0}^{M-1} \left\| x(t-\Delta_i) - \sum_{k\,\in\,\boldsymbol{\mathcal{S}}} \sqrt{2}\,\Re\left\{\widehat{X}_k\; e^{jk\omega_0(t-\Delta_i)}\right\} \right\|_2^2$$

Here $x(t)$ is a given polyphase waveform (i.e., a $1 \times m$ vector), and $\|\cdot\|_2$ denotes the standard (i.e., Euclidean) vector norm. Also, $\{\widehat{X}_k\,;\;k \in \boldsymbol{\mathcal{S}}\}$ are the estimated polyphase phasors, $\boldsymbol{\Delta} = \left\{\Delta_i\,;\;0 \le i \le M-1\right\}$ is a set of (distinct, non-negative) inter-sample shifts, with $\Delta_0 = 0$, and $\boldsymbol{\mathcal{S}} = \{k_\ell\,;\;1 \le \ell \le L\}$ is a set of selected harmonic indices, which always includes the fundamental harmonic. Notice that the solution of this least-squares problem is a function of the sliding-window set of polyphase samples $\{x(t), x(t-\Delta_1), \ldots, x(t-\Delta_{M-1})\}$ and thus depends, in general, on the time-instant "t." We will use the notation $\widehat{X}_k(t)$ to highlight the time dependence of the resulting estimated phasors.

We focus here on the over-determined case (i.e., $M \ge 2L$) and derive conditions for existence of a unique solution (Sect. G.2). We demonstrate that the solution of this least-squares problem is, in fact, a linear time-invariant map that can be conveniently characterized in terms of its frequency response (Sect. G.1). We use this filtering interpretation to characterize several important properties of the resulting dynamic phasor estimates (see Theorems G.2, G.3, G.4).

G.1 Matrix Formulation of the Least-Squares Problem

This linear least-squares optimization problem can be rewritten, for any time-instant "t," in a compact form using matrix notation, viz.,

$$\min_{\{\widehat{X}_k\,;\ k\in\boldsymbol{\mathcal{S}}\}} \left\| \boldsymbol{\chi}(t) - 2\,\Re\left\{\boldsymbol{\Psi}\,\mathbf{H}(t)\right\}\right\|_F^2 \tag{G.1a}$$

where $\|\ \|_F$ denotes the Frobenius norm of a matrix, and notice that $\Delta_0 = 0$.

$$\boldsymbol{\chi}(t) \stackrel{\text{def}}{=} \begin{pmatrix} x(t) \\ x(t-\Delta_1) \\ \vdots \\ x(t-\Delta_{M-1}) \end{pmatrix}, \qquad \mathbf{H}(t) \stackrel{\text{def}}{=} \sqrt{\frac{M}{2}} \begin{pmatrix} \widehat{X}_{k_1}\, e^{jk_1\omega_0 t} \\ \widehat{X}_{k_2}\, e^{jk_2\omega_0 t} \\ \vdots \\ \widehat{X}_{k_L}\, e^{jk_L\omega_0 t} \end{pmatrix} \tag{G.1b}$$

The dimensions of the matrices $\boldsymbol{\chi}$ and $\mathbf{H}$ are $M \times m$ and $L \times m$, respectively, where m is the number of phases. The time-invariant $M \times L$ matrix $\boldsymbol{\Psi}$ is determined by the harmonic index set $\boldsymbol{\mathcal{S}}$ and the time shifts $\{\Delta_i\,;\ 1 \le i \le M-1\}$, viz.,

$$\boldsymbol{\Psi} \stackrel{\text{def}}{=} \begin{bmatrix} \psi_{k_1} & \psi_{k_2} & \dots & \psi_{k_L} \end{bmatrix} \tag{G.1c}$$

where, for each $k \in \boldsymbol{\mathcal{S}}$,

$$\psi_k \stackrel{\text{def}}{=} \frac{1}{\sqrt{M}} \begin{bmatrix} 1 & e^{-jk\omega_0\,\Delta_1} & \dots & e^{-jk\omega_0\,\Delta_{M-1}} \end{bmatrix}^{\top} \tag{G.1d}$$

and the superscript $^{\top}$ denotes transposition (without conjugation). We shall assume from now on that $M \ge 2L$, which is a necessary (but not sufficient) condition for (G.1) to have a unique solution.

Equivalently, we can rewrite (G.1) as a *structured* linear least-squares problem, viz.,

$$\min_{\mathbf{H}} \left\| \boldsymbol{\chi}(t) - \boldsymbol{\mathcal{W}} \begin{pmatrix} \mathbf{H} \\ \mathbf{H}^* \end{pmatrix} \right\|_F^2, \qquad \boldsymbol{\mathcal{W}} \stackrel{\text{def}}{=} [\,\boldsymbol{\Psi} \;\; \boldsymbol{\Psi}^*\,] \tag{G.2}$$

where the superscript * denotes conjugation (without transposition). Notice that $\boldsymbol{\chi}(t)$ is real-valued, while $\mathbf{H}$ and $\boldsymbol{\Psi}$ are complex. Also notice that solving (G.2) for $\mathbf{H}$ is equivalent to solving (G.1) for the complex sub-cycle phasors $\{\widehat{X}_k\,;\ k \in \boldsymbol{\mathcal{S}}\}$, because the scaling factors $e^{jk_\ell\omega_0 t}$ in (G.1b) are always invertible.

Theorem G.1 (Structured Least Squares): *When the $M \times 2L$ matrix $\mathcal{W}$ has full column rank, the (unique) optimizing solution of the structured least-squares problem (G.2) is*

$$\mathbf{H}_{opt.}(t) = \begin{bmatrix} \mathbf{I}_L & \mathbf{0}_L \end{bmatrix} \left(\mathcal{W}^H \mathcal{W} \right)^{-1} \mathcal{W}^H \chi(t) \tag{G.3}$$

Proof: First, consider the *unstructured* linear least-squares problem

$$\min_{\mathbf{C}} \left\| \chi - \mathcal{W}\, \mathbf{C} \right\|_F^2 \tag{G.4a}$$

where $\mathbf{C}$ is any complex matrix of size $2L \times m$. Notice that

$$\left\| \chi - \mathcal{W}\, \mathbf{C} \right\|_F^2 = \sum_{p=1}^{m} \left\| \chi^{(p)} - \mathcal{W}\, \mathbf{C}^{(p)} \right\|_2^2$$

where $\chi^{(p)}$ and $\mathbf{C}^{(p)}$ denote the p-th column of χ and $\mathbf{C}$, respectively. Thus the solution of (G.4a) can be determined individually for each phase, namely, the optimizing $2L \times 1$ vector $\mathbf{C}^{(p)}_{opt.}$ is the solution of the single-phase least-squares problem $\min_{\mathbf{C}^{(p)}} \left\| \chi^{(p)} - \mathcal{W}\, \mathbf{C}^{(p)} \right\|_2^2$, whose well-known solution is [295, Ch. 9]

$$\mathbf{C}^{(p)}_{opt.} = \left(\mathcal{W}^H \mathcal{W} \right)^{-1} \mathcal{W}^H \chi^{(p)}$$

Consequently, the unique solution of (G.4a) is

$$\mathbf{C}_{opt.} = \left(\mathcal{W}^H \mathcal{W} \right)^{-1} \mathcal{W}^H \chi \tag{G.4b}$$

We shall presently show that the matrix $\mathbf{C}_{opt.}$ has indeed the structure needed for the solution of (G.2), namely

$$\mathbf{C}_{opt.} = \begin{pmatrix} \mathbf{H} \\ \mathbf{H}^* \end{pmatrix} \quad \text{for some } L \times m \text{ matrix } \mathbf{H} \tag{G.5}$$

Consequently, the top half of $\mathbf{C}_{opt.}$ is the (unique) solution of (G.2), which establishes (G.3).

To establish the special structure of $\mathbf{C}_{opt.}$, note first that this structural constraint is equivalent to the property

$$\tilde{\mathbf{I}}\, \mathbf{C}_{opt.} = \mathbf{C}^*_{opt.} \quad , \quad \tilde{\mathbf{I}} \stackrel{\text{def}}{=} \begin{pmatrix} \mathbf{0}_L & \mathbf{I}_L \\ \mathbf{I}_L & \mathbf{0}_L \end{pmatrix}$$

Next, observe that

$$\mathcal{W}\,\widetilde{\mathbf{I}} \equiv \begin{bmatrix} \boldsymbol{\Psi} & \boldsymbol{\Psi}^* \end{bmatrix} \widetilde{\mathbf{I}} = \begin{bmatrix} \boldsymbol{\Psi}^* & \boldsymbol{\Psi} \end{bmatrix} \equiv \mathcal{W}^*$$

and use the fact that $\mathcal{X}$ is real-valued, as well as the identity $\widetilde{\mathbf{I}} = \widetilde{\mathbf{I}}^{-1} = \widetilde{\mathbf{I}}^H$, to modify the expression (G.4b), viz.,

$$\begin{aligned} \widetilde{\mathbf{I}}\,\mathbf{C}_{opt.} &= \left[\widetilde{\mathbf{I}}\left(\mathcal{W}^H \mathcal{W}\right)^{-1} \widetilde{\mathbf{I}}\right]\left[\widetilde{\mathbf{I}}\,\mathcal{W}^H\right]\mathcal{X} = \left((\mathcal{W}^*)^H (\mathcal{W}^*)\right)^{-1} (\mathcal{W}^*)^H \mathcal{X} \\ &= \mathbf{C}^*_{opt.} \end{aligned}$$

This establishes (G.5) and completes the proof of the theorem. ■

Theorem G.1 provides an explicit expression for the sub-cycle polyphase phasors $\{\,\widehat{X}_k(t)\,;\ k \in \mathcal{S}\,\}$ in terms of the waveform samples $\{\,x(t-\Delta_i)\,;\ 0 \le i \le M-1\}$ (recall also (3.18a)), viz.,

$$\begin{pmatrix} \widehat{X}_{k_1}(t)\, e^{jk_1\omega_0 t} \\ \widehat{X}_{k_2}(t)\, e^{jk_2\omega_0 t} \\ \vdots \\ \widehat{X}_{k_L}(t)\, e^{jk_L\omega_0 t} \end{pmatrix} = \sqrt{\frac{2}{M}} \begin{bmatrix} \mathbf{I}_L & \mathbf{0}_L \end{bmatrix} \left(\mathcal{W}^H \mathcal{W}\right)^{-1} \mathcal{W}^H \begin{pmatrix} x(t) \\ x(t-\Delta_1) \\ \vdots \\ x(t-\Delta_{M-1}) \end{pmatrix} \tag{G.6}$$

where we assume that $\mathcal{W}$ has full column rank as a result of an appropriate choice of the harmonic index set $\mathcal{S}$ and the inter-sample shifts $\{\Delta_i\}$.

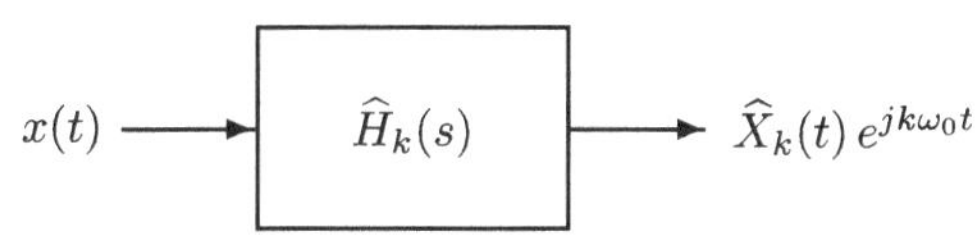

Fig. G.1 Filtering interpretation of the waveform-to-phasor map for a single $k \in \mathcal{S}$.

This waveform-to-phasor map is, in fact, a linear time-invariant filter-bank, with a single (polyphase) input $x(t)$ and L (polyphase) outputs, one for each $k \in \mathcal{S}$ (Fig. G.1). Notice that each phase of the row vector $x(t)$ is mapped into its own phasor (i.e., there is no coupling between phases). The transfer function $\widehat{H}_k(s)$ determines the steady-state and transient properties of the estimated phasor $\widehat{X}_k(t)$. In particular, $\widehat{H}_k(j\omega)$ has a *steady-state harmonics-blocking property* as described in the following theorem.

Theorem G.2 (Steady-State Harmonics-Blocking Property): *Consider the transfer function* $\widehat{H}_k(s)$ *associated with the waveform-to-phasor map (G.6). For every* $k \in \mathcal{S}$ *and* $\ell \in \mathcal{S}$

$$\begin{aligned}\widehat{H}_k(j\ell\omega_0) &= \sqrt{2}\,\delta_{k,\ell} \\ \widehat{H}_k(-j\ell\omega_0) &= 0\end{aligned} \tag{G.7}$$

Proof: Let k_ℓ and k_r denote the ℓ-th and r-th elements of the harmonic index set $\mathcal{S}$, respectively. Starting with the transfer function expression (3.18b), we evaluate the frequency response $\widehat{H}_{k_\ell}(j\omega)$, at the (harmonic) frequency $\omega = k_r\omega_0$, viz.,

$$\widehat{H}_{k_\ell}(jk_r\omega_0) = \sqrt{\frac{2}{M}}\, e_\ell \left(\mathcal{W}^H \mathcal{W}\right)^{-1} \mathcal{W}^H \begin{pmatrix} 1 \\ e^{-jk_r\omega_0\,\Delta_1} \\ e^{-jk_r\omega_0\,\Delta_2} \\ \vdots \\ e^{-jk_r\omega_0\,\Delta_{M-1}} \end{pmatrix}, \qquad 1 \le \ell \le L$$

where $e_\ell \overset{\text{def}}{=} [\,\underbrace{0\;\ldots\;0\;1}_{\ell}\;0\;\ldots\;0\,]$. Next, we observe (from (3.17a,b)) that

$$\begin{pmatrix} 1 \\ e^{-jk_r\omega_0\,\Delta_1} \\ e^{-jk_r\omega_0\,\Delta_2} \\ \vdots \\ e^{-jk_r\omega_0\,\Delta_{M-1}} \end{pmatrix} = \sqrt{M}\,\psi_{k_r} = \sqrt{M}\,\mathcal{W}\, e_r^\top$$

and thus

$$\begin{aligned}\widehat{H}_{k_\ell}(jk_r\omega_0) &= \sqrt{2}\left[e_\ell \left(\mathcal{W}^H \mathcal{W}\right)^{-1} \mathcal{W}^H\right] \mathcal{W}\, e_r^\top = \sqrt{2}\, e_\ell\, e_r^\top = \sqrt{2}\,\delta_{\ell,r} \\ &= \sqrt{2}\,\delta_{k_\ell,k_r}\end{aligned}$$

which establishes the first part of (G.7). Similarly, $\widehat{H}_{k_\ell}(-jk_r\,\omega_0)$ involves $\psi_{k_r}^* = \mathcal{W}\, e_{r+L}^\top$, so that

$$\widehat{H}_{k_\ell}(-jk_r\,\omega_0) = \sqrt{2}\left[e_\ell \left(\mathcal{W}^H \mathcal{W}\right)^{-1} \mathcal{W}^H\right] \mathcal{W}\, e_{r+L}^\top = \sqrt{2}\, e_\ell\, e_{r+L}^\top = 0$$

which establishes the second part of (G.7). The derivation given here is an extension to the case $M \ge 2L$ of the proof provided in [78] for the special case $M = 2L$, known as *sub-cycle*. ■

G.2 Uniformly-Shifted Waveform Samples

The complexity and computational cost of the general expression (G.3) can be significantly reduced by using uniformly-shifted waveform samples, namely

$$\Delta_i = i\ \Delta \ , \qquad i = 1, 2, \ldots, M-1 \tag{G.8a}$$

The choice (G.8a) makes $\mathcal{W}$ into a *Vandermonde matrix*: its columns are given by

$$\psi_k \ = \ \frac{1}{\sqrt{M}} \begin{bmatrix} 1 & \alpha^k & \alpha^{2k} & \ldots & \alpha^{(M-1)k} \end{bmatrix}^{\mathrm{T}} \tag{G.8b}$$

where $\alpha \stackrel{\text{def}}{=} e^{-j2\pi\xi}$, with $\xi = \frac{\Delta}{T} > 0$, a normalized inter-sample shift. This rectangular Vandermonde matrix has full column rank (recall that $M \geq 2L$) if, and only if, its columns are all *distinct*. To be specific, this means that, for every $k \in \mathcal{S}$ and every $\ell \in \mathcal{S}$: (i) $\psi_k \neq \psi_\ell$ (when $k \neq \ell$) and (ii) $\psi_k \neq \psi_\ell^*$ (including the case $k = \ell$). In view of (G.8b), these constraints are equivalent to

$$\left.\begin{array}{l} \text{(i)}\ \alpha^{(k-\ell)} \neq 1 \ \text{(when } k \neq \ell\text{)} \\ \text{(ii)}\ \alpha^{(k+\ell)} \neq 1 \end{array}\right\} \quad \text{for every } k \in \mathcal{S}\ ,\ \ell \in \mathcal{S} \tag{G.9}$$

where we used the fact that $\alpha^* = \alpha^{-1}$.

In sampled-data (i.e., digital) implementations, the value of the normalized inter-sample shift ξ is always rational, namely, $\xi = m/D$ where m and D are co-prime positive integers. Consequently, the sequence $\alpha^k = e^{-jk2\pi\xi} \equiv e^{-jk2\pi m/D}$ is periodic in k with period D. The periodicity property $\alpha^k = \alpha^{k+D}$ can be equivalently expressed as $\alpha^k = \alpha^{\{k \bmod D\}}$, where $k \bmod D$ denotes the remainder of dividing k by D. To highlight the fact that the period D always depends on ξ, we will use in the sequel the explicit notation $D(\xi)$. We now observe that, for rational values of ξ, the non-singularity constraint (G.9) induces an equivalent constraint on the harmonic indices k and ℓ, viz.,

$$\left.\begin{array}{l} \text{(i)}\ (k-\ell) \bmod D(\xi) \ \neq\ 0 \ \text{(when } k \neq \ell\text{)} \\ \text{(ii)}\ (k+\ell) \bmod D(\xi) \ \neq\ 0 \end{array}\right\} \quad \text{for every } k \in \mathcal{S}\ ,\ \ell \in \mathcal{S} \tag{G.10}$$

In conclusion, the matrix $\mathcal{W}$ has full column rank if, and only if, the condition (G.10) is satisfied. For instance, the choice $\mathcal{S} = \{1, 5\}$ with $\xi = \frac{1}{8}$ results in a full rank $\mathcal{W}$, since $D(\frac{1}{8}) = 8$ with $(5-1) \bmod 8 \neq 0$, $(1+1) \bmod 8 \neq 0$, $(5+5) \bmod 8 \neq 0$, and $(1+5) \bmod 8 \neq 0$. On the other hand, selecting the value $\xi = \frac{1}{10}$ results in a rank-deficient $\mathcal{W}$: now $D(\frac{1}{10}) = 10$ and $(5+5) \bmod 10 = 0$, which violates the second part of (G.10).

One simple way of satisfying the full-rank constraint (G.10) is to limit the range of indices in $\mathcal{S}$ to a single period, say $0 \le k \le D(\xi) - 1$. With this special choice, the first part of (G.10) is automatically satisfied, while the second part can be replaced by the *non-complementarity* restriction

$$k + \ell \neq D(\xi) \quad \text{for every } k \in \mathcal{S}\,,\ \ell \in \mathcal{S}$$

Since typical power system (voltage and current) waveforms contain primarily odd harmonics, it makes sense to choose $\mathcal{S}$ as a subset (of size L) of the odd harmonic indices in the range $\left[0 \quad D(\xi)-1\right]$, subject to the non-complementarity constraint. In addition, the index set $\mathcal{S}$ should always include the fundamental harmonic. There is a limited number of such choices, which we call *primary*. For instance consider the case $M = 4 = 2L$, with $\xi = \frac{1}{8}$, so that $D(\xi) = 8$, and the collection of possible (odd) index choices is $\{1, 3, 5, 7\}$. The set $\mathcal{S} = \{k_1, k_2\}$ is constructed by selecting two of these, subject to the constraint $k_1 + k_2 \bmod 8 \neq 0$ or, equivalently, $k_1 + k_2 \neq 8$. Thus the only valid *primary* choices for $\mathcal{S}$ are $\{1, 3\}$ or $\{1, 5\}$, and their complementary duals are $\{3, 7\}$ and $\{5, 7\}$ (see Table G.1 in Sect. G.6).

The transfer function $\widehat{H}_k(s)$ associated with the waveform-to-phasor map (G.6) is obtained by setting $\Delta_i = i\Delta$ and applying a Laplace transform, resulting in (also, recall (3.26))

$$\begin{pmatrix} \widehat{H}_{k_1}(s) \\ \widehat{H}_{k_2}(s) \\ \vdots \\ \widehat{H}_{k_L}(s) \end{pmatrix} = \sqrt{\frac{2}{M}} \begin{bmatrix} \mathbf{I}_L & \mathbf{0}_L \end{bmatrix} \left(\mathcal{W}^H \mathcal{W}\right)^{-1} \mathcal{W}^H \begin{pmatrix} 1 \\ e^{-s\Delta} \\ \vdots \\ e^{-s(M-1)\Delta} \end{pmatrix}$$

Notice that each transfer function $\widehat{H}_k(s)$ is now a *polynomial of degree* $M - 1$ in the complex variable $z = e^{-s\Delta}$, with $2L - 1$ of its zeros explicitly specified by the harmonics-blocking theorem.

In the special case when $M = 2L$, these observations make it possible to obtain a simplified expression for the frequency response $\widehat{H}_k(j\omega)$ for every $k \in \mathcal{S}$ [78,79], viz.,

$$\begin{aligned} \widehat{H}_k(j\omega) = {} & \sqrt{2}\, e^{-j(\frac{\omega}{\omega_0}-k)(M-1)\pi\xi} \, \frac{\sin(\frac{\omega}{\omega_0}+k)\pi\xi}{\sin(2\pi k\,\xi)} \\ & \times \prod_{\ell\in\mathcal{S},\ \ell\neq k} \frac{\sin(\frac{\omega}{\omega_0}+\ell)\pi\xi\ \sin(\frac{\omega}{\omega_0}-\ell)\pi\xi}{\sin(k+\ell)\pi\xi\ \sin(k-\ell)\pi\xi} \end{aligned} \tag{G.11}$$

where $\xi \stackrel{\text{def}}{=} \frac{\Delta}{T} > 0$ is a normalized inter-sample shift parameter. Notice that the denominator terms of $\widehat{H}_k(j\omega)$ can never vanish when $\mathcal{W}$ has full column rank. For instance, the term $\sin(k+\ell)\pi\xi$ can vanish only if $\xi = \frac{r}{k+\ell}$ for some integer r, which means that ξ must be rational. In this case, $D(\xi) = \frac{k+\ell}{m}$, where $m = \gcd(k+\ell, r)$ or, equivalently, $k+\ell = m\,D(\xi)$, which violates the second part of the full-rank condition (G.10). Similar analysis reveals that the remaining denominator terms, namely $\sin(2\pi k\,\xi)$ and $\sin(k-\ell)\pi\xi$, also do not vanish when $\mathcal{W}$ has full column rank.

The explicit expression (G.11) implies that, in the uniformly-shifted case, the group-delay associated with $\widehat{H}_k(j\omega)$ is frequency-independent and proportional to the inter-sample shift Δ, namely,

$$\text{group-delay} \stackrel{\text{def}}{=} -\frac{d}{d\omega}\left\{\arg\,\widehat{H}_k(j\omega)\right\} = \Delta\left(\frac{M-1}{2}\right) \tag{G.12}$$

This means that reduction of group-delay can be achieved by reducing either the inter-sample shift Δ, or the number of waveform samples (i.e., M) used in the waveform-to-phasor map.

G.3 Column-Orthogonality of $\mathcal{W}$

The $M \times 2L$ matrix $\mathcal{W}$ is called *column-orthogonal* when $\mathcal{W}^H\mathcal{W} = \mathbf{I}_{2L}$. Under column-orthogonality, the waveform-to-phasor map (G.3) reduces to

$$\mathbf{H}_{opt.}(t) = \mathbf{\Psi}^H\,\boldsymbol{\chi}(t) \tag{G.13a}$$

which provides a simple explicit expression for the sub-cycle phasors $\widehat{X}_k(t)$, viz.,

$$\widehat{X}_k(t)\,e^{jk\omega_0 t} = \sqrt{\frac{2}{M}}\;\psi_k^H \begin{pmatrix} x(t) \\ x(t-\Delta_1) \\ \vdots \\ x(t-\Delta_{M-1}) \end{pmatrix} \tag{G.13b}$$

The following theorem[3] provides a complete characterization of column-orthogonality in terms of the value of ξ and the content of the harmonic index set $\mathcal{S}$.

Theorem G.3 (Column-Orthogonality of the Matrix $\mathcal{W}$): *A* necessary condition *for mutual orthogonality of the columns of the matrix* $\mathcal{W}$, *subject to the fractionality constraint (3.23c), is that either* $\xi = \frac{1}{M}$ *or* $\xi = \frac{1}{2M}$. *In addition, when* $\mathcal{W}$ *has full column rank:*

[3] A more restricted version of this theorem was presented in [78] for the special case $M = 2L$.

- *The full-cycle choice* $\xi = \frac{1}{M}$ *results in mutual orthogonality of all the columns of* $\mathcal{W}$, *regardless of the content of the harmonic index set* $\mathcal{S}$.
- *The half-cycle choice* $\xi = \frac{1}{2M}$ *ensures column-orthogonality of* $\mathcal{W}$ *when the index set* $\mathcal{S}$ *consists entirely of odd harmonics.*

These serve as a ξ*-dependent* sufficient condition *(when* $\xi \le \frac{1}{M}$*) for the column-orthogonality of* $\mathcal{W}$.

Corollary: *Notice that according to this theorem the columns of* $\mathcal{W}$ *can never be mutually orthogonal when* $0 < \xi < \frac{1}{2M}$ *or* $\frac{1}{2M} < \xi < \frac{1}{M}$.

Proof: Since the columns of the matrix $\mathcal{W}$ always include ψ_1 and ψ_1^*, a necessary condition for column-orthogonality of $\mathcal{W}$ is that $\psi_1 \perp \psi_1^*$. Thus, since $\alpha \neq 1$ due to the fractionality constraint $0 < \xi \le \frac{1}{M}$, it follows that

$$[\psi_1^*]^H \psi_1 \equiv \psi_1^\top \psi_1 = \frac{1}{M}\sum_{i=0}^{M-1} \alpha^{2i} = \frac{1}{M}\cdot\frac{1-\alpha^{2M}}{1-\alpha^2}$$

which means that $\psi_1 \perp \psi_1^*$ if, and only if, $\alpha^{2M} = 1 = e^{-j4\pi\xi M}$. This can happen only if $2\xi M$ is an integer or, equivalently, if ξ is an integral multiple of $\frac{1}{2M}$. There are only two such values in the range $0 < \xi \le \frac{1}{M}$, namely $\xi = \frac{1}{2M}$ and $\xi = \frac{1}{M}$. This completes the proof of the necessary condition.

In order to establish the first ξ-dependent sufficient condition (i.e., when $\xi = \frac{1}{M}$), we need to prove that all the columns of $\mathcal{W} = [\boldsymbol{\Psi} \quad \boldsymbol{\Psi}^*]$ are mutually orthogonal, namely,

$$\psi_\ell^H \psi_k = \delta_{k,\ell} \quad \text{and} \quad [\psi_\ell^*]^H \psi_k = 0$$

for every $k \in \mathcal{S}$ and $\ell \in \mathcal{S}$. Using the expressions (3.24a,b), we find that

$$\psi_\ell^H \psi_k = \frac{1}{M}\sum_{i=0}^{M-1}\left[\alpha^k(\alpha^*)^\ell\right]^i = \frac{1}{M}\sum_{i=0}^{M-1}\left[e^{-j(k-\ell)2\pi\xi}\right]^i$$

$$(\psi_\ell^*)^H \psi_k = \frac{1}{M}\sum_{i=0}^{M-1}\left[\alpha^k\alpha^\ell\right]^i = \frac{1}{M}\sum_{i=0}^{M-1}\left[e^{-j(k+\ell)2\pi\xi}\right]^i$$

Both of these expressions involve summation of a finite geometric series. We recall that, in general,

$$\frac{1}{M}\sum_{i=0}^{M-1}\Gamma^i = \begin{cases} 1 & \Gamma = 1 \\ \dfrac{1-\Gamma^M}{M(1-\Gamma)} & \Gamma \neq 1 \end{cases}$$

which implies that this sum vanishes if (and only if) *both* $\Gamma \neq 1$ *and* $\Gamma^M = 1$.

Applying this observation to the first ξ-dependent sufficient condition (i.e., $\xi = \frac{1}{M}$) we find that $D(\xi) = M$, and, by virtue of the full-rank property (3.25):

$$(\bigstar)\quad e^{-j(k-\ell)2\pi\xi} = e^{-j(\frac{k-\ell}{M})2\pi} \neq 1 \text{ because } (k-\ell) \bmod M \neq 0 \text{ (for } k \neq \ell)$$
$$(\bigstar)\quad e^{-j(k+\ell)2\pi\xi} = e^{-j(\frac{k+\ell}{M})2\pi} \neq 1, \text{ because } (k+\ell) \bmod M \neq 0 \text{ (for all } k,\ell).$$

In addition,

$$\left(e^{-j(\frac{k\pm\ell}{M})2\pi}\right)^M = e^{-j(k\pm\ell)2\pi} = \left(e^{-j2\pi}\right)^{k\pm\ell} = 1$$

regardless of the values of k, ℓ. This completes the proof that $\psi_\ell^H \psi_k = \delta_{k,\ell}$ and $[\psi_\ell^*]^H \psi_k = 0$ when $\xi = \frac{1}{M}$, for all $k \in \mathcal{S}$ and $\ell \in \mathcal{S}$.

Next, in order to establish the second ξ-dependent sufficient condition (i.e., when $\xi = \frac{1}{2M}$), we need to prove that, for every $k \in \mathcal{S}$ and $\ell \in \mathcal{S}$,

$$\psi_\ell^H \psi_k = \delta_{k,\ell} \quad \text{and} \quad \left[\psi_\ell^*\right]^H \psi_k = 0$$

when $\mathcal{S}$ consists *entirely of odd harmonics*. In this (half-cycle) case, $D(\xi) = 2M$, and thus, by virtue of the full-rank property (3.25):

$$(\bigstar)\quad e^{-j(k-\ell)2\pi\xi} = e^{-j(\frac{k-\ell}{2M})2\pi} \neq 1 \text{ because } (k-\ell) \bmod 2M \neq 0 \text{ (for } k \neq \ell).$$
$$(\bigstar)\quad e^{-j(k+\ell)2\pi\xi} = e^{-j(\frac{k+\ell}{2M})2\pi} \neq 1 \text{ because } (k+\ell) \bmod 2M \neq 0 \text{ (for all } k,\ell).$$

In addition,

$$\left(e^{-j(\frac{k\pm\ell}{2M})2\pi}\right)^M = e^{-j(k\pm\ell)\pi} = \left(e^{-j\pi}\right)^{k\pm\ell} = \left(-1\right)^{k\pm\ell} = 1$$

because both k and ℓ are odd. This completes the proof that $\psi_\ell^H \psi_k = \delta_{k,\ell}$ and $[\psi_\ell^*]^H \psi_k = 0$ when $\xi = \frac{1}{2M}$, for all $k \in \mathcal{S}$ and $\ell \in \mathcal{S}$, subject to the added constraint that $\mathcal{S}$ consists entirely of odd harmonics. ■

G.4 Orthogonality of $\mathcal{W}$ and the Parseval Property

The sub-cycle scheme of phasor estimation, as described in Sect. 3.3, involves the added constraint $M = 2L$, so that the $M \times 2L$ matrix $\mathcal{W}$ becomes square. In this case column-orthogonality induces row orthogonality, namely

$$\mathcal{W}^H \mathcal{W} = \mathbf{I}_M = \mathcal{W}\mathcal{W}^H$$

which makes it possible to establish a *Parseval-type relation* between the time-domain samples of voltage and current waveforms and the associated sub-cycle phasor arrays $\widehat{\mathbf{V}}(t)$ and $\widehat{\mathbf{I}}(t)$, which are defined in (9.16a).

Theorem G.4 (Sub-cycle Parseval Relation): *When* $M = 2L$ *and the square matrix* $\mathcal{W}$ *is orthogonal, the estimated phasors associated with the polyphase waveforms* $x(t)$ *and* $y(t)$ *satisfy the Parseval-type identity*

$$\sum_{k\in\mathcal{S}} \Re\left\{\widehat{X}_k(t)\,\widehat{Y}_k^H(t)\right\} \;=\; \frac{1}{M}\sum_{i=0}^{M-1} x(t-\Delta_i)\,y^{\top}(t-\Delta_i) \tag{G.14}$$

Proof: Our first step toward establishing this relation is to observe that orthogonality of $\mathcal{W}$ implies the identities (recall that $M = 2L$)

$$\begin{bmatrix}\boldsymbol{\Psi} & \boldsymbol{\Psi}^*\end{bmatrix}^H \begin{bmatrix}\boldsymbol{\Psi} & \boldsymbol{\Psi}^*\end{bmatrix} \;=\; \mathbf{I}_{2L} \;=\; \begin{bmatrix}\boldsymbol{\Psi} & \boldsymbol{\Psi}^*\end{bmatrix}\begin{bmatrix}\boldsymbol{\Psi} & \boldsymbol{\Psi}^*\end{bmatrix}^H$$

which, in turn, results in two important properties of the $M \times L$ matrix $\boldsymbol{\Psi}$. The first identity implies that

$$\boldsymbol{\Psi}^H\boldsymbol{\Psi} \;=\; \mathbf{I}_L \qquad\Longrightarrow\qquad \psi_k^H\psi_\ell \;=\; \delta_{k,\ell} \tag{G.15a}$$

namely, the columns of $\boldsymbol{\Psi}$ are mutually orthonormal. The second identity implies that $\boldsymbol{\Psi}\boldsymbol{\Psi}^H + (\boldsymbol{\Psi}\boldsymbol{\Psi}^H)^* = \mathbf{I}_M$ or, equivalently,

$$\Re\left\{\boldsymbol{\Psi}\,\boldsymbol{\Psi}^H\right\} \;=\; \sum_{k\in\mathcal{S}}\Re\left\{\psi_k\,\psi_k^H\right\} \;=\; \tfrac{1}{2}\,\mathbf{I}_M \tag{G.15b}$$

Combining this identity with the waveform-to-phasor map (G.13b) results in a Parseval-type relation between the set of sub-cycle phasors $\{\widehat{X}_k(t)\,;\; k\in\mathcal{S}\}$ and the set of waveform samples $\{x\,(t - i\,\frac{T}{2M})\,;\; 0 \le i \le M-1\}$. First, notice that (G.13b) induces the bilinear relation (recall that $\Delta_i = i\,\frac{T}{2M}$)

$$\sum_{k\in\mathcal{S}}\widehat{X}_k(t)\,\widehat{Y}_k^H(t) = \frac{1}{L}\sum_{k\in\mathcal{S}}\left[\psi_k^H\begin{pmatrix} x(t)\\ x(t-\Delta_1)\\ \vdots\\ x(t-\Delta_{M-1})\end{pmatrix}\right]\left[\psi_k^H\begin{pmatrix} y(t)\\ y(t-\Delta_1)\\ \vdots\\ y(t-\Delta_{M-1})\end{pmatrix}\right]^H$$

$$= \frac{1}{L}\,\mathrm{tr}\left\{\left[\sum_{k\in\mathcal{S}}\psi_k\,\psi_k^H\right]\begin{pmatrix} x(t)\\ x(t-\Delta_1)\\ \vdots\\ x(t-\Delta_{M-1})\end{pmatrix}\begin{pmatrix} y(t)\\ y(t-\Delta_1)\\ \vdots\\ y(t-\Delta_{M-1})\end{pmatrix}^{\top}\right\}$$

where $\{\widehat{Y}_k(t)\,;\; k \in \boldsymbol{\mathcal{S}}\}$ is the set of sub-cycle phasors associated with the waveform $y(t)$, and tr{ } denotes the trace of a square matrix. Next, using the fact that $\sum_{k\in\boldsymbol{\mathcal{S}}} \Re\{\psi_k\, \psi_k^H\} = \frac{1}{2}\,\mathbf{I}_M$, we get

$$\sum_{k\in\boldsymbol{\mathcal{S}}} \Re\left\{\widehat{X}_k(t)\, \widehat{Y}_k^H(t)\right\} = \frac{1}{2L}\,\mathrm{tr}\left\{\begin{pmatrix} x(t) \\ x(t-\Delta_1) \\ \vdots \\ x(t-\Delta_{M-1}) \end{pmatrix}\begin{pmatrix} y(t) \\ y(t-\Delta_1) \\ \vdots \\ y(t-\Delta_{M-1}) \end{pmatrix}^{\top}\right\}$$

Finally, since $M = 2L$, we obtain the sub-cycle Parseval relation

$$\sum_{k\in\boldsymbol{\mathcal{S}}} \Re\left\{\widehat{X}_k(t)\, \widehat{Y}_k^H(t)\right\} = \frac{1}{M}\,\mathrm{tr}\left\{\boldsymbol{\mathcal{X}}(t)\, \boldsymbol{\mathcal{Y}}^{\top}(t)\right\} \equiv \frac{1}{M}\sum_{i=0}^{M-1} x(t-\Delta_i)\, y^{\top}(t-\Delta_i)$$

where $\Delta_i = i\,\frac{T}{2M}$, and

$$\boldsymbol{\mathcal{X}}(t) = \begin{pmatrix} x(t) \\ x(t-\Delta_1) \\ \vdots \\ x(t-\Delta_{M-1}) \end{pmatrix}, \quad \boldsymbol{\mathcal{Y}}(t) = \begin{pmatrix} y(t) \\ y(t-\Delta_1) \\ \vdots \\ y(t-\Delta_{M-1}) \end{pmatrix} \tag{G.16}$$

This completes the proof of the sub-cycle Parseval relation. ■

This relation is the sub-cycle equivalent of the (full-cycle) Parseval relation (1.9c). It establishes the equality of sub-cycle averaging (= discrete-time-domain inner-product) and phasor array multiplication (= frequency-domain inner-product). In particular, we have the relations

$$P(t) \stackrel{\text{def}}{=} \Re\left\{\mathbf{V}(t)\, \mathbf{I}^H(t)\right\} = \frac{1}{M}\sum_{i=0}^{M-1} p(t-\Delta_i)\;, \quad \Delta_i = i\,\frac{T}{2M} \tag{G.17a}$$

$$V_{rms}^2(t) \stackrel{\text{def}}{=} \mathbf{V}(t)\, \mathbf{V}^H(t) = \frac{1}{M}\sum_{i=0}^{M-1} v(t-\Delta_i)\, v^{\top}(t-\Delta_i) \tag{G.17b}$$

$$I_{rms}^2(t) \stackrel{\text{def}}{=} \mathbf{V}(t)\, \mathbf{I}^H(t) = \frac{1}{M}\sum_{i=0}^{M-1} i(t-\Delta_i)\, i^{\top}(t-\Delta_i) \tag{G.17c}$$

where $p(t) \stackrel{\text{def}}{=} v(t)\, i^{\top}(t)$ denotes the instantaneous power, and we use the $1 \times mL$ phasor arrays (with L polyphase phasors of length m each)

$$\mathbf{V}(t) \stackrel{\text{def}}{=} \left[\widehat{V}_{k_1}(t) \;\ldots\; \widehat{V}_{k_L}(t) \right] , \qquad \mathbf{I}(t) \stackrel{\text{def}}{=} \left[\widehat{I}_{k_1}(t) \;\ldots\; \widehat{I}_{k_L}(t) \right] \tag{G.17d}$$

To complete the analogy with full-cycle power metrics, we now define a sub-cycle counterpart of Budeanu's reactive power, namely,

$$Q(t) \stackrel{\text{def}}{=} \Im\{\mathbf{V}(t)\, \mathbf{I}^H(t)\} \equiv \sum_{k\in\mathcal{S}} \Im\left\{\widehat{V}_k(t)\, \widehat{I}_k^H(t)\right\} \tag{G.18}$$

It turns out that this phasor-based expression also has a time-domain equivalent, albeit a more complicated one, as shown in the following section.

G.5 The Sub-cycle Hilbert Transform

In the special case discussed in the preceding section, i.e., a uniformly-shifted samples scheme with $M = 2L$ and $\xi = \frac{1}{2M}$, the columns of $\boldsymbol{\Psi}$ are given by (recall (G.8b))

$$\psi_k = \frac{1}{\sqrt{M}} \left[1 \quad \alpha^k \quad \alpha^{2k} \quad \ldots \quad \alpha^{(\frac{M}{2}-1)k} \;\middle|\; \alpha^{(\frac{M}{2})k} \quad \ldots \quad \alpha^{(M-1)k} \right]^{\top},$$
$$\alpha = e^{-j\pi/M}$$

We notice that the bottom half of ψ_k is a scaled version of the top half, viz.,

$$\psi_k = \begin{pmatrix} a_k \\ (-j)^k\, a_k \end{pmatrix} , \qquad a_k \stackrel{\text{def}}{=} \frac{1}{\sqrt{M}} \left[1 \quad \alpha^k \quad \alpha^{2k} \quad \ldots \quad \alpha^{(\frac{M}{2}-1)k} \right]^{\top}$$

because $\alpha^{(\frac{M}{2})k} = e^{-jk\pi/2} = (-j)^k$. Consequently,

$$\begin{pmatrix} \mathbf{0}_L & (-1)^k \mathbf{I}_L \\ \mathbf{I}_L & \mathbf{0}_L \end{pmatrix} \psi_k = \begin{pmatrix} j^k a_k \\ a_k \end{pmatrix} = j^k\, \psi_k$$

Since we consider only odd harmonics as members of the index set $\mathcal{S}$, this equation reduces to

$$\begin{pmatrix} \mathbf{0}_L & -\mathbf{I}_L \\ \mathbf{I}_L & \mathbf{0}_L \end{pmatrix} \psi_k = j^k\, \psi_k \ , \qquad k \text{ is odd} \tag{G.19}$$

Combining this result with (G.13b), we conclude that

$$(-j)^k\ \widehat{X}_k(t)\ e^{jk\omega_0 t} \;=\; \frac{1}{\sqrt{L}}\left(j^k\ \psi_k\right)^H\ \mathcal{X}(t) \;=\; \frac{1}{\sqrt{L}}\ \psi_k^H \begin{pmatrix} \mathbf{0}_L & \mathbf{I}_L \\ -\mathbf{I}_L & \mathbf{0}_L \end{pmatrix} \mathcal{X}(t)$$

where $\mathcal{X}(t)$ is as defined in (G.16), with $\Delta = \frac{T}{2M}$. Thus, the k-th harmonic phasor associated with the modified set of input samples

$$\mathcal{H}\left\{\mathcal{X}(t)\right\} \;\stackrel{\text{def}}{=}\; \begin{pmatrix} \mathbf{0}_L & \mathbf{I}_L \\ -\mathbf{I}_L & \mathbf{0}_L \end{pmatrix} \mathcal{X}(t) \tag{G.20}$$

is simply $(-j)^k\widehat{X}_k(t)$. For instance, when $k = 1$, the fundamental harmonic phasor corresponding to $\mathcal{H}\left\{\mathcal{X}(t)\right\}$ is $-j\,\widehat{X}_1(t)$. This observation motivates us to denote $\mathcal{H}\left\{\mathcal{X}(t)\right\}$ as the *sub-cycle Hilbert transform* of $x(\cdot)$.

Notice that since we consider only odd harmonics, the phasor associated with $\mathcal{H}\left\{\mathcal{X}(t)\right\}$ is either $-j\widehat{X}_k(t)$ (when $k = 4\ell + 1$) or $j\widehat{X}_k(t)$ (when $k = 4\ell + 3$). Thus, in the *special case* when every $k \in \mathcal{S}$ is of the form $k = 4\ell + 1$, the sub-cycle reactive power $Q(t)$ can be expressed in the time-domain using the sub-cycle Hilbert transform $\mathcal{H}\{\ \}$, viz.,

$$Q(t) \;=\; \frac{1}{M}\ \text{tr}\left\{\mathcal{H}\left\{\mathcal{V}(t)\right\}\mathcal{I}^\top(t)\right\} \tag{G.21}$$

This is so because, according to the sub-cycle Parseval relation (G.14),

$$\frac{1}{M}\ \text{tr}\left\{\mathcal{H}\left\{\mathcal{V}(t)\right\}\mathcal{I}^\top(t)\right\} \;=\; \sum_{k\in\mathcal{S}} \Re\left\{\left[-j\,\widehat{V}_k(t)\right]\widehat{I}_k^H(t)\right\} \;=\; \sum_{k\in\mathcal{S}} \Im\left\{\widehat{V}_k(t)\,\widehat{I}_k^H(t)\right\}$$

which matches the definition (G.18) of the sub-cycle reactive power $Q(t)$. In this special case, the effect of the time-domain transformation (G.20) is precisely the same as that of the classical Hilbert transform: it imparts a $-90°$ phase-shift to all the sub-cycle phasors derived from $\mathcal{H}\left\{\mathcal{V}(t)\right\}$.

However, when $\mathcal{S}$ contains harmonics of both types (i.e., $k = 4\ell + 1$ and $k = 4\ell + 3$), the expression (G.21) no longer holds, so that $Q(t)$ does not admit a simple time-domain-averaging representation.

G.6 Valid $\mathcal{S}$-Choices for Orthogonal Sub-cycle Schemes

We examine here in some detail the $\mathcal{S}$-choices available in orthogonal sub-cycle schemes, which means we limit our discussion to the case where $M = 2L$ and $\xi = \frac{1}{2M}$. This means that the corresponding ψ_k-vectors are periodic in k with period $P = 2M$, which limits our choices to the range $0 \le k \le 2M - 1$. We provide below a summary of the valid $\mathcal{S}$-choices for $L = 2, 3, 4, 5$ (see Tables G.1 and G.2). Recall that $\mathcal{S}$-choices are designated as "primary" when they contain the fundamental harmonic (i.e., the index 1).

For instance, when $L = 3$, we have $2M \equiv 4L = 12$, and the collection of possible (odd) index choices is $\{1, 3, 5, 7, 9, 11\}$. The set $\mathcal{S} = \{k_1, k_2, k_3\}$ is constructed by selecting three of these, subject to the constraints $k_1 + k_2 \neq 12$, $k_1 + k_3 \neq 12$, and $k_2 + k_3 \neq 12$. The valid primary and dual choices are shown in Table G.1. Notice that for each one of these choices, the columns of the square (6×6) matrix $\mathcal{W} = \begin{bmatrix} \mathbf{\Psi} & \mathbf{\Psi}^* \end{bmatrix}$ are simply a permutation of the column set $\{\psi_1\,, \psi_3\,, \psi_5\,, \psi_7\,, \psi_9\,, \psi_{11}\}$.

The valid primary and dual choices for $L = 4$ $(2M \equiv 4L = 16)$ are also shown in Table G.1. All previously observed properties still hold in this case. In addition, we notice several choices that contain many of the harmonics (odd and indivisible by 3) typically found in real-life power system waveforms, which for $0 \leq k \leq 15$ are $\{1, 5, 7, 11, 13\}$. In particular, each one of the two preferred primary choices (shown in boldface) contains all but one of these typical harmonics.

Finally, the valid primary and dual choices for $L = 5$ $(2M \equiv 4L = 20)$ are shown in Table G.2. Again, all previously observed properties hold in this case as well. The harmonics typically found in real-life power system waveforms, in the range $0 \leq k \leq 19$, are $\{1, 5, 7, 11, 13, 17, 19\}$. Again, we observe two preferred primary choices (shown in boldface) that contain all but two of these typical harmonics.

In summary, we observe that in the general case (i.e., arbitrary L) there are 2^L valid choices for $\mathcal{S}$, half of which are primary.

Table G.1 Valid $\mathcal{S}$-choices for $L = 2, 3$ and 4

L	Primary choice	Dual choice
2	{1,3}	{5,7}
	{1,5}	{3,7}
3	{1,3,5}	{7,9,11}
	{1,3,7}	{5,9,11}
	{1,5,9}	{3,7,11}
	{1,7,9}	{3,5,11}
4	{1,3,5,7}	{9,11,13,15}
	{1,3,5,9}	{7,11,13,15}
	{1,3,7,11}	{5,9,13,15}
	{1,3,9,11}	{5,7,13,15}
	{1,5,7,13}	{3,9,11,15}
	{1,5,9,13}	{3,7,11,15}
	{1,7,11,13}	{3,5,9,15}
	{1,9,11,13}	{3,5,7,15}

Table G.2 Valid $\mathcal{S}$-choices for $L = 5$

L	Primary choice	Dual choice
5	{1,3,5,7,9}	{11,13,15,17,19}
	{1,3,5,7,11}	{9,13,15,17,19}
	{1,3,5,9,13}	{7,11,15,17,19}
	{1,3,5,11,13}	{7,9,15,17,19}
	{1,3,7,9,15}	{5,11,13,17,19}
	{1,3,7,11,15}	{5,9,13,17,19}
	{1,3,9,13,15}	{5,7,11,17,19}
	{1,3,11,13,15}	{5,7,9,17,19}
	{1,5,7,9,17}	{3,11,13,15,19}
	{1,5,7,11,17}	{3,9,13,15,19}
	{1,5,9,13,17}	{3,7,11,15,19}
	{1,5,11,13,17}	{3,7,9,15,19}
	{1,7,9,15,17}	{3,5,11,13,19}
	{1,7,11,15,17}	{3,5,9,13,19}
	{1,9,13,15,17}	{3,5,7,11,19}
	{1,11,13,15,17}	{3,5,7,9,19}

Appendix H: Cycle Averaging of Instantaneous Metrics

Purely instantaneous power metrics in an m-phase system satisfy the key relation (recall (9.7))

$$\left[v(t)\,v^{\top}(t)\right]\left[i(t)\,i^{\top}(t)\right] \;=\; p^2(t) \,+\, \tfrac{1}{2}\,\|\,\boldsymbol{Q}(t)\,\|_F^2$$

where $p(t) \stackrel{\text{def}}{=} v(t)\,i^{\top}(t)$ and $\boldsymbol{Q}(t) \stackrel{\text{def}}{=} i^{\top}(t)v(t) - v^{\top}(t)\,i(t)$. We have introduced in Sect. 9.1 the dynamic full-cycle metrics $V_{rms}^2(t)$, $I_{rms}^2(t)$, and $P(t)$, which are obtained by cycle averaging over the interval $(t - T \quad t]$ of the instantaneous metrics $v(\cdot)v^{\top}(\cdot)$, $i(\cdot)i^{\top}(\cdot)$, and $p(\cdot)$, respectively. In general, we define cycle averaging of a polyphase waveform $x(\cdot)$ as a special case of the generalized averaging operation (2.2c), viz.,

$$\langle\, x\,\rangle_0\,(t) \;\stackrel{\text{def}}{=}\; \frac{1}{T}\int_{t-T}^{t} x(s)\,ds \tag{H.1a}$$

We now turn to study the properties of the elements $Q_{r\ell}(t)$ obtained by cycle averaging of the strictly upper-diagonal elements of the matrix $\boldsymbol{Q}(\cdot)$, viz.,

$$Q_{r\ell}(t) \;\stackrel{\text{def}}{=}\; \langle\, q_{r\ell}\,\rangle_0\,(t) \;\equiv\; \frac{1}{T}\int_{t-T}^{t} q_{r\ell}(s)\,ds \qquad \left(r < \ell\right) \tag{H.1b}$$

where $1 \le r \le m$ and $1 \le \ell \le m$.

H.1 Coordinate-Transformed Q-Matrix

A *coordinate transformation,* such as the Clarke transform, replaces the polyphase $v(t)$ and $i(t)$ waveforms by $v(t)\mathcal{M}$ and $i(t)\mathcal{M}$, respectively, where $\mathcal{M}$ could be any real-valued orthogonal matrix. The resulting $\boldsymbol{Q}$-matrix is

$$\boldsymbol{Q}^{\mathcal{M}}(t) \stackrel{\text{def}}{=} \Big(i(t)\mathcal{M}\Big)^{\top}\Big(v(t)\mathcal{M}\Big) - \Big(v(t)\mathcal{M}\Big)^{\top}\Big(i(t)\mathcal{M}\Big) = \mathcal{M}^{\top}\boldsymbol{Q}(t)\,\mathcal{M}$$

Notice that this coordinate transformation does not affect the Frobenius norm of the $\boldsymbol{Q}$-matrix. Indeed, using the fact that $\| A \|_F^2 \equiv \operatorname{tr}\{AA^{\top}\}$ for any real-valued square matrix A, we obtain

$$\begin{aligned}
\| \boldsymbol{Q}^{\mathcal{M}}(t) \|_F^2 &= \operatorname{tr}\Big\{\Big[\mathcal{M}^{\top}\boldsymbol{Q}(t)\,\mathcal{M}\Big]\Big[\mathcal{M}^{\top}\boldsymbol{Q}(t)\,\mathcal{M}\Big]^{\top}\Big\} \\
&= \operatorname{tr}\Big\{\boldsymbol{Q}(t)\,\mathcal{M}\mathcal{M}^{\top}\boldsymbol{Q}^{\top}(t)\,\mathcal{M}\mathcal{M}^{\top}\Big\} \\
&= \operatorname{tr}\Big\{\boldsymbol{Q}(t)\,\boldsymbol{Q}^{\top}(t)\Big\} \\
&= \| \boldsymbol{Q}(t) \|_F^2
\end{aligned}$$

where we used the orthogonality property $\mathcal{M}\mathcal{M}^{\top} = I$. In other words, while each element $q_{r\ell}^{\mathcal{M}}(t)$ of the transformed matrix $\boldsymbol{Q}^{\mathcal{M}}(t)$ is dependent on the choice of the transformation matrix $\mathcal{M}$, the sum $\sum_{r<\ell}\big[q_{r\ell}^{\mathcal{M}}(t)\big]^2$ is independent of $\mathcal{M}$.

The elements of $\boldsymbol{Q}^{\mathcal{M}}(t)$ can be expressed directly in terms of the original (instantaneous) voltage and current waveforms. First, notice that

$$q_{r\ell}^{\mathcal{M}}(t) = e_r^{\top}\boldsymbol{Q}^{\mathcal{M}}(t)\,e_\ell = e_r^{\top}\mathcal{M}^{\top}\boldsymbol{Q}(t)\,\mathcal{M}\,e_\ell = \mu_r^{\top}\boldsymbol{Q}(t)\,\mu_\ell$$

where $\mu_r \stackrel{\text{def}}{=} \mathcal{M}\,e_r$, and

$$e_r \stackrel{\text{def}}{=} [\underbrace{0\ \ldots\ 0\ 1}_{r}\ 0\ \ldots 0]^{\top} \qquad\qquad (1 \le r \le m)$$

is the r-th $(m \times 1)$ unit vector. Next, using the definition of $\boldsymbol{Q}(t)$, we find that

$$\begin{aligned}
\mu_r^{\top}\boldsymbol{Q}(t)\,\mu_\ell &= \Big[\mu_r^{\top}i^{\top}(t)\Big]\Big[v(t)\mu_\ell\Big] - \Big[\mu_r^{\top}v^{\top}(t)\Big]\Big[i(t)\mu_\ell\Big] \\
&= \Big[v(t)\mu_\ell\Big]\Big[\mu_r^{\top}i^{\top}(t)\Big] - \Big[v(t)\mu_r\Big]\Big[\mu_\ell^{\top}i^{\top}(t)\Big] \\
&= v(t)\Big[\mu_\ell\,\mu_r^{\top} - \mu_r\,\mu_\ell^{\top}\Big]i^{\top}(t)
\end{aligned}$$

where we used the facts that (i) each expression in square brackets is a scalar, so they all commute, and (ii) each one of these expressions is a product between a row vector "x" and a column vector "y" and thus can be written either as $x\,y$ or as $y^\top x^\top$. In summary, we obtain an explicit expression for $q_{r\ell}^{\mathcal{M}}(t)$, viz.,

$$q_{r\ell}^{\mathcal{M}}(t) \;=\; v(t)\,\Omega_{r\ell}\,i^\top(t) \;, \qquad \text{where} \qquad \Omega_{r\ell} \stackrel{\text{def}}{=} \mu_\ell\,\mu_r^\top - \mu_r\,\mu_\ell^\top \tag{H.2}$$

In particular, when $\mathcal{M}$ is the Clarke transform matrix, i.e.,

$$\mathcal{M} \;=\; \frac{1}{\sqrt{3}} \begin{pmatrix} \sqrt{2} & 0 & 1 \\ -\frac{\sqrt{2}}{2} & \sqrt{\frac{3}{2}} & 1 \\ -\frac{\sqrt{2}}{2} & -\sqrt{\frac{3}{2}} & 1 \end{pmatrix} \stackrel{\text{def}}{=} \mathcal{M}_{Clarke} \tag{H.3}$$

we have

$$\mu_1 \;=\; \frac{1}{\sqrt{6}} \begin{pmatrix} 2 \\ -1 \\ -1 \end{pmatrix} \;, \quad \mu_2 \;=\; \frac{1}{\sqrt{2}} \begin{pmatrix} 0 \\ 1 \\ -1 \end{pmatrix} \;, \quad \mu_3 \;=\; \frac{1}{\sqrt{3}} \begin{pmatrix} 1 \\ 1 \\ 1 \end{pmatrix}$$

and thus

$$\Omega_{12} \;=\; \mu_2\,\mu_1^\top - \mu_1\,\mu_2^\top \;=\; \frac{1}{\sqrt{3}} \begin{pmatrix} 0 & -1 & 1 \\ 1 & 0 & -1 \\ -1 & 1 & 0 \end{pmatrix}$$

$$\Omega_{13} \;=\; \mu_3\,\mu_1^\top - \mu_1\,\mu_3^\top \;=\; \frac{1}{\sqrt{2}} \begin{pmatrix} 0 & -1 & -1 \\ 1 & 0 & 0 \\ 1 & 0 & 0 \end{pmatrix}$$

$$\Omega_{23} \;=\; \mu_3\,\mu_2^\top - \mu_2\,\mu_3^\top \;=\; \frac{1}{\sqrt{6}} \begin{pmatrix} 0 & 1 & -1 \\ -1 & 0 & -2 \\ 1 & 2 & 0 \end{pmatrix}$$

H.2 Cycle-Averaged Q-Matrix

Applying the cycle-averaging operation (H.1b) to the expression (H.2), we obtain

$$\begin{aligned} Q_{r\ell}^{\mathcal{M}}(t) = \frac{1}{T}\int_{t-T}^{t} q_{r\ell}^{\mathcal{M}}(s)\,ds \;&=\; \frac{1}{T}\int_{t-T}^{t} v(s)\,\Omega_{r\ell}\,i^\top(s)\,ds \\ &\equiv\; \langle\, v(\cdot)\,\Omega_{r\ell}\,,\, i(\cdot)\,\rangle\,(t) \end{aligned}$$

where we used the definition (9.1) of a time-dependent inner-product. This cycle-averaged metric can be expressed, via the Parseval identity (9.3), directly in terms of the full-cycle dynamic voltage and current phasors $V_k(t)$, $I_k(t)$, viz.,

$$Q_{r\ell}^{\mathcal{M}}(t) = \sum_{k=0}^{\infty} \Re\left\{V_k(t)\, \Omega_{r\ell}\, I_k^H(t)\right\} \tag{H.4}$$

This should be compared with $P(t) = \sum_{k=0}^{\infty} \Re\left\{V_k(t)\, I_k^H(t)\right\}$, the well-known expression for the full-cycle dynamic real power (recall (9.4)).

In particular, in a 3-phase system, choosing $\mathcal{M}$ as the Clarke transform (H.3) leads to useful compact expressions for $\{Q_{r\ell}^{\mathcal{M}}(t)\,;\; 1 \le r \le 3, 1 \le \ell \le 3\}$ in terms of the dynamic symmetrical sequence components of voltage and current. First, recall that the full-cycle dynamic symmetrical sequence phasors are given by (see, e.g., (2.18a))

$$\begin{aligned} \left[\, V_{k,0}(t) \quad V_{k,+}(t) \quad V_{k,-}(t)\,\right] &\stackrel{\text{def}}{=} V_k^{(ss)}(t) = V_k(t)\, \mathcal{M}_{sc} \\ \left[\, I_{k,0}(t) \quad I_{k,+}(t) \quad I_{k,-}(t)\,\right] &\stackrel{\text{def}}{=} I_k^{(ss)}(t) = I_k(t)\, \mathcal{M}_{sc} \end{aligned} \tag{H.5a}$$

where

$$\mathcal{M}_{sc} = \frac{1}{\sqrt{3}} \begin{pmatrix} 1 & 1 & 1 \\ 1 & \alpha & \alpha^2 \\ 1 & \alpha^2 & \alpha \end{pmatrix} \quad , \qquad \alpha \stackrel{\text{def}}{=} e^{j2\pi/3} \tag{H.5b}$$

Combining these relations with (H.4), we obtain an explicit expression for the Clarke-transformed $Q_{r\ell}^{\mathcal{M}}(t)$ in terms of the dynamic symmetrical sequence component voltage and current phasors, viz.,

$$Q_{r\ell}^{\mathcal{M}}(t) = \sum_{k=0}^{\infty} \Re\left\{V_k^{(ss)}(t) \left(\mathcal{M}_{sc}^H\, \Omega_{r\ell}\, \mathcal{M}_{sc}\right) \left[I_k^{(ss)}(t)\right]^H\right\}$$

Next, we observe that

$$\mathcal{M}_{sc}^H\, \Omega_{12}\, \mathcal{M}_{sc} = j \begin{pmatrix} 0 & 0 & 0 \\ 0 & -1 & 0 \\ 0 & 0 & 1 \end{pmatrix}$$

$$\mathcal{M}_{sc}^H\, \Omega_{13}\, \mathcal{M}_{sc} = \frac{1}{\sqrt{2}} \begin{pmatrix} 0 & 1 & 1 \\ -1 & 0 & 0 \\ -1 & 0 & 0 \end{pmatrix}$$

$$\mathcal{M}_{sc}^H\, \Omega_{23}\, \mathcal{M}_{sc} = \frac{j}{\sqrt{2}} \begin{pmatrix} 0 & 1 & -1 \\ 1 & 0 & 0 \\ -1 & 0 & 0 \end{pmatrix}$$

which generates a dynamic version of a well-known result, viz.,

$$\begin{aligned} Q_{12}^{\mathcal{M}}(t) &= \Im\Big\{ \sum_k V_{k,+}(t)\, I_{k,+}^*(t) - \sum_k V_{k,-}(t)\, I_{k,-}^*(t) \Big\} \\ &= Q_{B,+}(t) - Q_{B,-}(t) \end{aligned} \tag{H.6a}$$

as well as the less-known expressions

$$\begin{aligned} Q_{13}^{\mathcal{M}}(t) &= \tfrac{1}{\sqrt{2}}\,\Re\Bigg\{ \sum_k V_{k,0}(t)\Big[I_{k,+}(t) + I_{k,-}(t)\Big]^* \\ &\qquad - \sum_k \Big[V_{k,+}(t) + V_{k,-}(t)\Big] I_{k,0}^*(t) \Bigg\} \\ Q_{23}^{\mathcal{M}}(t) &= \tfrac{1}{\sqrt{2}}\,\Im\Bigg\{ \sum_k V_{k,0}(t)\Big[I_{k,-}(t) - I_{k,+}(t)\Big]^* \\ &\qquad + \sum_k \Big[V_{k,-}(t) - V_{k,+}(t)\Big] I_{k,0}^*(t) \Bigg\} \end{aligned} \tag{H.6b}$$

Finally, when the zero-sequence voltage and current vanish, $q_{12}(t)$ is the only non-vanishing $q_{r\ell}(t)$ component, which we denoted by $q_{\alpha\beta}(t)$ (in Sect. 9.2). In this case

$$\langle\, q_{\alpha\beta}\,\rangle_0\,(t) \;=\; Q_{B,+}(t) - Q_{B,-}(t)$$

is the only non-vanishing cycle-averaged reactive power component.

H.3 Compensation-Oriented Cycle-Averaged Metrics

We have presented in Sect. 9.2.3 the concept of instantaneous compensation, which eliminates the instantaneous power quality gap $s^2(t) - p^2(t)$ by injecting a suitable shunt current $i_x(t)$. The dynamic rms value of this current, evaluated by cycle averaging, is

$$I_{x,rms}^2(t) \stackrel{\text{def}}{=} \frac{1}{T}\int_{t-T}^{t} i_x(s)\, i_x^{\top}(s)\, ds \;=\; \frac{1}{T}\int_{t-T}^{t} \sum_{r<\ell} \left(\frac{q_{r\ell}^2(s)}{v(s)\, v^{\top}(s)} \right) ds \tag{H.7}$$

where we used the explicit expression (9.14) for the product $i_x(s)\, i_x^{\top}(s)$.

Recall, also, that $i_a(s)\, i_x^{\top}(s) = 0$ for all s, so that averaging over the cycle-long interval $(t-T \quad t]$ leads to the conclusion that $\langle\, i_a\, i_x^{\top}\,\rangle_0\,(t) = 0$. Consequently,

$$i(s)\, i^{\top}(s) \;=\; i_a(s)\, i_a^{\top}(s) + i_x(s)\, i_x^{\top}(s)$$

and averaging over the same interval results in the Pythagorean identity (for all t)

$$I_{rms}^2(t) \;=\; I_{a,rms}^2(t) \,+\, I_{x,rms}^2(t) \tag{H.8a}$$

where $I_{x,rms}(t)$ is defined in (H.7), $I_{rms}(t)$ is the full-cycle dynamic rms current, as defined in Sect. 9.1, and, similarly,

$$I_{a,rms}^2(t) \;\stackrel{\text{def}}{=}\; \frac{1}{T}\int_{t-T}^{t} i_a(s)\, i_a^{\top}(s)\, ds \;=\; \frac{1}{T}\int_{t-T}^{t} \frac{p^2(s)}{v(s)\, v^{\top}(s)}\, ds \tag{H.8b}$$

Multiplying the Pythagorean identity (H.8a) by $V_{rms}^2(t)$ produces a 2-component decomposition of the full-cycle dynamic apparent power, viz.,

$$S^2(t) \;=\; S_a^2(t) \,+\, Q_x^2(t) \tag{H.9a}$$

where

$$S_a(t) \;\stackrel{\text{def}}{=}\; V_{rms}(t)\, I_{a,rms}(t) \;\;, \qquad Q_x(t) \;\stackrel{\text{def}}{=}\; V_{rms}(t)\, I_{x,rms}(t) \tag{H.9b}$$

Here $Q_x(t)$ is the part of the full-cycle dynamic apparent power $S(t)$ that can be compensated without memory (i.e., by using only instantaneous operations to reduce $\sum_{r<\ell} q_{r\ell}^2(t)$ to zero).

Notice that $S_a(t)$ is always larger than the dynamic (full-cycle) real power $P(t)$. This property can be established by observing that

$$\begin{aligned} P^2(t) \;&=\; \left[\frac{1}{T}\int_{t-T}^{t} p(s)\, ds\right]^2 = \left[\frac{1}{T}\int_{t-T}^{t} \left(\frac{p(s)}{\sqrt{v(s)\, v^{\top}(s)}}\right) \sqrt{v(s)\, v^{\top}(s)}\; ds\right]^2 \\ &\leq\; \left(\frac{1}{T}\int_{t-T}^{t} \frac{p^2(s)}{v(s)\, v^{\top}(s)}\, ds\right)\left(\frac{1}{T}\int_{t-T}^{t} v(s)\, v^{\top}(s)\, ds\right) \;=\; I_{a,rms}^2(t)\; V_{rms}^2(t) \end{aligned}$$

where we used the Cauchy–Schwarz inequality. Since $S_a(t) \stackrel{\text{def}}{=} I_{a,rms}(t)\, V_{rms}(t)$, it follows that $S_a^2(t) \geq P^2(t)$. Allowing memory and/or energy storage facilitates compensation of a portion of the power quality gap $S_a^2(t) - P^2(t)$.

In the special case of periodic waveforms, the 2-component decomposition (H.9a) can be refined by splitting the term S_a^2 into two separate metrics, viz.,

(continued)

$$S^2 = \underbrace{P^2 + Q_y^2}_{S_a^2} + Q_x^2$$

showing explicitly the real (time-invariant) power P, and the part Q_y that can always be eliminated by using energy-storing devices [29].

Combining the definition of $Q_x(t)$ with (H.7) results in a direct relation between the purely instantaneous reactive power metrics $q_{r\ell}(t)$ and the full-cycle metric $Q_x(t)$, namely,

$$Q_x^2(t) = V_{rms}^2(t) \left[\frac{1}{T} \int_{t-T}^{t} \sum_{r<\ell} \left(\frac{q_{r\ell}^2(s)}{v(s)\, v^\top(s)} \right) ds \right] \tag{H.10}$$

This expression makes it evident that elimination of the instantaneous power gap $\sum_{r<\ell} q_{r\ell}^2(t) \equiv s^2(t) - p^2(t)$ results in $Q_x(t) = 0$ and vice versa.

H.4 Bounds on Cycle-Averaged Power Metrics

Jensen's Inequality

The cycle-averaging operation (H.1a) satisfies an associated inequality, which is a special case of the well-known Jensen's inequality, viz.,

$$\left| \langle x \rangle_0 (t) \right|^2 \le \langle x^2 \rangle_0 (t) \tag{H.11}$$

In particular, this means that

$$V_{rms}^2(t) \le \langle v\, v^\top \rangle_0 (t) \quad , \qquad I_{rms}^2(t) \le \langle i\, i^\top \rangle_0 (t)$$

as well as

$$P^2(t) \le \langle p^2 \rangle_0 (t) \quad , \qquad Q_{r\ell}^2(t) \le \langle q_{r\ell}^2 \rangle_0 (t)$$

so that (recall (9.7))

$$\begin{aligned} \langle s^2 \rangle_0 (t) \equiv \langle [v\, v^\top][i\, i^\top] \rangle_0 (t) &= \langle p^2 \rangle_0 (t) + \sum_{r<\ell} \langle q_{r\ell}^2 \rangle_0 (t) \\ &\ge P^2(t) + \sum_{r<\ell} Q_{r\ell}^2(t) \end{aligned}$$

Notice that since the (full-cycle) dynamic apparent power is defined as $S(t) = V_{rms}(t)\, I_{rms}(t)$, the quantity $\langle\, s^2 \,\rangle_0\,(t)$ is not related to $S(t)$.

On the other hand, the cycle-averaged power quality gap $\langle\, s^2 - p^2 \,\rangle_0\,(t)$ is directly related to a compensation-oriented metric $Q_x(t)$, as we have demonstrated in Sect. H.3.

Compensation-Oriented Bounds

The close inter-dependence between $\sum_{r<\ell} q_{r\ell}^2(t)$ and $Q_x(t)$ (recall (H.10)) can also be expressed in the form of upper and lower bounds. For instance, using the Cauchy–Schwarz inequality, we find that

$$
\begin{aligned}
\frac{1}{T}\int_{t-T}^{t}\Big[\sum_{r<\ell} q_{r\ell}^2(s)\Big]^{1/2} ds &= \frac{1}{T}\int_{t-T}^{t}\Big[v(s)\,v^{\top}(s)\Big]^{1/2}\Big[\frac{\sum_{r<\ell} q_{r\ell}^2(s)}{v(s)\,v^{\top}(s)}\Big]^{1/2} ds \\
&\le \Big[\frac{1}{T}\int_{t-T}^{t} v(s)\,v^{\top}(s)\,ds\Big]^{1/2} \\
&\quad \times \Big[\frac{1}{T}\int_{t-T}^{t}\frac{\sum_{r<\ell} q_{r\ell}^2(s)}{v(s)\,v^{\top}(s)}\,ds\Big]^{1/2} \\
&= V_{rms}(t)\, I_{x,rms}(t)
\end{aligned}
$$

which results in the upper bound

$$
\Big\langle \Big[\textstyle\sum_{r<\ell} q_{r\ell}^2\Big]^{1/2} \Big\rangle_0 (t) \;\le\; Q_x(t) \tag{H.12a}
$$

Similarly, let

$$
\nu(t) \;\stackrel{\text{def}}{=}\; \frac{\sqrt{v(s)\,v^{\top}(s)}}{V_{rms}(t)} \quad , \qquad \zeta(t) \;\stackrel{\text{def}}{=}\; \frac{\sqrt{i_x(s)\,i_x^{\top}(s)}}{I_{x,rms}(t)}
$$

and observe that $\langle\, \nu^2 \,\rangle_0\,(t) = 1 = \langle\, \zeta^2 \,\rangle_0\,(t)$, so that

$$
\Big\langle (\nu^2 - 1)(\zeta^2 - 1) \Big\rangle_0 (t) \;=\; \langle\, \nu^2 \zeta^2 \,\rangle_0\,(t) - 1 \;=\; \frac{\langle \sum_{r<\ell} q_{r\ell}^2 \rangle_0(t)}{Q_x^2(t)} \;-\; 1
$$

where, again, we used the identity (9.14). Next, we use a Cauchy–Schwarz inequality to find an upper bound for this expression, viz.,

$$\left|\left\langle (\nu^2-1)(\zeta^2-1)\right\rangle_0(t)\right| \equiv \left|\frac{1}{T}\int_{t-T}^{t}\left(\nu^2(s)-1\right)\left(\zeta^2(s)-1\right)ds\right|$$
$$\leq \left[\frac{1}{T}\int_{t-T}^{t}\left(\nu^2(s)-1\right)^2 ds\right]$$
$$\times\left[\frac{1}{T}\int_{t-T}^{t}\left(\zeta^2(s)-1\right)^2 ds\right]$$

Combining this inequality with the identities

$$\frac{1}{T}\int_{t-T}^{t}\left(\nu^2(s)-1\right)^2 ds = \langle \nu^4\rangle_0(t) - 2\langle \nu^2\rangle_0(t)+1 = \langle \nu^4\rangle_0(t)-1$$
$$\frac{1}{T}\int_{t-T}^{t}\left(\zeta^2(s)-1\right)^2 ds = \langle \zeta^4\rangle_0(t) - 2\langle \zeta^2\rangle_0(t)+1 = \langle \zeta^4\rangle_0(t)-1$$

generates the upper bound

$$\left|\frac{\left\langle \sum_{r<\ell} q_{r\ell}^2\right\rangle_0(t)}{Q_x^2(t)} - 1\right| \leq \sqrt{\left(\gamma_v(t)-1\right)\left(\gamma_i(t)-1\right)} \tag{H.12b}$$

where

$$\gamma_v(t) \stackrel{\text{def}}{=} \frac{\left\langle [v\,v^\top]^2\right\rangle_0(t)}{V_{rms}^4(t)}\ ,\qquad \gamma_i(t) \stackrel{\text{def}}{=} \frac{\left\langle [i_x\,i_x^\top]^2\right\rangle_0(t)}{I_{x,rms}^4(t)} \tag{H.12c}$$

These results extend steady-state bounds obtained in [29] to the case of non-periodic waveforms.

In the special case of a three-phase system with vanishing zero-sequence voltage and current, the only non-vanishing (Clarke-transformed) $q_{r\ell}(t)$ component is $q_{\alpha\beta}(t)$. Consequently, $\sum_{r<\ell} q_{r\ell}^2(t) \equiv q_{\alpha\beta}^2(t)$, and the inequalities (H.12) reduce to

$$\langle q_{\alpha\beta}\rangle_0(t) \leq \langle |q_{\alpha\beta}|\rangle_0(t) \leq Q_x(t) \tag{H.13a}$$

and

$$\left|\frac{\left\langle q_{\alpha\beta}^2\right\rangle_0(t)}{Q_x^2(t)} - 1\right| \leq \sqrt{\left(\gamma_v(t)-1\right)\left(\gamma_i(t)-1\right)} \tag{H.13b}$$

H.5 Averaging the Generalized Instantaneous Reactive Power

In addition to the purely instantaneous $m \times m$ reactive power matrix $\boldsymbol{Q}(t)$, we have reviewed in Sect. 9.2 several other proposed definitions of instantaneous reactive power, which are all special cases of the generalized definition (9.32a), viz.,

$$\breve{q}(t) \stackrel{\text{def}}{=} \frac{1}{2}\left[v_h(t)\, i_g^\top(t) - v_g(t)\, i_h^\top(t) \right] \tag{H.14}$$

where (recall Fig. 9.19)

$$\begin{aligned} v_h(t) &= h(t) \circledast v(t) \;, \quad v_g(t) = g(t) \circledast v(t) \;, \\ i_h(t) &= h(t) \circledast i(t) \;, \quad i_g(t) = g(t) \circledast i(t) \end{aligned}$$

The corresponding cycle-averaged metric is

$$\begin{aligned} \breve{Q}(t) \stackrel{\text{def}}{=} \langle \breve{q} \rangle_0(t) &\equiv \frac{1}{2T} \int_{t-T}^{t} \left[v_h(s)\, i_g^\top(s) - v_g(s)\, i_h^\top(s) \right] ds \\ &= \frac{1}{2}\left[\langle\, v_h(\cdot)\,,\, i_g(\cdot)\,\rangle(t) - \langle\, v_g(\cdot)\,,\, i_h(\cdot)\,\rangle(t) \right] \end{aligned} \tag{H.15}$$

where we used the time-variant inner-product notation, as defined in (2.4).

In particular, the choice $h(t) = \frac{1}{\pi t}$ (the impulse response associated with the Hilbert transform filter) and $g(t) = \delta(t)$ results in $v_h(t) = \{\mathcal{H}v\}(t)$ and similarly for the Hilbert-transformed current $i_h(t)$, so that

$$\begin{aligned} \breve{Q}(t) &= \frac{1}{2}\left[\langle\, \mathcal{H}v\,(\cdot)\,,\, i(\cdot)\,\rangle(t) - \langle\, v(\cdot)\,,\, \mathcal{H}i\,(\cdot)\,\rangle(t) \right] = \langle\, \mathcal{H}v\,(\cdot)\,,\, i(\cdot)\,\rangle(t) \\ &\equiv Q_B(t) \end{aligned}$$

where we relied on the fact that the Hilbert transform is a skew-Hermitian operator, i.e., $\mathcal{H}^* = -\mathcal{H}$ or, equivalently, $\langle\, v(\cdot)\,,\, \mathcal{H}i\,(\cdot)\,\rangle(t) \equiv \langle\, \mathcal{H}^* v\,(\cdot)\,,\, i(\cdot)\,\rangle(t) = -\langle\, \mathcal{H}v\,(\cdot)\,,\, i(\cdot)\,\rangle(t)$.

Similarly, the choice $h(t) = \omega_0\, u(t) \circledast \left[\delta(t) - u_T(t)\right]$ and $g(t) = \delta(t)$ generates the "instantaneous integral reactive power" $q^\alpha(t)$ of [277], which we discussed in Sect. 9.6. Applying cycle averaging results in an average reactive power metric $Q^\alpha(t)$. In steady-state, this metric captures energy storage in reactive components [277]: (a) $Q^\alpha = 2\omega_0 \left(\frac{1}{2} L\, I_{rms}^2\right)$ for an inductor and (b) $Q^\alpha = -2\omega_0 \left(\frac{1}{2} C\, V_{rms}^2\right)$ for a capacitor.

Appendix I: Tellegen's Theorem and Network-Conservative Power Metrics

Tellegen's theorem, in its most basic form, states that the instantaneous power $p(t)$ is *network-conservative,* namely, the summation of $p(t)$ over all branches of any electrical network adds up to zero, for every individual time-instant t. The theorem relies only on Kirchhoff's laws: it holds for any set of branch voltages that satisfy Kirchhoff's voltage law (KVL) on a given network, and any set of branch currents that satisfy Kirchhoff's current law (KCL) on the same network.

I.1 Tellegen's Theorem [296]

A single-phase electric circuit that consists of interconnected two-terminal lumped components and sources (i.e., an electrical network) can always be described by a finite simple graph, with M branches and N nodes, where each branch corresponds to a component or a source, and each node to a point of interconnection between branches. It is customary to assign a reference direction to each branch, so that the graph becomes *directed*. The physical nature of branch components can be completely ignored for the purpose of invoking Tellegen's theorem: only the topology of the graph, as captured, for instance, by the $M \times N$ incidence matrix $\mathbf{A}$, is used in establishing the theorem. The elements of the incidence matrix are defined as follows:

$$a_{mn} = \begin{cases} +1 & \text{if branch } m \text{ leaves node } n \\ -1 & \text{if branch } m \text{ enters node } n \\ 0 & \text{if branch } m \text{ is not adjacent to node } n \end{cases} \tag{I.1}$$

The statement of Tellegen's theorem assumes that a pair of (real) numbers (v_m, i_m), labeled "voltage" and "current," is associated with each branch of the graph: these numbers can represent any two (possibly unrelated) quantities, not necessarily physical voltage or current. Since the characteristics of the actual branch component are not used in formulating the theorem, the values of v_m and i_m need not reflect any physical reality. The only requirement is that they satisfy Kirchhoff's voltage and current laws.

The $M \times 1$ vector of branch "currents" $\mathbf{i} = [i_1 \;\; i_2 \;\; \ldots i_M]^\top$ is said to satisfy Kirchhoff's current law (KCL) when $\mathbf{i}^\top \mathbf{A} = 0$ (i.e., the sum of currents entering the node equals the sum of currents leaving the node). Notice that the transposed current vector $\mathbf{i}^\top$ must be a left null-vector of the incidence matrix $\mathbf{A}$. Next, the $M \times 1$ vector of branch "voltages" $\mathbf{v} = [v_1 \;\; v_2 \;\; \ldots v_M]^\top$ is said to satisfy Kirchhoff's voltage law (KVL) when the equation $\mathbf{v} = \mathbf{A}\boldsymbol{\phi}$ has a solution $\boldsymbol{\phi}$. The elements of the vector $\boldsymbol{\phi}$ (i.e., the so-called node potentials) must be unique up to a common additive constant: this is so because the vector $[1 \;\; 1 \;\; \ldots \;\; 1]^\top$ is always a right null-vector of the incidence matrix $\mathbf{A}$.

A current vector $\mathbf{i}$ that satisfies KCL will be called *feasible*. Similarly, a voltage vector $\mathbf{v}$ will be called *feasible* if it satisfies KVL. Combining the KCL and KVL constraints, we conclude that

$$\mathbf{i}^\top \mathbf{v} = \mathbf{i}^\top \mathbf{A}\boldsymbol{\phi} = 0$$

or, equivalently, $\sum_{m=1}^{M} v_m\, i_m = 0$. In summary, Tellegen's theorem states that, for any feasible set of branch voltages $\{v_m\,;\ 1 \leq m \leq M\}$, and any feasible set of branch currents $\{i_m\,;\ 1 \leq m \leq M\}$, the *network-conservativity* property $\sum_m v_m\, i_m = 0$ always holds.

The voltages and currents in a network can be time-variant: in this case, the theorem holds for every pair $(t\,,s)$ of time-instants, provided that KCL and KVL hold for each individual time-instant. This means that Tellegen's theorem can be stated as

$$\sum_m v_m(t)\, i_m(s) = 0 \tag{I.2}$$

where t and s are any two time-instants (including the special case $t = s$).

Extending the notion of feasibility to network waveforms, we will say that a set of voltage waveforms $\{v_m(t)\,;\ 1 \leq m \leq M\}$ is feasible if it satisfies KVL for every time-instant t. Similarly, a set of current waveforms $\{i_m(s)\,;\ 1 \leq m \leq M\}$ will be called feasible if it satisfies KCL for every time-instant s. The result (I.2) states that, for a given incidence matrix $\mathbf{A}$, every product of a feasible voltage set and a feasible current set is network-conservative.

I.2 Tellegen's Theorem for Filtered Waveforms

A Kirchhoff voltage operator is any operator (not necessarily linear) that preserves voltage feasibility (i.e., preserves KVL for all t) [297]. In other words, if the voltage waveforms $\{v_m(t)\,;\ 1 \leq m \leq M\}$ are feasible for a given incidence matrix, and the transformed voltage waveforms $\{\mathcal{F}\, v_m(t)\,;\ 1 \leq m \leq M\}$ are also feasible for the same incidence matrix, then $\mathcal{F}$ is called a Kirchhoff voltage operator. Similarly, a Kirchhoff current operator is any operator that preserves current feasibility (i.e., preserves KCL for all t) [297]. Notice that every network-invariant[4] linear operator is both a Kirchhoff voltage operator and a Kirchhoff current operator.

For our purposes, it will be sufficient to restrict our discussion to linear operators defined by the input–output relationship of linear time-invariant (LTI) filters (see Fig. I.1). These include as (very) special cases the Hilbert transform, the derivative, the anti-derivative (i.e., the indefinite integral), as well as the short-time Fourier

[4] An operator on a function space (a.k.a. waveform space or signal space) is called network-invariant if its definition does not depend on the network location of the waveform to which this operator is applied.

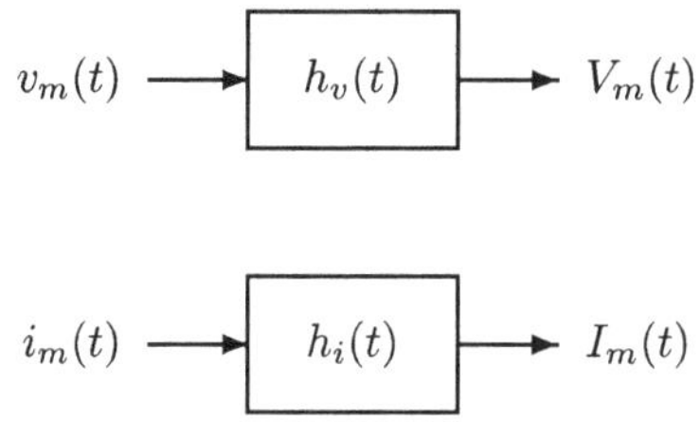

Fig. I.1 Generation of filtered voltage and current waveforms.

transform (STFT). Notice that a single LTI filter $h_v(t)$ is used for all voltage waveforms $\{v_m(t)\,;\ 1 \leq m \leq M\}$, and another single LTI filter $h_i(t)$, which could be the same as $h_v(t)$, is used for all current waveforms $\{i_m(t)\,;\ 1 \leq m \leq M\}$.

When the waveform sets $\{v_m(t)\,;\ 1 \leq m \leq M\}$ and $\{i_m(t)\,;\ 1 \leq m \leq M\}$ satisfy KVL and KCL, respectively (for all t), the same holds true for the filtered waveform sets $\{V_m(t)\,;\ 1 \leq m \leq M\}$ and $\{I_m(t)\,;\ 1 \leq m \leq M\}$. Since these waveform sets are feasible, the generalized version (I.2) of Tellegen's theorem applies, namely, the identity

$$\sum_m V_m(t)\ I_m(s) \ = \ 0 \tag{I.3}$$

holds for any two (possibly identical) time-instants t and s. This result can be extended to the case when the output waveforms are complex-valued (i.e., the LTI filters have a complex-valued impulse response). Indeed, splitting $V_m(t)$ into its real and imaginary parts, and similarly for $I_m(s)$, we obtain two sets of branch voltages – $\Re\{V_m(t)\}$ and $\Im\{V_m(t)\}$ – both of which satisfy KVL, and two sets of branch currents – $\Re\{I_m(s)\}$ and $\Im\{I_m(s)\}$ – both of which satisfy KCL. Since Tellegen's theorem applies to any pairing of feasible voltages and feasible currents, we find that

$$\sum_m \Re\{V_m(t)\}\,\Re\{I_m(s)\} = 0 \ , \qquad \sum_m \Re\{V_m(t)\}\,\Im\{I_m(s)\} = 0$$

as well as

$$\sum_m \Im\{V_m(t)\}\,\Re\{I_m(s)\} = 0 \ , \qquad \sum_m \Im\{V_m(t)\}\,\Im\{I_m(s)\} = 0$$

In particular, these identities imply that

$$\sum_m V_m(t)\ I_m(s) \ = \ 0 \tag{I.4a}$$

and

$$\sum_m V_m(t)\, I_m^*(s) \;=\; 0 \tag{I.4b}$$

for any two (possibly identical) time-instants t and s, and for any choice of the complex-valued impulse responses $h_v(t)$ and $h_i(t)$. In other words, both $V_m(t)\; I_m(s)$ and $V_m(t)\; I_m^*(s)$ are network-conservative, for any choice of the time-instants t and s.

Since the preceding discussion involves only two time-instants (i.e., t and s), there is no need for an explicit specification of a signal space for the input and output waveforms. In practice, all waveforms of interest are bounded (i.e., belong to L^∞), and the LTI filters used are BIBO stable.

I.3 Tellegen's Theorem for Filter-Banks

The general network-conservativity result (I.4) can be applied to the case where we use a filter-bank to map a waveform $x(t)$ into its subband components – a.k.a. generalized dynamic phasors (see Fig. I.2).

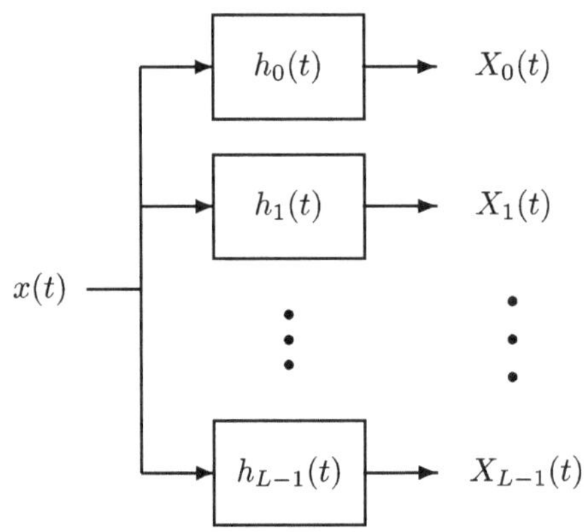

Fig. I.2 Filter-bank generation of generalized dynamic phasors.

To be specific, we can use a filter-bank of choice to map every pair of branch voltage and branch current in an electric network into their respective subband components, viz.,

$$v_m(\cdot) \;\Longrightarrow\; \left\{V_0^{(m)}(\cdot)\,,\; V_1^{(m)}(\cdot)\,,\; \ldots\,,\; V_{L-1}^{(m)}(\cdot)\right\}$$

$$i_m(\cdot) \;\Longrightarrow\; \left\{I_0^{(m)}(\cdot)\,,\; I_1^{(m)}(\cdot)\,,\; \ldots\,,\; I_{L-1}^{(m)}(\cdot)\right\}$$

The results of the preceding section now imply that the products $V_k(t)\, I_\ell(s)$ and $V_k(t)\, I_\ell^*(s)$ are both network-conservative, namely

$$\sum_m V_k^{(m)}(t)\, I_\ell^{(m)}(s) = 0 \tag{I.5a}$$

and

$$\sum_m V_k^{(m)}(t) \left[I_\ell^{(m)}(s) \right]^* = 0 \tag{I.5b}$$

for every k and ℓ in the range $[0\ \ L-1]$ (including $k = \ell$), and for every t and every s (including $t = s$).

In particular, setting $k = \ell$ and $t = s$ in (I.5b), we conclude that the dynamic phasor product $V_k(t)\, I_k^*(t)$ is network-conservative, for each k. Consequently, the dynamic real power $P(t) \stackrel{\text{def}}{=} \sum_k \Re\{V_k(t)\, I_k^*(t)\}$ and the dynamic reactive power $Q(t) \stackrel{\text{def}}{=} \sum_k \Im\{V_k(t)\, I_k^*(t)\}$ are also network-conservative, in both full-cycle and sub-cycle schemes (see Sect. 9.6 for further details).

Appendix J: Fundamentals of Euclidean and Hilbert Spaces

Euclidean and Hilbert spaces are the two prominent examples of *inner-product spaces*: this means they are *linear spaces* for which an operation called *inner-product* has been defined. A linear space is a collection of objects, known as elements, that is closed under the operations of scaling (by a real number) and addition, i.e., the result of such an operation is another element in the same collection. An inner-product associates a real-valued scalar with every pair of elements in this collection: inner-products can be used to define length of elements and angles between them.

As an example, the space R^N of real-valued N-tuples is evidently a linear space. If we represent the elements of this space by row vectors of length N, viz.,

$$\mathbf{x} = \begin{bmatrix} x_1 & x_2 & \dots & x_N \end{bmatrix} \,, \qquad \mathbf{y} = \begin{bmatrix} y_1 & y_2 & \dots & y_N \end{bmatrix}$$

then the inner-product is defined as

$$\langle\, \mathbf{x}\,,\, \mathbf{y} \,\rangle \stackrel{\text{def}}{=} \mathbf{x}\mathbf{y}^T = \sum_{i=1}^{N} x_i\, y_i \tag{J.1a}$$

which makes R^N an inner-product space. The inner-product also induces a norm in R^N, viz.,

$$\|\, \mathbf{x} \,\| \stackrel{\text{def}}{=} \sqrt{\langle\, \mathbf{x}\,,\, \mathbf{x} \,\rangle} \equiv \sqrt{\mathbf{x}\mathbf{x}^T} = \sum_{i=1}^{N} x_i^2 \tag{J.1b}$$

The angle between two elements $\mathbf{x} \in R^N$ and $\mathbf{y} \in R^N$ is defined in terms of its cosine, viz.,

$$\cos\phi \stackrel{\text{def}}{=} \frac{\mathbf{x}\,\mathbf{y}^T}{\sqrt{\mathbf{x}\,\mathbf{x}^T}\,\sqrt{\mathbf{y}\,\mathbf{y}^T}} \equiv \frac{\langle\, \mathbf{x}\,,\, \mathbf{y}\,\rangle}{\|\,\mathbf{x}\,\|\,\|\,\mathbf{y}\,\|} \tag{J.1c}$$

In particular, when $\mathbf{x}\,\mathbf{y}^T = 0$, we say that $\mathbf{x}$ and $\mathbf{y}$ are orthogonal. Orthogonality and the definition of a norm make it possible to construct orthonormal bases in R^N.

Another example is provided by the space $\mathcal{L}^2(t_0, t_1)$ of real-valued waveforms $\mathbf{x} \equiv \{x(t)\,;\; t_0 \leq t \leq t_1\}$ that are square-integrable over the interval $[\,t_0 \;\; t_1\,]$. This space is clearly linear: scaling and/or adding waveforms from $\mathcal{L}^2(t_0, t_1)$ produces another waveform in the same space. The inner-product in this space can be defined as

$$\langle\, \mathbf{x}\,,\, \mathbf{y}\,\rangle \stackrel{\text{def}}{=} \int_{t_0}^{t_1} x(t)\, y(t)\, dt \tag{J.2a}$$

which makes $\mathcal{L}^2(t_0, t_1)$ an inner-product space. This inner-product induces a norm in $\mathcal{L}^2(t_0, t_1)$, viz.,

$$\|\,\mathbf{x}\,\| \stackrel{\text{def}}{=} \sqrt{\langle\, \mathbf{x}\,,\, \mathbf{x}\,\rangle} \equiv \int_{t_0}^{t_1} x^2(t)\, dt \;<\; \infty \tag{J.2b}$$

which is finite by virtue of the square-integrability assumption. The angle between two elements $\mathbf{x} \in \mathcal{L}^2(t_0, t_1)$ and $\mathbf{y} \in \mathcal{L}^2(t_0, t_1)$ is defined in terms of its cosine, viz.,

$$\cos\phi \stackrel{\text{def}}{=} \frac{\int_{t_0}^{t_1} x(t)\, y(t)\, dt}{\sqrt{\int_{t_0}^{t_1} x^2(t)\, dt}\,\sqrt{\int_{t_0}^{t_1} y^2(t)\, dt}} \equiv \frac{\langle\, \mathbf{x}\,,\, \mathbf{y}\,\rangle}{\|\,\mathbf{x}\,\|\,\|\,\mathbf{y}\,\|} \tag{J.2c}$$

In particular, when $\int_{t_0}^{t_1} x(t)\, y(t)\, dt = 0$, we say that the waveforms $\mathbf{x}$ and $\mathbf{y}$ are orthogonal. Orthogonality and the definition of a norm make it possible to introduce the concept of an orthonormal basis in the space $\mathcal{L}^2(t_0, t_1)$. These concepts and definitions can be readily extended to *polyphase waveforms,* as demonstrated in Sect. 1.2.

Additional examples of inner-product spaces include: (a) the collection of all real-valued matrices of a given size, with the inner-product $\mathrm{tr}\{A\,B^T\}$; (b) the collection of real-valued zero-mean finite-variance random variables, with the inner-product $E\{\mathbf{x}\,\mathbf{y}\}$; (c) the collection ℓ^2 of all real-valued square-summable sequences $\{x_i\,,\;\; 1 \leq i < \infty\}$, with the inner-product $\sum_i x_i\, y_i$; and many more [298, 299].

J.1 Linear Space

A *linear space* (a.k.a. vector space) is a collection of objects, known as *elements* (also *vectors*) that can be (a) scaled (i.e., multiplied by a real number), and (b) added together. The result of such an operation is another element in the same collection. For instance, the collection of all waveforms on a given time interval is a linear space because the sum of two waveforms is itself a waveform defined on the same interval, and so is a scaled version of any waveform. This property is known as *closure*: we say that the linear space is closed under the operations of scaling and addition.

A formal (a.k.a. "axiomatic") definition of a linear space involves no assumptions about the nature of the elements in the space, or the specifics of the scaling and addition operations. This allows for many distinct specific examples to be covered by the same definition, and thus share the resulting properties. The formal definition, lists, however, certain requirements that must be met, such as associativity and commutativity of addition, the existence of a neutral element (with respect to addition), etc. A complete listing of these requirements is provided in the box below. The formal definition also allows for generalized scaling, using scalars chosen from any desired *field* [300]. However, for our purposes, it will be sufficient to restrict our discussion to scaling by real numbers. In this appendix we will use lower case boldface letters to denote elements of the linear space and lower case Greek letters to denote scalars.

A linear space over the real scalars is a set $\mathcal{V}$ with two operations:

- A scaling operation $\mathbf{x} \rightarrow \alpha\mathbf{x}$, where α is a real-valued number, and for every $\mathbf{x} \in \mathcal{V}$ it follows that $\alpha\mathbf{x} \in \mathcal{V}$.
- An addition operation $\mathbf{z} = \mathbf{x} + \mathbf{y}$, such that for every $\mathbf{x} \in \mathcal{V}$ and $\mathbf{y} \in \mathcal{V}$, it follows that $\mathbf{z} \in \mathcal{V}$ as well.

These two operations must satisfy the following requirements for all $\mathbf{x} \in \mathcal{V}$, $\mathbf{y} \in \mathcal{V}$, $\mathbf{z} \in \mathcal{V}$, and all scalars α, β:

- Commutativity of addition: $\mathbf{x} + \mathbf{y} = \mathbf{y} + \mathbf{x}$.
- Associativity of addition: $\mathbf{x} + (\mathbf{y} + \mathbf{z}) = (\mathbf{x} + \mathbf{y}) + \mathbf{z}$.
- Neutral element of addition: there exists an element $\mathbf{0} \in \mathcal{V}$ such that $\mathbf{x} + \mathbf{0} = \mathbf{x}$.
- Inverse element of addition: for every $\mathbf{x} \in \mathcal{V}$, there exists an element $\widetilde{\mathbf{x}} \in \mathcal{V}$ such that $\mathbf{x} + \widetilde{\mathbf{x}} = \mathbf{0}$.
- Scaling by unity: $1\,\mathbf{x} = \mathbf{x}$.
- Associativity of scaling: $\alpha(\beta\mathbf{x}) = (\alpha\beta)\mathbf{x}$.
- Distributivity of scaling with respect to addition: $\alpha(\mathbf{x} + \mathbf{y}) = \alpha\mathbf{x} + \alpha\mathbf{y}$, as well as $(\alpha + \beta)\mathbf{x} = \alpha\mathbf{x} + \beta\mathbf{x}$.

In view of these properties, it follows that $\widetilde{\mathbf{x}} = (-1)\mathbf{x} \stackrel{\text{def}}{=} -\mathbf{x}$.

An element of $\mathcal{V}$ of the form $\alpha\mathbf{x}+\beta\mathbf{y}+\gamma\mathbf{z}$ is known as a linear combination of the linear space elements $\mathbf{x}$, $\mathbf{y}$, and $\mathbf{z}$. A non-empty subset of $\mathcal{V}$ that is closed under the operation of linear combination of its elements is called a *linear subspace.* In particular, the collection of all linear combinations of a given finite set of elements in $\mathcal{V}$, say $\{\mathbf{x}_1, \mathbf{x}_2, \ \ldots \ , \mathbf{x}_N\}$, is a linear subspace, denoted by *span* $\{\mathbf{x}_1, \mathbf{x}_2, \ \ldots \ , \mathbf{x}_N\}$. Notice, however, that the span of an infinite set of elements $\mathcal{S} \subset \mathcal{V}$ is defined as the set of all *finite linear combinations* of elements of $\mathcal{S}$. Consequently, the construction of subspaces in an infinite-dimensional inner-product space involves an additional constraint known as completeness (see below).

J.2 Inner-Product Space

An *inner-product* associates a real-valued scalar with any pair of elements $\mathbf{x}, \mathbf{y}$ in a linear space $\mathcal{V}$. This makes it possible to define lengths of elements and angles between them. In particular, orthogonality of elements and of subspaces in $\mathcal{V}$ can be defined in terms of the inner-product. In addition, convergence of a sequence of elements in $\mathcal{V}$ can be defined in terms of the norm that is induced by the inner-product (see below). A linear space $\mathcal{V}$ with an inner-product is called an *inner-product space.* Several examples of inner-product spaces were mentioned at the beginning of this appendix. In particular, the spaces $\mathcal{L}^2(t_0, t_1)$ and R^N are used repeatedly throughout this book.

An *inner-product,* denoted by $\langle\, \mathbf{x}\, ,\, \mathbf{y}\, \rangle$, is an operation that maps a pair of elements $\mathbf{x}, \mathbf{y}$ of a linear space $\mathcal{V}$ into a (real-valued) scalar. The inner-product operation must satisfy the following requirements for all $\mathbf{x} \in \mathcal{V}$, $\mathbf{y} \in \mathcal{V}, \mathbf{z} \in \mathcal{V}$, and all scalars α:

- $\langle\, \mathbf{y}\, ,\, \mathbf{x}\, \rangle = \langle\, \mathbf{x}\, ,\, \mathbf{y}\, \rangle$.
- $\langle\, \alpha\mathbf{x}\, ,\, \mathbf{y}\, \rangle = \alpha\langle\, \mathbf{x}\, ,\, \mathbf{y}\, \rangle$.
- $\langle\, \mathbf{x}+\mathbf{y}\, ,\, \mathbf{z}\, \rangle = \langle\, \mathbf{x}\, ,\, \mathbf{z}\, \rangle + \langle\, \mathbf{y}\, ,\, \mathbf{z}\, \rangle$.
- $\langle\, \mathbf{x}\, ,\, \mathbf{x}\, \rangle > 0$ for every $\mathbf{x} \neq \mathbf{0}$ in $\mathcal{V}$.

The *norm* of an element $\mathbf{x} \in \mathcal{V}$ is defined as $\|\, \mathbf{x}\, \| \stackrel{\text{def}}{=} \sqrt{\langle\, \mathbf{x}\, ,\, \mathbf{x}\, \rangle}$. It follows from the above requirements that $\|\, \mathbf{x}\, \| = 0$ if, and only if, $\mathbf{x} = \mathbf{0}$.

As mentioned earlier, the inner-product can be used to define the angles between elements of an inner-product space. Two elements in an inner-product space are said to be *orthogonal* when the angle between them is a right angle (i.e., the cosine of this angle vanishes). Orthogonality is central to the construction of orthonormal bases in an inner-product space.

The angle between any two elements $\mathbf{x} \in \mathcal{V}$ and $\mathbf{y} \in \mathcal{V}$ is defined in terms of its cosine, viz.,

$$\cos \phi \stackrel{\text{def}}{=} \frac{\langle \mathbf{x}, \mathbf{y} \rangle}{\| \mathbf{x} \| \, \| \mathbf{y} \|}$$

The two elements $\mathbf{x}$ and $\mathbf{y}$ are said to be *orthogonal* when $\langle \mathbf{x}, \mathbf{y} \rangle = 0$. An *orthonormal family* is a subset $\mathcal{B} \subset \mathcal{V}$ such that:

- Every two distinct elements in $\mathcal{B}$ are orthogonal.
- Every element in $\mathcal{B}$ has unit norm.

An orthonormal family $\mathcal{B}$ is called *complete* if it is not contained in a larger orthonormal family. Equivalently, completeness means that the only element of $\mathcal{V}$ that is orthogonal to every element of $\mathcal{B}$ is the zero element.

The *dimension* of an inner-product space $\mathcal{V}$ is determined by the cardinality (i.e., "size") of its complete orthonormal families. When a complete orthonormal family is finite, the inner-product space is called *Euclidean* and has finite dimension (see below). Otherwise, the space $\mathcal{V}$ is infinite-dimensional and is known as a *pre-Hilbert space*: additional regularity conditions (i.e., completeness) need to be satisfied in order to qualify for the title *Hilbert space* (see below).

J.3 Euclidean Space

When a complete orthonormal family in an inner-product space $\mathcal{V}$ is finite, say $\mathcal{B} = \{b_k\,,\ 1 \le k \le N\}$, the dimension of the space $\mathcal{V}$ is defined as the size N of this orthonormal family. In this case every complete orthonormal family in $\mathcal{V}$ has the same size N, and can serve as a *basis* of $\mathcal{V}$, in the sense that every element of $\mathcal{V}$ can be represented as a finite linear combination of the elements of $\mathcal{B}$, namely $\mathcal{V} = span\ \mathcal{B}$.

A finite-dimensional inner-product space is known as a *Euclidean space.* Two Euclidean spaces of the same dimension, say $\mathcal{V}_1$ and $\mathcal{V}_2$, are always *isometric* to each other. Isometry means that there exists a one-to-one correspondence between elements of $\mathcal{V}_1$ and elements of $\mathcal{V}_2$, such that norms and angles are preserved: when this correspondence matches $\mathbf{x}_1 \in \mathcal{V}_1$ with $\mathbf{x}_2 \in \mathcal{V}_2$, and $\mathbf{y}_1 \in \mathcal{V}_1$ with $\mathbf{y}_2 \in \mathcal{V}_2$, then $\|\mathbf{x}_1\| = \|\mathbf{x}_2\|$ and $\|\mathbf{y}_1\| = \|\mathbf{y}_2\|$, as well as $\langle \mathbf{x}_1, \mathbf{y}_1 \rangle = \langle \mathbf{x}_2, \mathbf{y}_2 \rangle$. In particular, every N-dimensional Euclidean space is isometric to R^N.

The discipline of classical physics relies heavily on the description of physical quantities such as forces, velocities, accelerations, and electromagnetic fields as elements in the Euclidean space R^3. Another example is the depiction of m-phase *instantaneous* voltage and current samples as elements in the Euclidean space R^m, namely as $1 \times m$ arrays (see, e.g., Sect. 9.2).

J.4 Hilbert Space

A pre-Hilbert space that has the *completeness* property is called a Hilbert space. Completeness guarantees that the limit element of any convergent sequence in $\mathcal{V}$ is contained in the same space. For instance, the collection $\mathcal{C}(t_0, t_1)$ of *continuous* waveforms defined over the interval $[t_0 \quad t_1]$, together with the inner-product (J.2a), is an inner-product space. However, a sequence of continuous waveforms can converge to a limit waveform (such as a step function) that contains point discontinuities and is thus outside of $\mathcal{C}(t_0, t_1)$.

Convergence in an inner-product space is defined in terms of the norm $\| \cdot \|$: a sequence $\{\mathbf{x}_k\,;\; k = 1, 2, \ldots\}$ of elements of $\mathcal{V}$ is said to converge to a limit $\mathbf{x}_\infty$ if $\lim_{k\to\infty} \| \mathbf{x}_k - \mathbf{x}_\infty \| = 0$. Notice that this definition does not require that $\mathbf{x}_\infty$ is an element of $\mathcal{V}$: only the difference $\mathbf{x}_k - \mathbf{x}_\infty$ has to belong in the space $\mathcal{V}$.

A sequence $\{\mathbf{x}_k\,;\; k = 1, 2, \ldots\}$ of elements of $\mathcal{V}$ is called a *Cauchy sequence* if the norm $\| \mathbf{x}_k - \mathbf{x}_\ell \|$ converges to zero as $k, \ell \to \infty$. A more rigorous definition is provided in the box below.

A sequence $\{\mathbf{x}_k\,;\; k = 1, 2, \ldots\}$ of elements of $\mathcal{V}$ is called a *Cauchy sequence* if for every $\varepsilon > 0$ there exists a positive integer n_0 such that $\| \mathbf{x}_k - \mathbf{x}_\ell \| < \varepsilon$ holds for every $k \geq n_0$ and every $\ell \geq n_0$.

A pre-Hilbert space $\mathcal{V}$ is said to be *complete* when every Cauchy sequence of elements from this space converges to an element of the same space.

Every complete orthonormal family $\mathcal{B}$ in a Hilbert space $\mathcal{V}$ can serve as a basis for this space. This means that every Hilbert space element $\mathbf{x}$ can be represented as a linear combination of (possibly infinitely many) basis elements. When the basis $\mathcal{B}$ is countable, the Hilbert space is called *separable*. In particular, the space $\mathcal{L}^2(t_0, t_0 + T)$ is separable: one choice of an orthonormal basis for this space is the (countable) set of harmonics $\{(\cos k\omega_0 t\,,\; \sin k\omega_0 t);\; 1 \leq k < \infty\}$, where

$\omega_0 \overset{\text{def}}{=} 2\pi/T$, together with the constant (DC) waveform $x(t) = \{1;\ t_0 \le t \le t_0 + T\}$. Another example of a separable Hilbert space is $\ell^2(R)$, the collection of all (real-valued) square-summable sequences. All separable Hilbert spaces have the same dimension, known as $\aleph_0$ (pronounced "aleph naught"), and are thus all isometric to each other.

J.5 Direct Sum of Orthogonal Subspaces

If a non-empty subset $\mathcal{S}$ of a Euclidean space $\mathcal{V}$ is itself a Euclidean space (with the same inner-product as the one defined in $\mathcal{V}$), it is called a *subspace.* Subspaces of a Euclidean space are necessarily finite-dimensional. Similarly, a subspace of a Hilbert space $\mathcal{V}$ is a non-empty subset $\mathcal{S} \subset \mathcal{V}$ that is itself a Hilbert space (with the same inner-product as the one defined in $\mathcal{V}$). Subspaces of a Hilbert space can be finite-dimensional or infinite-dimensional.

Two subspaces, say $\mathcal{S}_1$ and $\mathcal{S}_2$, of a Euclidean/Hilbert space $\mathcal{V}$ are said to be *orthogonal* when $\langle\, \mathbf{x}, \mathbf{y} \,\rangle = 0$ for every $\mathbf{x} \in \mathcal{S}_1$ and every $\mathbf{y} \in \mathcal{S}_2$. The only element shared by a pair of orthogonal subspaces is the zero element, i.e., $\mathcal{S}_1 \cap \mathcal{S}_2 = \mathbf{0}$. We say that $\mathcal{V}$ is an *orthogonal direct sum* of the subspaces $\{\mathcal{S}_i\,;\ \mathcal{S}_i \subset \mathcal{V},\ i = 1, 2, 3\ \ldots\}$ when:

- Every element $\mathbf{x} \in \mathcal{V}$ can be expressed (uniquely) in the form $\mathbf{x} = \sum_i \mathbf{x}_i$, where $\mathbf{x}_i \in \mathcal{S}_i$.
- The subspaces $\mathcal{S}_i$ are all mutually orthogonal.

One of the key properties of an orthogonal direct sum of subspaces is additivity of dimension, namely, $\dim \mathcal{V} = \sum_i \dim \mathcal{S}_i$. Thus, decomposing a Euclidean/Hilbert space $\mathcal{V}$ into an orthogonal direct sum of subspaces makes it possible to study the properties of the space $\mathcal{V}$ by examining the properties of the individual subspaces that make up the direct sum.

For instance, we introduced in Sect. 1.2.2 the notion of (k-th harmonic/p-th phase) waveforms, i.e., waveforms described by the expression

$$x_k^{(p)}(t) = \sqrt{2}\,\Re\left\{X_k^{(p)} e^{jk\omega_0 t}\right\} \eta_p \ , \qquad \eta_p = [\,\underbrace{0 \ldots 0\ 1}_{p}\ 0 \ldots 0\,]$$

where $X_k^{(p)}$ is a complex scalar. The collection of all such waveforms, obtained by considering every possible choice (for fixed k and p) of the complex scalar $X_k^{(p)}$, is a two-dimensional subspace of the infinite-dimensional Hilbert space of all T-periodic, m-phase waveforms. We demonstrated in Sect. 1.2.2 that these subspaces are all mutually orthogonal. Since every m-phase, T-periodic waveform can be (uniquely) written as a sum of (k-th harmonic/p-th phase) components (recall (1.11c)), we conclude that the Hilbert space of all such waveforms is an *orthogonal direct sum* of its (k-th harmonic/p-th phase)-subspaces.

Appendix K: State-Space Model for Symmetrical Sequence Components

The phase-domain state-space model (3.5)–(3.6) involves no assumptions about the physical meaning of the elements of the state-vector $x(\cdot)$. In most cases, we may be interested in estimating the symmetrical sequence components associated with only a *subset* of the elements of $x(\cdot)$, such as certain 3-phase current or voltage waveforms. We present in this appendix the details of a transformation that converts this phase-domain state-space model into its *symmetrical-component* (SC) equivalent.

For the sake of clarity, we consider first the case where $x(\cdot)$ contains only a single 3-phase waveform of interest, whose symmetrical components we wish to estimate. However, the same approach can be extended without difficulty to handle any number of 3-phase waveforms of interest, as we demonstrate in Sect. K.2.

K.1 A Single 3-Phase Waveform of Interest

In this section, we assume that the $n \times 1$ state-vector $x(t)$ in (3.5)–(3.6) consists of two parts, viz.,

$$x(t) = \begin{pmatrix} i^\top(t) \\ \xi(t) \end{pmatrix}$$

where the 1×3 vector $i(t)$ is a 3-phase waveform (say, current) whose symmetrical sequence components we wish to estimate. The rest of the state variables are contained in the $(n-3) \times 1$ vector $\xi(t)$.

Thus the (passband) phasors $\mathcal{X}_k(t)$ that make up the phase-domain state-vector in (3.5)–(3.6) exhibit the same structural decomposition, viz.,

$$\mathcal{X}_k(t) = \begin{pmatrix} \mathcal{I}_k^\top(t) \\ \Xi_k(t) \end{pmatrix}$$

where $\mathcal{I}_k(t)$ and $\Xi_k(t)$ are the passband phasors associated with $i(t)$ and $\xi(t)$, respectively. The corresponding SC version is obtained by replacing $\mathcal{I}_k(t)$ with its SC-domain equivalent, using the mapping (2.18a) as a point of departure.

To be specific, we need to convert the definition (2.18a) of the symmetrical components associated with $I_k(t)$ into a *passband format*, written in terms of *column vectors*. For $k \neq 0$ this results in the relation

$$\mathcal{I}_k^{(sc)}(t) = e^{jk\omega_0 t}\Big[I_k(t)\mathcal{M}_{sc}\Big]^\top = \mathcal{M}_{sc}^\top\, \mathcal{I}_k^\top(t)$$

or, equivalently,

$$\mathcal{I}_k^\top(t) = \mathcal{M}_{sc}^{-\top}\,\mathcal{I}_k^{(sc)}(t) = \mathcal{M}_{sc}^{*}\,\mathcal{I}_k^{(sc)}(t)$$

where the asterisk $*$ denotes complex conjugation (without transposition). We also introduced the 1×3 passband phasor $\mathcal{I}_k(t) \stackrel{\text{def}}{=} I_k(t)\,e^{jk\omega_0 t}$ and the 3×1 (column) vector

$$\mathcal{I}_k^{(sc)}(t) \stackrel{\text{def}}{=} \begin{pmatrix} I_{k,0}(t) \\ I_{k,+}(t) \\ I_{k,-}(t) \end{pmatrix} e^{jk\omega_0 t} \tag{K.1}$$

and used the unitarity property $\mathcal{M}_{sc}\,\mathcal{M}_{sc}^{H} = I$. Separating the real and imaginary parts produces an equivalent dual-channel relation (for $k \neq 0$), viz.,

$$\begin{pmatrix} \Re\{\mathcal{I}_k^\top(t)\} \\ \Im\{\mathcal{I}_k^\top(t)\} \end{pmatrix} = \underbrace{\begin{pmatrix} \Re\{\mathcal{M}_{sc}\} & \Im\{\mathcal{M}_{sc}\} \\ -\Im\{\mathcal{M}_{sc}\} & \Re\{\mathcal{M}_{sc}\} \end{pmatrix}}_{\Theta_{sc}} \begin{pmatrix} \Re\{\mathcal{I}_k^{(sc)}(t)\} \\ \Im\{\mathcal{I}_k^{(sc)}(t)\} \end{pmatrix} \tag{K.2a}$$

where we recall that

$$\Re\{\mathcal{M}_{sc}\} = \frac{1}{\sqrt{3}} \begin{pmatrix} 1 & 1 & 1 \\ 1 & -\frac{1}{2} & -\frac{1}{2} \\ 1 & -\frac{1}{2} & -\frac{1}{2} \end{pmatrix}, \quad \Im\{\mathcal{M}_{sc}\} = \frac{1}{2} \begin{pmatrix} 0 & 0 & 0 \\ 0 & 1 & -1 \\ 0 & -1 & 1 \end{pmatrix} \tag{K.2b}$$

The 6×6 matrix Θ_{sc} is orthogonal (= real-valued unitary), i.e., $\Theta_{sc}\,\Theta_{sc}^\top = I$.

For the special case $k = 0$, we use the dynamic version of the real-valued mapping (1.16) to obtain the relation

$$\begin{pmatrix} I_{0,0}(t) \\ \sqrt{2}\,\Re\{I_{0,+}(t)\} \\ \sqrt{2}\,\Im\{I_{0,+}(t)\} \end{pmatrix} = \Theta_0^\top\, I_0^\top(t)$$

where, as defined in Sect. 1.3,

$$\Theta_0 = \frac{1}{\sqrt{3}} \begin{pmatrix} 1 & \sqrt{2} & 0 \\ 1 & -\frac{\sqrt{2}}{2} & \sqrt{\frac{3}{2}} \\ 1 & -\frac{\sqrt{2}}{2} & -\sqrt{\frac{3}{2}} \end{pmatrix} \tag{K.3a}$$

Since the matrix Θ_0 is *orthogonal*, i.e., $\Theta_0\,\Theta_0^\top = I$ or, equivalently, $\Theta_0^{-1} = \Theta_0^\top$, we conclude that

$$I_0^\top(t) = \Theta_0 \underbrace{\begin{pmatrix} I_{0,0}(t) \\ \sqrt{2}\,\Re\{I_{0,+}(t)\} \\ \sqrt{2}\,\Im\{I_{0,+}(t)\} \end{pmatrix}}_{I_0^{(sc)}(t)} \tag{K.3b}$$

Notice that these mappings are all time-invariant, so that the resulting SC-domain state-space model will have time-invariant coefficients, as we demonstrate in Sect. 3.1.3.

Next, we assemble these component-wise mappings into a single transformation that relates (for $k \neq 0$) the real-valued $n \times 1$ vectors $\Re\{\mathcal{X}_k(t)\}$ and $\Im\{\mathcal{X}_k(t)\}$ to their SC-domain versions, viz.,

$$\begin{pmatrix} \Re\{\mathcal{X}_k(t)\} \\ \Im\{\mathcal{X}_k(t)\} \end{pmatrix} = \underbrace{\begin{pmatrix} I_3 & 0 & 0 & 0 \\ 0 & 0 & I_{n-3} & 0 \\ - & - & - & - \\ 0 & I_3 & 0 & 0 \\ 0 & 0 & 0 & I_{n-3} \end{pmatrix}}_{\mathcal{P}_1} \begin{pmatrix} \Re\{\mathcal{I}_k^\top(t)\} \\ \Im\{\mathcal{I}_k^\top(t)\} \\ - \\ \Re\{\Xi_k(t)\} \\ \Im\{\Xi_k(t)\} \end{pmatrix}$$

$$= \mathcal{P}_1 \begin{pmatrix} \Theta_{sc} & 0 \\ 0 & I_{2n-6} \end{pmatrix} \begin{pmatrix} \Re\{\mathcal{I}_k^{(sc)}(t)\} \\ \Im\{\mathcal{I}_k^{(sc)}(t)\} \\ - \\ \Re\{\Xi_k(t)\} \\ \Im\{\Xi_k(t)\} \end{pmatrix}$$

where we used (K.2a). The $2n \times 2n$ matrix $\mathcal{P}_1$ is a *permutation matrix*: like all permutation matrices, it satisfies the orthogonality property (i.e., $\mathcal{P}_1\,\mathcal{P}_1^\top = I$).

Now, since

$$\begin{pmatrix} \Re\{\mathcal{I}_k^{(sc)}(t)\} \\ \Im\{\mathcal{I}_k^{(sc)}(t)\} \\ - \\ \Re\{\Xi_k(t)\} \\ \Im\{\Xi_k(t)\} \end{pmatrix} = \mathcal{P}_1^\top \begin{pmatrix} \Re\{\mathcal{X}_k^{(sc)}(t)\} \\ \Im\{\mathcal{X}_k^{(sc)}(t)\} \end{pmatrix}$$

where

$$\mathcal{X}_k^{(sc)}(t) \stackrel{\text{def}}{=} \begin{pmatrix} \mathcal{I}_k^{(sc)}(t) \\ ----- \\ \Xi_k(t) \end{pmatrix} \tag{K.4a}$$

it follows that, for every $k \neq 0$,

$$\begin{pmatrix} \Re\{\mathcal{X}_k(t)\} \\ \Im\{\mathcal{X}_k(t)\} \end{pmatrix} = \underbrace{\mathcal{P}_1 \begin{pmatrix} \Theta_{sc} & 0 \\ 0 & I_{2n-6} \end{pmatrix} \mathcal{P}_1^\top}_{\Phi_1} \begin{pmatrix} \Re\{\mathcal{X}_k^{(sc)}(t)\} \\ \Im\{\mathcal{X}_k^{(sc)}(t)\} \end{pmatrix} \tag{K.4b}$$

Notice that Φ_1 is the product of three orthogonal matrices and thus inherits the orthogonality property. We can also multiply out the matrices that make up this orthogonal transformation to obtain the explicit expression

$$\Phi_1 = \begin{pmatrix} \Re\{\mathcal{M}_{sc}\} & 0 & \Im\{\mathcal{M}_{sc}\} & 0 \\ 0 & I_{n-3} & 0 & 0 \\ -\Im\{\mathcal{M}_{sc}\} & 0 & \Re\{\mathcal{M}_{sc}\} & 0 \\ 0 & 0 & 0 & I_{n-3} \end{pmatrix} \tag{K.4c}$$

When $k = 0$, we need to replace the mapping (K.4) by

$$X_0(t) \equiv \begin{pmatrix} I_0^\top(t) \\ ----- \\ \Xi_0(t) \end{pmatrix} = \underbrace{\begin{pmatrix} \Theta_0 & 0 \\ 0 & I_{n-3} \end{pmatrix}}_{\Phi_{1,dc}} X_0^{(sc)}(t) \tag{K.5a}$$

where we used the relation (K.3), and introduced the shorthand notation

$$X_0^{(sc)}(t) \stackrel{\text{def}}{=} \begin{pmatrix} I_{0,0}(t) \\ \sqrt{2}\,\Re\{I_{0,+}(t)\} \\ \sqrt{2}\,\Im\{I_{0,+}(t)\} \\ ----- \\ \Xi_0(t) \end{pmatrix} \tag{K.5b}$$

Again, we observe that the matrix $\Phi_{1,dc}$ inherits the orthogonality property of Θ_0.

Finally, the combined similarity transformation $\mathbf{X}(t) = \mathcal{T}\mathbf{X}^{(sc)}(t)$ maps the state-vector $\mathbf{X}(t) = \{a_k\, \mathcal{X}_k(t)\,;\; k \in \boldsymbol{\mathcal{S}}\}$ into its SC-domain version $\mathbf{X}^{(sc)}(t) = \{a_k\, \mathcal{X}_k^{(sc)}(t)\,;\; k \in \boldsymbol{\mathcal{S}}\}$ (recall (3.4)), where

$$\mathcal{T} = \begin{pmatrix} \mathcal{T}_{k_1} & & \boldsymbol{O} \\ & \mathcal{T}_{k_2} & \\ \boldsymbol{O} & & \ddots \\ & & & \mathcal{T}_{k_L} \end{pmatrix} \tag{K.6a}$$

and

$$\mathcal{T}_k = \begin{cases} \Phi_{1,dc} & k = 0 \\ \Phi_1 & k \neq 0 \end{cases} \tag{K.6b}$$

Notice that the mappings Φ_1 and $\Phi_{1,dc}$ depend only on the structure of the state-vector $x(\cdot)$, and nothing else. Also notice that $\mathcal{T}$ is an orthogonal matrix, since each one of its diagonal blocks is orthogonal.

K.2 Multiple 3-Phase Waveforms of Interest

When the state-vector $x(t)$ contains several (say, p) 3-phase waveforms of interest, as well as $n-3p$ other state variables, the k-th harmonic passband phasor consists of $p+1$ parts, viz.,

$$\mathcal{X}_k(t) = \begin{pmatrix} \left[\mathcal{I}_k^{(1)}(t)\right]^\top \\ \left[\mathcal{I}_k^{(2)}(t)\right]^\top \\ \vdots \\ \left[\mathcal{I}_k^{(p)}(t)\right]^\top \\ - - - - \\ \Xi_k(t) \end{pmatrix}$$

Consequently, the relation (K.4) is extended to accommodate multiple waveforms of interest, viz.,

$$\begin{pmatrix} \Re\{\mathcal{X}_k(t)\} \\ \Im\{\mathcal{X}_k(t)\} \end{pmatrix} = \mathcal{P}_p \begin{pmatrix} \Theta_{sc} & & & & O \\ & \Theta_{sc} & & & \\ & & \ddots & & \\ & & & \Theta_{sc} & \\ O & & & & I_{2n-6p} \end{pmatrix} \begin{pmatrix} \Re\{\mathcal{I}_k^{(1,sc)}(t)\} \\ \Im\{\mathcal{I}_k^{(1,sc)}(t)\} \\ \Re\{\mathcal{I}_k^{(2,sc)}(t)\} \\ \Im\{\mathcal{I}_k^{(2,sc)}(t)\} \\ \vdots \\ \Re\{\mathcal{I}_k^{(p,sc)}(t)\} \\ \Im\{\mathcal{I}_k^{(p,sc)}(t)\} \\ - - - - \\ \Re\{\Xi_k(t)\} \\ \Im\{\Xi_k(t)\} \end{pmatrix}$$

$$= \mathcal{P}_p \begin{pmatrix} \Theta_{sc} & & & & \\ & \Theta_{sc} & & \mathbf{O} & \\ & & \ddots & & \\ & \mathbf{O} & & \Theta_{sc} & \\ & & & & I_{2n-6p} \end{pmatrix} \mathcal{P}_p^\top \begin{pmatrix} \Re\{\mathcal{X}_k^{(sc)}(t)\} \\ \Im\{\mathcal{X}_k^{(sc)}(t)\} \end{pmatrix}$$

where

$$\mathcal{X}_k^{(sc)}(t) \stackrel{\text{def}}{=} \begin{pmatrix} \mathcal{I}_k^{(1,sc)}(t) \\ \mathcal{I}_k^{(2,sc)}(t) \\ \vdots \\ \mathcal{I}_k^{(p,sc)}(t) \\ \hline \Xi_k(t) \end{pmatrix} \tag{K.7}$$

and $\mathcal{I}_k^{(\ell,sc)}(t)$ represents the symmetrical components associated with the k-th harmonic of the 3-phase current $i^{(\ell)}(t)$ (for $1 \le \ell \le p$), as defined in (K.1).

Here $\mathcal{P}_p$ is a $2n \times 2n$ permutation matrix, extending the definition of $\mathcal{P}_1$. For instance,

$$\mathcal{P}_2 = \begin{pmatrix} I_3 & 0 & 0 & 0 & 0 & 0 \\ 0 & 0 & I_3 & 0 & 0 & 0 \\ 0 & 0 & 0 & 0 & I_{n-6} & 0 \\ \hline 0 & I_3 & 0 & 0 & 0 & 0 \\ 0 & 0 & 0 & I_3 & 0 & 0 \\ 0 & 0 & 0 & 0 & 0 & I_{n-6} \end{pmatrix}$$

is the permutation matrix for the case of $p = 2$ waveforms of interest. In general, $\mathcal{P}_p$ can be obtained by:

(a) Partitioning the identity matrix I_{2n} into $2p + 2$ diagonal blocks ($2p$ blocks of size 3, and two blocks of size $n - 3p$), viz.,

$$I_{2n} = \begin{pmatrix} I_{2p} \otimes I_3 & 0 \\ 0 & I_2 \otimes I_{n-3p} \end{pmatrix} \equiv \operatorname{diag}\Big\{ \underbrace{I_3\,,\, I_3\,,\, \ldots\,,\, I_3}_{2p}\,,\, I_{n-3p}\,,\, I_{n-3p} \Big\}$$

(b) Numbering the resulting $2p + 2$ *block rows* as $\{1, 2, 3, \ldots, 2p + 1, 2p + 2\}$, and

(c) Reordering the block rows so that the odd-numbered block rows come first, namely $\{1, 3, 5, \ldots, 2p+1, 2, 4, 6, \ldots, 2p+2\}$

The combined similarity transformation $\mathbf{X}(t) = \mathcal{T}\mathbf{X}^{(sc)}(t)$ maps the state-vector $\mathbf{X}(t) = \{a_k\,\mathcal{X}_k(t)\,;\; k \in \mathcal{S}\}$ into its SC-domain version $\mathbf{X}^{(sc)}(t) = \{a_k\,\mathcal{X}_k^{(sc)}(t)\,;\; k \in \mathcal{S}\}$ (recall (3.4)). The transformation matrix $\mathcal{T}$ has the same global structure as (K.6), namely

$$\mathcal{T} = \begin{pmatrix} \mathcal{T}_{k_1} & & & \boldsymbol{O} \\ & \mathcal{T}_{k_2} & & \\ & & \ddots & \\ \boldsymbol{O} & & & \mathcal{T}_{k_L} \end{pmatrix} \tag{K.8a}$$

but with the p-dependent blocks

$$\mathcal{T}_k = \begin{cases} \Phi_{p,dc} & k = 0 \\ \Phi_p & k \neq 0 \end{cases} \tag{K.8b}$$

where the $n \times n$ transformation matrix $\Phi_{p,dc}$ is an extension of $\Phi_{1,dc}$, viz.,

$$\Phi_{p,dc} = \begin{pmatrix} \Theta_0 & & & & \boldsymbol{O} \\ & \Theta_0 & & & \\ & & \ddots & & \\ & & & \Theta_0 & \\ \boldsymbol{O} & & & & I_{n-3p} \end{pmatrix} \equiv \operatorname{diag}\Big\{ \underbrace{\Theta_0\,, \Theta_0\,, \;\ldots\;, \Theta_0}_{p}\,, I_{n-3p} \Big\} \tag{K.9a}$$

and the $2n \times 2n$ transformation matrix Φ_p is an extension of Φ_1, viz.,

$$\Phi_p = \mathcal{P}_p \begin{pmatrix} \Theta_{sc} & & & & \boldsymbol{O} \\ & \Theta_{sc} & & & \\ & & \ddots & & \\ & & & \Theta_{sc} & \\ \boldsymbol{O} & & & & I_{2n-6p} \end{pmatrix} \mathcal{P}_p^{\top} \tag{K.9b}$$

$$\equiv \mathcal{P}_p \begin{pmatrix} I_p \otimes \Theta_{sc} & 0 \\ 0 & I_{2n-6p} \end{pmatrix} \mathcal{P}_p^\top$$

$$\equiv \mathcal{P}_p \ \text{diag}\Big\{ \underbrace{\Theta_{sc}\,, \Theta_{sc}\,, \ \ldots\ , \Theta_{sc}}_{p}\,, I_{2n-6p} \Big\} \ \mathcal{P}_p^\top$$

Notice that the mappings Φ_p and $\Phi_{p,dc}$ depend only on the structure of the state-vector $x(\cdot)$, and nothing else. Also notice that $\mathcal{T}$ is an orthogonal matrix, since each one of its diagonal blocks is orthogonal.

This transformation is used in Sect. 3.1.3 to map the phase-domain state-space model (3.5)–(3.6) into the equivalent SC-domain version (3.14), namely

$$\frac{d}{dt}\,\mathbf{X}^{(sc)}(t) \;=\; \big(\mathcal{T}^{-1}\mathcal{A}\mathcal{T}\big)\,\mathbf{X}^{(sc)}(t) + \big(\mathcal{T}^{-1}\mathcal{B}\mathcal{T}\big)\,\mathbf{U}^{(sc)}(t) \tag{K.10a}$$

where $\mathbf{U}^{(sc)}(t) \stackrel{\text{def}}{=} \mathcal{T}^{-1}\mathbf{U}(t)$, and

$$\mathbf{y}(t) \;=\; \begin{pmatrix} x(t) \\ x(t-h) \\ \vdots \\ x(t-(M-1)h) \end{pmatrix} = \big(\mathcal{C}\mathcal{T}\big)\,\mathbf{X}^{(sc)}(t) \;+\; \begin{pmatrix} \eta(t) \\ \eta(t-h) \\ \vdots \\ \eta(t-(M-1)h) \end{pmatrix} \tag{K.10b}$$

where we recall that $\mathcal{C} = C \otimes I_n$, with C given by (3.6).

Appendix L: Example Scenarios

We provide in this appendix detailed information about the several steady-state and transient-event scenarios used as examples in this book.

L.1 Steady-State Scenarios

L.1.1 Switched Single-Phase Resistive Load

A single-phase system, consisting of a switched purely resistive load (with duty-cycle factor D), driven by a sinusoidal voltage, as shown in Fig. L.1. The voltage waveform is given by $v(t) = V_{max}\,\sin\omega_0 t$, and a single period of the current waveform is given by

$$i(t) \;=\; \begin{cases} 0 & |\,t\,| \;\le\; \frac{T}{4}\,(1-D) \\ I_{max}\,\sin\omega_0 t & \frac{T}{4}\,(1-D) \;\le\; |\,t\,| \;\le\; \frac{T}{4}\,(1+D) \\ 0 & \frac{T}{4}\,(1+D) \;\le\; |\,t\,| \le \frac{T}{2} \end{cases}$$

where D is the duty-cycle factor. Since the voltage is purely sinusoidal and single phase, it follows that Ω_v consists solely of the fundamental frequency, so that N_g and N_b vanish, and $P = \Re\{V_1 I_1^*\}$, $Q_B = \Im\{V_1 I_1^*\}$. The rms voltage phasor is $V_1 = \frac{V_{max}}{\sqrt{2}j}$, so $V_{rms} = |V_1|$. In contrast, the current consists of multiple harmonics: its fundamental (rms) phasor is $I_1 = \frac{I_{max}}{\sqrt{2}j}\left[D+\frac{1}{\pi}\sin\pi D\right]$, resulting in a vanishing Q_B. All higher current harmonics belong in the out-of-band component. Also, $I_{rms} = \frac{I_{max}}{\sqrt{2}}\sqrt{D+\frac{1}{\pi}\sin\pi D}$: consequently,

$$S = \frac{V_{max}\, I_{max}}{2}\sqrt{D + \frac{\sin\pi D}{\pi}} \; , \quad P = \frac{V_{max}\, I_{max}}{2}\left(D + \frac{\sin\pi D}{\pi}\right) \; ,$$

$$S_\perp = \sqrt{S^2 - P^2}$$

In this case, the inactive power is entirely due to the out-of-band current harmonics.

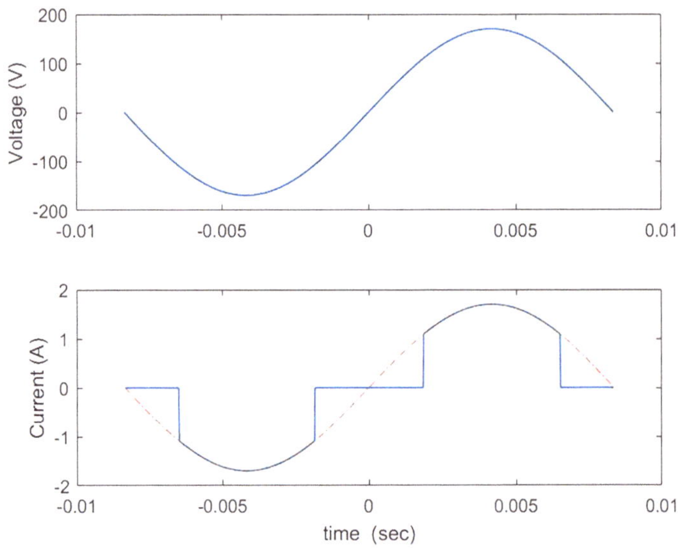

Fig. L.1 Switched resistive load: voltage (top) and current (bottom, solid line).

L.1.2 Single-Phase Purely-Reactive Load

A single-phase, ideal linear inductor of unit reactance at the fundamental frequency, driven by fundamental and 5-th harmonic voltages, of magnitudes 1 and 0.1, respectively. The corresponding voltage and current phasors are

$$V_1 = 1 \;\; , \quad I_1 = -j \;\; , \quad V_5 = 0.1 \;\; , \quad I_5 = -0.02j$$

so that $g_k = 0$ for all k, and $b_1 = 1$, $b_5 = 0.2$.

L.1.3 RLC Load, Balanced Sinusoidal Excitation [252]

A three-phase steady-state system, driven by balanced voltages (of unit magnitude) at the fundamental frequency. The unbalanced load consists of a resistor in phase a, an inductor in phase b, and a capacitor in phase c, each drawing a unit magnitude current. The corresponding voltage and current phasors are summarized in Table L.1.

Table L.1 RLC load, sinusoidal excitation: fundamental harmonic voltage and current phasors with circuit diagram

	a	b	c
$V_1^{(p)}$	1	$e^{-j2\pi/3}$	$e^{j2\pi/3}$
$I_1^{(p)}$	1	$e^{j5\pi/6}$	$e^{-j5\pi/6}$
$Y_1^{(p)}$	1	$-j$	j
$g_1^{(p)}$	1	0	0
$b_1^{(p)}$	0	1	-1

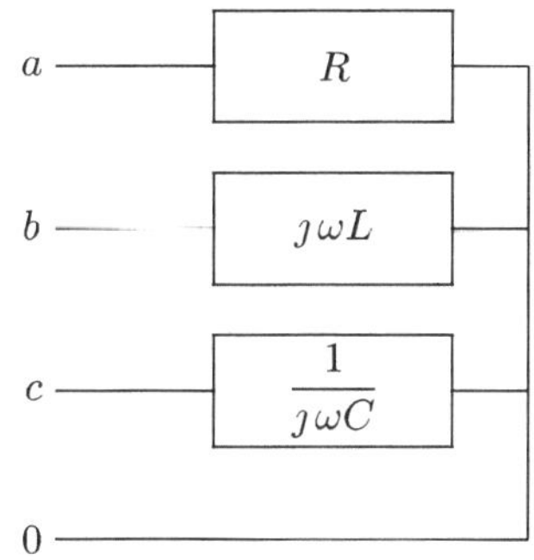

L.1.4 RLC Load, Two Harmonics

A three-phase steady-state system, driven by balanced fundamental and fifth harmonic voltages, with magnitudes 1 and 0.1, respectively, so that the in-band index set is $\Omega_v = \{ (k, p) ; \ k = 1, 5 \text{ and } p = a, b, c \}$. As in the previous scenario (Sect. L.1.3), the unbalanced load consists of a resistor in phase a, an inductor in phase b, and a capacitor in phase c, each drawing a unit magnitude current at the fundamental harmonic. The corresponding voltage and current phasors are summarized in Tables L.1 and L.2.

Table L.2 RLC load, two harmonics: fifth harmonic voltage and current phasors

	a	b	c
$V_5^{(p)}$	0.1	$0.1\, e^{-j2\pi/3}$	$0.1\, e^{j2\pi/3}$
$I_5^{(p)}$	0.1	$0.02\, e^{j5\pi/6}$	$0.5\, e^{-j5\pi/6}$
$Y_5^{(p)}$	1	$-0.2j$	$5j$
$g_5^{(p)}$	1	0	0
$b_5^{(p)}$	0	0.2	-5

L.1.5 RLC Load, Multiple Higher Harmonics

A three-phase steady-state system, driven by balanced fundamental, seventh and eleventh harmonic voltages, with (rms) magnitudes $V_{1,rms}$, $V_{7,rms}$, and $V_{11,rms}$, respectively. As in the previous two scenarios (Sects. L.1.3–L.1.4), the unbalanced load consists of a resistor in phase a, an inductor in phase b, and a capacitor in phase c, each drawing a unit magnitude current at the fundamental harmonic. In all examples, we use $V_{1,rms} = \sqrt{1000}$ V, so that $P \approx 1$ kW. The (much smaller) magnitudes of the higher harmonics are selected individually for each specific example: (a) in Sect. 3.3.1, we use $V_{7,rms} = V_{11,rms} = 0.005\, V_{1,rms}$, and (b) in Sect. 9.5.1, we use $V_{7,rms} = 0.05\, V_{1,rms}$ and $V_{11,rms} = 0$.

L.1.6 Strongly-Unbalanced Load, Balanced Sinusoidal Excitation

A three-phase $\Delta - \text{to} - Y$ transformer with 1:1 winding ratio, driven on the primary side by a balanced fundamental harmonic voltage with magnitude 1 (line to neutral). The load is an impedance of $Z = \sqrt{3} + j$, connected between phase a and the neutral terminal on the secondary side. A real-life example of this scenario is provided by an electric train, which loads only a single phase of a three-phase system, and is supplied through such a $\Delta - \text{to} - Y$ transformer [265].[5] The corresponding voltage and current phasors (at the primary terminals) are summarized in Table L.3. The circuit diagram describes the equivalent circuit on the primary side.

Table L.3 Strongly unbalanced load: fundamental harmonic voltage and current phasors with diagram of equivalent circuit

	a	b	c
$V_1^{(p)}$	1	$e^{-j2\pi/3}$	$e^{j2\pi/3}$
$I_1^{(p)}$	$\frac{\sqrt{3}}{2}$	$-\frac{\sqrt{3}}{2}$	0
$Y_1^{(p)}$	$\frac{\sqrt{3}}{2}$	$\frac{\sqrt{3}}{2}e^{-j\pi/3}$	0
$g_1^{(p)}$	$\frac{\sqrt{3}}{2}$	$\frac{\sqrt{3}}{4}$	0
$b_1^{(p)}$	0	$\frac{3}{4}$	0

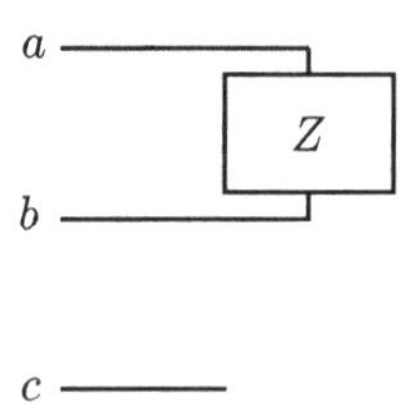

[5] The same example, but with different parameter values, is presented in [265, Fig. 10].

Table L.4 Adjustable-speed drive: voltage phasors

	Phase a		Phase b		Phase c	
Harmonic index	Magn. (V)	Phase (rad)	Magn. (V)	Phase (rad)	Magn. (V)	Phase (rad)
1	221.4	− 0.0032	223.0	− 2.0873	224.2	2.0905
3	0.5600	1.9100	0.5600	− 2.2788	0.5600	− 0.1844
5	3.744	− 2.4884	4.252	− 0.4832	4.324	1.7550
7	1.243	0.4647	1.693	− 2.2310	0.7836	1.6643
9	0.2252	− 2.2100	0.2252	− 0.1156	0.2252	1.9788
11	0.4042	0.9500	0.4042	3.0440	0.4042	− 1.1444

Table L.5 Adjustable-speed drive: current phasors

	Phase a		Phase b		Phase c	
Harmonic index	Magn. (V)	Phase (rad)	Magn. (V)	Phase (rad)	Magn. (V)	Phase (rad)
1	342.8	− 0.2269	342.8	− 2.3213	342.8	1.8675
3	3.428	0.5760	3.428	− 1.5184	3.428	2.6704
5	13.37	− 0.2793	13.37	− 2.3736	13.37	1.8151
7	7.199	1.3439	7.199	− 0.7505	7.199	3.4383
9	0.3428	0.8552	0.3428	− 1.2392	0.3428	2.9496
11	4.456	2.5307	4.456	0.4363	4.456	4.6251
13	2.057	− 2.8623	2.057	− 4.9567	2.057	− 0.7679
15	0.3428	− 0.7854	0.3428	− 2.8798	0.3428	1.3090
17	7.199	− 0.7854	7.199	− 2.8798	7.199	1.309
19	5.142	− 0.8901	5.142	− 2.9845	5.142	1.2043

L.1.7 Adjustable-Speed Drive [252, 301]

A 350-HP industrial adjustable-speed drive (ASD), operating in parallel with a linear unbalanced load. The voltages contain (odd) harmonics up to the 11-th, while current has harmonics up to the 19-th. The corresponding voltage and current phasors are summarized in Tables L.4 and L.5, which are based on the values listed in [301].

L.1.8 Hydro-Ottawa Distribution Grid, Steady-State Operation

This real-world steady-state case was obtained from Hydro-Ottawa, the largest local distribution company in eastern Ontario, Canada. Hydro-Ottawa customer base consists of 303,571 residential homes, 28,193 commercial, and 13 large users. This dataset was recorded by a Schneider Electric ION 86xx meter at 128 samples/cycle (= 7680 samples/sec), on Jan. 7, 2017 at Hydro-Ottawa Site 100. Voltage and current waveforms for this case are shown in Fig. L.2.

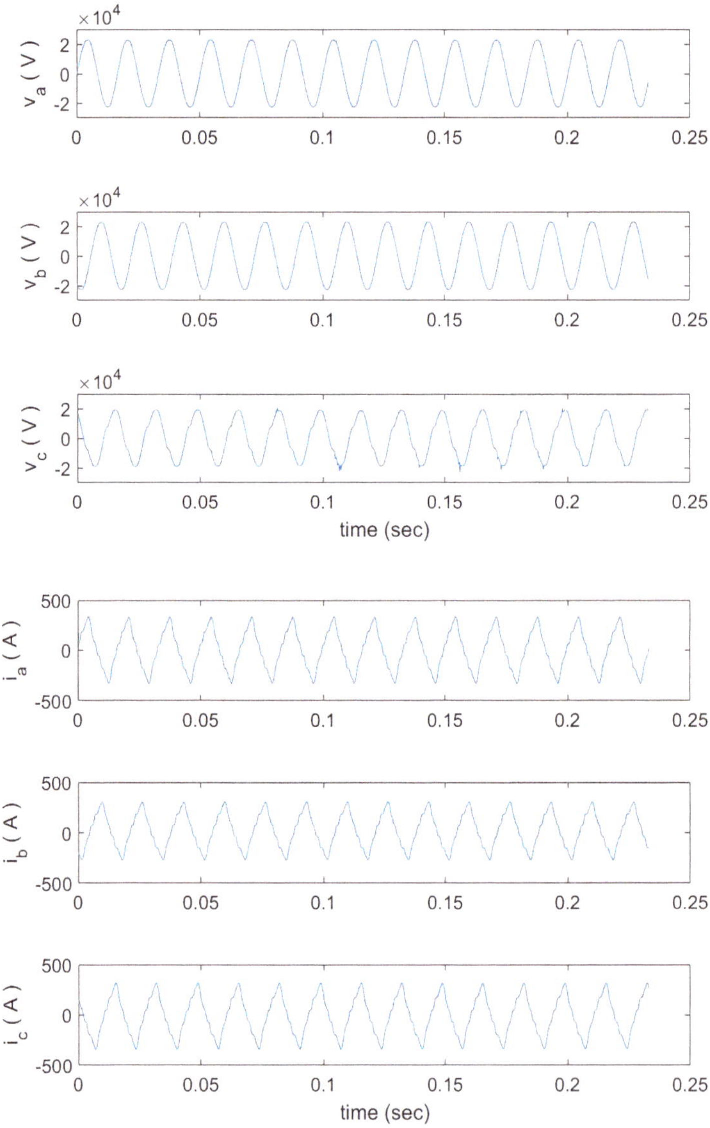

Fig. L.2 Hydro-Ottawa steady-state dataset: voltages (top) and currents (bottom).

L.2 Transient Scenarios

L.2.1 Synthetic Short Transient

A three-phase system, driven by balanced fundamental and fifth harmonic voltages, and an injected 15-th harmonic transient, limited to the interval from $t = 0.03$ sec to $t = 0.035$ sec, viz.,

$$\left.\begin{aligned} v_{transient}(t) &= 5.5\, e^{-t/0.0018} \cos\big(15\,\omega_0(t-0.03)\big) \\ i_{transient}(t) &= 0.9\, e^{-t/0.0018} \cos\big(15\,\omega_0(t-0.03)-1.4\big) \end{aligned}\right\}\quad 0.03 \le t < 0.035$$

Voltage and current waveforms are shown in Fig. L.3. The unbalanced load consists of resistive–inductive impedances $Z_1 = Z_2 = Z_3 = (1 + j\,0.4\,k)\,\Omega$, where k denotes the harmonic index. The fundamental frequency is 60 Hz, so that $\omega_0 = 120\,\pi$ rad/sec. Plots in this book were generated using 256 samples/cycle (= 15,360 samples/sec). The corresponding steady-state voltage and current phasors are summarized in Table L.6.

Table L.6 Synthetic short transient: steady-state voltage and current phasors

	a	b	c
$V_1^{(p)}$	$1.0\,e^{j\pi/3}$	$1.0\,e^{-j\pi/3}$	$1.0\,e^{j\pi}$
$V_5^{(p)}$	0.8	$0.8\,e^{-j2\pi/3}$	$0.8\,e^{j2\pi/3}$
$I_1^{(p)}$	$0.9285\,e^{j0.6667}$	$0.9285\,e^{j(0.6667-2\pi/3)}$	$0.9285\,e^{j(0.6667+2\pi/3)}$
$I_5^{(p)}$	$0.3578\,e^{j5.1761}$	$0.3578\,e^{j(5.1761-2\pi/3)}$	$0.3578\,e^{j(5.1761+2\pi/3)}$

This transient belongs in the "wide-band" category, based on its time–frequency representation (Fig. 5.2).

L.2.2 Wind-Farm Transient [303]

This real-world transient case was provided by the National Renewable Energy Laboratory (NREL) and collected from the Trent Mesa Wind Farm in Texas. A short transient in the current waveform is caused by switching in a bank of shunt capacitors. This dataset was recorded at 128 samples/cycle (= 7680 samples/sec). Voltage and current waveforms for this case are shown in Fig. L.4.

This transient belongs in the "wide-band" category, based on its time–frequency representation (Fig. 4.5).

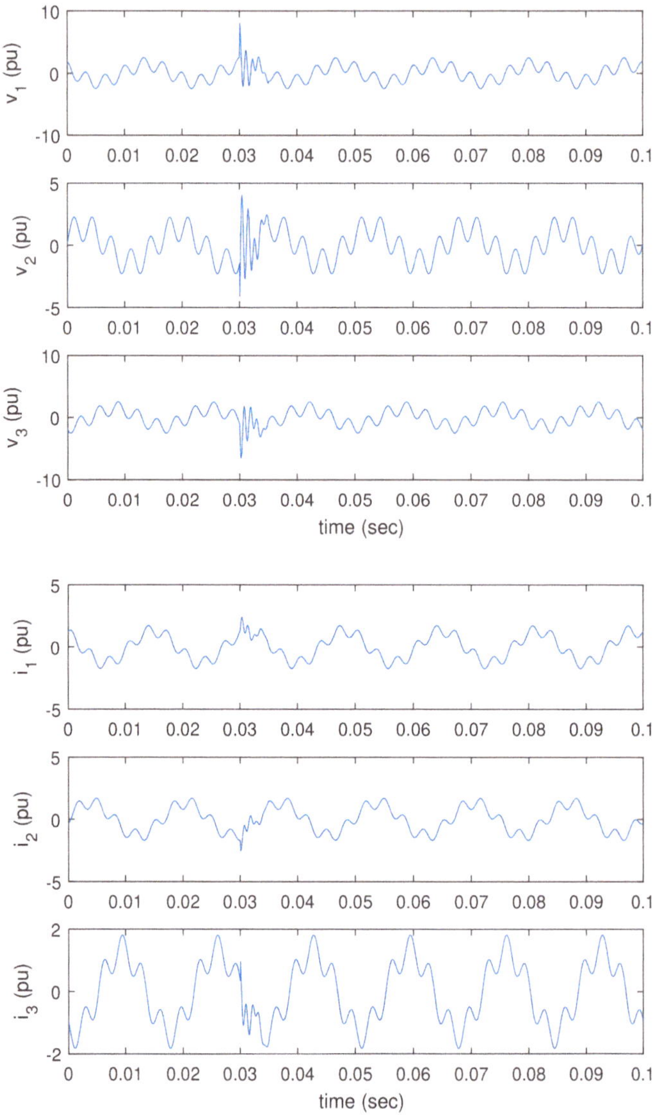

Fig. L.3 Synthetic short transient: voltages (top) and currents (bottom).

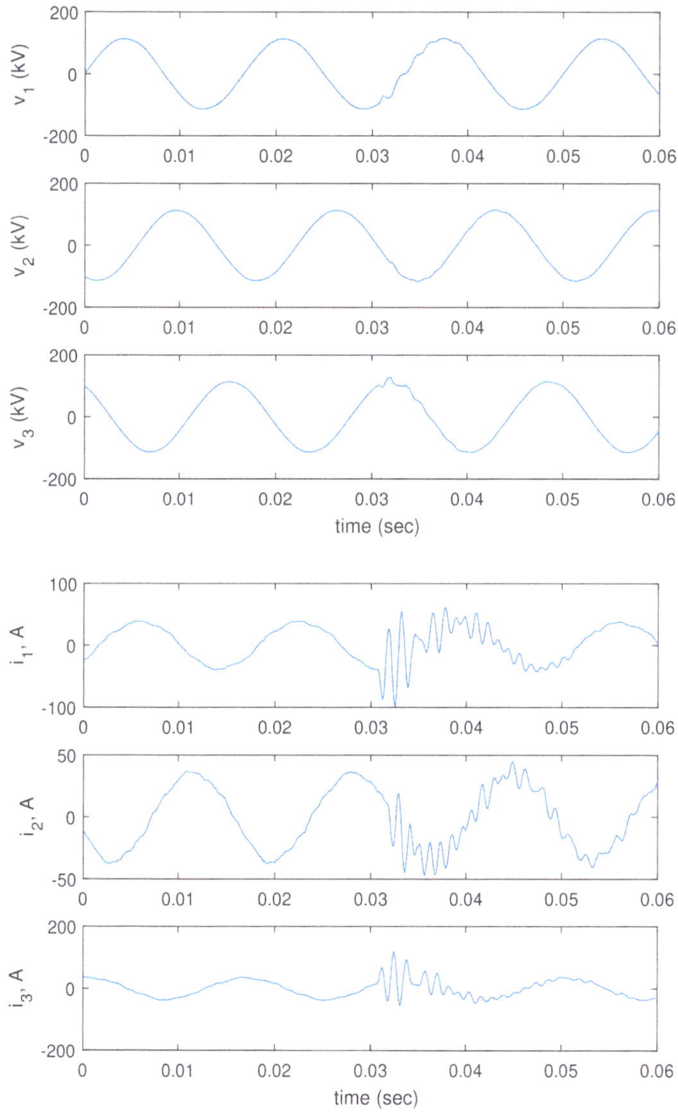

Fig. L.4 Wind-farm transient dataset: voltages (top) and currents (bottom).

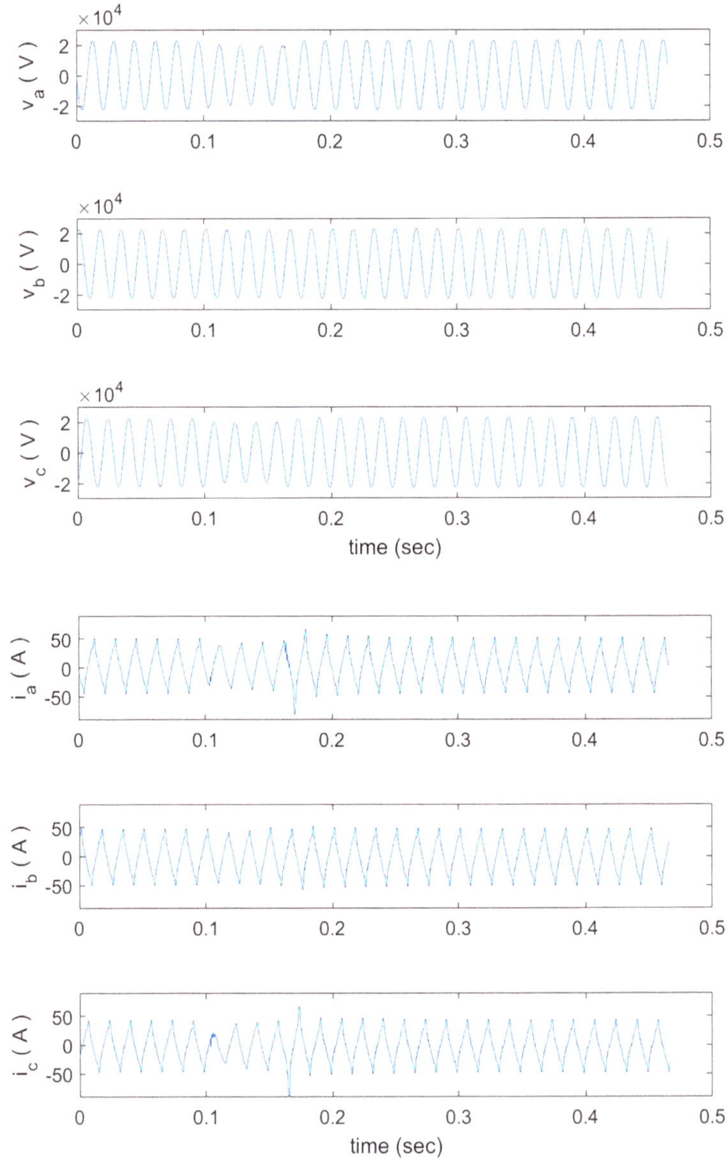

Fig. L.5 Hydro-Ottawa transient dataset: voltages (top) and currents (bottom).

L.2.3 Hydro-Ottawa Distribution Grid, Transient Operation

This real-world transient case was obtained from Hydro-Ottawa, the largest local distribution company in eastern Ontario, Canada. Hydro-Ottawa customer base consists of 303,571 residential homes, 28,193 commercial, and 13 large users. This dataset was recorded by Schneider Electric ION 86xx meters at 128 samples/cycle (= 7680 samples/sec), on June 26, 2017 at Hydro-Ottawa Site 10. Voltage and current waveforms for this case are shown in Fig. L.5.

This transient belongs in the "in-band" category, based on its time–frequency representation (Fig. 4.1).

L.2.4 Paper-Mill Transient [302]

This real-world transient case was extracted from a one-year power quality study (2004–2005) in the Mazandaran Wood and Paper Industries (MWPI), the largest paper manufacturer in Iran with a production capacity of 175,000 tons of paper per year, including 90,000 tons of newsprint, printing, and writing paper and 85,000 tons of fluting paper. The total power consumed by plant is 35 MW, and the fundamental frequency is 50 Hz. This dataset was recorded at 140 samples/cycle (= 7000 samples/sec). Voltage and current waveforms for this case are shown in Fig. L.6.

This transient belongs in the "in-band" category, based on its time–frequency representation (Fig. 4.3).

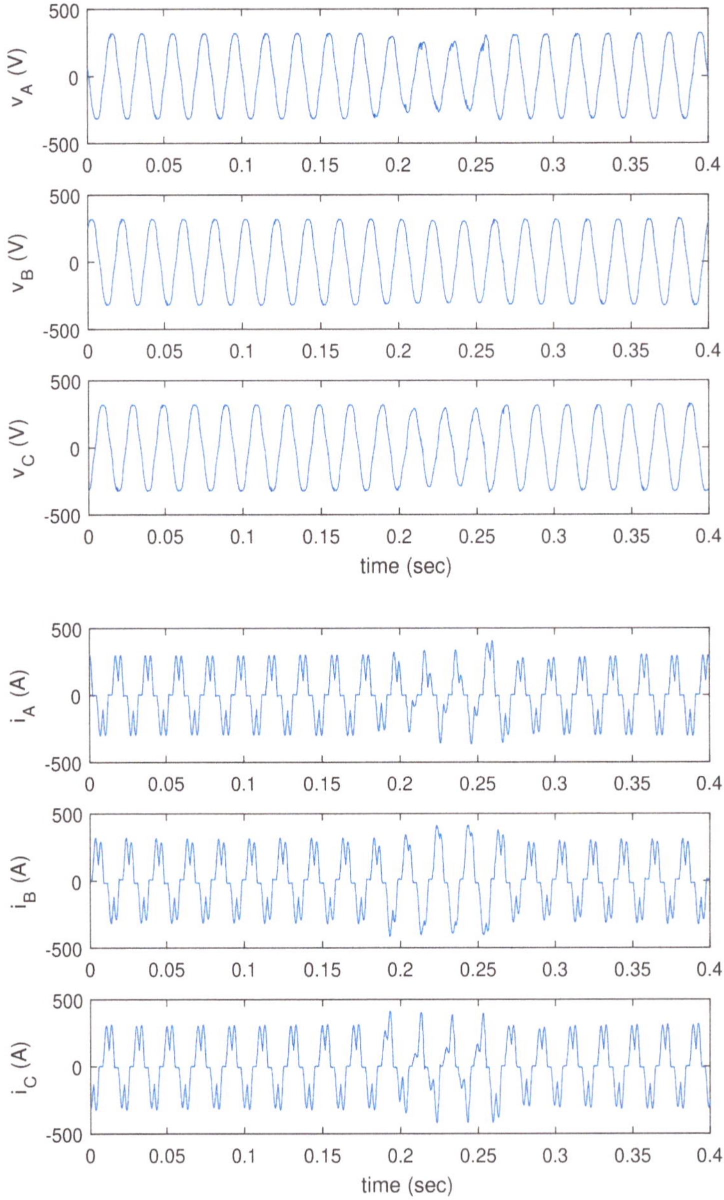

Fig. L.6 Paper-mill transient dataset: voltages (top) and currents (bottom).

Bibliography

1. G.C. Verghese, "Getting to the Gray Box: Some Challenges for Model Reduction," *2009 American Control Conference Plenary Talk*, https://acc2009.a2c2.org/plenary.htm
2. A.V. Pavlov, N. van de Wouw and H. Nijmeijer, *Uniform Output Regulation of Nonlinear Systems: A convergent Dynamics Approach In Systems & Control*, Birkhauser, Boston, 2005.
3. A.V. Pavlov, N. van de Wouw and H. Nijmeijer, "Frequency response functions for nonlinear convergent systems," *IEEE Trans. on Automatic Control*, Vol. 52, No. 6, pp. 1159–1165, June 2007.
4. C.L. DeMarco and G.C. Verghese, "Bringing phasor dynamics into the power system load flow," *North American Power Symposium*, Washington, D.C., pp. 463–469, Oct. 1993.
5. S.G. Jalali, R.A. Hedin, M. Pereira and K. Sadek, "A stability model for the advanced series compensator (ASC)," *IEEE Trans. on Power Delivery*, Vol. 11, No. 2, pp. 1128–1137, Apr. 1996.
6. H.K. Khalil, *Nonlinear systems*, 2nd edition, Prentice Hall, 1996.
7. N. Fenichel, "Geometric Singular Perturbation Theory for Ordinary Differential Equations," *Journal of Differential Equations*, Vol. 31, No. 1, pp. 53–98, Jan. 1979.
8. J.H. Taylor and B.H. Wilson, "A frequency domain model order reduction algorithm for nonlinear systems," *IEEE Conference on Control Applications*, pp. 1053–1058, 1995.
9. S.R. Sanders, "On limit cycles and the describing function method in periodically switched circuits," *IEEE Trans. on Circuits and Systems – Part I*, Vol. 40, No. 9, pp. 564–572, Sept. 1993.
10. I. Dobson, "Stability of ideal thyristor and diode switching circuits," *IEEE Trans. on Circuits and Systems-I*, Vol. 42, No. 9, pp. 517–529, Sept. 1995.
11. A. Brambilla, D. D'Amore and M. Pillan, "Convergence improvements of the harmonic balance method," *International Symposium on Circuits and Systems*, Vol. 4, pp. 2482–2485, 1993.
12. M. Green and D.J.N. Limebeer, *Linear Robust Control*, Prentice Hall, 1994.
13. G. Escobar, A. A. Valdez, J. Leyva-Ramos and P. Mattavelli, "Repetitive-Based Controller for a UPS Inverter to Compensate Unbalance and Harmonic Distortion," *IEEE Trans. Industrial Electronics*, Vol. 54, No. 1, pp. 504–510, Feb. 2007.
14. P. Mattavelli, G. Escobar and A.M. Stanković, "Dissipativity-Based Adaptive and Robust Control of UPS," *IEEE Trans. Industrial Electronics*, Vol. 48, No. 2, pp. 334–343, Apr. 2001.
15. IEEE PES Task Force, "Modeling and simulation of the propagation of harmonics in electric power networks – parts I and II," *IEEE PES Winter Meeting*, Nos. 264-2 and 265-9 PWRD, 1995.
16. Probabilistic Aspects Task Force of the Harmonics Working Group Subcommittee of the Transmission and Distribution Committee, "Time-Varying Harmonics: Part I – Characterizing Measured Data," *IEEE Transactions on Power Delivery*, Vol. 13, No. 3, pp. 938–944, July 1998.

H. Lev-Ari, A. M. Stanković, *Dynamic Phasors in Energy Processing Systems*, Power Electronics and Power Systems, https://doi.org/10.1007/978-3-031-50020-6

17. A.E. Emanuel, "The Randomness Power: An Other New Quantity to be Considered," *IEEE Transactions on Power Delivery*, Vol. 22, No. 3, pp. 1304–1308, July 2007.
18. A.E. Emanuel, "An Introduction to the Concept of Randomness Power," *Proceedings of 14th International Conference on Harmonics and Quality of Power (ICHQP 2010)*, Bergamo, Italy, Sept. 2010.
19. C. Slivinsky, S. Stanton, and J. Esztergalyos, "Phase angle measurement applications – a new transducer that measures state variables in real time," *19-th Ann. Western Protective Relay Conference*, 1992.
20. T. Berry, D. Giannakis, and J. Harlim, "Bridging Data Science and Dynamical Systems Theory," *Notices of the American Mathematical Society*, Vol. 67, No. 9, pp. 1336–1348, Oct. 2020.
21. Y. Katznelson, *An introduction to Harmonic Analysis*, Dover, 1976.
22. G.P. Tolstov, *Fourier Series*, Dover, 1962.
23. H. Kwakernaak and R. Sivan, *Modern Signals and Systems*, Prentice Hall, 1991.
24. A.V. Oppenheim and A.S. Willsky with S. Nawab, *Signals & Systems*, 2nd edition, Prentice-Hall 1997.
25. C. Phillips, J. Parr and E. Riskin, *Signals, Systems and Transforms*, 5th Edition, Pearson, 2013.
26. D.W. Novotny and T.A. Lipo, *Vector Control and Dynamics of AC Machines*, Oxford University Press, 1996.
27. S. Fryze, "Active, Reactive and Apparent Power in Networks with Nonsinusoidal Waveforms of Voltage and Current," (in German) *Electrotechnische Zeitschrift*, Vol. 53, No. 25–27, pp. 596–599, 625–627, 700–702, 1932.
28. W. Quade, "Neue Darstellung der Verzerrungsleistung eines Wechselstroms mit Hilfe des Funktionenraums," *Archiv für Elektrotechnik*, No. 5., pp. 277–305, 1939.
29. H. Lev-Ari, A.M. Stanković, "Hilbert Space Techniques for Modeling and Compensation of Reactive Power in Energy Processing Systems," *IEEE Trans. Circuits & Systems*, Vol. 50, No. 4, pp. 540–556, Apr. 2003.
30. J.D. Glover, M.S. Sarma and T.J. Overbye, *Power System Analysis and Design*, 5th edition, Cengage Learning, 2012.
31. K.K. Sen. "Static synchronous compensator – theory, modeling and applications," *IEEE Trans. on Power Delivery*, Vol. 13, No. 1, pp. 241–246, Jan. 1998.
32. J.G. Proakis and D. Manolakis, *Digital Signal Processing: Principles, Algorithms and Applications*, 4th edition, Prentice Hall, 2007.
33. R. Cristi, *Modern Digital Signal Processing*, Thomson (Brooks/Cole), 2004.
34. M.K. Penshanwar, M. Gavande and M.F.A.R. Satarkar, "Phasor Measurement Unit Technology and its Applications – A Review," *International Conference on Energy Systems and Applications (ICESA)*, Nov. 2015.
35. Powerside microPMU & LV Data Sheet. URL: https://powerside.com/wp-content/uploads/2020/12/MicroPMU-LV-Data-Sheet.V1_En.pdf
36. Qualitrol Q-PMU 9/18/36 brochure. URL: https://2hfybu1lrdue3x9wnu1dvw7s-wpengine.netdna-ssl.com/wp-content/uploads/2016/09/IP-P33-01L-03E_Q_PMU_9_18_36.pdf
37. Vizimax PMU Data Sheet. URL: https://www.vizimax.com/products-services/phasor-measurement-unit
38. K.J. Astrom and B. Wittenmark, *Computer Controlled Systems*, 3rd edition, Dover, 2011.
39. W. Xu, Z. Huang, X. Xie and C. Li, "Synchronized Waveforms – a Frontier of Data-Based Power System and Apparatus Monitoring, Protection and Control," *IEEE Transactions on Power Delivery*, Vol. 37, No. 1, pp. 3–17, Feb. 2022.
40. A.C. Gilbert, M.J. Strauss and J.A. Tropp, "A tutorial on fast Fourier sampling," *IEEE Signal Processing Magazine*, Vol. 25, No. 2, pp. 57–66, March 2008.
41. A.C. Gilbert, P. Indyk, M. Iwen and L. Schmidt, "Recent developments in the sparse Fourier transform: A compressed Fourier transform for big data," *IEEE Signal Processing Magazine*, Vol. 31, No. 5, pp. 91–100, Sept. 2014.

42. A.G. Phadke and J.S. Thorp, *Synchronized Phasor Measurements and Their Applications*, 2nd edition, Springer, 2017.
43. A. Gelb and W.E. van der Velde, *Multiple Input Describing Functions and Nonlinear Systems Design*, McGraw-Hill, 1968.
44. A.I. Mees and A.R. Bergen, "Describing function revisited," *IEEE Transactions on Automatic Control*, Vol. 20, No. 4, pp. 473–478, Sept. 1975.
45. S.R. Sanders, J.M. Noworolski, X.Z. Liu and G.C. Verghese, "Generalized averaging method for power conversion circuits," *IEEE Transactions on Power Electronics*, Vol. 6, No. 2, pp. 251–259, Apr. 1991.
46. V.A. Caliskan, G.C. Verghese and A.M. Stanković, "Multifrequency averaging of dc/dc converters," *IEEE Transactions on Power Electronics*, Vol. 14, No. 1, pp. 124–133, Jan. 1999.
47. S. Yamamura, "Spiral vector theory of ac circuits and machines," *Proc. Japan Acad.*, Vol. 65 (Ser. B), pp. 142–145, 1989.
48. W.V. Lyon, *Transient Analysis of Alternating Current Machinery*, Wiley, 1954.
49. P. Kundur, *Power System Stability and Control*, McGraw-Hill, 1994.
50. Q. Huang and V. Vittal, "Advanced EMT and Phasor-Domain Hybrid Simulation With Simulation Mode Switching Capability for Transmission and Distribution Systems," *IEEE Transactions on Power Systems*, Vol. 33, No. 6, pp. 6298–6308, Nov. 2018
51. D. Shu, X. Xie, V. Dinavahi, C. Zhang, X. Ye and Q. Jiang, "Dynamic Phasor Based Interface Model for EMT and Transient Stability Hybrid Simulations," *IEEE Transactions on Power Systems*, Vol. 33, No. 4, pp. 3930–3939, July 2018
52. B.C. Lesieutre, J.A. Mohamed and A.M. Stanković, "Analysis of Ferroresonance in Three-phase Transformers," *Proceedings of the International Conference on Power System Technology (PowerCon 2000)*, Perth, Australia, pp. 1013–1018, Dec. 2000.
53. M. Dragomir, A. Iamandi, M. Istrate and A. Dragomir, "A Review of Phasor Estimation Algorithms for Power System Applications," *IEEE PES Innovative Smart Grid Technologies Europe (ISGT-Europe)*, Bucharest, Romania, pp. 1–5, Sept. 2019.
54. A.G. Phadke and T. Bi, "Phasor measurement units, WAMS, and their applications in protection and control of power systems," *Springer Journal of Modern Power Systems and Clean Energy*, Vol. 6, No. 4, pp. 619–629, July 2018.
55. Y. Yeung and C. K. Lau, "Exploring the Application of Phasor Measurement Units in the Distribution Network," *IEEE International Conference on Power and Energy (PECon)*, Penang, Malaysia, pp. 299–303, Dec. 2020.
56. A. G. Kasembe, Z. Muller, J. Svec, J. Tlusty and V. Valouch, "Synchronous Phasors Monitoring System Application Possibilities," *IEEE 27th Convention of Electrical and Electronics Engineers in Israel*, Eilat, Israel, pp. 1–3, Nov. 2012.
57. V. Vittal, "Application of Phasor Measurements for Dynamic Security Assessment using Decision Trees," *IEEE Power and Energy Society General Meeting*, San Diego, CA, pp. 1–3, July 2012.
58. J.A. de la O Serna, and J. Rodriguez-Maldonado, "Instantaneous Oscillating Phasor Estimates with Taylor^K-Kalman Filters," *IEEE Transaction on Power Systems*, Vol. 26, No. 4, pp. 2336–2344, Nov. 2011.
59. J.A. de la O Serna, and J. Rodriguez-Maldonado, "Taylor-Kalman-Fourier Filters for Instantaneous Oscillating Phasor and Harmonic Estimates," *IEEE Transaction on Instrumentation and Measurement*, Vol. 61, No. 4, pp. 941–951, Apr. 2012.
60. J. Liu, F. Ni, J. Tang, F. Ponci and A. Monti, "A Modified Taylor-Kalman Filter for Instantaneous Dynamic Phasor Estimation," *3rd IEEE PES Innovative Smart Grid Technologies Europe (ISGT Europe)*, Berlin, Germany, pp. 1–7, Oct. 2012.
61. J. Khodaparast and M. Khederzadeh, "Least Square and Kalman Based Methods for Dynamic Phasor Estimation: a review," *Springer Open: Protection and Control of Modern Power Systems*, Vol. 2, Article 1, pp. 1–18, Jan. 2017.
62. A.G. Phadke and J.S. Thorp, *Computer Relaying for Power Systems*, 2nd edition, Wiley, 2009.

63. L. Chen, W. Zhao, Y. Yu and S. Huang, "Interharmonic and Fundamental Phasor Estimator for Smart Grid Applications," *2018 Conference on Precision Electromagnetic Measurements (CPEM 2018)*, Paris, France, pp. 1–2, July 2018.
64. A. Silverstein and J. Follum, "High-Resolution, Time-Synchronized Grid Monitoring Devices," *North American Synchrophasor Initiative (NASPI)*, Technical Report, March 2020. URL: https://naspi.org/node/819
65. B. Picinbono and P. Chevalier, "Widely Linear Estimation with Complex Data," *IEEE Trans. Signal Processing*, Vol. 43, No. 8, pp. 2030–2033, Aug. 1995.
66. D.H. Dini and D.P. Mandic, "Class of Widely Linear Complex Kalman Filters," *IEEE Trans. Neural Networks and Learning Systems*, Vol. 23, No. 5, pp. 775–786, May 2012.
67. W. Dang and L.L. Scharf, "Extensions to the Theory of Widely Linear Complex Kalman Filtering," *IEEE Trans. Signal Processing*, Vol. 60, No. 12, pp. 6669–6674, Dec. 2012.
68. R.A. Horn and C.R. Johnson, *Topics in Matrix Analysis*, Cambridge University Press, 2008.
69. A.M. Stanković, H. Lev-Ari and M.M. Perišić, "Analysis and Implementation of Model-Based Linear Estimation of Dynamic Phasors," *IEEE Trans. Power Systems*, Vol. 19, No. 4, pp. 1903–1910, Nov. 2004.
70. T. Kailath, *Linear Systems*, Prentice Hall, 1980.
71. A. Rahmati and R. Adhami, "A Fault Detection and Classification Technique Based on Sequential Components," *IEEE Trans. Industry Applications*, Vol. 50, No. 6, pp. 4202–4909, Nov. 2014.
72. N.H. van der Blij, L.M. Ramirez-Elizondo, M.T.J. Spaan and P. Bauer, "Symmetrical Component Decomposition of DC Distribution Systems," *IEEE Trans. Power Systems*, Vol. 33, No. 3, pp. 2733–2741, May 2018.
73. D. Yazdani, M. Mojiri, A. Bakhshai and G. Joos, "A Fast and Accurate Synchronization Technique for Extraction of Symmetrical Components," *IEEE Trans. Power Electronics*, Vol. 24, No. 3, pp. 674–684, March 2009.
74. M. Karimi-Ghartemani and H. Karimi, "Processing of Symmetrical Components in Time-Domain," *IEEE Trans. Power Systems*, Vol. 22, No. 2, pp. 572–579, May 2007.
75. E. Karami, G.B. Gharehpetian, M. Madrigal and J. de Jesus Chavez, "Dynamic Phasor-Based Analysis of Unbalanced Three-Phase Systems in Presence of Harmonic Distortion," *IEEE Trans. Power Systems*, Vol. 33, No. 6, pp. 6642–6654, Nov. 2018.
76. C.J. Lawson and R.J. Hanson, *Solving Least Squares Problems*, Society for Industrial and Applied Mathematics (SIAM), 1995.
77. H. Lev-Ari and A.M. Stanković, "A Sparse Sampling Approach to Dynamic Sub-Cycle Decomposition of Apparent Power in General Polyphase Networks," *Proceedings of the International School on Nonsinusoidal Currents and Compensation* (ISNCC 2015), Lagów, Poland, June, 2015.
78. A. Ghanavati, H. Lev-Ari and A.M. Stanković, "Sub-Cycle Dynamic Phasors with Adjustable Transient Response," *IEEE Transactions on Power Delivery*, Vol. 35, No. 4, pp. 1886–1894, Aug. 2020.
79. A. Ghanavati, "A Sub-Cycle Approach to Dynamic Phasors with Application to Dynamic Power Quality Metrics," PhD Dissertation, Dept. of Electrical and Computer Engineering, Northeastern University, Boston, MA, May 2018.
80. A. Ghanavati, H. Lev-Ari and A.M. Stanković, "A Sub-Cycle Approach to Dynamic Phasors with Application to Dynamic Power Quality Metrics," *IEEE Transactions on Power Delivery*, Vol. 33, No. 5, pp. 2217–2225, Oct. 2018.
81. J.H.R. Enslin and P.J.M. Heskes, "Harmonic Interaction Between a Large Number of Distributed Power Inverters and the Distribution Network," *IEEE Transactions on Power Electronics*, Vol. 19, No. 6, pp. 1586–1593, Nov. 2004.
82. S.L. Hahn, *Hilbert Transforms in Signal Processing*, Artech House, 1996.
83. R.N. Bracewell, *The Fourier Transform and Its Applications*, 2nd Edition, McGraw-Hill, 1978.
84. F. Hlawatsch and F. Auger, *Time-Frequency Analysis*, Wiley, 2013.
85. B. Picinbono, "On Instantaneous Amplitude and Phase of Signals," *IEEE Transactions on Signal Processing*, Vol. 45, No. 3, pp. 552–560, March 1997.

86. D. Gabor, "Theory of communication," *Journal of the Institution of Electrical Engineers*, Vol. 93, pp. 429–457, Nov. 1946.
87. E. Bedrosian, "The Analytic Signal Representation of Modulated Waveforms," *Proceedings of the IRE*, Vol. 50, No. 10, pp. 2071–2076, Oct. 1962.
88. J.G. Proakis and M. Salehi, *Communication System Engineering*, 2nd edition, Prentice Hall, 2002.
89. S.L. Hahn, "The History of Applications of Analytic Signals in Electrical and Radio Engineering," *EUROCON 2007 – The International Conference on "Computer as a Tool"*, Warsaw, Poland, Sept. 2007.
90. B. Boashash, "Estimating and Interpreting the Instantaneous Frequency of a Signal — Part 1: Fundamentals," *Proceedings of the IEEE*, Vol. 80, No. 4, pp. 520–538, Apr. 1992.
91. L. Cohen, *Time-Frequency Analysis,* Prentice Hall, 1995.
92. E. Bedrosian, "A product theorem for Hilbert transforms," *Proceedings of the IEEE*, Vol. 51, No. 5, pp. 868–869, May 1963.
93. M. Schwartz, W.R. Bennett and S. Stein, *Communication Systems and Techniques*, McGraw-Hill, 1966.
94. L. Cohen, P. Loughlin and D. Vakman, "On an Ambiguity in the Definition of the Amplitude and Phase of a Signal," *Signal Processing*, Vol. 79, pp. 301–307, 1999.
95. N.E. Huang et al., "The Empirical Mode Decomposition and the Hilbert Spectrum for Nonlinear and Non-Stationary Time Series Analysis," *Proceedings of the Royal Society of London*, Series A, Vol. 454, pp. 903–995, 1998.
96. N. Senroy, S. Suryanarayanan and P.F. Ribeiro, "An Improved Hilbert-Huang Method for Analysis of Time-Varying Waveforms in Power Quality," *IEEE Transactions on Power Systems*, Vol. 22, No. 4, pp. 1843–1850, Nov. 2007.
97. M.J. Afroni, D. Sutanto and D. Stirling, "Analysis of Nonstationary Power-Quality Waveforms Using Iterative Hilbert Huang Transform and SAX Algorithm," *IEEE Transactions on Power Delivery*, Vol. 28, No. 4, pp. 2134–2144, Oct. 2013.
98. K. Dragomiretskiy and D. Zosso, "Variational Mode Decomposition," *IEEE Transactions on Signal Processing*, Vol. 62, No. 3, pp. 531–544, Feb. 2014.
99. T. Croes, C. Gherasim, J. Van den Keybus, J. Ghijselen, J. Driesen and R. Belmans, "Power Measurement using the Wavelet Transform of Analytic Signals," *11th International Conference on Harmonics and Quality of Power*, Lake Placid, NY, Sept. 2004.
100. S. Qian, *Introduction to Time-Frequency and Wavelet Transforms*, Prentice Hall, 2002.
101. B. Boashash, "Estimating and Interpreting the Instantaneous Frequency of a Signal — Part 2: Algorithms and Applications," *Proceedings of the IEEE*, Vol. 80, No. 4, pp. 540–568, Apr. 1992.
102. L. Zhan, Y. Liu and Y. Liu, "A Clarke Transformation-Based DFT Phasor and Frequency Algorithm for Wide Frequency Range," *IEEE Transactions on Smart Grid*, Vol. 9, No. 1, pp. 67–77 , Jan. 2018.
103. S. Tretter, "Estimating the frequency of a noisy sinusoid by linear regression," *IEEE Transactions on Information Theory*, Vol. 31, No. 6, pp. 832–835, Nov. 1985.
104. S. Kay, "Statistically/computationally efficient frequency estimation," *IEEE International Conference on Acoustics, Speech and Signal Processing*, New York, NY, Apr. 1988, pp. 2292–2295.
105. C.F. Ten and P.A. Crossley, "Evaluation of Rocof Relay Performances on Networks with Distributed Generation," *IET 9th International Conference on Developments in Power System Protection (DPSP 2008)*, Glasgow, UK, March 2008.
106. B. Porat, *Digital Processing of Random Signals: Theory and Methods*, Prentice Hall, 1994.
107. P. Stoica and R.L. Moses, *Introduction to Spectral Analysis*, Prentice Hall, 1997.
108. P.B. Petrović, "Power Harmonics Estimation Based on Analytical Signal Concept," *2017 International Symposium on Power Electronics (Ee)*, pp. 1–6, Novi Sad, Serbia, Oct. 2017.
109. S.S. Kelkar, L.L. Grigsby and J. Langsner, "An Extension of Parseval's Theorem and Its Use in Calculating Transient Energy in the Frequency Domain," *IEEE Transactions on Industrial Electronics*, Vol. IE-30, No. 1, pp. 42–45, Feb. 1983.

110. H.-H. Chang, K.-L. Lian, Y.-C. Su and W.-J. Lee, "Power-Spectrum-Based Wavelet Transform for Nonintrusive Demand Monitoring and Load Identification," *IEEE Transactions on Industry Applications*, Vol. 50, No. 3, pp. 2081–2089, May-June 2014.
111. IEEE Standards Association, "IEEE Recommended Practice for Monitoring Electric Power Quality," *IEEE Standard 1159-2019*, Aug. 2019. Available at: https://standards.ieee.org/ieee/1159/6124/
112. M.H.J. Bolen and I.Y.H. Gu, *Signal Processing of Power Quality Events*, Wiley, 2006.
113. K. Takagi and H. Fujita, "A Three-Phase Grid-Connected Inverter Equipped With a Shunt Instantaneous Reactive Power Compensator," *IEEE Transactions on Industry Applications*, Vol. 55, No. 4, pp. 3955–3966, July/Aug. 2019.
114. D. Wiot, "A New Adaptive Transient Monitoring Scheme for Detection of Power System Events," *IEEE Transactions on Power Delivery*, Vol. 19, No. 1, pp. 42–48, Jan. 2004.
115. M. Simonov, G. Chicco and G. Zanetto, "Real-Time Event-Based Energy Metering," *IEEE Transactions on Industrial Informatics*, Vol. 13, No. 6, pp. 2813–2823, Dec. 2017.
116. R. Yadav, A.K. Pradhan and I. Kamwa, "Real-Time Multiple Event Detection and Classification in Power System Using Signal Energy Transformations," *IEEE Transactions on Industrial Informatics*, Vol. 15, No. 3, pp. 1521–1531, March 2019.
117. O.P. Mahela, B. Khan, H.H. Alhelou and P. Siano, "Power Quality Assessment and Event Detection in Distribution Network With Wind Energy Penetration Using Stockwell Transform and Fuzzy Clustering," *IEEE Transactions on Industrial Informatics*, Vol. 16, No. 11, pp. 6922–6932, Nov. 2020.
118. F. Zhang, H. Kodituwakku, J.W. Hines and J. Coble, "Multilayer Data-Driven Cyber-Attack Detection System for Industrial Control Systems Based on Network, System, and Process Data," *IEEE Transactions on Industrial Informatics*, Vol. 15, No. 7, pp. 4362–4369, July 2019.
119. P. Mattavelli and A.M. Stanković, "Dynamical Phasors in Modeling and Control of Active Filters," *IEEE International Symposium on Circuits and Systems*, Vol. 5, pp. 278–282, May 1999.
120. G. Escobar, A.M. Stanković, P. Mattavelli, "An adaptive controller in stationary reference frame for D-Statcom in unbalanced operation," *IEEE Transactions on Industrial Electronics*, Vol. 51 , No. 2, pp. 401–409, April 2004.
121. G. Escobar, P. Mattavelli, A.M. Stanković, A.A. Valdez and J. Leyva-Ramos, "An Adaptive Control for UPS to Compensate Unbalance and Harmonic Distortion Using a Combined Capacitor/Load Current Sensing," *IEEE Transactions on Industrial Electronics*, Vol. 54, No. 2, pp. 839–847, April 2007.
122. M. Karimi-Ghartemani and M.R. Iravani, "A Method for Synchronization of Power Electronic Converters in Polluted and Variable-Frequency Environments," *IEEE Transactions on Power Systems*, Vol. 19, No. 3, pp. 1263–1270, Aug. 2004.
123. M.R. Schroeder and B.S. Atal, "Generalized Short-Time Power Spectra and Autocorrelation Functions," *Journal of the Acoustical Society of America*, Vol. 34, No. 11, pp. 1679–1683, Nov. 1962.
124. J.L. Flanagan, D.I.S. Meinhart, R.M. Golden and M.M. Sondhi, "Phase Vocoder," *Journal of the Acoustical Society of America*, Vol.38, No. 5, pp. 939–940, Nov. 1965.
125. J.L. Flanagan and R.M. Golden, "Phase Vocoder," *Bell Systems Technical Journal*, Vol. 45, pp. 1493–1509, Nov. 1966.
126. G. Gambardella, "A Contribution to the Theory of Short-Time Spectral Analysis with Nonuniform Bandwidth Filters," *IEEE Transactions on Circuit Theory*, Vol. CT-18, No. 4, pp. 455–460, July 1971.
127. J.E. Youngberg and S.F. Boll, "Constant-Q Signal Analysis and Synthesis," *IEEE International Conference on Acoustics, Speech and Signal Processing*, pp. 375–378, Apr. 1978.
128. F. Hlawatsch and G.F. Boudreaux-Bartels, "Linear and Quadratic Time-Frequency Signal Representations," *IEEE Signal Processing Magazine*, Vol. 9, No. 2, pp. 21–60, Apr. 1992.
129. C.W. Helstrom, "An Expansion of a Signal in Gaussian Elementary Signals," *IEEE Transactions on Information Theory*, Vol 12, No. 1, pp. 81–82, Jan. 1966.

130. L.K. Montgomery and I.S. Reed, "A Generalization of the Gabor-Helstrom Transform," *IEEE Transactions on Information Theory*, Vol 13, No. 2, pp. 344–345, Apr. 1967.
131. Z. Cvetković and M. Vetterli, "Oversampled Filter Banks," *IEEE Transactions on Signal Processing*, Vol. 46, No. 5, pp. 1245–1255, May 1998.
132. M. Vetterli and J. Kovačević, *Wavelets and Subband Coding*, Prentice Hall, 1995.
133. P.P. Vaidyanathan, *Multirate Systems and Filter Banks*, Prentice Hall, 1993.
134. H. Lev-Ari and A.M. Stankovć, "Phasor Banks: Customizable Filter Banks for Robust Dynamic Time-Frequency Analysis," *40-th North American Power Symposium*, Calgary, Canada, Sept. 2008.
135. P. Argyropoulos, H. Lev-Ari and A.M. Stanković, "IIR Phasor-Banks: Causal, Delay-Free, Numerically Robust, Customizable Uniform-DFT-Like Perfect Reconstruction Filter Banks," *IEEE International Conference on Acoustics, Speech and Signal Processing*, Vancouver, BC, May 2013, pp. 5613–5617.
136. F.J. Harris, "On the Use of Windows for Harmonic Analysis with the Discrete Fourier Transform," *Proceedings of the IEEE*, Vol. 66, No. 1, pp. 51–83 , Jan. 1978.
137. B. Porat, *A Course in Digital Signal Processing*, Wiley, 1997.
138. *Testing and Measurement Techniques – General Guide on Harmonics and Interharmonics Measurements and Instrumentation, for Power Supply Systems and Equipment Connected Thereto*, International Electrotechnical Commission standard IEC 61000–4–7:2002 (updated in 2008).
139. *Testing and Measurement Techniques – Power Quality Measurement Methods*, International Electrotechnical Commission standard IEC 61000–4–30:2003 (updated in 2015, 2021).
140. A. Testa, D. Gallo and R. Langella, "On the Processing of Harmonics and Interharmonics: Using Hanning Window in Standard Framework," *IEEE Transactions on Power Delivery*, Vol. 19, No. 1, pp. 28–34, Jan. 2004.
141. J. Kovačević and A. Chebira, "Life Beyond Bases: The Advent of Frames (Part I)," *IEEE Signal Processing Magazine*, Vol. 24, No. 4, pp. 86–104, July 2007.
142. J. Kovačević and A. Chebira, "Life Beyond Bases: The Advent of Frames (Part II)," *IEEE Signal Processing Magazine*, Vol. 24, No. 5, pp. 115–125, Sept. 2007.
143. V.K. Goyal, J. Kovačević, and J.A. Kelner, "Quantized frame expansions with erasures," *Journal of Applied and Computational Harmonic Analysis*, Vol. 10, No. 3, pp. 203–233, May 2001.
144. P.P. Vaidyanathan and P.-Q. Hoang, "Lattice Structures for Optimal Design and Robust Implementation of Two-Channel Perfect-Reconstruction QMF Banks," *IEEE Transactions on Acoustics, Speech and Signal Processing*, Vol. 36, No. 1, pp. 81–94, Jan. 1988.
145. F.J. Harris, "High-Resolution Spectral Analysis with Arbitrary Spectral Centers and Adjustable Spectral Resolutions," *Computers & Electrical Engineering*, Vol. 3, No. 2, pp. 171–191, June 1976.
146. J.C. Brown, "Calculation of a Constant-Q Spectral Transform," *Journal of the Acoustical Society of America*, Vol. 89, No. 1, pp. 425–433, Jan. 1991.
147. H. Bo, H. Li, L. Ma and B. Yu, "A Constant-Q Transform Based Approach for Robust EEG Spectral Analysis," *2014 International Conference on Audio, Language and Image Processing*, pp. 58–63, July 2014.
148. P.S. Addison, *The Illustrated Wavelet Transform Handbook*, 2nd edition, CRC Press, 2017.
149. "CWT-Based Time-Frequency Analysis," MathWorks Help Center, URL: https://www.mathworks.com/help/wavelet/ug/cwt-based-time-frequency-analysis.html
150. "Choose a Wavelet," MathWorks Help Center, URL: https://www.mathworks.com/help/wavelet/gs/choose-a-wavelet.html
151. R.G. Stockwell, "A basis for efficient representation of the S-transform," *Digital Signal Processing*, Vol. 17, No. 1, pp. 371–393, Jan. 2007.
152. R.G. Stockwell, L. Mansinha and R.P. Lowe, "Localization of the Complex Spectrum: The *S* Transform," *IEEE Transactions on Signal Processing*, Vol. 44, No. 4, pp. 998–1001, Apr. 1996.

153. A. Bruns, "Fourier-, Hilbert- and wavelet-based signal analysis: are they really different approaches?" *Journal of Neuroscience Methods*, Vol. 137, No. 2, pp. 321–332, Aug. 2004.
154. IEEE Standards Association, "IEEE Recommended Practice and Requirements for Harmonic Control in Electric Power Systems," *IEEE Standard 519-2014*, June 2014. Available at: https://standards.ieee.org/ieee/519/3710/
155. J.M. Lilly and S.C. Olhede, "Higher-Order Properties of Analytic Wavelets," *IEEE Transactions on Signal Processing*, Vol. 57, No. 1, pp. 146–160, Jan. 2009.
156. H. Ryan, "Ricker, Ormsby, Klauder, Butterworth – A Choice of Wavelets," *CSEG (Canadian Society of Exploration Geophysicists) Recorder*, Vol. 19, No. 7, pp. 8–9, Sept. 1994.
157. "Continuous and Discrete Wavelet Transform," MathWorks Help Center, URL: https://www.mathworks.com/help/wavelet/gs/continuous-and-discrete-wavelet-transforms.html
158. J.M. Lilly and S.C. Olhede, "Analysis of Modulated Multivariate Oscillations," *IEEE Transactions On Signal Processing*, Vol. 60, No. 2, pp. 600–612, Feb. 2012.
159. L. Stanković, D. Mandić, M. Daković and M. Brajović, "Time-frequency decomposition of multivariate multicomponent signals," *Signal Processing*, Vol. 142, pp. 468–479, Jan. 2018.
160. C. R. Pinnegar and L. Mansinha, "The S-transform with Windows of Arbitrary and Varying Shape," *Geophysics*, Vol. 68, No. 1, pp. 381–385, Jan. 2003.
161. N. Singh and P.M. Pradhan, "Efficient Discrete S-Transform Based on Optimally Concentrated Window," *IEEE Signal Processing Letters*, Vol. 26, No. 1, pp. 14–18, Jan. 2019.
162. N. Singh and P.M. Pradhan, "Sharp Detection of Event's Onset in Seismic Signals With Asymmetrical Kaiser Window-Based S-Transform," *IEEE Geoscience and Remote Sensing Letters*, Vol. 16, No. 10, pp. 1620–1624, Oct. 2019.
163. I. Daubechies, *Ten Lectures on Wavelets*, SIAM, 1992.
164. G. Strang and T. Nguyen, *Wavelets and Filter Banks*, Wellesley-Cambridge Press, 1996.
165. P. Abry and P. Flandrin, "On the Initialization of the Discrete Wavelet Transform Algorithm," *IEEE Signal Processing Letters*, Vol. 1, No. 2, pp. 32–34, Feb. 1994.
166. X.-G. Xia, C.-C.J. Kuo and Z. Zhang, "Wavelet Coefficient Computation with Optimal Prefiltering," *IEEE Transactions on Signal Processing*, Vol. 42, No. 8, pp. 2191–2197, Aug. 1994.
167. D. Veitch, M.S. Taqqu and P. Abry, "Meaningful MRA Initialization for Discrete Time Series," *Signal Processing*, Vol. 80, No. 9, pp. 1971–1983, Sept. 2000.
168. Y. Wang and J. Orchard, "Fast Discrete Orthonormal Stockwell Transform," *SIAM Journal on Scientific Computing*, Vol. 31, No. 5, pp. 4000–4012, 2009.
169. I.W. Selesnick. "Wavelet Transform With Tunable Q-Factor," *IEEE Transactions on Signal Processing*, Vol. 59, No. 8, pp. 3560–3575, Aug. 2011.
170. S.L. Marple, "Time-Frequency Signal Analysis: Issues and Alternative Methods," *Proceedings of the IEEE-SP International Symposium on Time-Frequency and Time-Scale Analysis*, Pittsburgh, PA, Oct. 1998, pp. 329–332.
171. L. Cohen, "Time-Frequency Distributions – A Review," *Proceedings of the IEEE*, Vol. 77, No. 7, pp. 941–981, July 1989.
172. P. Flandrin, "Time-frequency and time-scale," *IEEE Fourth ASSP Workshop on Spectrum Estimation and Modeling*, Minneapolis, MN, Aug. 1988, pp. 77–80.
173. C.S. Burrus, R.A. Gopinath and H. Guo, *Introduction to Wavelets and Wavelet Transforms*, Prentice Hall, 1998.
174. S.G. Jalali, R.H. Lasseter and I. Dobson, "Dynamic response of a thyristor-controlled switched capacitor," *IEEE Trans. on Power Delivery*, Vol. 9, No. 3, pp. 1609–1615, July 1994.
175. P.C. Stefanov and A.M. Stanković, "Modeling of UPFC Operation Under Unbalanced Conditions with Dynamic Phasors," IEEE Transactions on Power Systems, Vol. 17, No. 2, pp. 395–403, May 2002.
176. R. Caldon, P. Mattavelli and B.M. Han, "Dynamic analysis and control of upfc using transient simulation," *Intl. Conf. on Power System Transients – IPST'99*, pp. 527–532, 1999.
177. L. Gyugyi, "A unified power flow control concept for flexible ac transmission systems," *IEE Proceedings – Part C*, Vol. 139, No. 4, pp. 323–331, July 1992.

178. N.G. Hingorani and L. Gyugyi, *Understanding FACTS – Concepts and Technology of Flexible AC Transmission Systems*, IEEE Press, New York, 2000.
179. Y.H. Song and A.T. Johns (Eds.), *Flexible AC Transmission Systems (FACTS)*, IEE Press, London, 2000.
180. H. Schauder and H. Mehta, "Vector analysis and control of advanced static var compensators," *IEE Proceedings-C*, Vol. 140, No. 4, pp. 299–305, July 1993.
181. K.K. Sen and E.J. Stacey, "UPFC – unified power flow controller: Theory, modeling, and applications," *IEEE Trans. on Power Delivery*, Vol. 13, No. 4, pp. 1453–1460, October 1998.
182. J.G. Kassakian, M.F. Schlecht and G.C. Verghese, *Principles of Power Electronics*, Addison–Wesley, 1991.
183. V. Caliskan, G.C. Verghese and A.M. Stanković, "Multi-frequency averaging of dc-dc converters," *IEEE 5th Workshop on Computers in Power Electronics*, pp. 113–119, 1996.
184. C.R. Giardina and P.M. Chirlian, "Bounds on the Truncation Error of Periodic Signals," *IEEE Transactions on Circuit Theory*, Vol. 19, No. 2, pp. 206–207, March 1972.
185. I.J. Perez-Arriaga, G.C. Verghese, F.L. Pagola, J.L. Sancha and F.C. Schweppe, "Developments in selective modal analysis of small-signal stability in electric power systems," *Automatica*, Vol. 26, No. 2, pp. 215–231, 1990.
186. P.V. Kokotović, H.K. Khalil and J. O'Reilly, *Singular Perturbation Methods in Control: Analysis and Design*, Academic Press, London, 1986.
187. B.A. Lehman and R.M. Bass, "Switching frequency dependent averaged models for pwm dc-dc converters," *IEEE Trans. on Power Electronics*, Vol. 11, No. 1, pp. 89–98, Jan. 1996.
188. P. Mattavelli, G.C. Verghese and A.M. Stanković, "Phasor dynamics of thyristor-controlled series capacitor systems," *IEEE Trans. on Power Systems*, Vol. 12, No. 3, pp. 1259–1267, Aug. 1997.
189. P. Mattavelli, A.M. Stanković and G.C. Verghese, "SSR Analysis with Dynamic Phasor Model of Thyristor–Controlled Series Capacitor," *IEEE Transactions on Power Systems*, Vol. 14, No. 1, pp. 200–208, Feb. 1999.
190. IEEE PES Committee Report, "Terms, definitions, and symbols for subsynchronous resonance oscillations," *IEEE Trans. on Power Apparatus and Systems*, Vol. 104, No. 6, pp. 1326–1334, June 1985.
191. IEEE PES Committee Report, "Reader's guide to subsynchronous resonance," *IEEE Trans. on Power Systems*, Vol. 7, No. 1, pp. 150–157, Feb. 1992.
192. J. Urbanek, R.J. Piwko, E.V. Larsen, B.L. Damsky, B.C. Furumasu, W. Mittelstadt and J.D. Eden, "Thyristor controlled series capacitor – prototype installation at the Slatt 500 kv substation," *IEEE Trans. on Power Delivery*, Vol. 9, No. 2, pp. 992–1000, Apr. 1994.
193. K. Renz, G. Thumm and S. Weiss, "Thyristor controlled reactor enhances series capacitor applications," *6th European Conf. on Power Electronics and Applications*, pp. 911–914, 1995.
194. N. Christl, R. Hedin, R. Johnson, P. Krause and A. Montoya, "Power system studies and modeling for the Kayenta 230kv substation advanced series compensation," *IEE Proc. Conf. Fifth International Conference on AC and DC Power Transmission AC-DC Transmission Systems*, pp. 33–37, Apr. 1991.
195. W. Zhu, R. Spee, R.R. Moler, G.C. Alexander, W.A. Mittelstadt and D. Maratukulam, "An EMTP study of SSR mitigation using the thyristor controlled series capacitor," *IEEE Trans. on Power Delivery*, Vol. 10, No. 3, pp. 1479–1485, July 1995.
196. J.F. Hauer, W.A. Mittelstadt, R.J. Piwko, B.L. Damsky and J.D. Eden, "Modulation and SSR tests performed on the BPA 500kv thyristor controlled series capacitor unit at Slatt substation," *IEEE Trans. on Power Systems*, Vol. 11, No. 2, pp. 801–806, May 1996.
197. R.A. Hedin, S. Weiss, D. Torgerson and L.E. Eits, "SSR characteristics of alternative types of series compensation schemes," *IEEE Trans. on Power Systems*, Vol. 10, No. 2, pp. 845–850, May 1995.
198. D.L. Trudnowski, M.K.Donnelly and J.F. Hauer, "Estimating damping effectiveness of BPA's thyristor controlled series capacitor by applying time and frequency domain methods to measured response," *IEEE Trans. on Power Systems*, Vol. 11, No. 2, pp. 761–766, May 1996.

199. J.J. Paserba, N.W. Miller, E.V. Larsen and R.J. Piwko, "A thyristor controller series compensation model for power system stability analysis," *IEEE Trans. on Power Delivery*, Vol. 10, No. 3, pp. 1471–1478, July 1995.
200. IEEE Power Engineering Society, *Facts Overview*, Vol. TP-108, IEEE Press, 1995.
201. IEEE Power Engineering Society, *Facts Applications*, Vol. TP-116, IEEE Press, 1996.
202. R. Rajaraman, I. Dobson, R.H. Lasseter and Y. Shern, "Computing the damping of subsynchronous oscillation due to a thyristor controlled series capacitor," *IEEE Trans. on Power Delivery*, Vol. 11, No. 2, pp. 1120–1127, Apr. 1996.
203. R. Rajaraman and I. Dobson, "Damping estimates of subsynchronous and power swing oscillations in power systems with thyristor switching devices," *IEEE Trans. on Power Systems*, Vol. 11, No. 4, pp. 1926–1930, Nov. 1996.
204. P.M. Anderson, B.L. Agrawal and J.E. Van Ness, *Subsynchronous Resonance in Power Systems*, IEEE Press, 1990.
205. A.M. Stanković, B.C. Lesieutre and T. Aydin, "Modeling and Analysis of Single–Phase Induction Machines with Dynamic Phasors," *IEEE Transactions on Power Systems*, Vol. 14, No. 1, pp. 9–14, Feb. 1999.
206. A.E. Fitzgerald, C. Kingsley and S. D. Umans, *Electric Machinery*, McGraw-Hill, New York, Fourth Ed., 1983.
207. G.R. Slemon and A. Straughen, *Electric Machines*, Addison-Wesley, Reading, MA, 1980.
208. S.J. Chapman, *Electric Machinery Fundamentals*, McGraw-Hill, New York, Second Ed., 1991.
209. D.V. Richardson and A.J. Caisse, *Rotating Electric Machinery and Transformer Technology*, Prentice Hall, Fourth Edition, 1997.
210. P. Krause, O. Wasynczuk and S. Sudhoff, *Analysis of Electric Machinery*, IEEE Press, 1995.
211. E. Delaleau and A.M. Stanković, "Modeling and Simulation of the Induction Motor with Position-Dependent Load Torque," *IEEE Conference on Decision and Control*, Maui, HI, pp. 6212–6217, Dec. 2003.
212. D.W. Novotny and T.A. Lipo, *Vector Control and Dynamics of AC Drives*, Oxford University Press, 1996.
213. A.M. Stanković, S.R. Sanders and T. Aydin, "Dynamic Phasors in Analysis of Unbalanced Polyphase AC Machines," *IEEE Transactions on Energy Conversion*, Vol. 17, No. 1, pp. 107–113, March 2002.
214. S.M. Halpin, C.A. Gross and L.L. Grigsby, "An improved method of including detailed synchronous machine representations in large power system models for fault analysis," *IEEE Trans. Energy Conversion*, Vol. 8, No. 4, pp. 719–725, Aug. 1993.
215. A.M. Stanković and T. Aydin, "Analysis of Asymmetrical Faults in Power Systems Using Dynamic Phasors," *IEEE Transactions on Power Systems*, Vol. 15, No. 3, pp. 1062–1068, Aug. 2000.
216. B.C. Lesieutre, J.A. Mohamed and A.M. Stanković, "Analysis of ferroresonance in three-phase transformers," *Intl. Conf. on Power System Technology – Powercon*, Vol. 2, 2000.
217. B.C. Lesieutre, A.M. Stanković and J.R. Lacalle-Melero, "A study of state variable participation in nonlinear limit-cycle behavior," *Intl. Conf. on Control Applications*, 1995.
218. G. Escobar, A.M. Stanković, P. Mattavelli and R. Ortega, "A Comparative Study of Three Nonlinear Controllers for TCSC," *IEEE Powertech 2001*, Porto, Portugal, pp. 422–427, Sep. 2001.
219. G. Escobar, A.M. Stanković, R. Ortega and P. Mattavelli, "On the nonlinear control of TCSC," *35-th Hawaii International Conference on System Science*, Jan. 2002.
220. G. Escobar, A.M. Stanković, R. Ortega and A. Astolfi, "An adaptive controller for the TCSC based on the immersion and invariance design technique," *Power Systems Computation Conference*, Sevilla, Spain, pp. 345–349, June 2002.
221. P.A. Vela, *Averaging and Control of Nonlinear Systems*, PhD Dissertation, Caltech, 2003.
222. G.A. Leonov, I.M. Burkin and A.I. Shepeljavyi, *Frequency Methods in Oscillation Theory*, Kluwer, 1996.

223. D. Naunin, "The Calculation of the Dynamic Behavior of Electrical Machines by Space-Phasors," *Electric Machines & Power Systems*, Vol. 4, No. 1, pp. 31–45, 1979.
224. S. Yamamura, *Spiral Vector Theory of AC Circuits and Machines*, Oxford University Press, 1992.
225. V. Staudt, "AC Machine Dynamics Described by Space Vectors Using State-Space Notation", *European Transactions on Electric Power*, Vol. 9, No. 1, pp. 17–25, Jan. 1999.
226. A.C. Smith, R.C. Healey and S. Williamson, "A Transient Induction Motor Model Including Saturation and Deep Bar Effect," *IEEE Trans. on Energy Conversion*, Vol. 11, No. 1, pp. 8–16, March 1996.
227. A.M. Stanković, S.R. Sanders and T. Aydin, "Analysis of Unbalanced AC Machines with Dynamic Phasors," *Naval Symposium on Electric Machines*, pp. 219–224, Oct. 1998.
228. H. Lev-Ari and A.M. Stanković, "Phasor-Banks: Customizable Filter Banks for Robust Dynamic Time-Frequency Analysis," *40-th North American Power Symposium*, Sept. 2008.
229. T. Yang, S. Bozhko, J.-M. Le-Peuvedic, G. Asher and C.I. Hill, "Dynamic Phasor Modeling of Multi-Generator Variable Frequency Electrical Power Systems," *IEEE Trans. on Power Systems*, Vol. 31, No. 1, pp. 563–571, Jan. 2016.
230. O.C. Sakinci and Jef Beerten, "Generalized Dynamic Phasor Modeling of the MMC for Small-Signal Stability Analysis," *IEEE Trans. Power Delivery*, Vol. 34, No. 3, pp. 991–1000, June 2019.
231. J. Rupasinghe, S. Filizadeh, and Liwei Wang, "A Dynamic Phasor Model of an MMC With Extended Frequency Range for EMT Simulations," *IEEE Jour. Emerging and Selected Topics in Power Electronics*, Vol. 7, No. 1, pp. 30–40, Mar. 2019.
232. T.H. Demiray, *Simulation of power system dynamics using dynamic phasor models*, Ph.D. Dissertation, Swiss Federal Institute of Technology, Zurich, 2008.
233. K. Mudunkotuwa, S. Filizadeh and U. Annakkage, "Development of a hybrid simulator by interfacing dynamic phasors with electromagnetic transient simulation," *IET Generation, Transmission & Distribution*, Vol. 11, No. 12, pp. 2991–3001, 2017.
234. D. Shu, X. Xie, V. Dinavahi, C. Zhang, X. Ye and Q. Jiang, "Dynamic Phasor Based Interface Model for EMT and Transient Stability Hybrid Simulations," *IEEE Trans. Power Systems*, Vol. 33, No. 4, pp. 3930–3939, July 2018.
235. T. Yang, S. Bozhko and G. Asher, "Modelling of Electrical Power Systems with Dynamic Phasors in Modelica," *10-th International Modelica Conference*, Lund, Sweden, pp. 507–514, March 2014.
236. M. Mirz, S. Vogel, G. Reinke and A. Monti, "DPsim – A Dynamic Phasor Real-Time Simulator for Power systems," *Software X*, 100253, Vol. 10, April 2019.
237. T. Bi, H. Liu, Q. Feng, C. Qian and Y. Liu, "Dynamic Phasor Model-Based Synchrophasor Estimation Algorithm for M-Class PMU," *IEEE Transactions on Power Delivery*, Vol. 30, No. 3, pp. 1162–1171, June 2015.
238. X. Guo, Z. Lu, B. Wang, X. Sun, L. Wang and J.M. Guerrero, "Dynamic Phasors-Based Modeling and Stability Analysis of Droop-Controlled Inverters for Microgrid Applications," *IEEE Transactions on Smart Grid*, Vol. 5, No. 6, pp. 2980–2987, Nov. 2014.
239. M. Daryabak, S. Filizadeh, J. Jatskevich, A. Davoudi, M. Saeedifard, V. K. Sood, J. A. Martinez, D. Aliprantis, J. Cano and A. Mehrizi-Sani, "Modeling of LCC-HVDC Systems Using Dynamic Phasors," *IEEE Transactions on Power Delivery*, Vol. 29, No. 4, pp. 1989–1998, Aug. 2014.
240. O.B. Shallenberger, "The Distribution of Electricity by Alternating Current," *Electrical World*, March 3, 1888, pp. 114–115.
241. W. Stanley, Jr., "Phenomena of Retardation in the Induction Coil," *American Institute of Electrical Engineers*, Vol. V, No. 4, Jan. 1888, pp. 97–115.
242. C.P. Steinmetz, "Findet eine Phasenverschiebung im Wechselstromlichtbogen statt?" *Electrotechnische Zeitschrift*, Vol. 42, 1892, pp. 567–568,
243. C.P. Steinmetz, *Theory of Alternating Current Phenomena*, McGraw, New York, 1897.
244. Preliminary Report of Special Joint Committee, "Power Factor in Polyphase Circuits," *Journal of the AIEE*, vol. 39(2), July 1920, pp. 1449–50.

245. A.E. Knowlton, "Reactive Power Concepts in Need of Clarification," *Transactions of the American Institute of Electrical Engineers*, Vol. 52, No. 3, pp. 744–747, Sept. 1933.
246. C.L. Fortescue, Power, power factor and reactive volt amperes. *Electrical Engineering*, Vol. 52, No. 5, pp. 319–323, May 1933.
247. H.L. Curtis and F.B. Silsbee, "Definitions of Power and Related Quantities," *Transactions of the American Institute of Electrical Engineers*, Vol. 54, No. 4, pp. 394–404, Apr. 1935.
248. C.I. Budeanu, "Puisance Reactives et Fictives," *Institut National Romain pour l'Etude de l'Aménagement et de l'Utilisation des Sources de l'Energie*, Vol. 2, Bucharest, Romania, 1927.
249. W. Shepherd and P. Zakikhani, "Suggested Definition of Reactive Power for Nonsinusoidal Systems," Proc. IEE, Vol. 119, pp. 1361–1362, Sep. 1972.
250. D. Sharon, "Reactive Power Definition and Power Factor Improvement in Nonlinear Systems," *Proceedings of the Institution of Electrical Engineers*, Vol. 120, No. 6, pp. 704–706, July 1973.
251. L.S. Czarnecki, "Orthogonal Decomposition of the Currents in a 3-Phase Nonlinear Circuit with a Nonsinusoidal Voltage Source," *IEEE Transactions on Instrumentation and Measurement*, Vol. 37, No. 1, pp. 30–34, March 1988.
252. H. Lev-Ari, A.M. Stanković and S.J. Ceballos, "An Orthogonal Decomposition of Apparent Power with Application to an Industrial Load," *Proceedings of the Power Systems Computation Conference (PSCC)*, Liege, Belgium, Aug. 2005.
253. H. Lev-Ari and A.M. Stanković, "A Decomposition of Apparent Power in Polyphase Unbalanced Networks in Nonsinusoidal Operation," *IEEE Trans. on Power Systems*, Vol. 21, No. 1, pp. 438–440, Feb. 2006.
254. M. Depenbrock, pp. 15–25 in "Untersuchungen uber die Spannungs und Leistungsverh altnisse bei Urnrichtern ohne Energiespeicher," Ph.D. dissertation, Hannover Technical University, 1962.
255. M. Depenbrock, "The FBD-Method, a Generally Applicable Tool for Analyzing Power Relations," *IEEE Transactions on Power Systems*, Vol. 8, No. 2, pp. 381–387, May 1993.
256. C. Doran and A. Lasenby, *Geometric Algebra for Physicists*, Cambridge University Press, 2003.
257. V. Staudt, "Fryze – Buchholz – Depenbrock: A Time-Domain Power Theory," *International School on Nonsinusoidal Currents and Compensation*, Lagów, Poland, pp. 1–12, 2008.
258. A.E. Emanuel, "The Buchholz – Goodhue Apparent Power Definition: the Practical Approach for Nonsinusoidal and Unbalanced Systems," *IEEE Trans. on Power Delivery*, Vol. 13, No. 2, pp. 344–350, April 1998.
259. N. LaWhite and M.D. Ilić, "Vector Space Decomposition of Reactive Power for Periodic Nonsinusoidal Signals," *IEEE Trans. Circ. & Syst.–I: Fundamental Theory And Applications*, Vol. 44, No. 4, pp. 338–346, Apr. 1997.
260. R. A. Horn, "The Hadamard Product," in *Matrix Theory and Applications, Proceedings of Symposia in Applied Mathematics*, Vol. 40, C.R. Johnson (ed.), American Mathematical Society, Providence, RI, 1990.
261. L.S. Czarnecki, "Considerations on the Reactive Power in Nonsinusoidal Situations," *IEEE Trans. Instrumentation & Measurement*, Vol. 34, pp. 399–404, Sept. 1985.
262. L. S. Czarnecki, "What is wrong with the Budeanu concept of reactive and distortion power and why it should be abandoned," *IEEE Transactions on Instrumentation and Measurement*, Vol. IM-36, No. 3, pp. 834–837, Sep. 1987.
263. J.L. Willems, "Budeanu's Reactive Power and Related Concepts Revisited," *IEEE Transactions on Instrumentation and Measurement*, Vol. 60, No. 4, pp. 1182–1186, Apr. 2011.
264. "IEEE Standard Definitions for the Measurement of Electric Power Quantities Under Sinusoidal, Nonsinusoidal, Balanced, or Unbalanced Conditions," *IEEE Standard 1459-2010*, March 2010.
265. L.S. Czarnecki, "Currents' Physical Components (CPC) in Circuits with Nonsinusoidal Voltages and Currents, Part 2: Three-Phase Three-Wire Linear Circuits," *Electrical Power Quality and Utilisation Journal*, Vol. 12, No. 1, pp. 3–13, 2006.

266. L.S. Czarnecki, "Currents' Physical Components (CPC) in Circuits with Nonsinusoidal Voltages and Currents, Part 1: Single-Phase Linear Circuits," *Electrical Power Quality and Utilisation Journal*, Vol. 11, No. 2, pp. 3–14, 2005.
267. W. Mathis and W. Marten, "A Unified Concept of Electric Power," *Proc. IEEE ISCAS*, Portland, OR, pp. 499–502, May 1989.
268. E.F. Bolinder, "Clifford Algebra: What is it?," *IEEE Antenna and Propagation Society Newsletter*, Aug. 1987, pp. 18–23.
269. H. Lev-Ari and A.M. Stanković, "A Geometric Algebra Approach to Decomposition of Apparent Power in General Polyphase Networks," *41-st North American Power Symposium*, Starkville, MS, Oct. 2009.
270. H. Lev-Ari, A.M. Stanković and A. Ghanavati, "Dynamic Decomposition of Apparent Power in Polyphase Unbalanced Networks with Application to Transients in an Industrial Load," *40-th North American Power Symposium*, Sept. 2008.
271. H. Akagi, Y. Kanazawa and A. Nabae, "Instantaneous Reactive Power Compensators Comprising Switching Devices Without Energy Storage Components," *IEEE Trans. Industry Applications*, Vol. 20, No. 3, pp. 625–630, 1984.
272. H. Akagi, E.H. Watanabe and M. Aredes, *Instantaneous Power Theory and Applications to Power Conditioning*, Wiley, 2007.
273. L.S. Czarnecki, "Meta-Theory of Electric Powers and Present State of Power Theory of Circuits with Periodic Voltages and Currents," *Przeglad Elektrotechniczny*, Vol. 89, No. 6, pp. 26–31, 2013.
274. H. Lev-Ari and A.M. Stanković, "Instantaneous Power Quantities in Polyphase Systems – A Geometric Algebra Approach," *Energy Conversion Congress & Expo (ECCE)*, San Jose, CA, Sept. 2009, pp. 592–596.
275. R. S. Herrera, P. Salmeron, J. R. Vazquez, S. P. Litran and A. Perez, "Generalized Instantaneous Reactive Power Theory in Polyphase Power Systems," *Proceedings of the 13th European Conference on Power Electronics and Applications*, Barcelona, Spain, Sept. 2009, pp. 1–10.
276. J. Wyatt and M. Ilić, "Time-Domain Reactive Power Concepts for Nonlinear, Nonsinusoidal or Nonperiodic Network," *IEEE Intl. Symp. Circuits and Systems*, Volume 1, Issue 3, May 1990, pp. 387–390.
277. P. Tenti and P. Mattavelli, "A Time-domain Approach to Power Terms Definitions under Non-sinusoidal Conditions," *6th Int. Workshop on Power Definitions and Measurement under Non-Sinusoidal Conditions*, Milan, Italy, 2003.
278. P. Tenti, P. Mattavelli and E. Tedeschi, "Compensation Techniques based on Reactive Power Conservation," *Seventh International Workshop on Power Definitions and Measurements under Non-Sinusoidal Conditions*, Cagliari, Italy, July 2006.
279. F.Z. Peng and J.-S. Lai, "Generalized instantaneous reactive power theory for three-phase power systems," *IEEE Transactions on Instrumentation and Measurement*, Vol. 45, No. 1, pp. 293–297, Feb. 1996.
280. J.L. Willems, "Mathematical foundations of the instantaneous power concepts: a geometrical approach," *European Transactions on Electrical Power*, Vol. 6, pp. 299–304, Sep./Oct. 1996.
281. A. Ferrero, "Mathematical foundations of the instantaneous power concepts: an algebraic approach," *European Transactions on Electrical Power*, Vol. 6, pp. 305–309, Sep./Oct. 1996.
282. X. Dai, G. Liu and R. Gretsch, "Generalized theory of instantaneous reactive quantity for multiphase power system," *IEEE Trans. Power Delivery*, Vol. 19, No. 3, pp. 965–978, July 2004.
283. X. Dai and G. Liu, "Closure on 'Generalized theory of instantaneous reactive quantity for multiphase power system'," *IEEE Trans. Power Delivery*, Vol. 21, No. 1, pp. 542–543, Jan. 2006.
284. E. Clarke, *Circuit Analysis of A-C Power Systems*, Wiley, 1943.
285. W.C. Duesterhoeft, M.W. Schulz and E. Clarke, "Determination of Instantaneous Currents and Voltages by Means of Alpha, Beta, and Zero Components," *Transactions of the American Institute of Electrical Engineers*, Vol. 70, no. 2, pp. 1248–1255, July 1951.

286. A. Ferrero and G. Superti-Furga, "A new approach to the definition of power components in three-phase systems under nonsinusoidal conditions," *IEEE Trans. Instrumentation and Measurement*, Vol. 40, No. 3, pp. 568–577, June 1991.
287. H. Kim, F. Blaabjerg, B. Bak-Nielsen and J. Choi, "Instantaneous Power Compensation in Three-Phase Systems by Using p-q-r Theory," *IEEE Trans. Power Electronics*, Vol. 17, No. 8, pp. 701–710, Sept. 2002.
288. M. Depenbrock, V. Staudt and H. Wrede, "Concerning Instantaneous Power Compensation in Three-Phase Systems by Using p-q-r Theory," *IEEE Trans. Power Electronics*, Vol. 19, No. 4, pp. 1151–1152, July 2004.
289. E.H. Watanabe, M. Aredes and H. Akagi, "The p-q Theory for Active Filter Control: Some Problems and Solutions," *Revista Controle & Automação*, Vol.15, No.1, pp. 78–84, Jan. 2004.
290. M. Popescu, A. Bitoleanu and V. Suru, "A DSP-Based Implementation of the p-q Theory in Active Power Filtering Under Nonideal Voltage Conditions," *IEEE Trans. Industrial Informatics*, Vol. 9, No. 2, pp. 880–889, May 2013.
291. A. Ghanavati, H. Lev-Ari and A.M. Stanković, "An Efficient Sub-Cycle Alternative to Instantaneous Power Metrics," *IEEE Transactions on Industrial Informatics*, Vol. 18, No. 2, pp. 776–785, Feb. 2022.
292. A.O. Gelfand, *Calculus of Finite Differences*, Hindustan Publishing, Delhi, 1971.
293. "Window function," Wikipedia. URL: https://en.wikipedia.org/wiki/Window_function
294. M.R. Spiegel and L.J. Stephens, *Schaum's Outlines of Statistics*, 6th edition, McGraw-Hill, 2018.
295. S. Haykin, *Adaptive Filter Theory*, 5th ed., Prentice Hall, 2014.
296. B.D.H. Tellegen, "A General Network Theorem with Applications," *Philips Research Reports*, Vol. 7, pp. 259–269, Aug. 1951.
297. P. Penfield, R. Spence and S. Duinker, "A Generalized Form of Tellegen's Theorem," *IEEE Trans. Circuit Theory*, Vol. CT-17, No. 3, pp. 302–305, Aug. 1970.
298. "Hilbert space," Wikipedia. URL: https://en.wikipedia.org/wiki/Hilbert_space
299. "Inner product space," Wikipedia. URL: https://en.wikipedia.org/wiki/Inner_product_space
300. D.G. Luenberger, *Optimization by Vector Space Methods*, Wiley, 1969.
301. M.E. Ragnarsdottir, "Java Program that Calculates the Decomposition of Load Current (and consequently the apparent power) into Seven Orthogonal components," MS project report, Dept. of Electrical and Computer Engineering, Northeastern University, Boston, MA, 2008.
302. M. Radmehr, Sh. Farhangi and A. Nasiri, "The Power of Paper: Effects of power quality distortions on electrical drives and transformer life in paper industries," *IEEE Industry Applications Magazine*, Vol. 13 , No. 5, pp. 38–48, Sept.-Oct. 2007.
303. M. Islam, H.A. Mohammadpour, A. Ghaderi, C.W. Brice and Y.-J. Shin, "Time-Frequency-Based Instantaneous Power Components for Transient Disturbances According to IEEE Standard 1459," *IEEE Trans. on Power Delivery*, Vol. 30, No. 3, pp. 1288–1297, June 2015.

Index

H. Lev-Ari, A. M. Stanković, *Dynamic Phasors in Energy Processing Systems*, Power Electronics and Power Systems, https://doi.org/10.1007/978-3-031-50020-6

W

MIX
Papier aus verantwortungsvollen Quellen
Paper from responsible sources
FSC® C105338

If you have any concerns about our products,
you can contact us on
ProductSafety@springernature.com

In case Publisher is established outside the EU,
the EU authorized representative is:
Springer Nature Customer Service Center GmbH
Europaplatz 3, 69115 Heidelberg, Germany

Printed by Libri Plureos GmbH
in Hamburg, Germany